Lecture Notes in Computer Science 15885

Advanced Research in Computing and Software Science

Subline of Lecture Notes in Computer Science

More information about this series at https://link.springer.com/bookseries/558

Henning Fernau · Binhai Zhu
Editors

Combinatorial Algorithms

36th International Workshop, IWOCA 2025
Bozeman, MT, USA, July 21–24, 2025
Proceedings

Springer

Editors
Henning Fernau
University of Trier
Trier, Rheinland-Pfalz, Germany

Binhai Zhu
Montana State University
Bozeman, MT, USA

ISSN 0302-9743 ISSN 1611-3349 (electronic)
Lecture Notes in Computer Science
ISBN 978-3-031-98739-7 ISBN 978-3-031-98740-3 (eBook)
https://doi.org/10.1007/978-3-031-98740-3

This Springer imprint is published by the registered company Springer Nature Switzerland AG
The registered company address is: Gewerbestrasse 11, 6330 Cham, Switzerland

Preface

The 36th International Workshop on Combinatorial Algorithms (IWOCA 2025) was organized by Montana State University in Bozeman, MT, USA. In order to lower the carbon footprint of the conference, it was decided that the conference could be also attended online. The conference was scheduled on July 21–24, 2025, including some research workshop.

IWOCA is an annual conference series that started in 1989 as AWOCA (Australasian Workshop on Combinatorial Algorithms), and became an international conference in 2007, having been held (in alphabetical order) in Australia, Canada, Czech Republic, Finland, France, Germany, Indonesia, India, Italy, Japan, Singapore, South Korea, Taiwan, UK, and USA. In 2025, IWOCA returned to the USA again, for the first time since 2014. The conference brought together researchers on diverse topics related to combinatorial algorithms, as specified in the Call for Papers: Ad Hoc, Dynamic and Evolving Networks; Algorithms and Data Structures; Algorithms on Strings & Graphs; Algorithms for Big Data and Networks Analytics; Algorithmic Game Theory; Approximation Algorithms; Combinatorial Generation, Enumeration and Counting; Combinatorial Optimization; Complexity Theory; Combinatorics of Words; Computational Algebra and Geometry; Computational Biology; Cryptography and Information Security; Distributed and Parallel Algorithms; Experimental Evaluation of Algorithms; Fine-Grained Complexity; Foundations of Cloud Computing; Graph Algorithms for Social Network Analysis; Graph Drawing and Labelling; Graph Theory & Combinatorics; Mobile Agents; New Paradigms of Computation; Online Algorithms; Parameterized and Exact Algorithms; Probabilistic and Randomized Algorithms; Scheduling; and Streaming Algorithms.

The Program Committee (PC) of IWOCA 2025 received 78 abstract submissions; finally, 71 full papers were submitted. Each submission was reviewed by at least three PC members and some trusted external referees, and evaluated on its quality, originality, and relevance to the conference. The PC selected 32 papers for presentation at the conference and inclusion in the proceedings. Two invited talks were scheduled for IWOCA 2025, by Paweł Gawrychowski (University of Wrocław, Poland) and Saket Saurabh (Institute of Mathematical Sciences, India; and University of Bergen, Norway). Two short abstracts are included in the frontmatter.

The Program Committee also selected two papers to receive the Best Paper Award and the Best Student Paper Award, respectively. These awards were sponsored by Springer. The awardees are:

– Best Paper Award: Vahideh Keikha, Maarten Löffler, Maria Saumell, and Pavel Valtr. Guarding a 1.5D Terrain with Imprecise Viewpoints.
– Best Student Paper Award: Kan Shota and Kazuya Haraguchi. A Linear Delay Algorithm of Enumerating Strongly-Connected Induced Subgraphs Based on SSD Set System.

We would like to thank all invited speakers for agreeing to give a talk at the conference, the Program Committee members who graciously gave their time and energy, and 64 external reviewers for their expertise.

We thank Springer for publishing the proceedings of IWOCA 2025 in the ARCoSS subline of their LNCS series and for their financial support towards the best paper and the best student paper awards. Also, we were one of the last conferences collecting experiences with EquinOCS, a Springer system we used for managing the collection and editing of the papers.

Finally, we thank the Steering Committee for giving us the opportunity to serve as program chairs of IWOCA 2025 and for their continuous support.

May 2025

Henning Fernau
Binhai Zhu

Organization

Steering Committee

Maria Chudnovsky	Princeton University, USA
Bill Smyth	McMaster University, Canada
Ralf Klasing	CNRS and University of Bordeaux, France
Wing-Kin (Ken) Sung	Chinese University of Hong Kong, China
Henning Fernau	University of Trier, Germany

Program Committee

Cristina Bazgan	Université Paris Dauphine, France
Sergey Bereg	University of Texas-Dallas, USA
Laurent Bulteau	CNRS and Université Gustave Eiffel, France
Gruia Calinescu	Illinois Institute of Technology, USA
Zhi-Zhong Chen	Tyoko Denki University, Japan
Maria Chudnovsky	Princeton University, USA
Riccardo Dondi	University of Bergamo, Italy
Piotr Faliszewski	AGH University of Science and Technology, Poland
Henning Fernau (Co-chair)	Universität Trier, Germany
Pamela Fleischmann	Universität Kiel, Germany
Paweł Gawrychowski	University of Wrocław, Poland
Gregory Gutin	Royal Holloway, University of London, UK
Adriana Hansberg	UNAM, Mexico
Pavol Hell	Simon Fraser University, Canada
Yuya Higashikawa	University of Hyogo, Japan
Ling-ju Hung	National Taipei University of Business, Taiwan
Haitao Jiang	Shandong University, China
Dominik Köppl	University of Yamanashi, Japan
Christian Komusiewicz	Friedrich-Schiller-Universität Jena, Germany
Manuel Lafond	University of Sherbrooke, Canada
Guohui Lin	University of Alberta, Canada
Adiesha Liyanage	Montana State University, USA
Neeldhara Misra	Indian Institute of Technology Gandhinagar, India
Brendan Mumey	Montana State University, USA
Sang-il Oum	IBS and KAIST, South Korea

List of Subreviewers

Taro Abe
Junho Ahn
Duncan Adamson
Ridho Alfarisi
Djamal Belazzougui
Sriram Bhyravarapu
Theo Boury
Nicolas Bousquet
Richard Brewster
Sanjana Dey
Diego Diaz
Silvia Fernandez-Merchant
Carleb Eardley
William Evans
Dalibor Froncek
Jaroslav Garvardt
Mingyang Gong
Maximilian Gorsky
Andreas Grigorjew
Alexander Guterman
Hovhannes Harutyunyan
Daniel Harvey
Felix Hommelsheim
Medhi Hosseinzadeh
Haoqiang Huang
Shang-En Huang
Annika Huch
Sven Jäger
Hemanshu Kaul
Kamyar Khodamoradi
Ralf Klasing
Dušan Knop
Martin Koutecky
Grzegorz Lisowski
Weidong Luo
Bradley McCoy
Bertrand Marchand
David Mitchell
Hendrik Molter
Nils Morawietz
Max Wiedenhöft
Subhas Nandy
Jochim Orthaber
Kristýna Pekárková
Daniel Perz
Yuri Pirola
Alexander Popa
Jurek Rostalsky
Jack Ruder
Yoshifumi Sakai
Saket Saurabh
Masoud Seddighin
Florian Sikova
Frank Sommer
Braeden Sopp
Krzysztof Sornat
Lucas Staus
Gupta Sushmita
Junichi Teruyama
Yuki Tokuni
Carlo Tosoni
Kutsuhisa Yamanaka

Abstracts of Invited Talks

Algorithms and Combinatorics on Two-Dimensional Strings

Paweł Gawrychowski

Institute of Computer Science, University of Wrocław, Poland
gawry@cs.uni.wroc.pl

Abstract. There is a vast literature concerning algorithmic and combinatorial properties of sequences of characters, called strings in the algorithmic context and words in the more combinatorial setting. The usual questions concerning such objects are related to different notions of repetitions (such as palindromes or squares), effectively locating occurrences of one such object in another (pattern matching), and building a data structure capable of answering such queries on demand (string indexing). For classical one-dimensional strings, many such questions are by now well-understood. However, this is not the case for the natural generalisation of strings to two-dimensions. A two-dimensional string is simply a two-dimensional array of characters. The complexity of exact pattern matching for two-dimensional strings has been already considered in the late 70s, and in the early 90s Galil and Park published their alphabetindependent linear-time algorithm algorithm [SICOMP 1996]. The main tool that enabled achieving such time complexity irrespectively of the size of the alphabet was periodicity, which turns out to be significantly more challenging to work with in two dimensions than in one dimension. However, for approximate pattern matching an algorithm faster than $\mathcal{O}\left(kn^2\right)$ (for an $n \times n$ text) has been given only very recently by Ellert, Gawrychowski, Górkiewicz, and Starikovskaya [SODA 2025]. Similarly, an extension of the classical notion of squares to two-dimension has been given by Apostolico and Brimkov [TCS 2000] a while go, but only recently asymptotically tight bounds have been given. Finally, for indexing two-dimensional strings a classical data structure of Giancarlo [SICOMP 1995] is capable of handling square patterns effectively, but this leaves handling the more general case of arbitrary rectangles as an open question. I will survey those and other results concerning two-dimensional strings.

Keywords. Two-dimensional strings · Repetitions · Pattern matching · String indexing

FPT Approximation Algorithms for Coverage and Satisfiability Problems

Saket Saurabh[1,2]
[1] The Institute of Mathematical Sciences, HBNI, Chennai, India
saket@imsc.res.in
[2] University of Bergen, Bergen, Norway
saket.saurabh@uib.no

Abstract. The classical MAX COVERAGE problem is defined as follows: given a universe $\mathcal{U}$ of n elements, a family $\mathcal{F}$ of subsets of $\mathcal{U}$, and an integer k, select k sets from $\mathcal{F}$ to maximize the number of elements covered. A greedy algorithm achieves a $(1 - 1/e)$-approximation in polynomial time, and this bound is tight assuming $\mathsf{P} \neq \mathsf{NP}$. However, from the parameterized complexity viewpoint, MAX COVERAGE is W[2]-hard when parameterized by k, and it is known to be hard to approximate within any constant factor in time $f(k) \cdot n^{o(k)}$, assuming the Exponential Time Hypothesis (ETH). This rules out the possibility of constant-factor FPT approximation algorithms in general.

To circumvent this hardness, we explore approximation schemes tailored to well-structured instances of the problem. An *Efficient Polynomial-Time Approximation Scheme (EPAS)* is an algorithm that, for every $\varepsilon > 0$, returns a $(1 + \varepsilon)$-approximate solution in time $f(k, \varepsilon) \cdot n^{\mathcal{O}(1)}$. A *Parameterized Approximation Scheme (PAS)* also returns a $(1 + \varepsilon)$-approximate solution, but allows the dependence on ε in the exponent of n; that is, its running time is of the form $f(k, \varepsilon) \cdot n^{g(\varepsilon)}$ for some function g.

In this survey, we list recent developments in designing EPASes and PASes for MAX COVERAGE and its variants over well-structured set systems. These include families with bounded VC-dimension, bounded intersection complexity, or geometric and graph-based constraints. We describe algorithmic techniques that exploit such structures to obtain efficient approximation schemes. We also highlight connections to related problems in partial coverage, capacitated coverage, connected variants, and structurally restricted MAX- SAT.

Keywords. Maximum Coverage, FPT Approximation, Satisfiability

Contents

Computational Geometry and Graph Drawing

Guarding a 1.5D Terrain with Imprecise Viewpoints

Vahideh Keikha[1], Maarten Löffler[2], Maria Saumell[3](✉), and Pavel Valtr[4]

[1] The Czech Academy of Sciences, Institute of Computer Science, Prague, Czech Republic
keikha@cs.cas.cz

[2] Department of Information and Computing Sciences, Utrecht University, Utrecht, The Netherlands
m.loffler@uu.nl

[3] Department of Theoretical Computer Science, Faculty of Information Technology, Czech Technical University in Prague, Prague, Czech Republic
maria.saumell@fit.cvut.cz

[4] Department of Applied Mathematics, Faculty of Mathematics and Physics, Charles University, Prague, Czech Republic
valtr@kam.mff.cuni.cz

Abstract. Given an n-vertex 1.5D terrain $\mathcal{T}$ and a set of m edges of $\mathcal{T}$, we study the problem of placing one viewpoint on each edge so that the total length of the visible portions of the terrain is maximized. We present an $O(n + m \log\ m)$ time $\frac{1}{2}$-approximation algorithm for the general problem, and polynomial-time algorithms for the cases $m = 1$ and $m = 2$. Additionally, we show that the problem of computing a point on $\mathcal{T}$ maximizing the visible portion of $\mathcal{T}$ can be solved in $O(n^3)$ time.

Keywords: 1.5D terrain · Visibility · Multiple Viewpoints · Data imprecision

1 Introduction

Visibility problems on terrains have been thoroughly studied in the literature. In this paper, we focus on *1.5D terrains*, defined as x-monotone polygonal lines in $\mathbb{R}^2$. A point on a 1.5D terrain $\mathcal{T}$ is *visible* from a viewpoint if the entire straight-line segment connecting them lies on and above $\mathcal{T}$. Despite the extensive literature on this topic, a natural variant has not been considered: What happens if guards (also called *viewpoints*) are *imprecise*, i.e., their exact location is not known, but it is contained in prescribed regions of the terrain? Formally, in a more general setting, let R be a set of m regions, each associated with one imprecise viewpoint. A point set $\mathcal{P}$ is a *realization* of R if there exists a bijection between $\mathcal{P}$ and R such that each point in $\mathcal{P}$ is contained in the corresponding

H. Fernau and B. Zhu (Eds.): IWOCA 2025, LNCS 15885, pp. 3–16, 2025.
https://doi.org/10.1007/978-3-031-98740-3_1

region in R (the point is called the *representative* of the region). Since many different realizations are possible, natural optimization problems arise. The general problem we consider is:

IV-mL (m Imprecise Viewpoints on set L): Given a 1.5D terrain $\mathcal{T}$ and a set L of m edges of $\mathcal{T}$, find a realization $\mathcal{P}$ of L such that the total length of the portion of $\mathcal{T}$ visible by $\mathcal{P}$ is maximized.

Our Results. In this paper, we present an $O(n + m \log m)$ time $\frac{1}{2}$-approximation algorithm for *IV-mL*. Furthermore, we consider the special cases $m = 1$ and $m = 2$. Finally, we show that with our techniques we can solve a very natural and fundamental problem that does not seem to have been solved before: Computing a point on $\mathcal{T}$ maximizing the visible portion of $\mathcal{T}$.

Related Work. A common goal in this type of problems is to make the entire terrain visible. However, computing the minimum number of viewpoints on the terrain to achieve this is NP-hard [9].

If we have only one available viewpoint but it can be placed above the terrain (such a viewpoint is called a *watchtower*), computing a watchtower of minimum height that covers the entire terrain is an easy problem: It can be solved in $O(n)$ time using the fact that the kernel of a simple polygon can be computed in linear time [10]. The problem becomes more interesting when we can place two watchtowers. If the watchtowers are allowed to be placed anywhere on $\mathcal{T}$, finding the optimal placement takes $O(n^3 \alpha(n) \log^3 n)$ time, where $\alpha(n)$ is the inverse Ackermann function [1]. If one or two watchtowers are restricted to be among the vertices of $\mathcal{T}$, finding the optimal placement takes $O(n^2 \log^4 n)$ time [1]. If the watchtowers lie on a fixed horizontal segment above the terrain, the minimum number of watchtowers that cover the terrain can be found in linear time [13]. Finally, another version that has been considered is that of guarding a finite set of input points lying on a terrain (instead of the whole terrain) with two or more watchtowers; see, e.g., [8].

A different type of guard that has been considered is a segment. In particular, it is possible to find in $O(n)$ time a segment of fixed length over a terrain such that the entire terrain is visible from the segment and the segment satisfies some optimization criteria: the segment is horizontal and has minimum height [14] or the segment is not necessarily horizontal and the maximum y-coordinate of its endpoints is minimized [12].

Another problem closely related to ours is that of constructing a data structure for reporting the visible portion of a terrain (or, more generally, a simple polygon) from a moving point along a given trajectory. If the trajectory is a sequence of m line segments, this problem can be solved in $O(mn \log n)$ time [4]. Additionally, an optimization problem generalizing ours (for fixed h, compute the watchtower at height h maximizing the visible portion of $\mathcal{T}$) has been considered in [5]. However, the algorithm is not correct (see Sect. 3.3).

The model of imprecision we adopt in this paper is called *region-based* and is among the simplest and most studied models for geometric uncertainty. In this model, instead of assuming that the exact location of each point is known,

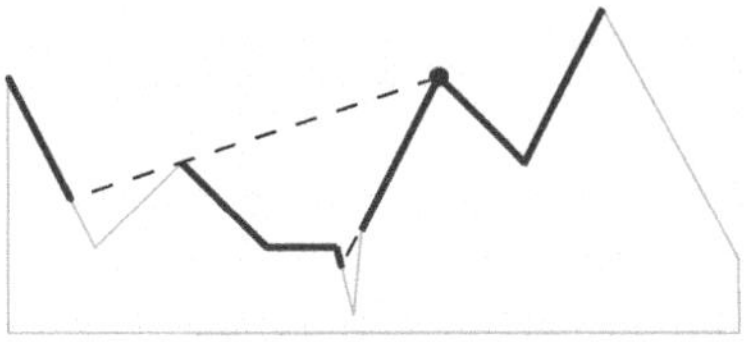

Fig. 1. The viewshed of a viewpoint. (For visualization purposes, in our figures we always "close" terrains by adding two vertical downward segments incident to v_1 and v_n, and one horizontal segment below $\mathcal{T}$.)

the point lies within a region (and its location can be anywhere in the region). Many problems in computational geometry have been studied in this model. Depending on the imprecision region (e.g., a line segment, a disk, etc.) and the studied problem, some variants can still be solved in polynomial time while some others become intractable even when the imprecise regions are simple.

Regarding previous research on terrains in the region-based model, there are some studies where the elevation of the vertices of $\mathcal{T}$ is imprecise. Then the aim might be to optimize some terrain parameter [11], or to compute a shortest watchtower guarding the entire terrain [12].

Preliminaries. We assume that $\mathcal{T}$ does not have three collinear vertices. Let $V = \{v_1, \ldots, v_n\}$ denote its vertex set ordered from left to right. The set of edges is $E = \{e_1, \ldots, e_{n-1}\}$, with $e_i = v_i v_{i+1}$. For each edge $e_i \in E$, let $e_i^l = v_i$ and $e_i^r = v_{i+1}$ denote the left and right endpoint of e_i, respectively.

For every vertex v_i, we say that v_i is *reflex* if the angle external to $\mathcal{T}$ between the edges incident to v_i is greater than π. If the angle is smaller than π, then the vertex is *convex*.

Two points on the terrain are said to be *visible* from each other if the line segment connecting them does not contain any point strictly below $\mathcal{T}$. The *viewshed* of a viewpoint p, denoted $\mathcal{V}_{\mathcal{T}}(p)$, is the union of all portions of $\mathcal{T}$ that are visible from p. See Fig. 1.

Omitted details and proofs will be given in the full version of this paper.

2 $\frac{1}{2}$-Approximation Algorithm

Let $\mathcal{P}$ be a set of viewpoints on $\mathcal{T}$. We denote the union of the viewsheds of all viewpoints in $\mathcal{P}$ by $\operatorname{Vis}(\mathcal{T}, \mathcal{P})$, and its length by $|\operatorname{Vis}(\mathcal{T}, \mathcal{P})|$.

We use $x(p)$ to denote the x-coordinate of a point p and $y(p)$ to denote its y-coordinate.

Lemma 1. *[Claim 2.1 in [2]] Let $a, b, c,$ and d be four points on $\mathcal{T}$ such that $x(a) < x(b) < x(c) < x(d)$. If a sees c and b sees d, then a sees d.*

We can use Lemma 1 to prove the following:

Lemma 2. *Let p_i be a point of e_i. Then $\mathcal{V}_\mathcal{T}(p_i) \subseteq \mathcal{V}_\mathcal{T}(e_i^l) \cup \mathcal{V}_\mathcal{T}(e_i^r)$.*

Proof. We prove that any point $q \in \mathcal{T}$ that is visible by p_i is also visible by e_i^l or e_i^r. If q lies on e_i, the claim is trivial. Suppose, without loss of generality, that q is to the right of e_i. If $p_i = e_i^l$ or $p_i = e_i^r$, the claim is again trivial. Otherwise, we use Lemma 1: since e_i^l sees e_i^r and p_i sees q, e_i^l also sees q. □

The following lemma is the base of our algorithm:

Lemma 3. *Let $\mathcal{P}_1$ and $\mathcal{P}_2$ be two realizations of L such that, for each l_i, in $\mathcal{P}_1$ the representative of l_i lies on one of the endpoints of that edge and in $\mathcal{P}_2$ the representative of l_i lies on the other endpoint. Let $P^* = \{p_1^*, p_2^*, \ldots, p_m^*\}$ denote an optimal realization. Then,* $\max\{|\operatorname{Vis}(\mathcal{T}, \mathcal{P}_1)|, |\operatorname{Vis}(\mathcal{T}, \mathcal{P}_2)|\} \geq \frac{1}{2}|\operatorname{Vis}(\mathcal{T}, P^*)|$.

Proof. By possibly renaming the edges of $\mathcal{T}$, we can assume that $l_i = e_i$ for $i = 1, 2, \ldots, m$. Then, $\operatorname{Vis}(\mathcal{T}, P^*) = \cup_{i=1}^m \mathcal{V}_\mathcal{T}(p_i^*)$. By Lemma 2, $\cup_{i=1}^m \mathcal{V}_\mathcal{T}(p_i^*) \subseteq \cup_{i=1}^m (\mathcal{V}_\mathcal{T}(e_i^l) \cup \mathcal{V}_\mathcal{T}(e_i^r)) = \operatorname{Vis}(\mathcal{T}, \mathcal{P}_1) \cup \operatorname{Vis}(\mathcal{T}, \mathcal{P}_2)$. Consequently,

$$
\begin{aligned}
|\operatorname{Vis}(\mathcal{T}, P^*)| &\leq |\operatorname{Vis}(\mathcal{T}, \mathcal{P}_1) \cup \operatorname{Vis}(\mathcal{T}, \mathcal{P}_2)| \\
&\leq |\operatorname{Vis}(\mathcal{T}, \mathcal{P}_1)| + |\operatorname{Vis}(\mathcal{T}, \mathcal{P}_2)| \\
&\leq 2\max\{|\operatorname{Vis}(\mathcal{T}, \mathcal{P}_1)|, |\operatorname{Vis}(\mathcal{T}, \mathcal{P}_2)|.
\end{aligned}
$$

□

The algorithm is a consequence of the previous lemma: Let $\mathcal{P}_1$ and $\mathcal{P}_2$ be two realizations of L as described in Lemma 3. We compute $|\operatorname{Vis}(\mathcal{T}, \mathcal{P}_1)|$ and $|\operatorname{Vis}(\mathcal{T}, \mathcal{P}_2)|$ in $O(n + m \log m)$ time using the algorithm from [6], and take the realization achieving the maximum among the two.

Theorem 1. *For the IV-mL problem, a $\frac{1}{2}$-approximation can be computed in $O(n + m \log m)$ time.*

3 Case $m = 1$

We observe that the problem is not trivial because the optimal realization is not necessarily at an endpoint of the edge; see Fig. 2 for an illustrative example.

3.1 Preliminaries

Let e be the edge of $\mathcal{T}$ equal to the imprecise viewpoint. We parametrize e by $\sigma \in [0, 1]$, where $\sigma = 0$ corresponds to e^l and $\sigma = 1$ corresponds to e^r. For each edge $e_x \in \mathcal{T}$ and each $\sigma \in [0, 1]$, we define $\mathcal{F}_x(\sigma) := |\mathcal{V}_\mathcal{T}(\sigma) \cap e_x|$.

Without loss of generality, we assume that e is to the left of e_x.

The following observation was made in [6] (proof of Theorem 1):

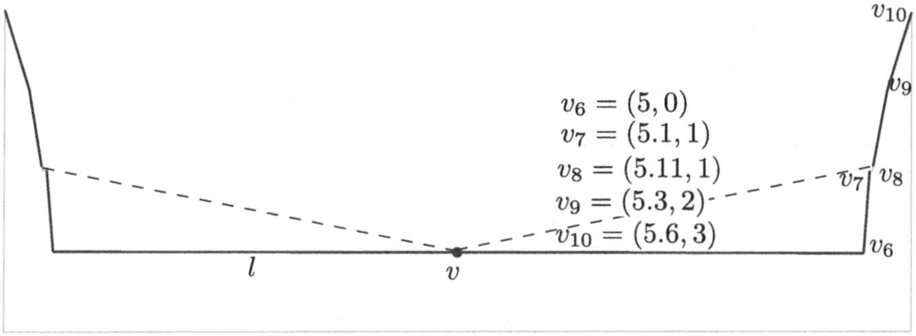

Fig. 2. Example with one imprecise viewpoint on edge l. For the coordinates of vertices $v_1, v_2, \ldots, v_5$, we note that the terrain is symmetric with respect to the y-axis. The point $v = (0, 0)$, which corresponds to the optimal placement inside l, lies in the interior of the edge.

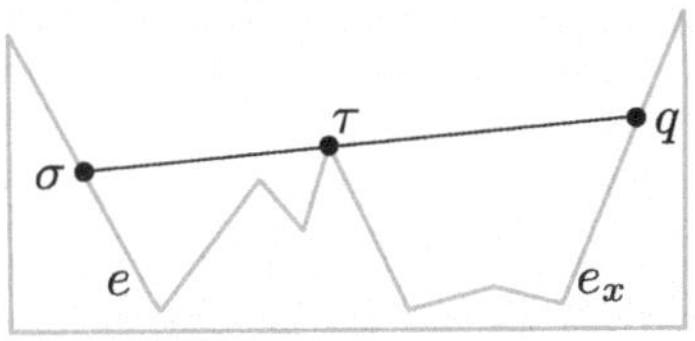

Fig. 3. Vertex τ obstructs the visibility of e_x from point σ in e.

Observation 1. *Consider an edge e_x of $\mathcal{T}$ and a point σ to the left of e_x. If σ sees some interior point q of e_x, then σ also sees all the segment from q to the rightmost endpoint of e_x.*

Lemma 4. *The function $\mathcal{F}_x(\sigma)$ is non-increasing in $[0, 1)$.*

Proof. Let $\sigma_1, \sigma_2 \in [0, 1)$ with $\sigma_1 < \sigma_2$. Let q be a point of e_x visible from σ_2. Since σ_2 is an interior point of e, by Observation 1 (on a mirror image of our situation), q sees all the segment from σ_2 to the leftmost endpoint of e. Thus, q also sees σ_1. In consequence, $\mathcal{V}_\mathcal{T}(\sigma_2) \cap e_x \subseteq \mathcal{V}_\mathcal{T}(\sigma_1) \cap e_x$ and $\mathcal{F}_x(\sigma_2) \leq \mathcal{F}_x(\sigma_1)$. $\square$

The following two claims are a direct consequence of Lemma 4:

Claim 1. *Let σ_1 be a value in $[0, 1)$ such that $\mathcal{F}_x(\sigma_1) = 0$. Then, $\mathcal{F}_x(\sigma_2) = 0$ for $\sigma_2 \in [\sigma_1, 1)$.*

Claim 2. *Let σ_2 be a value in $(0, 1)$ such that $\mathcal{F}_x(\sigma_2) = |e_x|$. Then, $\mathcal{F}_x(\sigma_1) = |e_x|$ for $\sigma_1 \in [0, \sigma_2)$.*

Definition 1. *Let e and e_x be edges of $\mathcal{T}$ such that e is to the left of e_x.*

- *Vertex τ* obstructs the visibility of e_x from point σ in e *if $\tau \notin \{e^r, e_x^l, e_x^r\}$ and the ray with origin at σ and passing through τ intersects e_x at a point q that is visible from σ.*

- *We denote by* n_x *the number of vertices* τ *of* $\mathcal{T}$ *such that there exists some* σ *in* e *satisfying that* τ *obstructs the visibility of* e_x *from point* σ.

An example of a vertex obstructing the visibility of an edge is given in Fig. 3.

Lemma 5. *Let* $\sigma_0, \sigma_1 \in [0,1]$, *with* $\sigma_0 < \sigma_1$, *be two points in* e *such that, for any* $\sigma \in [\sigma_0, \sigma_1]$, *the visibility of* e_x *from* σ *is obstructed by the same vertex* τ. *Then,*

$$\mathcal{F}_x(\sigma) = \frac{A\sigma + B}{C\sigma + D}, \text{ for } \sigma \in [\sigma_0, \sigma_1],$$

where A, B, C, D *are functions of the coordinates of* σ_0, σ_1, e_x *and* τ.

To fully characterize $\mathcal{F}_x(\sigma)$, we need some additional observations.

Claim 3. *Let* $\sigma_1, \sigma_2 \in [0,1)$, *with* $\sigma_1 < \sigma_2$, *be two points in* e *such that the visibility of* e_x *from* σ_1 *and* σ_2 *is obstructed by* τ_1 *and* τ_2, *respectively* ($\tau_2 \neq \tau_1$). *Then, for any* $\sigma_3 > \sigma_2$, *the visibility of* e_x *from* σ_3 *is not obstructed by* τ_1.

Lemma 6. *(a) If both endpoints of* e *see both endpoints of* e_x, *then every point of* e *sees every point of* e_x. *(b) If none of the endpoints of* e *sees any of the endpoints of* e_x, *then no point of* e *sees any point of* e_x.

We are now ready to characterize the function $\mathcal{F}_x(\sigma)$.

Lemma 7. $\mathcal{F}_x(\sigma)$ *is a piecewise-defined function. Each of the sub-functions has one of the following shapes:*

(i) ℓ, *with* $0 \leq \ell \leq |e_x|$;
(ii) $\frac{A\sigma+B}{C\sigma+D}$.

The number of pieces (i.e., sub-functions) of $\mathcal{F}_x(\sigma)$ *is at most* $n_x + 2$.

Proof. Let σ be a point in e. If σ sees no point of e_x or the entire e_x, we have $\mathcal{F}_x(\sigma) = 0$ or $\mathcal{F}_x(\sigma) = |e_x|$, respectively. If σ sees a portion of e_x of positive length that is not the whole e_x, this portion is a segment from some interior point q to e_x^r and σq passes through some vertex τ of $\mathcal{T}$. If $\tau = e^r$, $\mathcal{F}_x(\sigma') = \ell$, with $0 < \ell < |e_x|$, for σ' close enough to σ. Otherwise, by Lemma 5, $\mathcal{F}_x(\sigma') = \frac{A\sigma'+B}{C\sigma'+D}$, for $\sigma' \geq \sigma$ close enough to σ or for $\sigma' \leq \sigma$ close enough to σ, or for both (where A, B, C, D are functions of the coordinates of σ_0, σ_1, e_x and τ).

Let us next analyze the number of pieces of $\mathcal{F}_x(\sigma)$.

If e^r is reflex, we treat one edge separately: Let e_x be the edge containing the first intersection (if any) of $\mathcal{T}$ with the ray with origin at e^l and passing through e^r. Then, $\mathcal{F}_x(\sigma)$ is constant for all $\sigma \in [0,1)$. Consequently, the number of pieces of $\mathcal{F}_x(\sigma)$ is at most 2. Additionally, $n_x \leq 1$: Since $\mathcal{T}$ does not have three collinear vertices, a vertex τ might obstruct the visibility of e_x from a point σ only if $\sigma = e^r$. In the remainder of the proof, we assume that e_x is not this special edge.

Let σ_1 be the smallest value (if any) in $[0, 1)$ for which $\mathcal{F}_x(\sigma_1) = 0$. By Claim 1, $\mathcal{F}_x(\sigma_2) = 0$ for $\sigma_2 \in [\sigma_1, 1)$. Hence, there is at most one interval of $[0, 1]$ where $\mathcal{F}_x(\sigma) = 0$.

Similarly, we show that $\mathcal{F}_x(\sigma)$ is equal to $|e_x|$ at most at one interval. If $\mathcal{F}_x(\sigma) = |e_x|$ only at $\sigma = 1$, we are done. Otherwise, for any $\sigma_2 \in (0, 1)$ such that $\mathcal{F}_x(\sigma_2) = |e_x|$, Claim 2 implies that $\mathcal{F}_x(\sigma_1) = |e_x|$ for $\sigma_1 \in [0, \sigma_2)$. Finally, by Lemma 6a, the following situation is not possible: $\mathcal{F}_x(\sigma_1) = |e_x|$ for $\sigma_1 \in [0, \sigma_2]$, $\mathcal{F}_x(\sigma_3) < |e_x|$ for $\sigma_3 \in (\sigma_2, 1)$ and $\mathcal{F}_x(1) = |e_x|$.

Finally, suppose that there exist $\sigma \in [0, 1)$ and a vertex τ of $\mathcal{T}$ such that the visibility of e_x from σ is obstructed by τ. Let $[\sigma_1, \sigma_2]$ be the largest interval containing σ such that the visibility of e_x from any point of the interval is obstructed by τ. If $\sigma_2 \neq 1$, there exists some $\sigma_4 > \sigma_2$ such that the visibility of e_x from any $\sigma_3 \in (\sigma_2, \sigma_4]$ is not obstructed by τ. By Lemma 4, we have that either $\mathcal{F}_x(\sigma_4) = 0$ or the visibility of e_x from σ_4 is obstructed by some τ' with $\tau' \neq \tau$. In the first case, Lemma 4 yields that $\mathcal{F}_x(\cdot)$ is equal to zero for values in $(\sigma_4, 1)$. Additionally, it is easy to see that in this case the visibility of e_x from $\sigma = 1$ is not obstructed by τ. In the second case, by Claim 3, for any $\sigma_5 > \sigma_4$, the visibility of e_x from σ_5 is not obstructed by τ. In both cases, there is only one interval $[\sigma_1, \sigma_2] \subseteq [0, 1]$ such that the visibility of e_x from any point of the interval is obstructed by τ.

In conclusion, there might be at most one interval where $\mathcal{F}_x(\sigma) = 0$, at most one interval where $\mathcal{F}_x(\sigma) = |e_x|$ and, for every τ accounted in n_x, at most one interval such that the visibility of e_x from any point of the interval is obstructed by τ. This implies that the number of pieces of $\mathcal{F}_x(\sigma)$ is at most $n_x + 2$. □

Definition 2. *A* critical point *of e is a point on e that is collinear with and is visible from two different vertices τ and τ' of $\mathcal{T}$ simultaneously such that τ and τ' are on the same side of the line that contains e.*

Chen and Daescu [4] proved that the number of critical points of e is $O(n)$ (note that their result is for simple polygons). We use this result and our previous lemmas to prove the following:

Theorem 2. *The number of pieces of $\mathcal{F}_x(\sigma)$ added over all edges $e_x \in \mathcal{T}$ is $O(n)$.*

3.2 Computation of the Functions $\mathcal{F}_x(\sigma)$

We sketch how to compute the functions $\mathcal{F}_x(\sigma)$ for all e_x to the right of e.

First, we compute $\mathcal{V}_\mathcal{T}(e^l)$ and $\mathcal{V}_\mathcal{T}(e^r)$. For every edge e_x such that e^r or e^l sees a portion of e_x that is not the whole e_x, we store the vertex obstructing the visibility of e_x (or e^r/e^l_x).[1] Afterwards, we compute the set of critical points of e sorted along the edge. For every critical point we store the two vertices of $\mathcal{T}$ associated with it.

[1] There cannot be more than one vertex obstructing the visibility of e_x from e^r or e^l due to the fact that $\mathcal{T}$ does not have three collinear vertices.

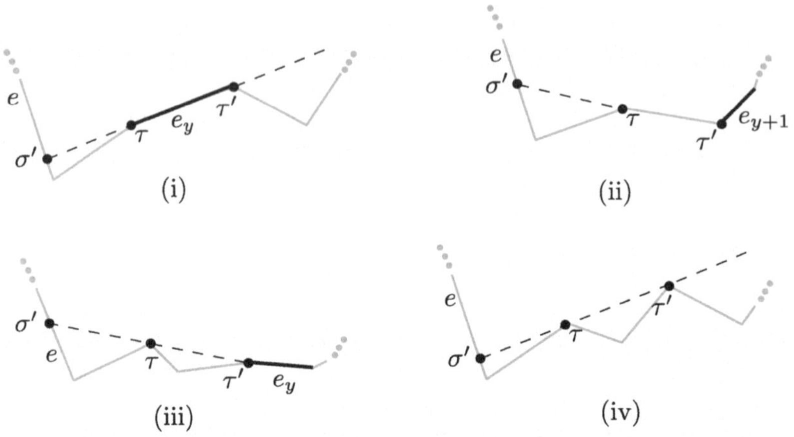

Fig. 4. The four different types of critical points.

We classify the critical points into four types (see Fig. 4). We denote by σ' the critical point, by τ the leftmost terrain vertex associated with σ', and by τ' the rightmost such terrain vertex. The four types are based on the answers to: Does there exist a portion of $\mathcal{T}$ of positive length starting at τ (i.e., having τ as its left endpoint) that is visible from σ'? And does there exist such a portion starting at τ'? If the answers are *yes, no*, the type is (i). If the answers are *yes, yes*, the type is (ii). If the answers are *no, yes*, the type is (iii). If the answers are *no, no*, the type is (iv).

In some cases, we can already determine some functions. In particular, for types (i) and (ii), τ and τ' are endpoints of the same edge e_y. Hence, we have that $\mathcal{F}_y(\sigma) = |e_y|$ for all $\sigma \in [0, \sigma']$ and $\mathcal{F}_y(\sigma) = 0$ for all $\sigma \in (\sigma', 1]$. For type (ii), since the portion of $\mathcal{T}$ immediately to the right of τ' is visible from σ', the whole edge e_{y+1} (which has τ' as left endpoint) also is. Thus, $\mathcal{F}_{y+1}(\sigma) = |e_{y+1}|$ for all $\sigma \in [0, \sigma']$. Finally, for values of σ infinitesimally greater than σ', the visibility of e_{y+1} from σ is obstructed by τ. We assign the critical point σ' to e_{y+1}.

In type (iii), τ' is the left endpoint of some edge e_y. Hence, we have that $\mathcal{F}_y(\sigma) = |e_y|$ for all $\sigma \in [0, \sigma']$. Furthermore, for values of σ infinitesimally greater than σ', the visibility of e_y from σ is obstructed by τ. We assign the critical point σ' to e_y.

In types (i) and (iv), we compute via a ray-shooting query the first intersection after τ, τ' between $\mathcal{T}$ and the ray with origin at σ' and passing through τ, τ', and we assign the critical point σ' to the intersected edge (if any).

After this preprocessing, we compute $\mathcal{F}_x(\sigma)$ for the remaining e_x. We start by sorting the critical points assigned to e_x. Note that $\mathcal{F}_x(0)$ and $\mathcal{F}_x(1)$ are obtained from $\mathcal{V}_\mathcal{T}(e^l)$ and $\mathcal{V}_\mathcal{T}(e^r)$, respectively. If $\mathcal{F}_x(0) = 0$, by Lemma 4 we are done. If e^r is reflex and e_x is the edge containing the first intersection of $\mathcal{T}$

with the ray with origin at e^l and passing through e^r, then $\mathcal{F}_x(\sigma)$ is constant for all $\sigma \in [0, 1)$.

Next, suppose that $\mathcal{F}_x(0) \in (0, |e_x|)$. Then there is a vertex τ obstructing the visibility of e_x from $\sigma = 0$ (τ has been found during the computation of $\mathcal{V}_\mathcal{T}(e^l)$). If e_x has critical points assigned to it, we take the first unprocessed one. For σ between 0 and the value of the critical point, the visibility of e_x is obstructed by τ. We compute $\mathcal{F}_x(\sigma)$ using Lemma 5. For values of σ infinitesimally greater than that of the critical point, the visibility of e_x is obstructed by the vertex associated to the critical point different from τ. Thus, we are in the same situation as in the beginning of this case. Otherwise, e_x has no unprocessed critical points assigned to it, so either the visibility of e_x is obstructed by τ until $\sigma = 1$, or the visibility of e_x is obstructed by τ until some value $\sigma' < 1$ and afterwards $\mathcal{F}_x(\cdot) = 0$. It is not difficult to distinguish between both cases and compute $\mathcal{F}_x(\sigma)$ in the rest of the domain.

The last case is when $\mathcal{F}_x(0) = |e_x|$. If $\mathcal{F}_x(1) = |e_x|$, it is easy to see that $\mathcal{F}_x(\sigma) = |e_x|$ for all $\sigma \in [0, 1]$. Otherwise, we deduce that there is a critical point σ' of type (ii) or (iii) whose two associated vertices are e_x^l and some vertex τ. The critical point corresponds to the first one assigned to e_x. For values of σ infinitesimally greater than σ', the visibility of e_x from σ is obstructed by τ. Thus, we are in the situation described in the previous paragraph.

We observe that the computation of the functions $\mathcal{F}_x(\sigma)$ for all e_x to the left of e is symmetric. We next analyze the total running time of the computation of all functions.

Computing $\mathcal{V}_\mathcal{T}(e^l)$ and $\mathcal{V}_\mathcal{T}(e^r)$ can be done in $O(n)$ time by turning the terrain into a simple polygon and using the algorithm from [7]. Additionally, computing the set of critical points of e sorted along the edge can be done in $O(n \log n)$ [4].

During the preprocessing of critical points, for every critical point of types (i) or (iv), we perform a ray-shooting query in $O(\log n)$ time [3]. All other operations performed take $O(1)$ time. Since the number of critical points is $O(n)$, the total cost of this phase is $O(n \log n)$.

Afterwards, the functions $\mathcal{F}_x(\sigma)$ are calculated. Apart from initialization steps and possibly a ray-shooting query, the cost of computing of $\mathcal{F}_x(\sigma)$ takes constant time per critical point assigned to e_x. We conclude the following:

Theorem 3. *The set of functions $\mathcal{F}_x(\sigma)$ can be computed in $O(n \log n)$ time.*

3.3 Optimization

By Theorem 2, $[0, 1]$ can be divided into a set $\mathcal{I}$ of $O(n)$ intervals such that, in every interval, each $\mathcal{F}_x(\sigma)$ is defined by only one piece. Let $[\sigma_0, \sigma_1] \in \mathcal{I}$. Since $\sum_{e_x \neq e} \mathcal{F}_x(\sigma)$ is potentially the sum of $\Theta(n)$ functions of type $\frac{A\sigma+B}{C\sigma+D}$, some of which are increasing and some of which are decreasing, it could in principle be possible that $\sum_{e_x \neq e} \mathcal{F}_x(\sigma)$ has $\Theta(n)$ local maxima in $\mathcal{I}$. By computing $(\mathcal{F}_x)''(\sigma)$, we can show that this is not the case.

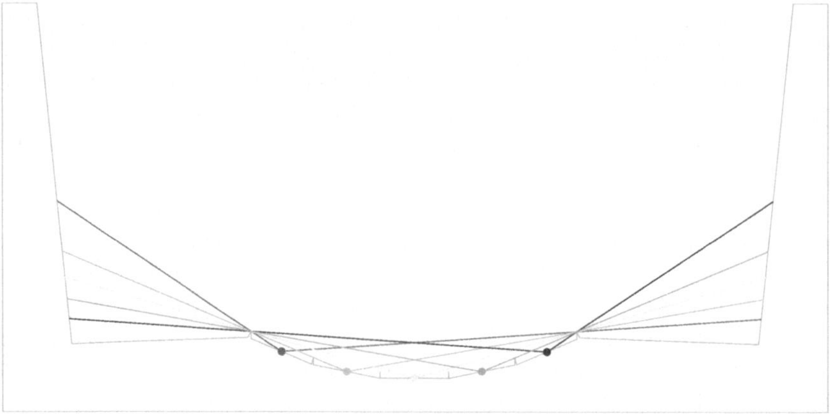

Fig. 5. A terrain with $\Theta(n)$ edges e such that $\sum_{e_x \neq e} \mathcal{F}_x(\sigma)$ have a local maximum.

Lemma 8. *$\sum_{e_x \neq e} \mathcal{F}_x(\sigma)$ is either constant or strictly concave in each $[\sigma_0, \sigma_1] \in \mathcal{I}$.*

By Lemma 8, $\sum_{e_x \neq e} \mathcal{F}_x(\sigma)$ has at most one local maximum in $[\sigma_0, \sigma_1]$, so the global maximum is attained at this local maximum, at $\sigma = \sigma_0$ or at $\sigma = \sigma_1$. Since there does not seem to be an easy geometrical characterization of the local maximum (recall the example in Fig. 2), we solve $(\mathcal{F}_x)'(\sigma) = 0$. We assume that our model of computation is able to compute roots of polynomials of degree d in $O(d)$ time. Thus, we can solve $(\mathcal{F}_x)'(\sigma) = 0$ in $[\sigma_0, \sigma_1]$ in $O(n)$ time. Repeating this procedure for every interval in $\mathcal{I}$, we obtain the following result:

Theorem 4. *The problem IV-1L can be solved in $O(n^2)$ time.*

If we run our algorithm for every edge e of the terrain, we can find a point on $\mathcal{T}$ that sees the largest portion of $\mathcal{T}$.

Theorem 5. *We can compute a point on $\mathcal{T}$ maximizing the visible portion of $\mathcal{T}$ in $O(n^3)$ time.*

Note 1. Computing such a point appears like a fundamental problem on visibility on terrains for which there only seems to be an incorrect algorithm: In [5], the authors study the following problem: Given a 1.5D terrain $\mathcal{T}$, find the location of a watchtower of given height h maximizing the visible portion of $\mathcal{T}$. Their approach is as follows: Let $\mathcal{T}'$ be the upward translation of $\mathcal{T}$ by height h. For each reflex vertex τ, a *grazing ray* is a ray originated at τ and extending upward following the direction of one of the two terrain edges incident to τ. The authors claim that the candidates points for the base of the optimal watchtower are the terrain vertices and the vertical projection on $\mathcal{T}$ of the intersections between $\mathcal{T}'$ and the grazing rays. However, if we consider the terrain in Fig. 2 and set $h = 0.1$ (or smaller), the base of the optimal watchtower is at the origin, but there is no candidate at the origin as defined in the algorithm.

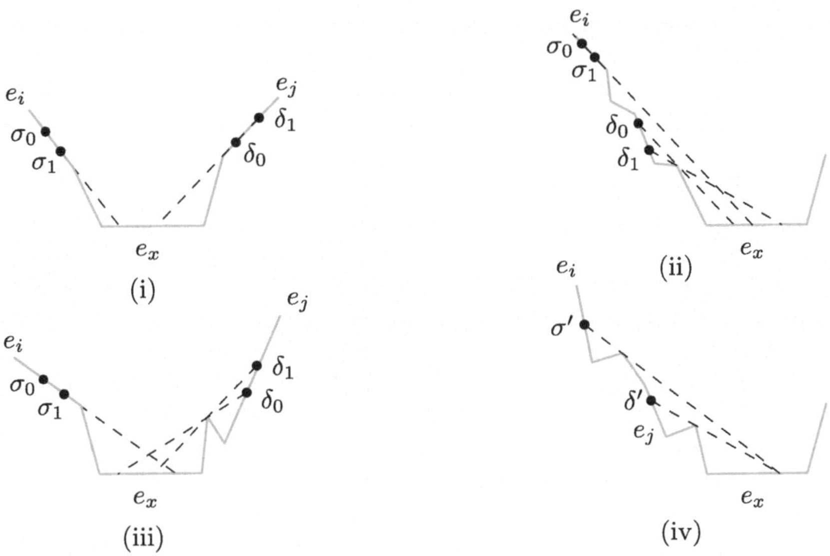

Fig. 6. (i) $\mathcal{F}_x^i(\sigma)$ and $\mathcal{F}_x^j(\delta)$ are constant in I and J, respectively, so $\mathcal{F}_x^{i,j}(\sigma,\delta)$ is constant (its value is $|e_x|$) in $I \times J$. (ii) $\mathcal{F}_x^i(\sigma)$ is constant and $\mathcal{F}_x^j(\delta)$ is a hyperbolic function. Let $\delta' \in [\delta_0, \delta_1]$ be the value such that $\mathcal{F}_x^j(\delta') = \mathcal{F}_x^i(\sigma_0)$. Then, below the line $\delta = \delta'$ we have $\mathcal{F}_x^{i,j}(\sigma,\delta) = \mathcal{F}_x^j(\delta)$ and above the line we have $\mathcal{F}_x^{i,j}(\sigma,\delta) = \mathcal{F}_x^i(\sigma)$. (iii) $\mathcal{F}_x^i(\sigma)$ is constant, $\mathcal{F}_x^j(\delta)$ is a hyperbolic function and $\mathcal{F}_x^{i,j}(\sigma,\delta) = \mathcal{F}_x^i(\sigma) + \mathcal{F}_x^j(\delta)$. (iv) $\mathcal{F}_x^i(\sigma)$ and $\mathcal{F}_x^j(\delta)$ are hyperbolic functions, and σ', δ' satisfy that $\mathcal{F}_x^i(\sigma') = \mathcal{F}_x^j(\delta')$. Then, for a small enough $\varepsilon > 0$, we have that $\mathcal{F}_x^{i,j}(\sigma', \delta' + \varepsilon) = \mathcal{F}_x^i(\sigma')$ and $\mathcal{F}_x^{i,j}(\sigma', \delta' - \varepsilon) = \mathcal{F}_x^j(\delta' - \varepsilon)$.

We finally show an example where there are many local maxima. The terrain, illustrated in Fig. 5, has $\Theta(n)$ edges e such that $\sum_{e_x \neq e} \mathcal{F}_x(\sigma)$ have a local maximum and each of these edges has $\Theta(n)$ pieces in the description of $\mathcal{F}_x$ added over all edges e_x.

4 IV-2L

We briefly sketch how to solve this variant in polynomial time.

Suppose that the imprecise viewpoints are located on edges e_i and e_j with $i < j$. We parameterize them by σ and δ, respectively. With slight abuse of notation, we will also use σ and δ to refer to the viewpoints themselves. Thus, every realization can be seen as a pair $(\sigma, \delta) \in [0,1] \times [0,1]$.

For each edge $e_x \in \mathcal{T}$ and each realization σ (resp., δ) of e_i (resp., e_j), we define $\mathcal{F}_x^i(\sigma)$ (resp., $\mathcal{F}_x^j(\delta)$) as the length of the portion of e_x that is visible from σ (resp., δ). So $\mathcal{F}_x^i(\sigma) = |\mathcal{V}_\mathcal{T}(\sigma) \cap e_x|$ and $\mathcal{F}_x^j(\delta) = |\mathcal{V}_\mathcal{T}(\delta) \cap e_x|$. Recall that $[0,1]$ can be divided into a set $\mathcal{I}$ of $O(n)$ intervals such that, in every interval, each $\mathcal{F}_x^i(\sigma)$ is defined by one sub-function. Analogously, $[0,1]$ can be divided into a set $\mathcal{J}$ of $O(n)$ intervals such that, in every interval, each $\mathcal{F}_x^j(\delta)$ is defined by one sub-function.

Let $I = [\sigma_0, \sigma_1] \subseteq [0,1]$ be an interval in $\mathcal{I}$ and $J = [\delta_0, \delta_1] \subseteq [0,1]$ be an interval in $\mathcal{J}$. We would like to describe the function $\mathcal{F}_x^{i,j}(\sigma,\delta) = |(\mathcal{V}_\mathcal{T}(\sigma) \cup \mathcal{V}_\mathcal{T}(\delta)) \cap e_x|$ giving the length of the portion(s) of e_x visible by σ or δ, for values $(\sigma,\delta) \in I \times J$. There are several cases:

1. If either $\mathcal{F}_x^i(\sigma)$ or $\mathcal{F}_x^j(\delta)$ is either 0 or $|e_x|$ in the whole domain, or if both functions are constant, then $\mathcal{F}_x^{i,j}(\sigma,\delta)$ has the same algebraic expression in the whole domain $I \times J$ (it might be a constant, or one of $\mathcal{F}_x^i(\sigma)$ or $\mathcal{F}_x^j(\delta)$ if the other one is 0). See Fig. 6 (i) for an example.
2. If $\mathcal{F}_x^i(\sigma) = \ell \in (0, |e_x|)$, $\mathcal{F}_x^j(\delta) = \frac{A\delta+B}{C\delta+D}$ and e_x is on the same side of e_i and e_j, then $\mathcal{F}_x^{i,j}(\sigma,\delta) = \mathcal{F}_x^i(\sigma)$, or $\mathcal{F}_x^{i,j}(\sigma,\delta) = \mathcal{F}_x^j(\delta)$, or there exists a horizontal line $\delta = \delta'$ such that for (σ,δ) above the line we have $\mathcal{F}_x^{i,j}(\sigma,\delta)$ is equal to one of $\mathcal{F}_x^i(\sigma)$ or $\mathcal{F}_x^j(\delta)$, for (σ,δ) below the line $\mathcal{F}_x^{i,j}(\sigma,\delta)$ is equal to the other function, and for (σ,δ) on the line $\mathcal{F}_x^{i,j}(\sigma,\delta)$ is equal to both functions because $\mathcal{V}_\mathcal{T}(\sigma) \cap e_x = \mathcal{V}_\mathcal{T}(\delta') \cap e_x$ for all $\sigma \in I$. See Fig. 6 (ii) for an example. The situation is similar if $\mathcal{F}_x^i(\sigma)$ is a piece of a hyperbolic function and $\mathcal{F}_x^j(\delta)$ is a constant.
3. If $\mathcal{F}_x^i(\sigma) = \ell \in (0, |e_x|)$, $\mathcal{F}_x^j(\delta) = \frac{A\delta+B}{C\delta+D}$ and e_x is to the right of e_i and to the left of e_j, then $\mathcal{F}_x^{i,j}(\sigma,\delta) = |e_x|$, $\mathcal{F}_x^{i,j}(\sigma,\delta) = \mathcal{F}_x^i(\sigma) + \mathcal{F}_x^j(\delta)$, or there exists a horizontal line $\delta = \delta'$ such that for (σ,δ) above the line we have $\mathcal{F}_x^{i,j}(\sigma,\delta) = |e_x|$ and for (σ,δ) below the line we have $\mathcal{F}_x^{i,j}(\sigma,\delta) = \mathcal{F}_x^i(\sigma) + \mathcal{F}_x^j(\delta)$. See Fig. 6 (iii) for an example. The situation is similar if $\mathcal{F}_x^i(\sigma)$ is a piece of a hyperbolic function and $\mathcal{F}_x^j(\delta)$ is a constant.
4. If $\mathcal{F}_x^i(\sigma) = \frac{A\sigma+B}{C\sigma+D}$ and $\mathcal{F}_x^j(\delta) = \frac{A'\delta+B'}{C'\delta+D'}$ and e_x is on the same side of e_i and e_j, then $\mathcal{F}_x^{i,j}(\sigma,\delta) = \mathcal{F}_x^i(\sigma)$, or $\mathcal{F}_x^{i,j}(\sigma,\delta) = \mathcal{F}_x^j(\delta)$, or there exists a σ-monotone (and δ-monotone) hyperbolic curve in $I \times J$ such that for (σ,δ) above the curve we have $\mathcal{F}_x^{i,j}(\sigma,\delta)$ is equal to one of $\mathcal{F}_x^i(\sigma)$ or $\mathcal{F}_x^j(\delta)$, for (σ,δ) below the curve $\mathcal{F}_x^{i,j}(\sigma,\delta)$ is equal to the other function, and for (σ,δ) on the curve $\mathcal{F}_x^{i,j}(\sigma,\delta)$ is equal to both functions because the curve visits precisely the values $(\sigma,\delta) \in I \times J$ such that $\mathcal{V}_\mathcal{T}(\sigma) \cap e_x = \mathcal{V}_\mathcal{T}(\delta) \cap e_x$. See Fig. 6 (iv) for an example.
5. If $\mathcal{F}_x^i(\sigma) = \frac{A\sigma+B}{C\sigma+D}$ and $\mathcal{F}_x^j(\delta) = \frac{A'\delta+B'}{C'\delta+D'}$ and e_x is to the right of e_i and to the left of e_j, then $\mathcal{F}_x^{i,j}(\sigma,\delta) = |e_x|$, or $\mathcal{F}_x^{i,j}(\sigma,\delta) = \mathcal{F}_x^i(\sigma) + \mathcal{F}_x^j(\delta)$, or there exists σ-monotone hyperbolic curve in $I \times J$ such that for (σ,δ) on or above the curve we have $\mathcal{F}_x^{i,j}(\sigma,\delta) = |e_x|$, and for (σ,δ) below the curve we have $\mathcal{F}_x^{i,j}(\sigma,\delta) = \mathcal{F}_x^i(\sigma) + \mathcal{F}_x^j(\delta)$.

In summary, for a fixed e_x, the domain $I \times J$ might get divided by a curve into at most two portions such that the function giving $\mathcal{F}_x^{i,j}(\sigma,\delta)$ within each portion is the same. We repeat this procedure for all edges e_x of $\mathcal{T}$ and consider the obtained arrangement of curves in $I \times J$. Since the curves are either axis-aligned segments or hyperbolic curves, every two curves intersect at most twice, and thus the arrangement has $O(n^2)$ faces.

To obtain the value maximizing $\sum_{e_x} \mathcal{F}_{i,j}^x(\sigma,\delta)$ in $I \times J$, we traverse the faces of the arrangement one by one. In each face, we know the algebraic expression of $\mathcal{F}_x^{i,j}(\sigma, \delta)$ for every e_x. Since this expression is $\mathcal{F}_x^i(\sigma)$, $\mathcal{F}_x^j(\delta)$, $|e_x|$ or $\mathcal{F}_x^i(\sigma) + \mathcal{F}_x^j(\delta)$, we can use Lemma 8 to derive that there are at most one value of σ and one value of δ (in the interior of the face of the arrangement) that must be checked when looking for local maxima. We also traverse the boundary of the face of the arrangement to see if the maximum of $\sum_{e_x} \mathcal{F}_{i,j}^x(\sigma, \delta)$ in the face is achieved on the boundary.

Since the domain $[0, 1] \times [0, 1]$ is partitioned into $O(n^2)$ rectangles of type $I \times J$, we conclude the following:

Theorem 6. *The problem IV-2L can be solved in polynomial time.*

5 Open Problems

The main open problems that arise from our work are related to the complexity of the problem IV-mL for general values of m: Is the problem NP-hard? Is it fixed parameter tractable?

Acknowledgement. V.K. was supported by the CAS PPPLZ grant L100302301, and the institutional support RVO: 67985807. M.S. and P.V. were supported by the Czech Science Foundation, grant number 23-04949X.

References

1. Agarwal, P.K., et al.: Guarding a terrain by two watchtowers. Algorithmica **58**(2), 352–390 (2010)
2. Ben-Moshe, B., Katz, M.J., Mitchell, J.: A constant-factor approximation algorithm for optimal 1.5D terrain guarding. SIAM J. Comput. **36**(6), 1631–1647 (2007)
3. Chazelle, B., et al.: Ray shooting in polygons using geodesic triangulations. Algorithmica **12**(1), 54–68 (1994)
4. Chen, D.Z., Daescu, O.: Maintaining visibility of a polygon with a moving point of view. Inf. Process. Lett. **65**(5), 269–275 (1998)
5. Gewali, L., Dahal, B.: Algorithms for tower placement on terrain. In: Latifi, S. (ed.) 16th International Conference on Information Technology-New Generations (ITNG 2019). AISC, vol. 800, pp. 551–556. Springer, Cham (2019). https://doi.org/10.1007/978-3-030-14070-0_77
6. Hurtado, F., et al.: Terrain visibility with multiple viewpoints. Int. J. Comput. Geom. Appl. **24**(04), 275–306 (2014)
7. Joe, B., Simpson, R.B.: Corrections to Lee's visibility polygon algorithm. BIT Numer. Math. **27**(4), 458–473 (1987)
8. Kang, B., Choi, J., Han, J., Ahn, H.-K.: Guarding points on a terrain by watchtowers. In: Proceedings of the 36th Canadian Conference on Computational Geometry, pp. 41–47 (2024)
9. King, J., Krohn, E.: Terrain guarding is NP-hard. SIAM J. Comput. **40**(5), 1316–1339 (2011)
10. Lee, D.-T., Preparata, F.P.: An optimal algorithm for finding the kernel of a polygon. J. ACM **26**(3), 415–421 (1979)

11. Lubiw, A., Stroud, G.: Computing realistic terrains from imprecise elevations. Comput. Geom. Topol. **2**(2), 3–1 (2023)
12. McCoy, B., Zhu, B., Dutt, A.: Guarding precise and imprecise polyhedral terrains with segments. In: Proceedings of the 17th International Conference on Combinatorial Optimization and Applications, pp. 323–336 (2023)
13. Nilsson, B.J.: Guarding art galleries: methods for mobile guards. Ph.D. thesis, Lund University (1996)
14. Seth, R., Maheshwari, A., Nandy, S.C.: Acrophobic guard watchtower problem. Comput. Geom. **109**, 101918 (2023)

Extending Simple Monotone Drawings

Jan Kynčl and Jan Soukup[(✉)]

Department of Applied Mathematics, Faculty of Mathematics and Physics, Charles University, Malostranské nám. 25, 118 00 Praha 1, Czech Republic
{kyncl,soukup}@kam.mff.cuni.cz

Abstract. We prove the following variant of Levi's Enlargement Lemma: for an arbitrary arrangement $\mathcal{A}$ of x-monotone pseudosegments in the plane and a pair of points a, b with distinct x-coordinates and not on the same pseudosegment, there exists a simple x-monotone curve with endpoints a, b that intersects every curve of $\mathcal{A}$ at most once. As a consequence, every simple monotone drawing of a graph can be extended to a simple monotone drawing of a complete graph. We also show that extending an arrangement of cylindrically monotone pseudosegments is not always possible; in fact, the corresponding decision problem is NP-hard.

Keywords: Simple topological graph · Monotone drawing · Cylindrically monotone drawing · Arrangement of pseudosegments

1 Introduction

Given $k \geq 1$, a finite set $\mathcal{A}$ of simple curves in the plane is called an *arrangement of k-strings* if every pair of the curves of $\mathcal{A}$ intersects at most k times, and every intersection point is either a proper crossing or an endpoint of at least one of the curves of the intersecting pair. Multiple k-strings can intersect at a common point, an endpoint of one curve can lie on another, but two curves cannot touch in their inner points. An arrangement of 1-strings is also called an *arrangement of pseudosegments*, and each curve in the arrangement is called a *pseudosegment*. In this paper, we represent simple curves as subsets of the plane (or another surface) that are homeomorphic images of a closed interval. When necessary, we also specify one endpoint of a curve as the starting point to introduce an orientation to the curve.

A simple curve γ in the plane is *x-monotone*, shortly *monotone*, if γ intersects every line parallel to the y-axis at most once.

Let $\mathcal{F}$ be a family of arrangements of pseudosegments in the plane, and let a, b be a pair of points in the plane. An arrangement $\mathcal{A}$ from $\mathcal{F}$ is *(a, b)-extendable in $\mathcal{F}$* if there exists a simple curve α with endpoints a, b so that $\mathcal{A} \cup \{\alpha\} \in \mathcal{F}$. The arrangement $\mathcal{A}$ is *extendable in $\mathcal{F}$* if it is (a, b)-extendable in $\mathcal{F}$ for all possible choices of a and b with distinct x-coordinates and not on the same pseudosegment. We omit the family $\mathcal{F}$ from the notation whenever it is clear from context. The family $\mathcal{F}$ is *extendable* if all its elements are extendable in $\mathcal{F}$.

Our main result is the following.

H. Fernau and B. Zhu (Eds.): IWOCA 2025, LNCS 15885, pp. 17–30, 2025.
https://doi.org/10.1007/978-3-031-98740-3_2

Theorem 1. *The family of all arrangements of monotone pseudosegments in the plane is extendable.*

Moreover, we show that the proof of Theorem 1 can be turned into an efficient algorithm, in particular, given a suitable representation of an arrangement of n monotone pseudosegments where m pairs of pseudosegments intersect we can find the new pseudosegment extending the arrangement in time $O(n+m)$.

A drawing of a graph in the plane is *simple* if every pair of edges has at most one common point, either a common endpoint or a proper crossing. A drawing of a graph is *monotone* if every edge is drawn as a monotone curve and no two vertices share the same x-coordinate. We have the following direct consequence of Theorem 1, illustrated in Fig. 1.

Corollary 2. *Every simple monotone drawing of a graph in the plane can be extended to a simple monotone drawing of the complete graph with the same set of vertices.*

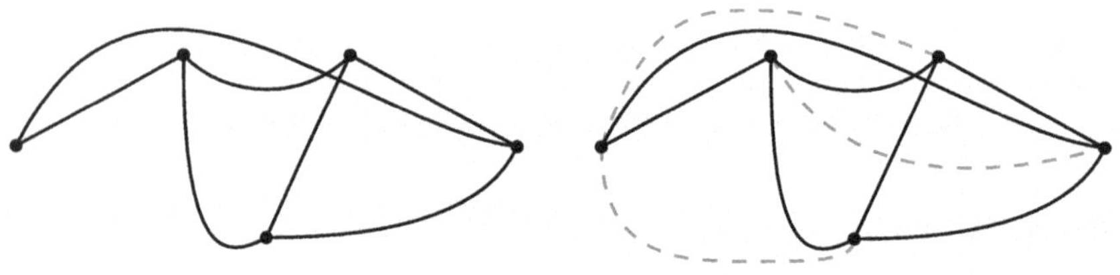

Fig. 1. Left: a simple monotone drawing of a graph. Right: an extension of the drawing on the left to a simple monotone drawing of a complete graph. The added edges are dashed.

1.1 Drawings on a Cylinder

A cylinder can be represented as the surface $S^1 \times \mathbb{R}$ embedded in $\mathbb{R}^3$, where S^1 is the unit circle in the xy-plane. A simple curve γ on the cylinder is *cylindrically monotone* if γ intersects every line parallel to the z-axis at most once. A finite system of circular arcs on the circle S^1 is *normal* if no pair of arcs covers the whole circle [13]. An arrangement of cylindrically monotone pseudosegments is *normal* if the projections of its curves to S^1 form a normal system of circular arcs. Drawings of graphs whose edges are cylindrically monotone have also been called *angularly monotone* [8], and in the case when the edges form a normal arrangement they have also been called *strongly c-monotone* [1].

Let $\mathcal{F}$ be a family of arrangements of pseudosegments on the cylinder, and let a, b be a pair of points on the cylinder. An arrangement $\mathcal{A}$ from $\mathcal{F}$ is (a, b)-*extendable in* $\mathcal{F}$ if there exists a simple curve α with endpoints a, b so that $\mathcal{A} \cup \{\alpha\} \in \mathcal{F}$. The arrangement $\mathcal{A}$ is *extendable in* $\mathcal{F}$ if it is (a, b)-extendable in $\mathcal{F}$ for all possible choices of a and b not on the same vertical line and not on the same pseudosegment. We omit the family $\mathcal{F}$ from the notation whenever it is clear from context. The family $\mathcal{F}$ is *extendable* if all its elements are extendable in $\mathcal{F}$.

Theorem 1 generalizes in a straightforward way to normal arrangements of cylindrically monotone pseudosegments:

Corollary 3. *The family of all normal arrangements of cylindrically monotone pseudosegments is extendable.*

It may be helpful to consider an alternative definition of normal arrangements as those where the curves are drawn the "shorter way" around the cylinder. The following result implies that these two definitions are combinatorially equivalent.

Proposition 4. *Given a normal system $\mathcal{C}$ of circular arcs on the circle S^1, there exists a homeomorphism of S^1 that maps each arc in $\mathcal{C}$ to an arc of length smaller than π.*

We suspect that Proposition 4 might be a known result, but we were not able to find it in the literature.

We show that further generalization of Corollary 3 to arbitrary arrangements of cylindrically monotone pseudosegments is not possible.

Proposition 5. *There exists an arrangement $\mathcal{A}$ of five cylindrically monotone pseudosegments and a pair of points a, b not on the pseudosegments of $\mathcal{A}$ and not on the same vertical line such that $\mathcal{A}$ is not (a, b)-extendable in the family of all arrangements of cylindrically monotone pseudosegments.*

Moreover, the decision problem of extendability of cylindrically monotone arrangements turns out to be NP-hard.

Theorem 6. *Given an arrangement $\mathcal{A}$ of cylindrically monotone pseudosegments and a pair of points a, b, it is NP-hard to decide whether $\mathcal{A}$ is (a, b)-extendable in the family of all arrangements of cylindrically monotone pseudosegments.*

The problem in Theorem 6 is, in fact, NP-complete: the membership in NP is rather straightforward since an arrangement of n cylindrically monotone pseudosegments can be encoded by a sequence consisting of $O(n)$ endpoints and $O(n^2)$ crossings, and by the above/below relations between the pseudosegments and the endpoints or crossings of other pseudosegments.

We prove Theorem 1 in Sect. 2 and we either prove or at least give the main idea of proofs of Corollary 3, Proposition 5 and Theorem 6 in Sect. 3. We omit the proof of Proposition 4 due to the space constraints.

1.2 Related Results

A *pseudoline* in the plane is an image of a Euclidean line under a homeomorphism of the plane; in other words, a pseudoline is a homeomorphic image of the set $\mathbb{R}$, unbounded in both directions. An *arrangement of pseudolines* is a finite set of pseudolines such that every pair of them has exactly one crossing, and no other common intersection point. Pseudolines are also often defined in the projective plane, as nonseparating simple closed curves.

Levi's Enlargement Lemma [12] states that for every arrangement of pseudolines and every pair of points a, b not on the same pseudoline, one can draw a new pseudoline through a and b, crossing every curve from the given arrangement exactly once. The lemma has several alternative proofs in the literature [5,14].

Snoeyink and Hershberger [15] generalized Levi's Lemma to a sweeping theorem for pseudoline arrangements, which allows "rotating" a new pseudoline through a given point, sweeping the whole plane in the process.

By a classical result of Goodman [9], [7, Theorem 5.1.4], every arrangement of pseudolines can be transformed by a homeomorphism of the plane into an arrangement of monotone pseudolines, or a so-called *wiring diagram*. Therefore, monotone arrangements of pseudosegments can be considered as a generalization of pseudoline arrangements. On the other hand, Fig. 2 shows an example that not every monotone arrangement of pseudosegments can be seen as a "restriction" of a pseudoline arrangement, and so Theorem 1 does not easily follow from Levi's Lemma. See Arroyo, Bensmail and Richter [3, Figure 2] for more examples. Since a pseudoline (in the projective plane) can be considered as a union of two internally disjoint pseudosegments (one bounded and the other one crossing the line at infinity), Theorem 1 can also be considered as a generalization of "a half" of Levi's Lemma.

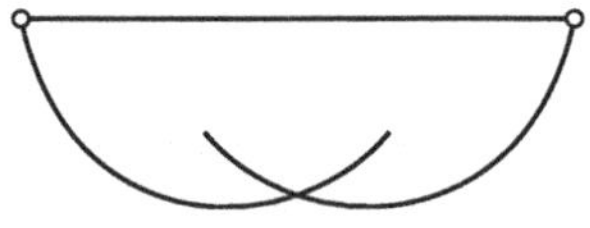

Fig. 2. An example of an arrangement of three pseudosegments that cannot be extended to pseudolines forming a pseudoline arrangement since any extension of the pseudosegments into pseudolines would contain a pair of pseudolines with two mutual intersections.

By $P_3 + P_3$ we denote the union of two vertex-disjoint paths of length 2. A simple drawing of $P_3 + P_3$ that cannot be extended to a simple drawing of a complete graph was constructed by Eggleton [6, Diagram 15(ii)] and later rediscovered by the first author [11, Figure 9]. Later a few more examples of non-extendable simple drawings were constructed [10, Figures 1, 10]. None of these drawings are homeomorphic to monotone drawings, which follows, for example, from Theorem 1, but some of them can easily be transformed into cylindrically monotone drawings; see Fig. 3.

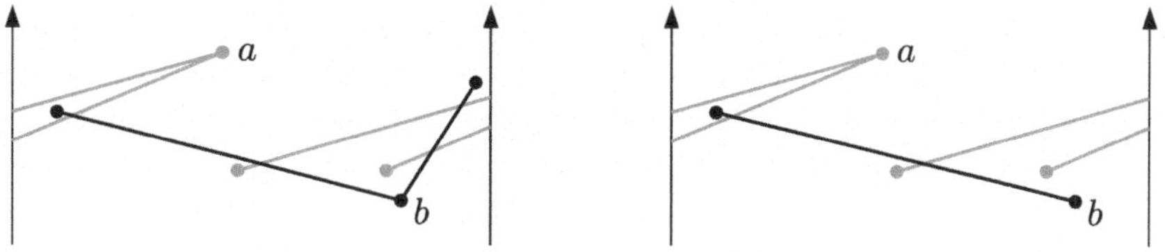

Fig. 3. Left: a simple cylindrically monotone drawing of $P_3 + P_3$ where the edge ab cannot be added [6,10,11] as a simple curve without crossing some edge of $P_3 + P_3$. Since all edges of $P_3 + P_3$ are incident to either a or b, the added edge ab would have at least two intersections with some edge of $P_3 + P_3$. Right: a simple cylindrically monotone drawing of $P_3 + P_2$ where the edge ab cannot be added as a cylindrically monotone curve without intersecting some edge of $P_3 + P_2$ twice.

The first author together with Pach, Radoičić and Tóth [10] proved that for every $k \geq 1$, for every arrangement $\mathcal{A}$ of k-strings and every pair of points a, b not on the same curve of $\mathcal{A}$, there exists a simple curve joining a and b intersecting every curve of $\mathcal{A}$ at most $2k$ times. They also constructed examples showing that the constant $2k$ cannot be improved.

Arroyo, Klute, Parada, Vogtenhuber, Seidel and Wiedera [4] showed that it is NP-hard to decide, given an arrangement $\mathcal{A}$ of pseudosegments and a pair of points a, b, whether a and b can be joined by a simple curve crossing each pseudosegment of $\mathcal{A}$ at most once. Our proof of Theorem 6 is a simple adaptation of this result to cylindrically monotone arrangements. Recently, Aichholzer Orthaber and Vogtenhuber [2] proved a result similar to Theorem 1 where they showed that the crossing-minimizing and the so-called separable drawings can be extended to simple drawings of the complete graph but not necessarily to the crossing-minimizing or separable drawings of the complete graph.

Arroyo, Bensmail and Richter [3] studied a slightly different extendability question: given an arrangement of pseudosegments, can it be extended to an arrangement of pseudolines by extending every given pseudosegment to a pseudoline? They determined the full infinite set of minimal obstructions, and found a polynomial time algorithm for detecting the obstructions and extending the arrangement.

2 Monotone Arrangements in the Plane

We start with a few definitions and tools for analyzing x-monotone arrangements. Given a pair of points a, b in the plane, we write $a \prec b$ if a has a smaller x-coordinate than b. Clearly, $\prec$ is a strict linear order on the points of any monotone curve.

We can naturally talk about objects lying "below" and "above" monotone curves. Let a, b be points such that $a \prec b$. For any monotone curve γ we denote by $\gamma[a,b]$ and $\gamma(a,b)$ the subset of γ formed by the points x of γ satisfying $a \preceq x \preceq b$ and $a \prec x \prec b$, respectively. Similarly, for an arrangement $\mathcal{A}$ of monotone pseudosegments we denote by $\mathcal{A}[a,b]$ the arrangement of pseudosegments where we replace each $\gamma \in \mathcal{A}$ by $\gamma[a,b]$.

By *consecutive intersections* of two monotone curves with finitely many intersections we mean consecutive intersections with respect to their x-coordinates. Let α, β be two monotone curves with finitely many intersections. Let a, b be two consecutive intersections of α, β such that $a \prec b$. Then the only intersections of $\alpha[a,b]$ with $\beta[a,b]$ are the points a and b. In this case we say that the curves α and β form a *bigon*. Furthermore, if $\alpha(a,b)$ lies above $\beta(a,b)$ we say that α and β form an *α-top*, or equivalently, a *β-bottom* bigon. Note that in general α and β can form both an α-bottom and a β-bottom bigon.

The *lower envelope* $\text{low}(\mathcal{U})$ of a set $\mathcal{U}$ of curves is the set of all points p of these curves such that no other point of any curve of $\mathcal{U}$ with the same x-coordinate as p is below p. Note that if $\mathcal{U}$ is an arrangement of monotone pseudosegments, then $\text{low}(\mathcal{U})$ is a finite union of connected parts of pseudosegments.

We continue by proving Theorem 1, followed by a discussion of the time complexity of extending the arrangement.

2.1 Proof of Theorem 1

Let $\mathcal{A}$ be an arrangement of monotone pseudosegments. Let a, b, with $a \prec b$, be points that are not on the same pseudosegment of $\mathcal{A}$. We need to find a monotone curve from a to b that intersects every curve of $\mathcal{A}$ at most once. Since every curve of $\mathcal{A}$ is monotone, we can without loss of generality assume that $\mathcal{A} = \mathcal{A}[a,b]$.

Let $\mathcal{A}'$ be an arrangement of monotone pseudosegments formed by all pseudosegments of $\mathcal{A}$ together with three new segments τ_1, τ_2, τ_3, defined as follows. The segment τ_1 is an almost vertical segment starting in a and ending in some new point to the right of a and above all pseudosegments of $\mathcal{A}$. Similarly, τ_3 is an almost vertical segment ending in b and starting in some new point to the left of b and above all pseudosegments of $\mathcal{A}$. Finally, τ_2 is a horizontal segment crossing τ_1 and τ_3, and lying entirely above all pseudosegments of $\mathcal{A}$; see Fig. 4. In this way, $\text{low}(\{\tau_1, \tau_2, \tau_3\})$ is a monotone curve connecting a and b "from above", so that every pseudosegment $\gamma \in \mathcal{A}$ intersects it at most twice. Furthermore no $\gamma \in \mathcal{A}$ forms a γ-top bigon with $\text{low}(\{\tau_1, \tau_2, \tau_3\})$ (it can only form a γ-bottom bigon).

In order to find an extending curve we do the following. We find a nonempty subset $\mathcal{U} \subseteq \mathcal{A}'$ of pseudosegments such that the lower envelope of $\mathcal{U}$ is a monotone curve connecting a to b, intersecting every pseudosegment of $\mathcal{A}' \setminus \mathcal{U}$ at most once. Furthermore, we find $\mathcal{U}$ so that no pseudosegment α touches $\text{low}(\mathcal{U})$ from below in an inner point of α. After finding such $\mathcal{U}$, a new pseudosegment connecting a and b can clearly be drawn slightly below the lower envelope of $\mathcal{U}$ and will indeed intersect every pseudosegment of $\mathcal{A}'$ at most once. Thus, if such $\mathcal{U}$ exists, $\mathcal{A}'$, and consequently $\mathcal{A}$, is (a,b)-extendable.

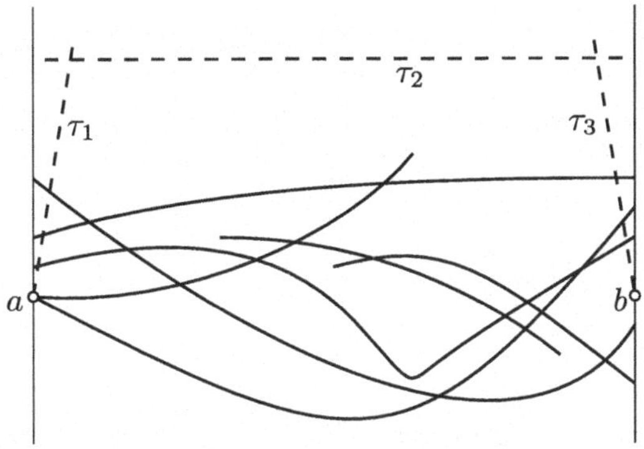

Fig. 4. An arrangement of monotone pseudosegments with three added segments τ_1, τ_2, τ_3 connecting points a, b "from above".

We find $\mathcal{U}$ inductively. We start with $\mathcal{U}_0 = \{\tau_1, \tau_2, \tau_3\}$ and always look at the lower envelope of $\mathcal{U}_i$. In the ith step we select an arbitrary pseudosegment γ_i of $\mathcal{A}' \setminus \mathcal{U}_{i-1}$ intersecting $\mathrm{low}(\mathcal{U}_{i-1})$ at least twice. If there is no such γ_i then $\mathcal{U} = \mathcal{U}_{i-1}$ and we are done. Otherwise, we set $\mathcal{U}_i = \mathcal{U}_{i-1} \cup \{\gamma_i\}$. The number of pseudosegments is finite, so this process finishes with a set $\mathcal{U}$ such that the lower envelope of $\mathcal{U}$ intersects every pseudosegment of $\mathcal{A}' \setminus \mathcal{U}$ at most once.

Additionally, we prove that the induction preserves the following invariants for every $\mathcal{U}_i$.

(I1) No pseudosegment α of $\mathcal{A}' \setminus \mathcal{U}_i$ forms an α-top bigon with $\mathrm{low}(\mathcal{U}_i)$.
(I2) No pseudosegment α of $\mathcal{A}' \setminus \mathcal{U}_i$ touches $\mathrm{low}(\mathcal{U}_i)$ from below in an inner point of α.
(I3) The lower envelope of $\mathcal{U}_i$ is connected and contains a and b.

In particular, by (I3), the lower envelope of $\mathcal{U}$ is a monotone curve connecting a to b and, by (I2), no pseudosegment α of $\mathcal{A}' \setminus \mathcal{U}$ touches $\mathrm{low}(\mathcal{U})$ from below in an inner point of α. Since $\mathrm{low}(\mathcal{U})$ intersects every pseudosegment of $\mathcal{A}' \setminus \mathcal{U}$ at most once by its construction, $\mathcal{A}$ is (a, b)-extendable by the previous discussion. Thus, it suffices to prove the correctness of these invariants to finish the proof.

The invariants hold for $\mathcal{U}_0$ by the construction of τ_1, τ_2 and τ_3. Suppose all invariants hold for $\mathcal{U}_{i-1}$. In particular, $\mathrm{low}(\mathcal{U}_{i-1})$ is a monotone curve connecting a to b by invariant (I3). We show that all invariants also hold for $\mathcal{U}_i$.

The pseudosegment γ_i intersects $\mathrm{low}(\mathcal{U}_{i-1})$ at least twice. We show that γ_i intersects $\mathrm{low}(\mathcal{U}_{i-1})$ exactly twice. Suppose, for contradiction, that there are three consecutive intersections c, d and e of γ_i with $\mathrm{low}(\mathcal{U}_{i-1})$ such that $c \prec d \prec e$. Then $\gamma_i[c, d]$ with $\mathrm{low}(\mathcal{U}_{i-1})[c, d]$ forms a bigon and so does $\gamma_i[d, e]$ with $\mathrm{low}(\mathcal{U}_{i-1})[d, e]$. By invariant (I1) both of these bigons must be $\mathrm{low}(\mathcal{U}_{i-1})$-top bigons. However, in this case γ_i touches $\mathrm{low}(\mathcal{U}_{i-1})$ from below in the point d. That is not possible by invariant (I2). Thus, γ_i intersects $\mathrm{low}(\mathcal{U}_{i-1})$ exactly twice. Furthermore, by invariant (I1), γ_i and $\mathrm{low}(\mathcal{U}_{i-1})$ form a γ_i-bottom bigon.

Let x and y be the two intersection points of γ_i and $\mathrm{low}(\mathcal{U}_{i-1})$. Refer to Fig. 5. Since γ_i and $\mathrm{low}(\mathcal{U}_{i-1})$ form a γ_i-bottom bigon, the only part of the curve γ_i that lies below $\mathrm{low}(\mathcal{U}_{i-1})$ is exactly $\gamma_i(x, y)$. Thus, the lower envelope of $\mathcal{U}_{i-1} \cup \{\gamma_i\}$ is a monotone curve connecting a and b. Therefore, invariant (I3) holds also for $\mathcal{U}_i$.

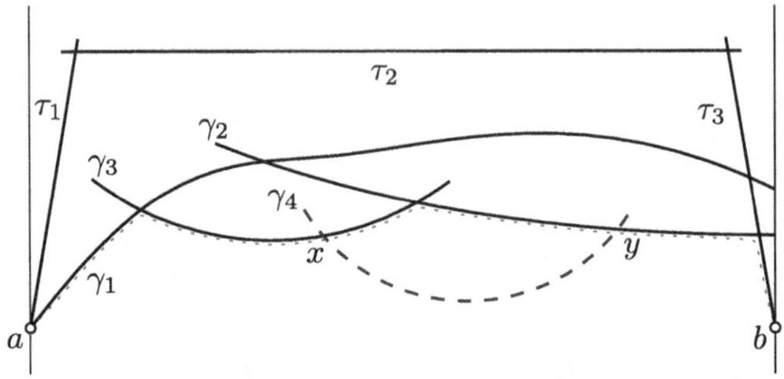

Fig. 5. Induction step in the proof of Theorem 1. In the ith step (fourth step in the figure) we add pseudosegment γ_i (dashed) intersecting the lower envelope (dotted) of the previous segments twice. The lower envelope remains a connected curve connecting a with b even after this addition.

Now, suppose that invariant (I2) does not hold, that is, there exists some pseudosegment β of $\mathcal{A}' \setminus \mathcal{U}_i$ that touches $\mathrm{low}(\mathcal{U}_i)$ from below in an inner point of β. Refer to Fig. 6. Since $\mathcal{U}_i = \mathcal{U}_{i-1} \cup \{\gamma_i\}$, the curve β has to touch γ_i or $\mathrm{low}(\mathcal{U}_{i-1})$ in an inner point of β, a contradiction. Hence, invariant (I2) also holds for $\mathcal{U}_i$. Note that the analogous statement for touchings from above does not hold, that is, there may exist some pseudosegment α of $\mathcal{A}' \setminus \mathcal{U}_i$ that both touches $\mathrm{low}(\mathcal{U}_i)$ from above in an inner point of α and touches neither γ_i nor $\mathrm{low}(\mathcal{U}_i)$ in an inner point of α.

Finally, suppose that invariant (I1) does not hold, that is, there exists some pseudosegment ρ of $\mathcal{A}' \setminus \mathcal{U}_i$ that together with $\mathrm{low}(\mathcal{U}_i)$ forms a ρ-top bigon. Call s and t the vertices of this bigon and assume $s \prec t$. See Fig. 7.

If s and t both lie on $\gamma_i[x, y]$, then ρ and γ_i intersect twice, a contradiction. Otherwise s or t does not lie on $\gamma_i[x, y]$. Without loss of generality assume that t does not lie on $\gamma_i[x, y]$ and $y \prec t$. Then s either lies on $\mathrm{low}(\mathcal{U}_{i-1})$ or below it. In both cases $\rho[s, t]$ intersects $\mathrm{low}(\mathcal{U}_{i-1})$ in some point other than t since $\rho[s, t]$ together with $\mathrm{low}(\mathcal{U}_i)$ forms a ρ-top bigon. Denote the rightmost intersection of $\rho[s, t]$ and $\mathrm{low}(\mathcal{U}_{i-1})$ other than t by u. Then $\rho(u, t)$ lies above $\mathrm{low}(\mathcal{U}_{i-1})$ and so $\rho[u, t]$ together with $\mathrm{low}(\mathcal{U}_i)$ forms a ρ-top bigon, a contradiction with invariant (I1) for $\mathcal{U}_{i-1}$. This concludes the proof of Theorem 1.

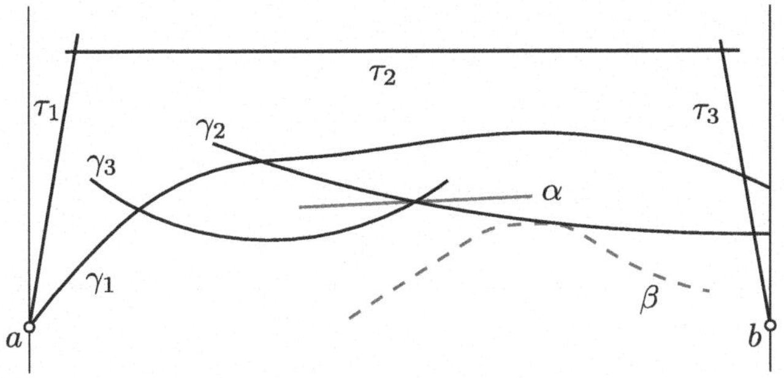

Fig. 6. Induction step in the proof of Theorem 1. During the selection of $\mathcal{U}$ some pseudosegments may touch low($\mathcal{U}$) from above but never from below. Pseudosegment α touches low($\{\tau_1, \tau_2, \tau_3, \gamma_1, \gamma_2, \gamma_3\}$) from above. On the other hand, β cannot be in the same arrangement of pseudosegments since it touches γ_2.

2.2 Complexity of Extending the Arrangement

We briefly discuss how our proof can be turned into an algorithm extending a given monotone arrangement. Due to the space constraints, we omit the technical part of choosing a suitable representation of an arrangement of monotone curves. We assume we can access all pseudosegments in constant time, every pseudosegment stores its incidences with its crossings and endpoints in the order of increasing x-coordinates, and every incidence of pseudosegments contains information on which pseudosegments are incident to it. As an output, we will find only low($\mathcal{U}$). Once again, we omit the exact representation of low($\mathcal{U}$) and the description of the new monotone pseudosegment that can be added slightly below it.

With these assumptions and ignoring the precise representation, we can see the algorithm already in the proof: We create $\mathcal{A}'$ from $\mathcal{A}$, set $\mathcal{U}_0 = \{\tau_1, \tau_2, \tau_3\}$ and iteratively find all $\mathcal{U}_i$ until we get $\mathcal{U}$. The only unexplained part is how to effectively find an arbitrary pseudosegment γ_i of $\mathcal{A}' \setminus \mathcal{U}_{i-1}$ in each step such that γ_i intersects low($\mathcal{U}_{i-1}$) at least twice.

From the proof, we know that any such γ_i has to intersect low($\mathcal{U}_{i-1}$) exactly twice. To find γ_i efficiently, we dynamically keep track of the intersections of each pseudosegment with the lower envelope low($\mathcal{U}_{i-1}$) and maintain a list of pseudosegments that intersect low($\mathcal{U}_{i-1}$) exactly twice. In this way, we can choose γ_i in constant time, but we need to find how many times each pseudosegment intersects the new lower envelope low($\mathcal{U}_i$). We know it already for low($\mathcal{U}_{i-1}$). These two lower envelopes differ only in the bigon formed by γ_i and low($\mathcal{U}_{i-1}$). Thus, we can look at all the intersections along this bigon to update our intersection information. Since the bigons from different steps of the algorithm are clearly internally disjoint, we access every intersection point at most a constant number of times for every pseudosegment incident to it. Accessing an intersection of k pseudosegments takes asymptotically time $O(k)$, so, in total, we spend time $O(k^2)$ in every k-intersection. That is asymptotically the same as the number of

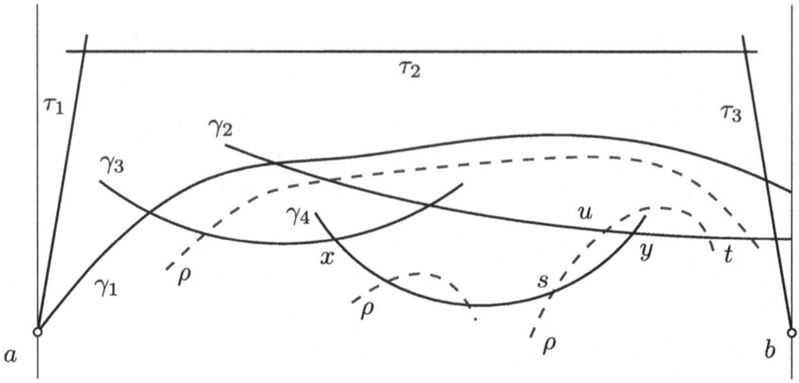

Fig. 7. Induction step in the proof of Theorem 1. If there was some pseudosegment ρ that together with the lower envelope ($\text{low}(\{\tau_1, \tau_2, \tau_3, \gamma_1, \gamma_2, \gamma_3, \gamma_4\})$ in the picture) formed a ρ-top bigon, it would either form a ρ-top bigon with the previous lower envelope or intersect twice the segment that was added as the last. In the picture, there are three such possible ρ's.

pairs of intersecting pseudosegments in the k-intersection. Therefore, the overall time complexity is asymptotically upper-bounded by the number of pseudosegments plus the number of pairs of intersecting pseudosegments.

3 Cylindrically Monotone Arrangements

Recall that we represent the cylinder as the surface $S^1 \times \mathbb{R}$. If we imagine cutting and unrolling the cylinder, we can represent it in the plane as a vertical strip whose left and right sides represent the same vertical line of the cylinder. We can also select the orientation so that the counter-clockwise direction on the cylinder corresponds to the left-to-right direction in the plane. For points $a, b \in S^1$ we denote by $[a, b]$ the counter-clockwise circular arc in S^1 from a to b and we call it an *interval*.

3.1 Normal Cylindrically Monotone Arrangements

We prove Corollary 3 using Proposition 4, but we omit the proof of the proposition due to space constraints.

Proof of Corollary 3. Let $\mathcal{A}$ be a normal arrangement of cylindrically monotone pseudosegments. Let a, b be a pair of points on the cylinder but not on the same vertical line and not on the same pseudosegment of $\mathcal{A}$. We need to show that $\mathcal{A}$ is (a, b)-extendable in the family of all normal arrangements of cylindrically monotone pseudosegments. That is, we need to find a cylindrically monotone simple curve γ with endpoints a, b that intersects every curve of $\mathcal{A}$ at most once such that $\mathcal{A} \cup \{\gamma\}$ is still normal.

Let π be the orthogonal projection from $S^1 \times \mathbb{R}$ to S^1. Let f be the homeomorphism from Proposition 4. We define a homeomorphism $g\colon S^1 \times \mathbb{R} \to S^1 \times \mathbb{R}$ as $g(x, y) = (f(x), y)$. The image $g(\mathcal{A}) = \{g[\alpha];\, \alpha \in \mathcal{A}\}$ of $\mathcal{A}$ under g is a homeomorphic arrangement of cylindrically monotone pseudosegments whose orthogonal projections to S^1 have lengths smaller than π. Furthermore, g maps every vertical line onto a vertical line. Thus, extending $\mathcal{A}$ by a pseudosegment from a to b is equivalent to extending $g(\mathcal{A})$ by a pseudosegment from $g(a)$ to $g(b)$.

We look at the orthogonal projections $\pi(g(a)), \pi(g(b))$ of $g(a), g(b)$ to S^1. Without loss of generality we may assume that the length of $[\pi(g(a)), \pi(g(b))]$ is smaller than π. Since orthogonal projections of all pseudosegments of $g(A)$ to S^1 have lengths smaller than π, then

$$\{g[\alpha] \cap ([\pi(g(a)), \pi(g(b))] \times \mathbb{R})\,;\, \alpha \in \mathcal{A}\} \setminus \{\emptyset\}$$

is an arrangement of cylindrically monotone pseudosegments on a cylindrical strip. This arrangement is homeomorphic to an arrangement of monotone pseudosegments in the plane (by unrolling the strip to the plane). Thus, by Theorem 1, we can extend the arrangement in the plane by a monotone pseudosegment β connecting the images of $g(a)$ and $g(b)$ after unrolling. Let β' be the cylindrically monotone curve on the cylinder corresponding to β. Clearly, β' is a curve connecting $g(a)$ and $g(b)$ and intersecting every pseudosegment from $g(\mathcal{A})$ at most once. Moreover, since the length of $[\pi(g(a)), \pi(g(b))]$ is smaller than π, the resulting arrangement is still normal. Thus, $\gamma = g^{-1}(\beta')$ is the desired curve extending the original arrangement $\mathcal{A}$.

Remark 1. Let us note that we could also find the extending curve γ without the help of Proposition 4 by "cutting" the cylinder along the vertical lines passing through a and b and showing that one of the resulting parts corresponds to the planar version.

3.2 General Cylindrically Monotone Arrangements

In this subsection, when we say that an arrangement is extendable, we mean that it is extendable in the family of all cylindrically monotone arrangements.

Unlike normal arrangements, general arrangements of cylindrically monotone pseudosegments are not always extendable, as stated in Proposition 5. Moreover, Theorem 6 states that in general, deciding whether a cylindrically monotone arrangement is (a, b)-extendable is NP-hard.

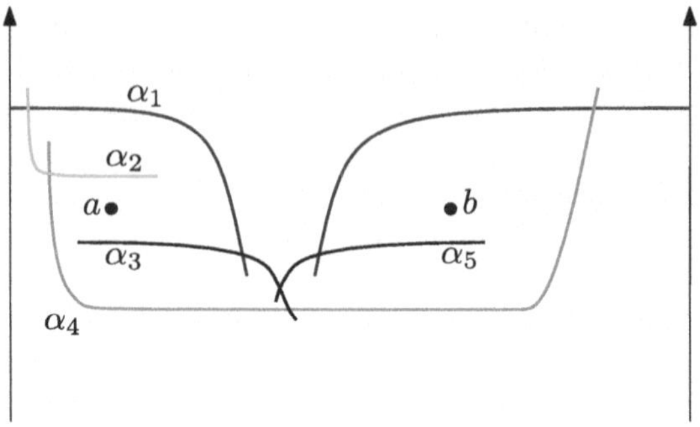

Fig. 8. An arrangement of five cylindrically monotone pseudosegments that is not (a, b)-extendable.

Proof of Proposition 5. Figure 8 provides an example of an arrangement $\mathcal{A}$ of five pseudosegments and a pair of points a, b such that every cylindrically monotone curve connecting a and b has to intersect some pseudosegment of $\mathcal{A}$ twice. This means that $\mathcal{A}$ is not (a, b)-extendable. This example is the smallest that we know of where the points a and b do not lie on any of the curves of $\mathcal{A}$.

Proof of Theorem 6. Our proof is heavily inspired by a recent proof by Arroyo et al. [4] showing that deciding the extendability of a simple drawing of a graph by one edge between a prescribed pair of points is NP-hard. We adapt their reduction from 3-SAT to the more restricted setting of cylindrically monotone drawings. We need the clauses to contain at least one positive and one negative literal, so first we apply logical transformations to make it so, for example, we replace the clause $x_1 \vee x_2 \vee x_3$ by $(x_2 \vee \neg y_1 \vee x_1) \wedge (y_1 \vee \mathtt{false_n} \vee x_3)$ where y_1 is a new variable, $\mathtt{false_n}$ is a logical constant `false`, and the subscript `n` indicates it counts as a negative literal. We omit the rest of the proof, but the main idea is illustrated in Fig. 9 where the reduction is applied to the formula $(x_1 \vee x_2 \vee x_3) \wedge (\neg x_1 \vee \neg x_2 \vee \neg x_4) \wedge (\neg x_1 \vee x_3 \vee \neg x_4)$.

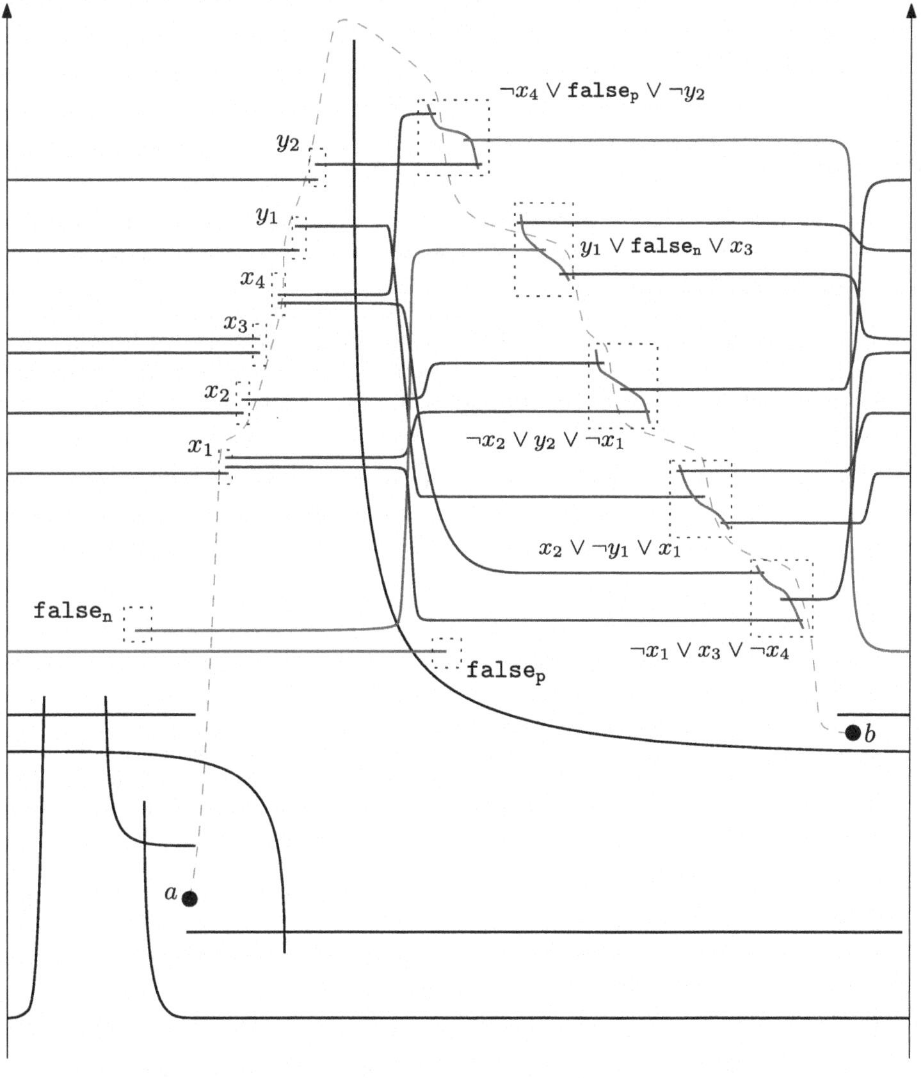

Fig. 9. Reduction of the satisfiability of the formula $(\neg x_1 \vee x_3 \vee \neg x_4) \wedge (x_2 \vee \neg y_1 \vee x_1) \wedge (\neg x_2 \vee y_2 \vee \neg x_1) \wedge (y_1 \vee \texttt{false}_\texttt{n} \vee x_3) \wedge (\neg x_4 \vee \texttt{false}_\texttt{p} \vee \neg y_2)$ to the problem of (a, b)-extendability of an arrangement of cylindrically monotone pseudosegments. Every occurrence of a literal corresponds to one pseudosegment in the arrangement. A new monotone pseudosegment connecting a and b (dashed green in the figure) can be added if and only if the formula is satisfiable. The highlighted labeled regions correspond to the variables and clauses of the formula. If a pseudosegment corresponding to an occurrence of a literal is not crossed in the variable region, it means it can be crossed in the clause region. Constants $\texttt{false}_\texttt{n}$ and $\texttt{false}_\texttt{p}$ can never be crossed in the clause regions. The dashed green line corresponds to the satisfying assignment $x_1, y_1, y_2 \longrightarrow \texttt{false}$ and $x_2, x_3, x_4 \longrightarrow \texttt{true}$.

Acknowledgement. Supported by project 23-04949X of the Czech Science Foundation (GAČR) and by the grant SVV2023260699. We thank the anonymous referees for helpful comments on the paper.

References

1. Aichholzer, O., Orthaber, J., Vogtenhuber, B.: Towards crossing-free Hamiltonian cycles in simple drawings of complete graphs. Comput. Geom. Topol. **3**(2), 30, Article no. 5 (2024)
2. Aichholzer, O., Orthaber, J., Vogtenhuber, B.: Separable drawings: extendability and crossing-free Hamiltonian cycles. In: Proceedings of the 32nd International Symposium on Graph Drawing and Network Visualization (GD 2024). Leibniz International Proceedings in Informatics (LIPIcs), vol. 320, pp. 34:1–34:17. Schloss Dagstuhl – Leibniz-Zentrum für Informatik (2024)
3. Arroyo, A., Bensmail, J., Richter, R.B.: Extending drawings of graphs to arrangements of pseudolines. J. Comput. Geom. **12**(2), 3–24 (2021)
4. Arroyo, A., Klute, F., Parada, I., Vogtenhuber, B., Seidel, R., Wiedera, T.: Inserting one edge into a simple drawing is hard. Discret. Comput. Geom. **69**(3), 745–770 (2023)
5. Arroyo, A., McQuillan, D., Richter, R.B., Salazar, G.: Levi's Lemma, pseudolinear drawings of K_n, and empty triangles. J. Graph Theory **87**(4), 443–459 (2018)
6. Eggleton, R.B.: Crossing numbers of graphs. Ph.D. thesis, University of Calgary (1973)
7. Felsner, S., Goodman, J.E.: Pseudoline arrangements. In: Goodman, J.E., O'Rourke, J., Tóth, C.D. (eds.) Handbook of Discrete and Computational Geometry, 3rd edn. Discrete Mathematics and its Applications (Boca Raton), CRC Press, Boca Raton, FL (2018). Electronic version. http://www.csun.edu/~ctoth/Handbook/HDCG3.html. ISBN 978-1-4987-1139. Accessed May 2023
8. Fulek, R., Ruiz-Vargas, A.J.: Topological graphs: empty triangles and disjoint matchings. In: Proceedings of the Twenty-Ninth Annual Symposium on Computational Geometry (SoCG 2013), pp. 259–266. ACM, New York (2013)
9. Goodman, J.E.: Proof of a conjecture of Burr, Grünbaum, and Sloane. Discret. Math. **32**(1), 27–35 (1980)
10. Kynčl, J., Pach, J., Radoičić, R., Tóth, G.: Saturated simple and k-simple topological graphs. Comput. Geom. **48**(4), 295–310 (2015)
11. Kynčl, J.: Improved enumeration of simple topological graphs. Discret. Comput. Geom. **50**(3), 727–770 (2013)
12. Levi, F.: Die Teilung der projektiven Ebene durch Gerade oder Pseudogerade. Berichte Math.-Phys. Kl. Sächs. Akad. Wiss. Leipzig **78**, 256–267 (1926)
13. Lin, M.C., Szwarcfiter, J.L.: Characterizations and recognition of circular-arc graphs and subclasses: a survey. Discret. Math. **309**(18), 5618–5635 (2009)
14. Schaefer, M.: A proof of Levi's Extension Lemma. arXiv:1910.05388v1 (2019)
15. Snoeyink, J., Hershberger, J.: Sweeping arrangements of curves. Discret. Comput. Geom. (New Brunswick, NJ, 1989/1990). DIMACS Ser. Discret. Math. Theor. Comput. Sci. **6**, 309–349 (1991)

Guarding Terrains with Guards on a Line

Byeonguk Kang[1(✉)], Hwi Kim[1], and Hee-Kap Ahn[2]

[1] Department of Computer Science and Engineering, Pohang University of Science and Technology, Pohang, Korea
{kbu417,hwikim}@postech.ac.kr

[2] Graduate School of Artificial Intelligence, Department of Computer Science and Engineering, Pohang University of Science and Technology, Pohang, Korea
heekap@postech.ac.kr

Abstract. Given an x-monotone polygonal chain T with n vertices, and an integer k, we consider the problem of finding the lowest horizontal line L lying above T with k point guards lying on L, so that every point on the chain is *visible* from some guard. A natural optimization is to minimize the y-coordinate of L. We present an algorithm for finding the optimal placements of L and k point guards for T in $O(k^2\lambda_{k-1}(n)\log n)$ time for even numbers $k \geq 2$, and in $O(k^2\lambda_{k-2}(n)\log n)$ time for odd numbers $k \geq 3$, where $\lambda_s(n)$ is the length of the longest (n, s)-Davenport-Schinzel sequence. We also study a variant with an additional requirement that T is partitioned into k subchains, each subchain is paired with exactly one guard, and every point on a subchain is visible from its paired guard. When L is fixed, we can place the minimum number of guards in $O(n)$ time. When the number k of guards is fixed, we can find an optimal placement of L with k point guards lying on L in $O(kn)$ time.

Keywords: Terrain guarding · Visibility problem · Contiguous variant

1 Introduction

We consider the problem of guarding a terrain. A (polygonal) *terrain* is a graph of a piecewise linear function $f : A \subset \mathbb{R} \to \mathbb{R}$ that assigns a height $f(p)$ to every point p in a compact domain A. In other words, a terrain is an x-monotone polygonal chain in the plane. A *guard* is a point lying on or above the terrain. A point on the terrain is *visible* from a guard if the line segment connecting the point and the guard does not cross the terrain. A subchain of the terrain is visible from a guard if every point on the subchain is visible from the guard. A set of guards *covers* a subchain if every point on the subchain is visible from some guard.

Given a terrain T with n vertices, an integer k, and a line L parallel to the x-axis and lying above T, we consider the problem of finding k point guards lying between T and L so that every point on T is visible from some guard. Observe that for $k \geq \lfloor n/2 \rfloor$, we can always cover T by placing guards at every other vertex along T. Thus, we consider the case for $k < \lfloor n/2 \rfloor$. Observe that this

H. Fernau and B. Zhu (Eds.): IWOCA 2025, LNCS 15885, pp. 31–43, 2025.
https://doi.org/10.1007/978-3-031-98740-3_3

problem is closely related to the SET COVER problem: the universe is the terrain T and the set corresponding to a guard consists of points on T visible from the guard. The goal is to select k sets (or the smallest collection of sets) whose union equals T. Observe that the sets corresponding to k guards are defined after the guards are placed.

Lemma 1. *Every point on T visible from a guard $u = (x, y)$ is also visible from a guard $u' = (x, y + \delta)$ for any $\delta > 0$.*

Since the terrain is x-monotone, Lemma 1 holds. By Lemma 1, a natural optimization is to minimize the y-coordinate (height) of the line L. Indeed, in many real-world applications such as security cameras mounted on top of poles, it is desirable to make the height of the cameras as low as possible so that they are sturdy against high wind and it is effective concerning the construction and the cost while the whole terrain is covered. We define two problems formally in the following.

Altitude Terrain Cover (ATC). Given a terrain T and an integer k, find the lowest horizontal line L with k point guards lying on L that cover T.

When $k = 1$, every segment of T must be visible from the unique guard. Thus, the optimal placement of L is determined by the intersection of $n - 1$ lines, each extending a segment of T. Specifically, it is the lowest point in the intersection of the (upper) half-planes, each bounded by one of the lines. Thus, this can be done in $O(n)$ time [15]. However, for the case $k \geq 2$, the optimal placement of L is not necessarily determined by the intersections of the extending lines. This is because a segment may not be visible from any single guard but it is visible from two or more guards in an optimal placement. Figure 1(a) illustrates a case for $k = 2$. The optimal line L is not at an intersection of the extending lines.

Bijective Altitude Terrain Cover (BATC). Given a terrain T and an integer k, find the lowest horizontal line L with k point guards lying on L such that T is partitioned into k subchains, each subchain is paired with exactly one guard, and every point on a subchain is visible from its paired guard.

We consider two cases. (1) Given a line L, place the minimum number of guards lying between T and L that together cover T. (2) Given an integer k, find the lowest horizontal line L and place k guards on L that together cover T. See Fig. 1(b) for an illustration for case (2) with $k = 2$.

1.1 Related Works

Daescu et al. [7] studied the problem of placing the minimum number of point guards on a fixed line so that every point on the terrain is visible from some guard. They presented a linear time algorithm for the problem. They also showed that the problem is NP-hard for a polyhedral terrain.

Katoh et al. [11] studied the problem of placing two horizontal segment guards on a fixed line such that the segment guards collectively cover the terrain

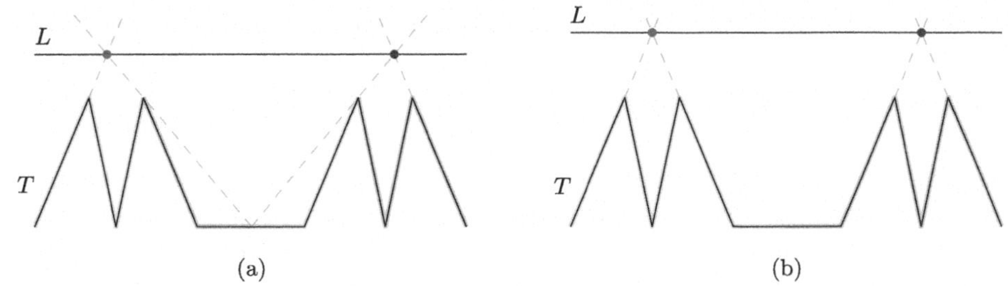

Fig. 1. Cases for $k = 2$. (a) The optimal placement of L such that T can be covered by two guards lying on L. (b) The optimal placement of L such that T is partitioned into two subchains: one of them is visible from the red guard and the other is visible from the blue guard. (Color figure online)

while minimizing the maximum length of the two segments. They presented an $O(n \log^2 n)$-time algorithm for the problem. Later, McCoy et al. [13] studied the problem of finding the lowest horizontal line on which two fixed length horizontal segment guards can be placed to collectively cover the terrain. They presented an $O(n^2 \log n)$-time algorithm for the problem.

There has been work on a variant in which the point guards must be placed on the terrain. Cole and Sharir [6] showed that the problem of covering a polyhedral terrain using the minimum number of guards lying on the terrain is NP-hard. For covering a polygonal terrain using the minimum number of guards lying on the terrain, Chen et al. [5] proposed an NP-completeness proof, but it was not completed. Later, King and Krohn [12] showed that the problem is NP-hard.

Another line of research on terrain guarding is to place k vertical line segments on top of the terrain such that the terrain is covered by them and the maximum length of the segments is minimized. It is known as the *k-watchtower problem*, and there is a fair amount of work on the k-watchtower problem for a polygonal terrain [1,3,9,16] and a polyhedral terrain [1,18]. A variant of the k-watchtower problem is to cover the given *point sites* lying on the terrain, instead of covering the whole terrain [9].

In the *boundary guarding problem*, we are given a simple polygon instead of a terrain. The goal is to minimize the number of point guards placed in the polygon to cover all its segments. Biniaz et al. [4] gave a polynomial-time exact algorithm and an $(\mathsf{OPT}+1)$-approximation algorithm under a contiguity restriction, where OPT is the number of guards in an optimal solution.

1.2 Our Results

Let $\lambda_s(n)$ denote the length of the longest (n, s)-Davenport-Schinzel sequence. See [17] for details about Davenport-Schinzel sequences. For the Altitude Terrain Cover problem, we give an algorithm for finding the optimal placements of L and k point guards for T in $O(k^2 \lambda_{k-1}(n) \log n)$ time for even numbers $k \geq 2$, and in $O(k^2 \lambda_{k-2}(n) \log n)$ time for odd numbers $k \geq 3$, where k is the number of guards. For the Bijective Altitude Terrain Cover problem, we give an algorithm for finding the minimum number of guards (case (1)) in $O(n)$ time, and an algorithm for finding the lowest horizontal line L (case (2)) in $O(kn)$ time.

Sketches of Our Algorithms. For the Altitude Terrain Cover problem, a naïve approach would be to apply Megiddo's parametric search technique [14]. It enables us to transform an $O(\mathsf{T_s})$-time decision algorithm that performs $O(\mathsf{C_s})$ comparisons into an $O(\mathsf{C_s T_s})$-time algorithm for its optimization version. In our problem, a decision algorithm should determine whether a given real value is greater than the optimal y-coordinate. Based on $O(n)$-time algorithm by Daescu et al. [7], we can design an $O(kn)$-time decision algorithm that performs $O(n)$ comparisons. By applying parametric search, we can obtain an $O(kn^2)$-time algorithm.

Our algorithm for the Altitude Terrain Cover problem finds the optimal y-coordinate of L, denoted by h^*, and the placements of k guards in an iterative manner. It first computes the set of *extreme* placements over all y-coordinates for the leftmost and the rightmost guards. From this, we compute a y-interval I that contains h^*. In more detail, it computes the rightmost placement for the leftmost guard over all y-coordinates. The set of placements forms a piecewise increasing function. Likewise, it computes the leftmost placement for the rightmost guard over all y-coordinates, which together form a piecewise decreasing function.

Then, our algorithm computes a y-interval I which contains the optimal y-coordinate h^* by binary search over the y-coordinates of the breakpoints (i.e. the endpoints of the pieces) of the two functions. Then it repeats the process by computing the functions for the second leftmost guard and the second rightmost guard, and by applying binary search over the y-coordinates of the breakpoints of the functions within I. Then we obtain a subinterval of I containing h^*.

For even numbers $k \geq 2$, we proceed until all guards ($k/2$ guards from the left and $k/2$ guards from the right) are placed. Let I^* be the final interval we have, which contains h^*. Let E be the set of edges of T lying between the two guards placed in the middle (their indices being $k/2$ and $k/2+1$ from left to right). Within I^*, we check for each edge in E whether every point on the edge is visible from any of the two guards, and if not, compute the minimum y-coordinate such that the edge is covered together by them. The optimal y-coordinate h^* is the maximum y-coordinate among the y-coordinates.

The time for computing the interval I^* dominates the time complexity of the algorithm, which is $O(k^2\lambda_{k-1}(n)\log n)$. This is because it takes $O(k)$ time to compute the intersection of two functions that intersect at most k times, the time to compute the two monotone functions (which are upper envelopes of rational functions defined by edges of T) is $O(k\lambda_{k-1}(n)\log n)$, and there are $O(k)$ guards.

For odd numbers k, we continue the process until we place $k-1$ guards ($\lfloor k/2 \rfloor$ guards from the left and $\lfloor k/2 \rfloor$ guards from the right). Instead of computing the edge set E and the corresponding y-coordinate for each edge in E as in the case for even numbers k, within I^* we find h^* in $O(kn)$ time using quasiconvex programming. The total time is again dominated by the time for computing I^*, which is $O(k^2\lambda_{k-2}(n)\log n)$.

For the Bijective Altitude Terrain Cover (BATC) problem, we first show that there is an optimal partition of T such that every endpoint of the subchains is a vertex of T. For a subchain, a guard must be placed at the intersection of the upper half-planes, each bounded by the line extending a segment of the subchain. Thus, we compute the lowest point of this intersection to guarantee the minimum y-coordinate for covering the subchain.

In case (1) of BATC, we place guards on L one by one from left to right in a greedy manner. We find the longest subchain starting from the leftmost vertex that can be covered by a single guard on L in time linear to the size of the subchain. We repeat this process for the remaining terrain until the whole terrain is covered. This can be done in $O(n)$ time.

In case (2) of BATC, there are $C(n-2, k-1) = O((n-2)^{k-1})$ partitions of T into k subchains. For $k = 2$, we can compute the lowest guard position for each subchain of an optimal partition over all $O(n)$ partitions in $O(n)$ time using an incremental algorithm. For $k > 2$, we place k guards one by one from left to right. We show that it suffices to consider $O(n)$ partitions, instead of $O((n-2)^{k-1})$ partitions. We design an $O(n)$-time incremental algorithm for placing a guard. We run the algorithm k times to place k guards, and obtain the optimal y-coordinate of L. Thus, it takes $O(kn)$ time in total.

Most proofs are omitted and can be found in the full version [10].

2 Preliminaries

For a point p in the plane, we use $x(p)$ and $y(p)$ to denote the x- and y-coordinate of p. For any two distinct points p and q in the plane, we denote by pq the line segment connecting p and q, and by $\overline{pq}$ the line passing through p and q. For a segment s, we use $\overline{s}$ to denote the line extending s. The *slope* of a line is the angle swept from the y-axis in the counterclockwise direction to the line, and it is thus in $[0, \pi)$. The slope of a line segment is the slope of the line that extends it. For a non-vertical line segment s, we use s^+ to denote the set of points in $\mathbb{R}^2$ that lie on or above $\overline{s}$. Thus, s^+ denotes the *upper half-plane* bounded by $\overline{s}$.

We use $T = \langle v_1, \dots, v_n \rangle$, a sequence of points with $x(v_i) < x(v_j)$ for any $1 \le i < j \le n$, to denote an x-monotone polygonal chain which we call *terrain*. We call every point v_i a *vertex* of T for $i = 1, \dots, n$, and every segment v_iv_{i+1} an *edge* of T for $i = 1, \dots, n-1$. We use $\tilde{T}$ to denote the set of points $u \in \mathbb{R}^2$ such that the vertical line through u intersects T at u' and $y(u) \ge y(u')$. A line segment on T is *fully visible* from a point if every point on the line segment is visible from the point. We let $y(T) = \max\{y(p) \mid p \in T\}$ denote the maximum y-coordinate among the points on T. For a real value $h \in [y(T), \infty)$, we call the horizontal line $L(h) := \{(x, y) \in \mathbb{R}^2 | y = h\}$ as the *altitude line*.

3 Altitude Terrain Cover

In this section, we present algorithms for the Altitude Terrain Cover (ATC) problem. Consider the case for $k = 1$. An edge e of T is fully visible from a point u only if $u \in e^+$. Since T is x-monotone, every edge of T is fully visible from the lowest point in $\bigcap_{e \subset T} e^+$. Thus, h^* is the y-coordinate of the lowest point in $\bigcap_{e \subset T} e^+$. We can compute the lowest point in linear time [15].

In Sect. 3.1, we give an $O(n \log n)$-time algorithm for $k = 2$. In Sect. 3.2, we present an $O(k\lambda_{k-1}(n) \log n)$-time algorithm for even numbers $k \geq 2$ of guards. In Sect. 3.3, we present an $O(k\lambda_{k-2}(n) \log n)$-time algorithm for odd numbers $k \geq 3$.

3.1 Two Guards

Clearly, $h^* \in I_0 := [y(T), \infty)$. Fix $h \in I_0$. For a point $p \in \tilde{T}$, let $f(p, h)$ be the rightmost point on $L(h)$ from which p is visible. For a line segment ℓ in $\tilde{T}$, let $f(\ell, h)$ (resp. $g(\ell, h)$) be the rightmost (resp. leftmost) point in $L(h)$ from which ℓ is fully visible. See Fig. 2(a). Let $f(h)$ be the leftmost point among $f(e, h)$ for all edges $e \subset T$. See Fig. 2(b). Similarly, let $g(h)$ be the rightmost point among $g(e, h)$ for all edges $e \subset T$.

Let u_1 and u_2 be two guards on $L(h)$ with $x(u_1) < x(u_2)$.

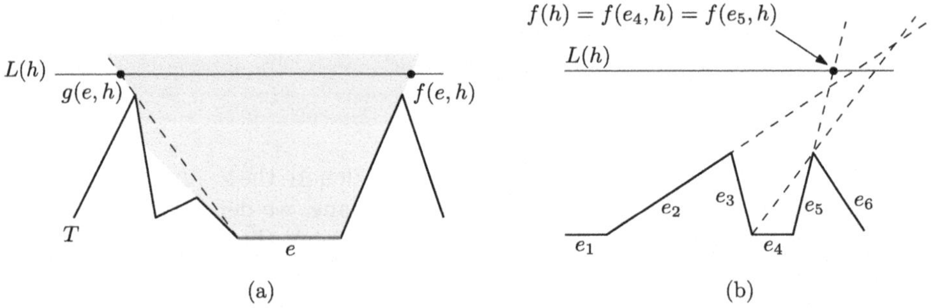

Fig. 2. (a) The gray region is the set of points visible from every point of e. $g(e, h)$ (resp. $f(e, h)$) is the leftmost (resp. rightmost) point of $L(h)$ in the region. (b) $f(h)$ is the leftmost point among $f(e, h)$ for all edges e of T.

Lemma 2. *Two guards u_1 and u_2 on $L(h)$ can guard T only if $x(u_1) \leq x(f(h))$ and $x(u_2) \geq x(g(h))$.*

By Lemma 2, for each $h \in I_0$, we must place u_1 at $f(h)$ or to its left, and u_2 at $g(h)$ or to its right. By the following lemma, u_1 placed at $f(h)$ and u_2 placed at $g(h)$ are indeed the best choice. For a point $p \in \tilde{T}$, let $\mathsf{Vis}(T, p)$ be the set of points of T that are visible from p. We simply use $\mathsf{Vis}(p)$ if it is clear from the context.

Lemma 3. *Vis*$(u) \subseteq$ *Vis*$(f(h))$ *for any point* $u \in L(h)$ *with* $x(u) < x(f(h))$*, and* *Vis*$(w) \subseteq$ *Vis*$(g(h))$ *for any point* $w \in L(h)$ *with* $x(w) > x(g(h))$.

By Lemma 3, for a fixed h, we find $f(h)$ and $g(h)$, and then check whether they together cover T. Using this, we find the minimum value of h such that T is covered. To find it efficiently, we show that both $f(h)$ and $g(h)$ are piecewise linear, and that they are xy-monotone curves in $\mathbb{R}^2$.

A *shortest path tree* of a polygon from a point p in the polygon is the tree rooted at p such that for each vertex u of the polygon, the path between p and u on the graph (as a geometric graph) coincides with the geodesic between p and u in the polygon [8]. See Fig. 3(a). It can be computed in linear time for any given point in the polygon. For a point $p \in \tilde{T}$, let $\mathsf{SPT}(p)$ be the shortest path tree from p to the vertices of T. Let $\mathsf{S}_1 = \mathsf{SPT}(v_1)$ and $\mathsf{S}_n = \mathsf{SPT}(v_n)$. For a node v in S_t, let $\pi_t(v)$ denote the parent node of v in S_t for $t = 1, n$.

Lemma 4. *Let* v *be a vertex of* T*. For any* $h \in I_0$, $f(v,h) = L(h) \cap \overline{v\pi_n(v)}$*, and* $g(v,h) = L(h) \cap \overline{v\pi_1(v)}$.

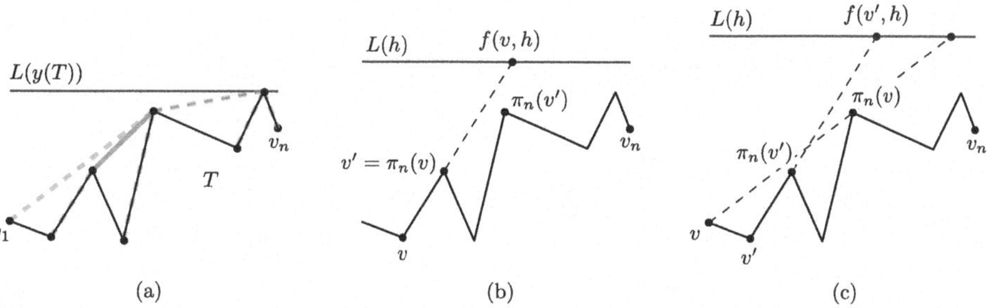

Fig. 3. (a) Red dashed segments are edges of S_n. For any fixed $h \in I_0$, (b) $f(vv',h) = f(v,h)$ if $x(\pi_n(v)) < x(\pi_n(v'))$. (c) $f(vv',h) = f(v',h)$ if $x(\pi_n(v)) \geq x(\pi_n(v'))$. (Color figure online)

Lemma 5. *Let* vv' *be an edge of* T *with* $x(v) < x(v')$*. For any* $h \in I_0$, $f(vv',h) = f(v,h)$ *if* $x(\pi_n(v)) < x(\pi_n(v'))$*, and* $f(vv',h) = f(v',h)$ *otherwise.*

Having computed $\mathsf{SPT}(v_1)$ and $\mathsf{SPT}(v_n)$ in linear time, we can compute $f(e,h)$ and $g(e,h)$ for any edge e of T and any fixed $h \in I_0$ in $O(1)$ time by Lemmas 4 and 5.

Lemma 6. $f(h)$ *is piecewise linear and monotone increasing, and* $g(h)$ *is piecewise linear and monotone decreasing. We can compute them in* $O(n)$ *time.*

By Lemma 6, we compute $f(h)$ and $g(h)$ in $O(n)$ time. The next step is to reduce the search space for h^* to the maximal subinterval $(y_0, y_0']$ of I_0 such that $h^* \in (y_0, y_0']$, and both $f(h)$ and $g(h)$ restricted to $(y_0, y_0']$ are linear functions.

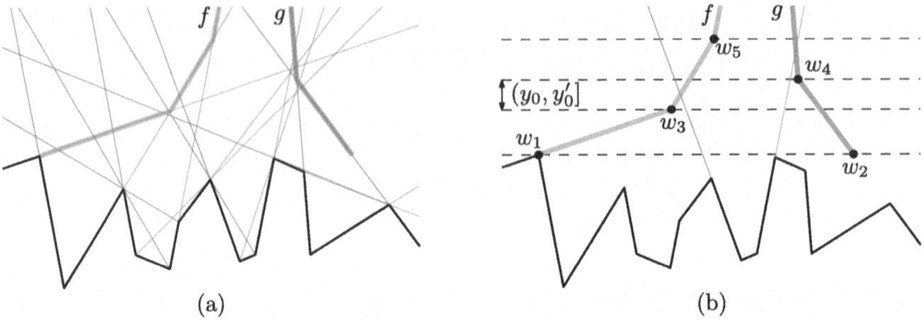

Fig. 4. (a) Red and blue half-line represents $f(e, h)$ and $g(e, h)$ for an edge e of T, respectively. (b) We have $W = \langle w_1, w_2, w_3, w_4, w_5 \rangle$, and $(y_0, y'_0] = (y(w_3), y(w_4)]$. (Color figure online)

Let $W = \langle w_1, \ldots, w_m \rangle$ be the sequence of vertices of $f(h)$ and $g(h)$, ordered by their y-coordinates, where $m = O(n)$. Let H denote the sequence $\langle y(w_1), y(w_2), \ldots, y(w_m) \rangle$.

The following lemma, Lemma 7, enables us to find $(y_0, y'_0]$ such that y_0 and y'_0 are consecutive in H using binary search. See Fig. 4(b).

Lemma 7. *We can compute the interval* $(y_0, y'_0]$ *in* $O(n \log n)$ *time.*

Let $q \in \tilde{T}$ and e be an edge of T that is not fully visible from q. Let e' be the maximal segment of e that is fully visible from q. Let p be the endpoint of e' that is not an endpoint of e. The *peak* of e for q, denoted by $\mathsf{peak}(e, q)$, is the set of vertices $v \in T$ such that $v \in pq$. See Fig. 5(b). For edges e with $e' = \emptyset$, we set $\mathsf{peak}(e, q) = \emptyset$.

Recall that u_1 and u_2 are the two guards. Let $U(h)$ be the set of fully visible edges from u_1 on $L(h)$. Let $V(h)$ be the set of pairs $(e, \mathsf{peak}(e, u_1))$ for edges e that are not fully visible from u_1 on $L(h)$, and $\mathsf{peak}(e, u_1) \neq \emptyset$. We similarly define $\overline{U}(h)$ and $\overline{V}(h)$ for u_2 on $L(h)$. Let (y_1, y'_1) be the maximal subinterval of $(y_0, y'_0]$ such that $h^* \in (y_1, y'_1]$, and the sets $U(h)$, $V(h)$, $\overline{U}(h)$, and $\overline{V}(h)$ remain unchanged for all $h \in (y_1, y'_1)$.

Lemma 8. *We can compute the interval* (y_1, y'_1) *in* $O(n \log n)$ *time.*

By Lemma 8, we can compute (y_1, y'_1) in $O(n \log n)$ time. For each edge e of T such that $e \notin U(y_1)$ and $e \notin \overline{U}(y_1)$, we compute the minimum y-coordinate h such that e is covered by u_1 and u_2 on $L(h)$ in $O(1)$ time. The optimal y-coordinate h^* is the maximum among the y-coordinates. Thus, we have the following theorem.

Theorem 1. *Given a terrain with* n *vertices, the Altitude Terrain Cover problem for* $k = 2$ *can be solved in* $O(n \log n)$ *time.*

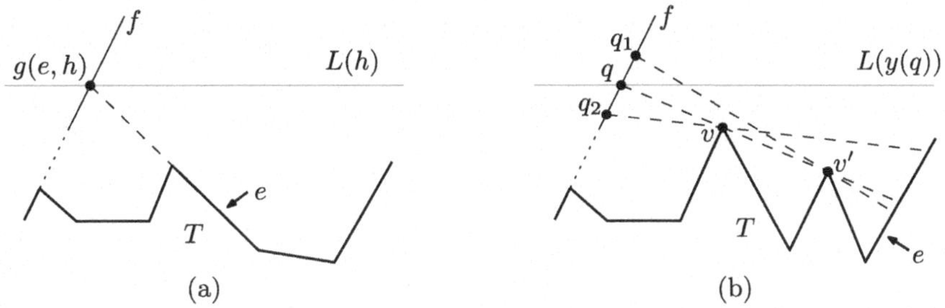

Fig. 5. (a) e is an edge of T. $f(h) = g(e,h)$. (b) vv' is an edge of S_1. We have that $\overline{vv'} \cap e$ is contained in the interior of e, $q = f(h) \cap \overline{vv'}$, and $\mathsf{peak}(e,q) = \{v, v'\}$ for $h = y(q)$. For arbitrary small $\epsilon > 0$, $\mathsf{peak}(e, q_1) = \{v'\}$ for $h = y(q) + \epsilon$, and $\mathsf{peak}(e, q_2) = \{v\}$ for $h = y(q) - \epsilon$.

3.2 Even Numbers of Guards

Here we consider even numbers $k > 2$. Let $u_1, \ldots, u_k$ be the guards on $L(h)$ from left to right. Our algorithm works in a greedy manner. It places u_1 to cover a subchain containing v_1, and places u_k to cover a subchain containing v_n. Let $\mathcal{T}_1$ and $\mathcal{T}_k$ be the set of maximal subchains of T covered by u_1 and u_k, respectively. The algorithm then places u_2 (and u_{k-1}) to cover the maximal subchains of $T \setminus (\mathcal{T}_1 \cup \mathcal{T}_k)$ containing the leftmost (and the rightmost) vertex of it. It continues to place the remaining guards in this manner until it places $u_{k/2}$ and $u_{k/2+1}$. The following lemma and corollary generalize Lemma 2, which gives a necessary condition on the position of guards.

Lemma 9. *Suppose that $u_1, \ldots, u_i$ and $u_{k-i+1}, \ldots, u_k$ are placed on $L(h)$ for $1 \le i < k/2$. Then T can be covered only if both of the followings hold.*

(a) $x(u_{i+1}) \le x(p)$, *where p is the leftmost point among $f(e,h)$'s for all edges e of $T \setminus \bigcup_{j=1}^{i} \big(\mathsf{Vis}(u_j) \cup \mathsf{Vis}(u_{k-j+1})\big)$.*
(b) $x(q) \le x(u_{k-i})$, *where q is the rightmost point among $g(e,h)$'s for all edges $e \subset T \setminus \bigcup_{j=1}^{i} \big(\mathsf{Vis}(u_j) \cup \mathsf{Vis}(u_{k-j+1})\big)$.*

Let $f_1(h) = f(h)$, $g_1(h) = g(h)$ and $T_1(h) = T$. Let $T_{i+1}(h) = T_i(h) \setminus \big(\mathsf{Vis}(f_i(h)) \cup \mathsf{Vis}(g_i(h))\big)$ be the parts of T not covered by any of $f_1(h), \ldots, f_i(h)$ and $g_1(h), \ldots, g_i(h)$. We define $f_i(h)$ as the leftmost point among $f(e,h)$ for all edges e of $T_i(h)$. Similarly, we define $g_i(h)$ as the rightmost point among $g(e,h)$ for all edges e of $T_i(h)$.

With Lemma 9, we place the k guards on $L(h)$ such that $u_i = f_i(h)$ for $i \le k/2$, and $u_i = g_{k+1-i}(h)$ for $k/2 < i \le k$.

Lemma 10. *Let i be an integer with $1 < i \le k/2$. For any point u on $L(h)$ with $x(f_{i-1}(h)) \le x(u) \le x(f_i(h))$, $T_i(h) \setminus \mathsf{Vis}(f_i(h)) \subseteq T_i(h) \setminus \mathsf{Vis}(u)$. For any point u' on $L(h)$ with $x(g_i(h)) \le x(u') \le x(g_{i-1}(h))$, $T_i(h) \setminus \mathsf{Vis}(g_i(h)) \subseteq T_i(h) \setminus \mathsf{Vis}(u')$.*

Lemma 10, which is a generalization of Lemma 3, indicates that this placement is indeed the best choice. For an edge $e \in T_i(h)$, we show that $f(e,h)$ is a piecewise rational function for any $i = 2, \ldots, k/2$. For a polynomial function ϕ, let $\deg(\phi)$ be the degree of ϕ. For two integers $i \geq 0$ and $j > 0$, let $\Pi_{i,j}$ be the set of rational functions $C(h)/D(h)$, where $C(h)$ and $D(h)$ are polynomials with $\deg(C(h)) \leq i$ and $\deg(D(h)) \leq j$.

Let $U_i(h)$ be the set of edges e of T such that every point of e is visible from some guard among $u_1 \ldots, u_i$ and $u_{k-i+1}, \ldots, u_k$ on $L(h)$. Let $V_i(h)$ be the set of pairs $(e, \mathsf{peak}(e, u_i))$ for edges e of T such that $e \notin U_i(h)$, for $1 \leq i \leq k/2$. Similarly, let $\overline{V}_i(h)$ be the set of pairs $(e, \mathsf{peak}(e, u_{k-i+1}))$ for edges e of T such that $e \notin U_i(h)$, for $1 \leq i \leq k/2$.

Let $I_i = (h_i, h_i')$ for $i = 1, \ldots, k/2$ be the maximal subinterval of I_{i-1} such that $h^* \in (h_i, h_i']$, the sets $V_i(h), \overline{V}_i(h)$, and $U_i(h)$ remain the same for all $h \in I_i$, and both $f_i(h)$ and $g_i(h)$ restricted to I_i are rational functions in $\Pi_{i,i-1}$.

For an edge e of T and an integer i with $1 < i \leq k/2$, let $e(i,h)$ denote the portion of e that is not visible from $f_{i-1}(h)$ or $g_{i-1}(h)$ on $L(h)$, and let $I_i(e) \subseteq I_{i-1}$ be a y-interval such that $h^* \in I_i(e)$ and $\mathsf{peak}(e, f(e(i,h),h))$ remains the same for all $h \in I_i(e)$.

Lemma 11. *For an integer i with $1 < i \leq k$, a y-coordinate $h \in I_i(e)$, and an edge e of T with $e \notin U_{i-1}(h)$, $f(e(i,h),h)$ is a function in $\Pi_{i,i-1}$.*

Lemma 11 shows that $f(e,h)$ restricted to $I_i(e)$ is a rational function. The following two lemmas show how to compute I_1 and I_i's for $i = 2, \ldots, k/2$ iteratively.

Lemma 12. *We can compute I_1 in $O(n \log n)$ time.*

After computing I_1 by Lemma 12, we compute I_i for $i = 2, \ldots, k/2$ in an iterative manner.

Lemma 13. *For any integer i with $1 < i \leq k/2$, we can compute I_i given I_{i-1} in $O(k\lambda_{2i-1}(n) \log n)$ time.*

By Lemma 13, we can compute $I_{k/2}$ in $O(k^2\lambda_{k-1}(n) \log n)$ time. Within the interval for each edge vv' of T such that $x(u_{k/2}) < x(v) < x(v') < x(u_{k/2+1})$ and it is not fully visible from $u_{k/2}$ or $u_{k/2+1}$, we compute the minimum y-coordinate h such that e is covered by $u_{k/2}$ and $u_{k/2+1}$ together on $L(h)$. The optimal y-coordinate h^* is then the maximum among all such values, which we can compute in $O(n)$ time using an argument similar to the proof of Theorem 1. Therefore, we conclude this section with Theorem 2.

Theorem 2. *For any even number $k > 2$, the Altitude Terrain Cover problem can be solved in $O(k^2\lambda_{k-1}(n) \log n)$ time.*

3.3 Odd Numbers of Guards

Consider odd numbers $k \geq 3$. As in Sect. 3.2, we place the guards $u_1, \ldots, u_{\lfloor k/2 \rfloor}$ at $f_1(h), \ldots, f_{\lfloor k/2 \rfloor}(h)$, and place $u_k, \ldots, u_{k-\lfloor k/2 \rfloor+1}$ at $g_1(h), \ldots, g_{\lfloor k/2 \rfloor}(h)$. Let

$m = \lceil k/2 \rceil$ be the middle index. By Lemma 9, the optimal y-coordinate h^* is the minimum value h such that $x(f_m(h)) = x(g_m(h))$, and we can compute it efficiently using quasiconvex programming [2].

Theorem 3. *For any odd number $k \geq 3$, the Altitude Terrain Cover problem can be solved in $O(k^2 \lambda_{k-2}(n) \log n)$ time.*

4 Bijective Altitude Terrain Cover

In this section, we present algorithms for the bijective altitude terrain cover problem. In Sect. 4.1, we present an $O(n)$-time algorithm for case (1). In Sect. 4.2, we present an $O(kn)$-time algorithm for case (2).

The following lemma shows that, for $k = 1$, there is an optimal placement of the guard u in the guard minimization version of the altitude terrain cover problem on the x-interval defined by the input terrain. Let $\mathcal{E} := \bigcap_{e \subset T} e^+$ be the intersection of the (closed) upper half-planes defined by the edges of T.

Lemma 14. *For any $h \in I_0$, if T is covered by a single guard u on $L(h)$, then there is a point u' on $L(h)$ with $x(v_1) \leq x(u') \leq x(v_n)$ that covers T.*

For any two points $p, q \in T$ with $x(p) \leq x(q)$, we use $T(p, q)$ to denote the subchain of T from p to q. Let $(u_1, T_1), \ldots, (u_k, T_k)$ be an optimal solution (*guard-subchain pairs*) of the bijective altitude terrain cover problem. By the above lemma, each u_i on $L(h)$ for $i = \{1, \ldots, k\}$ can be placed such that $x(a_i) \leq u_i \leq x(b_i)$ for $T_i = T(a_i, b_i)$.

Moreover, there always is an optimal solution corresponding to an *edge partition*: every endpoint of the k subchains is a vertex of T. This is verified in the following lemma.

Lemma 15. *For any optimal guard-subchain pair $(u_1, T_1), \ldots, (u_k, T_k)$, every endpoint of T_i, for each $i = 1, \ldots, k$, is a vertex of T.*

4.1 Guard Optimization

Similar to the algorithm by Daescu et al. [7] for the altitude terrain guarding problem, we greedily place guards on the given line $L(h)$ from left to right.

Let $\mathcal{E}(i, j) := \bigcap_{e \subset T(i,j)} e^+$ be the intersection of the upper half-planes defined by the consecutive edges $v_i v_{i+1}, \ldots, v_{j-1} v_j$ of T. For the leftmost guard, we find the largest index j_1 such that $\mathcal{E}(1, j_1) \cap L(h) \neq \varnothing$. By Lemma 15, we can optimally place u_1 on any point of $\mathcal{E}(1, j_1) \cap L(h)$ to cover $T(1, j_1)$. Likewise, for every $i = 2, \ldots, k$, we find the largest index j_i such that $\mathcal{E}(j_{i-1}, j_i) \cap L(h) \neq \varnothing$, and place u_i on any point of it. We repeat this process until the whole terrain T is covered by the guards.

By Lemmas 1 and 14, instead of computing each $\mathcal{E}(j_{i-1}, j_i)$, we compute its lowest vertex, which we denote by $w(j_{i-1}, j_i)$. For efficiency, our algorithm handles edges of positive slopes and edges of negative slopes separately while computing such lowest vertices. With edges of positive slope, it incrementally constructs the intersection $\mathcal{E}'$ of the corresponding half-planes. On the other hand, it handles edges of negative slopes one by one in order: compute the intersection of the edge and $\mathcal{E}'$, check if the lowest vertex changes, and update $\mathcal{E}'$ accordingly if so.

The following lemma shows how to efficiently compute such $w(j_{i-1}, j_i)$'s as mentioned above. Let $\mathcal{E}^+(i,j)$ and $\mathcal{E}^-(i,j)$ be the intersection of e^+'s for all edges e of T with positive and negative slopes, respectively.

Lemma 16. *For any indices i, j with $i < j$, we can compute $w(i,\ell)$'s for all $(i+1) \leq \ell \leq j$ in $O(j-i)$ time.*

Recall that our algorithm finds, for each guard u_i, the longest subchain (corresponding to the index j_i) starting from the leftmost edge of the remaining parts of T. By Lemma 16, we can find j_1 in $O(j_1)$ time, and each j_i in $O(j_i - j_{i-1})$ time. Thus, we can find an optimal set of guard-subchain pairs in $O(n)$ time.

Theorem 4. *Given a terrain with n vertices and a horizontal line L, we can place the minimum number of guards on L for case (1) of the Bijective Altitude Terrain Cover problem in $O(n)$ time.*

4.2 Altitude Optimization

Cases $k = 1, 2$ can be handled in $O(n)$ time as in Sect. 4.1.

Suppose there are more than two guards. By Lemmas 14 and 15, we have $h^* = \min_{1<i_1<\ldots i_{k-1}<n}\{\max\{y(w(1,i_1)), \ldots y(w(i_{k-1},n))\}\}$. We can compute $w(i,j)$ for all pairs (i,j) with $1 \leq i < j \leq n$ in $O(n^2)$ time using Lemma 16. Then h^* can be found in $O(n)$ time with a greedy approach as in Sect. 4.1. Finally, an optimal partition of T can be found in $O(kn^2)$ time using dynamic programming.

We can improve it to $O(kn)$ by not computing all $w(i,j)$'s but only $O(n)$ values for each guard. The idea is to maintain $O(n)$ subintervals of I_0 while placing the guards from left to right, such that the set of edges covered so far is the same for any y-coordinate in each interval.

For the leftmost guard, the algorithm computes the points $w(1,2), \ldots, w(1,n)$ in $O(n)$ time. The horizontal lines through these points define the initial $O(n)$ subintervals. For each remaining guard u_i, the algorithm computes $O(n)$ number of $w(i,j)$'s using the information for each subinterval. It further divides the subintervals using the newly computed $w(i,j)$'s. The number of subintervals is $O(n)$ for any u_i. The algorithm stops when it finds a lowest subinterval for the rightmost guard u_k that sees the rightmost edge of T.

Full details can be found in the full version [10].

Theorem 5. *Given a terrain with n vertices and a positive integer k, we can place the lowest horizontal line L and place k guards on L for case (2) of the Bijective Altitude Terrain Cover problem in $O(kn)$ time.*

References

1. Agarwal, P.K., et al.: Guarding a terrain by two watchtowers. Algorithmica **58**(2), 352–390 (2010)
2. Amenta, N., Bern, M., Eppstein, D.: Optimal point placement for mesh smoothing. J. Algorithms **30**(2), 302–322 (1999)
3. Bespamyatnikh, S., Chen, Z., Wang, K., Zhu, B.: On the planar two-watchtower problem. In: Wang, J. (ed.) COCOON 2001. LNCS, vol. 2108, pp. 121–130. Springer, Heidelberg (2001). https://doi.org/10.1007/3-540-44679-6_14
4. Biniaz, A., Maheshwari, A., Mitchell, J.S., Odak, S., Polishchuk, V., Shermer, T.: Contiguous boundary guarding. arXiv preprint arXiv:2412.15053 (2024)
5. Chen, D.Z., Estivill-Castro, V., Urrutia, J.: Optimal guarding of polygons and monotone chains. In: 7th Canadian Conference on Computational Geometry (CCCG 1995), pp. 133–138 (1995)
6. Cole, R., Sharir, M.: Visibility problems for polyhedral terrains. J. Symb. Comput. **7**(1), 11–30 (1989)
7. Daescu, O., Friedrichs, S., Malik, H., Polishchuk, V., Schmidt, C.: Altitude terrain guarding and guarding uni-monotone polygons. Comput. Geom. **84**, 22–35 (2019)
8. Guibas, L., Hershberger, J., Leven, D., Sharir, M., Tarjan, R.: Linear time algorithms for visibility and shortest path problems inside simple polygons. In: In 2nd Annual Symposium on Computational Geometry (SCG 1986), pp. 1–13 (1986)
9. Kang, B., Choi, J., Han, J., Ahn, H.K.: Guarding points on a terrain by watchtowers. In: 36th Canadian Conference on Computational Geometry (CCCG 2024), pp. 41–47 (2024)
10. Kang, B., Kim, H., Ahn, H.K.: Guarding terrains with guards on a line. arXiv preprint arXiv:2505.02373 (2025)
11. Katoh, N., Wang, W., Xu, Y., Zhu, B.: Parametric search: three new applications. Front. Math. China **5**, 65–73 (2010)
12. King, J., Krohn, E.: Terrain guarding is NP-hard. SIAM J. Comput. **40**(5), 1316–1339 (2011)
13. McCoy, B., Zhu, B., Dutt, A.: Guarding precise and imprecise polyhedral terrains with segments. In: International Conference on Combinatorial Optimization and Applications, pp. 323–336. Springer, Cham (2023)
14. Megiddo, N.: Applying parallel computation algorithms in the design of serial algorithms. J. ACM (JACM) **30**(4), 852–865 (1983)
15. Megiddo, N.: Linear programming in linear time when the dimension is fixed. J. ACM (JACM) **31**(1), 114–127 (1984)
16. Seth, R., Maheshwari, A., Nandy, S.C.: Acrophobic guard watchtower problem. Comput. Geom. **109**, 101918 (2023)
17. Sharir, M., Agarwal, P.K.: Davenport-Schinzel Sequences and Their Geometric Applications. Cambridge University Press (1995)
18. Tripathi, N., Pal, M., De, M., Das, G., Nandy, S.C.: Guarding polyhedral terrain by k-watchtowers. In: Chen, J., Lu, P. (eds.) FAW 2018. LNCS, vol. 10823, pp. 112–125. Springer, Cham (2018). https://doi.org/10.1007/978-3-319-78455-7_9

Minimum-Complexity Graph Simplification Under the Fréchet-Like Distance

Omrit Filtser[1], Majid Mirzanezhad[2(✉)], and Carola Wenk[3]

[1] Open University of Israel, Ra'anana 4353701, Israel
omritf@openu.ac.il
[2] Ohio University, Athens, OH 45701, USA
miirza@ohio.edu
[3] Tulane University, New Orleans, LA 70118, USA
cwenk@tulane.edu

Abstract. Simplifying granular geometric networks for future computations is vital, especially in geospatial data processing, where maps are frequently used. Given a geometric graph and an error bound, the objective is to compute an alternative graph of a minimum number of vertices and edges in total so that a "Fréchet-like" distance between the two graphs remains at most the error. As curve simplification has a cubic conditional lower bound under the Fréchet distance [9], it seems unlikely to achieve a fast polynomial-time algorithm for graphs under the same distance. In this paper, the Fréchet-like distance we consider between graphs is the *"Graph Distance"* introduced by Akitaya et al. [3]. Due to its recognized practice in GIS, we assume that the simplified graph is a subgraph of the original graph for which we prove the NP-hardness. Turning our attention to trees, the simplified subtree may not stay connected; hence, we slightly shift the setting to ensure the simplified tree selects its vertices from a subset of the original vertices. We propose two algorithms for tree simplification, including the case where the leaves of the simplified and original trees are mapped and enforced to correspond to one another under the graph distance.

Keywords: Graph simplification · Fréchet distance · Graph distance

1 Introduction

Unlike the curve simplification problem, there is much less known on simplifying more complicated input objects such as trees and graphs. Approximating geometric networks can be a crucial task in data analysis, e.g., in GIS, image processing, shape matching, mesh simplification, and molecular biology [8,16,22,28]. Broadly speaking, a practitioner may wish to obtain a coarse and simpler representation of a graph-like object such as cartographic maps, molecular structures, or neural and vascular systems while preserving the geometry and topology of

H. Fernau and B. Zhu (Eds.): IWOCA 2025, LNCS 15885, pp. 44–57, 2025.
https://doi.org/10.1007/978-3-031-98740-3_4

the underlying structure intact to a degree. This can bring the idea of approximating a network with another one of moderate complexity to the sphere of simplification problem.

1.1 Related Work

A very elementary variant of geometric simplification considers the input graph as a simple path (polygonal curve). There is a massive body of research on curve simplification problem under a variety of distances, notably the Fréchet and Hausdorff distances [1,2,5,7,10–15,17,21,24,25]. Intuitively, the former is the minimum-length of the leash connecting a man and dog walking along two curves, and the latter is the maximum of the nearest neighbor distances overall points from one curve to the other.

Recent works underlined the difference between *global* and *local* simplifications. in global simplification, a bound on a distance measure must be satisfied between the entire original and simplified curves, whereas in local simplification, the same bound holds between each original edge and the portion it surrounds on the simplified curve. Van Kreveld et al. [27] obtained a polynomial-time simplification algorithm under the Fréchet distance but NP-hardness under the Hausdorff distance. Later, a cubic algorithm under both distances and a cubic conditional lower bound were obtained [9]. Concurrently, Van Kerkhof et al. [26] considered various simplified vertex selections and provided bi-criteria approximation on the error bound and the number of simplified edges.

A few works studied approximating a planar subdivision of a map with the minimum number of links under topological constraints [18,23]. Most of the algorithms have applied to GIS data and are based on map schematization/generalization [30,31] in which, roughly speaking, the essential topological structure of the map remains the same and paths with vertices of degree two become simplified. A notable work on map schematization under some topological constraints, e.g., facet preserving, no self-intersecting boundary simplification, can be found in [20,29]. Simplifying a planar subdivision inside a polygon with another low-complexity planar subdivision homeomorphic to the original one has also been explored [18,23]. A heuristic algorithm for planar maps is proposed which keeps the boundary of the polygon simple. At the same time, it is impossible to give a polynomial-time algorithm within $n^{\frac{1}{5}-\epsilon}$ approximation factor for the problem, for any $\epsilon > 0$, assuming $\mathsf{P} \neq \mathsf{NP}$ [18].

In the context of simplification, one can consider different variants induced by the input parameters/constraints such as the distance measure, e.g., Fréchet, Hausdorff distances, its direction from input to output (if the distance is asymmetric), topological constraints (facet preservation), and others. Note that some distances are not symmetric. Therefore, it is crucial to determine in which direction the distance is applied. We define our problem of interest for graphs as follows, but it can apply to trees as well:

Problem: *Min-Complexity Graph Simplification (MCGS). Let* $\delta > 0$, $\mathsf{D}(\cdot,\cdot)$ *be a distance measure between graphs, and* $G = (V, E)$ *be a (connected) graph in* $\mathbb{R}^d$,

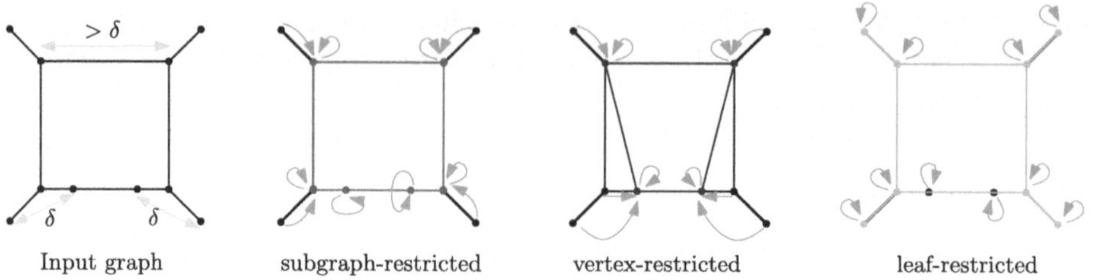

Fig. 1. Simplification of a graph with different vertex placement restrictions under the Graph distance mapping from input to output graph highlighted in green. (Color figure online)

whose edges in E are straight-line segments between the vertices in V. Compute an alternative (connected) graph G′, with the minimum number of edges and vertices in total satisfying $\mathsf{D}(G, G') \leq \delta$.

We use a Fréchet-like distance called *"Graph Distance"* introduced in [3]. The authors prove the NP-hardness for computing the graph distance between two graphs, but they give an $O(n^3)$-time algorithm if one graph is a tree, where n is the total number of vertices of the graphs. Under a *graph mapping*, a graph is mapped to a portion of the other so the edges are mapped to paths with a minimum Fréchet distance in the other graph. The *graph distance* is defined as the maximum of the Fréchet distances between the edges and the paths they are mapped to. The figure on the right shows the graph distance of δ under the optimal mapping $\mu : G_1 \to G_2$ with $\mu(a_1) = b_1$, $\mu(a_2) = b_2$, and $\mu(a_3) = b_3$.

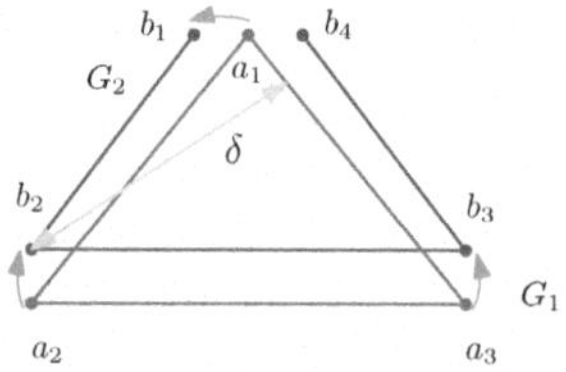

For the vertex placement of the simplified graph, we differentiate between *subgraph-restricted*, and *vertex-restricted.* In the subgraph-restricted simplification, the simplified graph is a subgraph of the input graph, whereas, in the vertex-restricted case, which is a less restrictive class compared to the former, a simplified graph selects its vertices from a subset of the input vertices. Another special case of the vertex-restricted setting is the *leaf-restricted* simplification, which requires the degree-one vertices of the output to be identical to a subset of degree-one vertices of the input graph. This might help capture the graph's structure, such as roads that terminate at cul-de-sacs (see e.g. [3]). In principle, we call a variant of MCGS with restriction $\mathcal{R}$ on the placement of vertices of the simplified graph that simplifies G_1 to a minimum-complexity simplified graph G_2, applying the distance $\overrightarrow{\mathsf{D}}(G_2, G_1)$ (from output G_2 to input G_1), an *$\mathcal{R}$-restricted min-complexity G_1-G_2 simplification under* $\overrightarrow{\mathsf{D}}(G_2, G_1) \leq \delta$. See Fig. 1 for a better understanding of different simplifications.

1.2 Our Results

Inspired by globally simplifying a curve in [2,6,9,26,27], we study globally approximating a graph under a "Fréchet-like" distance" with restrictions on the placement of the vertices and edges of the simplified graph. In this paper, we primarily study two restrictions: subgraph- and vertex-restricted. We show that the subgraph-restricted min-complexity graph-to-graph simplification is NP-hard (Theorem 1). Because of its NP-hardness, we explore trees as a viable option to approximate. Although the NP-hardness of the tree-tree simplification variant remains elusive, we give a fixed parameter algorithm that runs in $O(n^5+n^3\alpha^2 2^\alpha)$ time under the graph distance. Here, α is an implicit parameter that, roughly speaking, is the number of intersections between the simplified tree and the ball of radius δ around each vertex of the input tree (Theorem 2).

As the vertex-restricted variant appears to be hard to admit a fully polynomial time algorithm, we take our investigation further to the leaf-restricted variant. We show that the leaf-restricted tree-tree simplification when the distance is applied from input to output becomes NP-hard (Theorem 3); however, when the direction of the distance flips with the leaves of the output being identical to k leaves of the input, an $O(kn^5)$-time algorithm can be devised (Theorem 4).

An interesting pattern discerned from our results is that the difficulty of the problem variants gets counter-intuitively skewed depending on the direction of the distance measure and the vertex placement of the simplified graph. A more comprehensive version of our results in this paper can be found in [19].

2 Preliminaries

We first begin with introducing curves and the Fréchet distance between them. Let $P = \langle p_1, p_2, \cdots, p_n \rangle$ be a polygonal curve. We treat P as a continuous map $P : [1, n] \to \mathbb{R}^d$, where $P(i) = p_i$ for an integer i, and the i-th edge is linearly parameterized as $P(i+\lambda) = (1-\lambda)p_i + \lambda \cdot p_{i+1}$. We write $P[s,t]$ for the subcurve between $P(s)$ and $P(t)$ and $\langle P(s)P(t) \rangle$ for the line segment connecting the two points. Given two curves $P : [1, n] \to \mathbb{R}^d$ and $Q : [1, m] \to \mathbb{R}^d$, the Fréchet distance between P and Q is defined as:

$$\mathsf{F}(P,Q) = \inf_{f,g} \max_{t\in[0,1]} \|P(f(t)) - Q(g(t))\|,$$

where $f : [0, 1] \to [1, n]$ and $g : [0, 1] \to [1, m]$ are continuous non-decreasing functions. If f and g are not non-decreasing functions, then the obtained distance is called *the weak Fréchet distance* denoted by $\mathsf{wF}(P,Q)$. The *free space diagram* of the two curves P and Q of complexities of n and m, respectively, is denoted by $\mathsf{FSD}_\delta(P,Q)$. This diagram consists of $(n-1) \times (m-1)$ cells in the domain $[1, n] \times [1, m]$.

For any $\delta > 0$, the free space diagram $\mathsf{FSD}_\delta(P,Q)$ consists of *cells* and the boundary of each cell consists of four sides which each contains at most one *free space interval*. An interval is part of an edge of P that is within δ to a vertex of

Q and vice versa. A *monotone path* from $(1,1)$ to (n,m) that lies entirely within the free space corresponds to a pair of monotone reparameterizations (f,g) that witness $\mathsf{F}(P,Q) \le \delta$. Alt and Godau showed that such a *reachable path* can be computed in $O(mn)$ time by propagating reachable points across free space cell boundaries in a dynamic programming manner [4]. See Fig. 2 for an example of free space diagram for two curves.

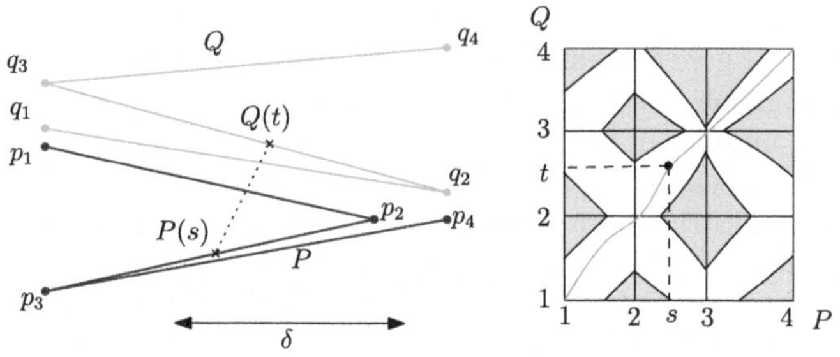

Fig. 2. (s,t) is a free point on a reachable path in $\mathsf{FSD}_\delta(P,Q)$. The free space is shown in white.

Now let $G = (V,E)$ be a graph *immersed* in $\mathbb{R}^d$, where $V(G) = \{v_1, v_2, \cdots, v_n\}$, v_i is embedded at a point $p \in \mathbb{R}^d$ for all $i \in \{1, \cdots, n\}$, and each edge $e = \langle uv \rangle \in E(G)$ is also linearly parameterized. Suppose μ is an arbitrary mapping from graph G_1 to graph G_2 that maps every vertex of G_1 to some point on G_2. In other words, $\mu(v) = p$ for all $v \in V(G_1)$, there exists some $p \in G_2$. Given an edge $\langle uv \rangle \in E(G_1)$, with a slight abuse of the notation, we have $\mu(e) = \mathcal{P}$, where $e = \langle uv \rangle$, and $\mathcal{P}$ is a path starting at $\mu(u)$ and ending at $\mu(v)$ on G_2. Now a *graph mapping* (introduced in [3]) is a function $\mu : G_1 \to G_2$ that (1) maps each vertex $v \in V(G_1)$ to a point $\mu(v)$ on an edge of G_2, and (2) maps each edge $e = \langle uv \rangle \in E(G_1)$ to a simple path $\mu(e)$, from $\mu(u)$ to $\mu(v)$, in G_2. The *directed (strong) graph distance* between G_1 and G_2 is:

$$\overrightarrow{\delta_G}(G_1, G_2) = \inf_{\mu: G_1 \to G_2} \max_{e \in E(G_1)} \mathsf{F}(e, \mu(e)).$$

Note that the path between $\mu(u)$ and $\mu(v)$ in G_2 may not exist which results in $\mathsf{F}(e, \mu(e)) = \infty$. This case can occur if G_2 is disconnected. The *directed (weak) graph distance* denoted by $\overrightarrow{\delta_{wG}}$ is obtained by replacing the F with wF.

3 Graph Simplification

In this section we prove that computing the subgraph-restricted minimum-complexity graph-graph simplification under $\overrightarrow{\delta_G}(G, G') \le \delta$ is NP-hard. We reduce from a specific variant of MAX-2SAT problem which is defined as follows: Given a set of variables $X = \{X_1, X_2, \cdots, X_n\}$ and a CNF formula $\mathcal{F}$ consisting of m disjunctive clauses each with at most two variables, find an assignment $\{\mathsf{T},\mathsf{F}\}^n$ to X such that the number of satisfied clauses is maximum. The decision version of this problem, MAX-2SAT$(X, \mathcal{F}, k)$, takes an integer $k > 0$ as an argument and asks whether the exists an assignment A to X under which the number of satisfied clauses in $\mathcal{F}$ is at least k or not. Given a formula $\mathcal{F}$, a *variable graph* $G_{\mathcal{F}}$ is a graph whose nodes are the variables in $\mathcal{F}$ and two nodes are connected by an edge if the respective variables of the nodes belong to the same clause in $\mathcal{F}$. A MAX-2SAT problem is said to be *BIPARTITE-MAX-2SAT* if the variable graph of its formula forms a bipartite graph. In the full version of this paper [19], we show that BIPARTITE-MAX-2SAT is NP-hard, and then we make our main reduction from BIPARTITE-MAX-2SAT.

Lemma 1. *BIPARTITE-MAX-2SAT is NP-hard.*

Proof. Replacing every clause $C = (x \vee y \vee z)$ in the 3-SAT formula with the following 16 clauses results in a bipartite variable graph: $\mathcal{F}_C = (\bar{x} \vee a) \wedge (\bar{x} \vee \bar{b}) \wedge (\bar{y} \vee b) \wedge (\bar{y} \vee \bar{c}) \wedge (\bar{z} \vee c) \wedge (\bar{z} \vee \bar{a}) \wedge (\bar{w} \vee d_1) \wedge (\bar{w} \vee d_2) \wedge (\bar{w} \vee d_3) \wedge (x \vee \bar{d_1}) \wedge (y \vee \bar{d_2}) \wedge (z \vee \bar{d_3}) \wedge (x) \wedge (y) \wedge (z) \wedge (w)$. It is not hard to see that C is satisfied if and only if 13 clauses of $\mathcal{F}_C$ are satisfiable (see [19] for further details).

We present our main reduction from BIPARTITE-MAX-2SAT to the min-complexity graph-graph simplification problem: set $\delta = 1$ and construct a graph G from $G_{\mathcal{F}}$ as follows: G consists of two types of gadgets; *variable* and *clause* gadgets. A variable gadget X_i has two vertices T_i and F_i representing the two possible assignments to X_i.

A binary clause gadget of $(X_i \vee X_j)$ connects two pairs of vertices of two variable gadgets using paths passing through some *hook* vertex h_{ij} that is located far away from the variables. The hook vertex controls the number of links such that a clause gadget can be simplified with exactly two links if it is satisfied and three links otherwise by the corresponding assignment to the variables in the clause. There are four interior vertices connected to the hook that are located close enough, by distance ϵ, to each other, where $0 < \epsilon \ll \delta = 1$. The *length* of the clause gadget is L_{ij} that is the difference between the y-coordinates of the two variable gadgets. Here we also set $L_{ij} \gg 1$. See Fig. 3(a), 3(b) for further illustration on two types of clause gadgets. The clause gadget for every unary clause (X_i) is similar to case (b) and involves one additional variable gadget X_t. To handle this, we remove the path between T_t and F_i, the path and F_t and T_i, from the original clause gadget in case (b).

Let $G_{\mathcal{F}} = (\mathcal{X}_1 \dot{\cup} \mathcal{X}_2, E)$, so $\mathcal{X}_1 \dot{\cup} \mathcal{X}_2$ is a partition of the variables, E set of edges/clause gadgets, and $G_{\mathcal{F}}$ is bipartite. Let ℓ_1 and ℓ_2 be two lines parallel to the x-axis, at vertical distance L. We place the variable gadgets belonging to $\mathcal{X}_1$ on ℓ_1 and those belonging to $\mathcal{X}_2$ on ℓ_2. Since $G_{\mathcal{F}}$ is bipartite, each clause gadget connects a variable from ℓ_1 to another variable from ℓ_2 (see Fig. 3(c)). Now, we need to make sure that the edges on the subgraph simplifying a clause gadget can be used only to *"cover"* the edges of that clause. By "cover", we mean that the graph distance from the clause gadget and the subgraph simplifying the clause gadget must be 1. For this, we choose $L \gg 1$ and the variable gadgets

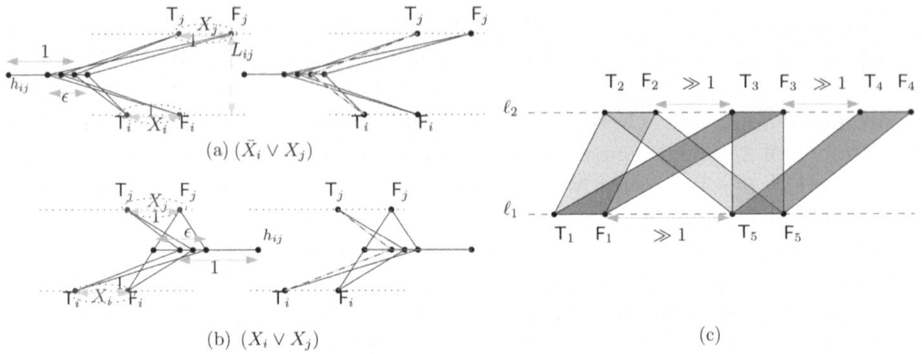

Fig. 3. (a) A clause gadget for $(\bar{X}_i \vee X_j)$. Note that a path with 2 edges, in green, from T_i to T_j simplifies the gadget while satisfying the clause and uses 3 edges in red, otherwise. (b) A clause gadget for $(X_i \vee X_j)$, that gets simplified similar to type (a). The constructions for $(X_i \vee \bar{X}_j)$ and $(\bar{X}_i \vee \bar{X}_j)$ are symmetric to types (a) and (b). (c) A schematic representation of the thickened embedded variable graph induced by $(X_1 \vee X_2) \wedge (X_1 \vee \bar{X}_3) \wedge (\bar{X}_2 \vee \bar{X}_5) \wedge (X_4 \vee \bar{X}_5) \wedge (X_3 \vee X_5)$. For a better presentation, types (a) and (b) are highlighted in orange and blue strips, respectively.

are spaced apart at distance larger than 1 along each ℓ_1 and ℓ_2. This way the clause gadgets of the same type (either type (a) or type (b)) do not overlay on top of each other and cannot *cover* for each other. The following lemmas imply our main theorem in this section.

Lemma 2. *A min-complexity simplified graph selects exactly one vertex per variable gadget.*

Lemma 3. *Let m be the number of clauses. An assignment satisfying at least $k < m$ clauses of* Bipartite-Max-2SAT *exists iff G' with $|E(G')| + |V(G')| \leq 6m - 2k + 1$ exists.*

Theorem 1. *The subgraph-restricted min-complexity graph-to-graph simplification under $\overrightarrow{\delta_G}(G, G') \leq \delta$ is NP-hard.*

4 Tree Simplification

In this section, we set out to give a fixed parameter algorithm for vertex-restricted min-complexity tree-tree simplification under the graph distance from input to output. The main idea is to consider the free space diagram between all edges of the input tree and all edges of the complete graph G induced by the vertices of T, i.e., $\mathsf{FSD}_\delta(G, T)$, and then find a reachable tree throughout $\mathsf{FSD}_\delta(G, T)$ that crosses the minimum number of vertices in G. It is not hard to see that there is always a tree $T' \subset G$ with minimum-complexity such that $\overrightarrow{\delta_G}(T, T') \leq \delta$. Given a tree $T = (V, E)$ in $\mathbb{R}^d$, we define the *shortcut graph* of T as $G = G(T) = (V(G), E(G))$, where $V(G) = V(T)$ and $E(G) = \{\langle uv \rangle \mid u, v \in V_G\}$.

Each edge $e = \langle uv \rangle \in E(G)$ is linearly parameterized. The parameter space of G is $E(G) \times [0,1]$ and the parameter space of T is $E(T) \times [0,1]$.

Now, let $\delta > 0$, and consider the joint parameter space $E(T) \times [0,1] \times E(G) \times [0,1]$ of T and G. Any $(e, \lambda, e', \gamma) \in E(T) \times [0,1] \times E(G) \times [0,1]$ is called *free* if $\|e(\lambda) - e'(\gamma)\| \leq \delta$, and the union of all free points are referred to as the *free space*. We consider a free space chunk, i.e., the free space diagram to be composed of an edge in T and the whole shortcut graph G. A chunk is comprised of two main components: (1) spine; the free space between an edge in T and a vertex in G, and (2) slice; the free space between a vertex in T and the entire graph G; see Fig. 4. For any $v \in V(T), v' \in V(G)$, $e \in E(T)$ and $e' \in E(G)$ we call $\mathsf{SP}(v') = e \times [0,1] \times v'$ a *spine*, and $\mathsf{SL}(v) = \cup_{e' \in E(G)} v \times e' \times [0,1]$ a *slice*. We also denote the free space within spines and slices as $\mathsf{SP}_\delta(v') = \{(\lambda, v') \mid 0 \leq \lambda \leq 1,\ \|e(\lambda) - v'\| \leq \delta\}$, and $\mathsf{SL}_\delta(v) = \{(v, \gamma) \mid 0 \leq \gamma \leq 1,\ \|v - e'(\gamma)\| \leq \delta\}$, respectively. For an edge $e = \langle uv \rangle \in E(T)$, it holds that $\mathsf{SL}(u), \mathsf{SL}(v) \subset \mathsf{FSD}_\delta(e, G)$ and $\mathsf{SL}(u)$ is a subset of all free spaces with respect to edges in T incident on u.

Definition 1 (Elementary Intervals). *For every $v \in V(T)$, a free space interval $I \in \mathsf{SL}_\delta(v)$ is called* elementary *if I lies completely within $v \times e' \times [0,1]$, for some $e' \in E(G)$; see Fig. 4.*

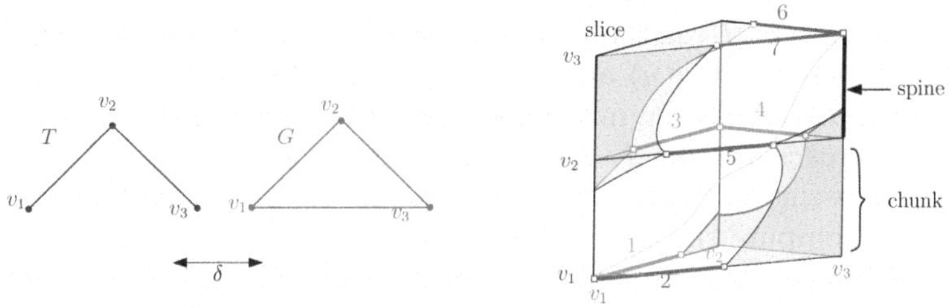

Fig. 4. The free space diagram between a tree T and its shortcut graph G consisting of seven elementary intervals. A reachable path in green uses a minimum number of spines traversing all edges in T. (Color figure online)

We use dynamic programming to compute an optimal solution to the problem. Let T be a rooted tree. In this setup every vertex $v \in V(T)$ is the parent of a set of vertices if (1) v is the neighbor of the vertices, and (2) takes fewer vertices on T to reach the root of T. The idea is to start constructing optimal trees rooted at elementary intervals of $\mathsf{SL}_\delta(v)$ for all $v \in V(T)$, where v is a leaf and then propagate the minimum-vertex rooted subtrees bottom-up towards the root in a dynamic programming fashion. For this, we associate a cost function $\psi : [0,1] \to \mathbb{N}$ with each elementary interval I, where $\psi(I)$ is the number of vertices in a minimum-vertex simplified subtree rooted at I. When a simplified tree T' *is rooted at an elementary interval* $I \in \mathsf{SL}_\delta(u)$ it means that T' simplifies the subtree of T rooted at u, and the mapping realizing $\overrightarrow{\delta_G}(T, T') \leq \delta$ matches u to a point $x \in I$. Let T'_I denote a simplified tree rooted at I and $\mathsf{W}(T'_I)$ be the *weight* of T'_I that indicates the number of vertices on it. Clearly, for an $I \in \mathsf{SL}_\delta(u)$, $\psi(I) = \min_{T'_I} \mathsf{W}(T'_I)$. We also extend our notations as follows: Suppose in a rooted tree T, u is the parent of v in T and $\mathsf{Children}(u) = \{v_1, \cdots, v_k\}$ is the set of children of u and correspondingly $\mathsf{Parent}(v_i) = u$, for all $i = 1, \cdots, k$.

Now suppose we are given $\mathsf{Children}(u) = \{v_1, \cdots, v_k\}$. Each $\mathsf{SL}_\delta(v_i)$ consists of a set of elementary intervals, for all $i \in \{1, \cdots, k\}$. Note that an elementary interval might belong to different sets of elementary intervals associated with different slices. In other words, each $\mathsf{SL}_\delta(v_i)$, for all $i \in \{1, \cdots, k\}$, has a set of elementary intervals and an elementary interval I can belong to $\{\mathsf{SL}_\delta(v_{i_1}), \cdots, \mathsf{SL}_\delta(v_{i_m})\}$ where $\{i_1, \cdots, i_m\} \subseteq \{1, \cdots, k\}$. At this point, we say that $I \in \mathsf{SL}_\delta(v_j)$ *"covers"* a subset of vertices $\{v_{i_1}, \cdots, v_{i_m}\}$ where $\{i_1, \cdots, i_m\} \subseteq \{1, \cdots, k\}$, and we denote this by $C(I) = \{v_{i_1}, \cdots v_{i_m}\}$ for some $j \in \{1, \cdots, k\}$. Let $\mathcal{I}^* = \{I_1, \cdots, I_\ell\}$ be the universal set of all elementary intervals over all $\mathsf{SL}_\delta(v_i)$ for all $i \in \{1, \cdots, k\}$. For any $I \in \mathsf{SL}_\delta(u)$ where u is a non-leaf vertex, we propose the following recursive formula:

$$\psi(I) = \min_{\mathcal{I} \subseteq \mathcal{I}^*} \sum_{I' \in \mathcal{I}} \big(\psi(I') + \kappa(I', I)\big), \tag{1}$$

where $\mathcal{I}$ ranges over all subsets of $\mathcal{I}^*$ such that $\cup_{I \in \mathcal{I}} C(I) = \mathsf{Children}(u) = \{v_1, \cdots, v_k\}$, i.e., all elementary intervals in $\mathcal{I}$ together cover all vertices in $\mathsf{Children}(u)$. Here, $\kappa(I, I')$ is the number of vertices in a minimum-vertex reachable path between $I \in \mathsf{SL}_\delta(u)$ and $I' \in \mathsf{SL}_\delta(v_i)$ in $\mathsf{FSD}_\delta(e, G)$ where $e = \langle uv_i \rangle$. In other words, $\kappa(I, I')$ is the minimum number of spines that a reachable path has to cross to reach I' from I. If u is a leaf then $\psi(I) = 1$.

We now present our dynamic programming algorithm as follows: Similar to [3], we first compute the connected components of all vertices in T (the part of G lying within the balls of radius δ around vertices of T). For two neighboring vertices u and v, we prune all invalid paths between u and v in G for which the Fréchet distance to $e = \langle uv \rangle$ is greater than δ. This gives us a *pruned graph* G'. Now we construct $\mathsf{FSD}_\delta(G', T)$ and define the slices with respect to G' and T. We perform a BFS search on T starting at the root and store the vertices into an auxiliary stack S in the order they are being encountered along the search. If a vertex u is a leaf then we set $\psi(I) = 1$ for all $I \in \mathsf{SL}_\delta(u)$ and $\psi(I) = \infty$, otherwise. Once we processed all the vertices in T, we pop the vertices from S. For every popped vertex u and every elementary interval $I \in \mathsf{SL}_\delta(u)$, we compute $\psi(I)$ using formula (1). We repeat the process until we reach the root r and I, for all $I \in \mathsf{SL}_\delta(r)$. In the end, we backtrack and find the elementary intervals in $\mathsf{FSD}_\delta(G', T)$ that contain the minimum ψ value. The vertices of the simplified tree are the endpoints of those intervals that end (or start) at their neighboring spines. The following lemmas lead to the main theorem:

Lemma 4. *Constructing G' takes $O(n^3)$ time under the weak graph distance and $O(n^5)$ time under the strong graph distance.*

Lemma 5. *For every $I \in \mathsf{SL}_\delta(u)$ and $I' \in \mathsf{SL}_\delta(v)$ where $v \in \mathsf{Children}(u)$, there exists a procedure that computes $\kappa(I, I')$ in $O(n^2)$ time under both the weak and strong graph distances.*

Theorem 2. *Let T be a tree in $\mathbb{R}^d$ with n vertices and $\delta > 0$. There exists an algorithm that computes a vertex-restricted min-complexity simplified tree T' in $O(\alpha^2 2^\alpha n^3)$ and $O(n^5 + \alpha^2 2^\alpha n^3)$ times fulfilling $\overrightarrow{\delta_{wG}}(T, T') \leq \delta$ and $\overrightarrow{\delta_G}(T, T') \leq \delta$, respectively, where α is the maximum number of elementary intervals over all slices in $\mathsf{FSD}_\delta(G', T)$.*

5 Leaf-Restricted Tree Simplification

We now consider the case where the leaves of T' are identical to subset of the leaves of T. We show that given a tree T in $\mathbb{R}^2$ and a threshold δ, it is NP-hard to compute a tree T' with a minimum complexity such that $\overrightarrow{\delta_G}(T, T') \leq \delta$, where every leaf in $V(T')$ is identical to some leaf in $V(T)$. The reduction is fairly straightforward as follows: Given a unit disk graph $G = (V, E)$ with $P = V(G)$, let $\delta = 1$ and B be the smallest bounding box of unit disks around the points in P. Construct T as follows: $V(T) = P \cup \{w\}$, and $E(T) = \{\langle pw \rangle \mid p \in P\}$, for some point w which lies far enough from B. We have the following theorem:

Theorem 3. *Let $\delta > 0$. Given a tree T in $\mathbb{R}^2$, computing a leaf-restricted min-complexity tree-tree simplification T' under $\overrightarrow{\delta_G}(T, T') \leq \delta$ is NP-Hard.*

Proof. Let $k > 0$ be an integer as the decision parameter to the Minimum Dominating Set in Unit Disk Graph (MDSUDG). We show that there exists a dominating set $S = \{s_1, \cdots, s_m\}$ with $m \leq k$ to the (MDSUDG) if and only if there exists a leaf-restricted simplification T' for T under $\overrightarrow{\delta_G}(T, T') \leq 1$ with at most $k + 1$ vertices. Let $S = \{s_1, \ldots, s_m\}$ with $m \leq k$ be a dominating set to G. Set $V(T') = \{w, s_1, \ldots, s_k\}$, and $E(T') = \{\langle s_i w \rangle \mid 1 \leq i \leq m\}$. Consider a mapping μ such that maps $\mu(w) = w$ and $\mu(p) = p'$ where $p \in V(T)$ and $p' \in V(T')$. Note that p is contained in the unit disk centered at p'. Now each edge $\langle xp \rangle \in E(T)$ is mapped to an edge $\langle wp' \rangle \in E(T')$, and thus $\overrightarrow{\delta_G}(T, T') \leq 1$.

To show the other direction, let T' be a simplified tree of number of vertices at most $k+1$. Let μ be the mapping realizing $\overrightarrow{\delta_G}(T, T') \leq 1$. Let $C = \{\mu(p) \mid p \in V(T) \setminus \{w\}\}$, then $C \subseteq V(T')$ because T' is a leaf-restricted simplification and thus μ matches each leaf p of T to a leaf $\mu(p)$ of T'. Also, C contains at most k vertices because w is far enough from B, so there must be at least one vertex of T' which lies outside of B such that none of the points of P is matched to it. Now consider the set centers $S = \{s_1, \cdots, s_m\}$ of the unit disks around $V(T')$ with $m \leq k$. Clearly, S is a dominating set for G.

5.1 An Output-Sensitive Algorithm

The leaf-restricted tree-tree simplification under $\overrightarrow{\delta_G}(T',T) \leq \delta$ (from output to input) allows us to obtain a polynomial-time algorithm. For a given subset $l = \{l_1, \cdots, l_k\}$ of leaves in T with $k \geq 1$, we require T' has k leaves identical to leaves in l that are given as part of the input. Note that the root r is given as an arbitrary node of T. Without any restriction on the leaves, the problem is trivial, as T' could consist of a single point. Similar to Sect. 4, let G be the shortcut graph. For a vertex $u \in V(T)$, let $\mathcal{I}_u$ be the set of all elementary intervals $I \in \mathsf{SL}_\delta(u)$. Note that for a tree T, leaf set l with root r and a leaf $l_1 \in l$, a solution to the leaf-restricted min-complexity tree simplification under $\overrightarrow{\delta_G}(T',T) \leq \delta$ when $k = 1$ is a reachable path across $\mathsf{FSD}_\delta(G,T)$ starting from an elementary interval $I \in \mathsf{SL}_\delta(r)$ to an elementary interval $l' \in \mathsf{SL}_\delta(l_1)$. In fact, the simplified tree for $k = 1$ is trivially the path from the respective leaf to the root. Note that due to the leaf restriction, the leaves and root of T' are identical to the leaves and root of T. Our algorithm simplifies the remaining vertices in T.

The Algorithm: We first compute the simple path from every leaf in l to the root. The union of these paths forms the *mapping subtree* M_T. A vertex where multiple paths merge/meet in M_T is called an *ancestor*. We associate a pointer with each leaf to point to the closest ancestor obtained along the path to r. We repeat the same process for the new ancestors unti we meet r. The ancestors and pointers together result in a tree called *ancestor tree* A_T. An ancestor u is the *parent* of v (v is a child) if it is pointed to by v in A_T. The idea is to use dynamic programming to propagate the optimal simplified tree rooted at every elementary interval I' of every child to the one rooted at every elementary interval I of the parent. For this, we associate a cost function $\psi : [0,1] \rightarrow \mathbb{N}$ with each edge I, where $\psi(I)$ is the number of vertices in a minimum-vertex simplified subtree rooted at I. Similar to Sect. 4, we again recall that a simplified tree T' *is rooted at an elementary interval* $I \in \mathsf{SL}_\delta(u)$ if T' simplifies the subtree of T rooted at u and the mapping realizing $\overrightarrow{\delta_G}(T',T) \leq \delta$ matches u to a point $x \in I$. For an elementary interval $I \in \mathsf{SL}_\delta(u)$ and $I' \in \mathsf{SL}_\delta(v)$, where u is the parent of v in A_T, we consider the following recursive formula:

$$\psi(I) = \min_{\mathcal{E}} \sum_{I' \in \mathcal{E}} (\psi(I') + \gamma(I', I)), \tag{2}$$

where $\mathcal{E} = \{\mathcal{E}_1, \mathcal{E}_2, \cdots, \mathcal{E}_{k'}\}$ and each elementary interval $\mathcal{E}_i$ belongs to the elementary interval set $\mathcal{I}_{v_i}$, for all $1 \leq i \leq k'$, where k' is the number of children of u in A_T. Here, $\gamma(I', I)$ is the number of spines hit by the reachable path in $\mathsf{FSD}_\delta(M_T, G)$ starting at I and ending at I'. Note that the path from u to v_i in M_T forms a path, thus the simplification of the path $P[u, v_i] \in M_T$ can be computed by the algorithm proposed in [26]. Also for every leaf $u \in l$ we set $\psi(I) = 1$. Note that unlike the formula in Sect. 4, $\mathcal{E}_i \cap \mathcal{E}_j = \emptyset$ for all $\mathcal{E}_i, \mathcal{E}_j \in \mathcal{E}$ with $i \neq j$, otherwise a point $x \in \mathcal{E}_i \cap \mathcal{E}_j$ would be mapped to two different v_i and

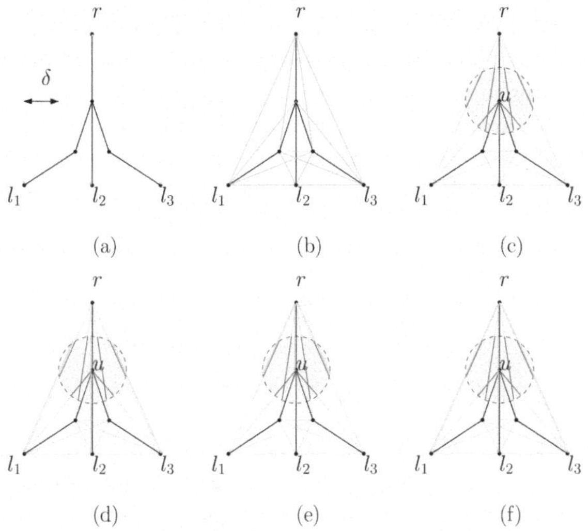

Fig. 5. An illustration of the algorithm. (a) The rooted tree, $\delta > 0$, and $k = 3$ (b) The shortcut graph G in pink. (c) Computing common ancestor u of the leaves $\{l_1, l_2, l_3\}$, identifying the elementary intervals of $\mathsf{SL}_\delta(u)$ in red. (d) Min-vertex curve simplification of $P[l_i, u]$ in green, starting at l_i and ending at some elementary interval in $\mathsf{SL}_\delta(u)$, for all $i = 1, 2, 3$. (e) The continuation of the min-vertex path simplification of $P[u, r]$ starting at elementary intervals in $\mathsf{SL}_\delta(u)$ and ending at r. (f) The resulting simplified tree in green. (Color figure online)

v_j in M_T which is not possible under the mapping realizing the graph distance from T' to T. Our proposed algorithm can run in polynomial time relying on such a fact; see Fig. 5.

Lemma 6. *M_T is the subtree of T that T' is mapped to under $\overrightarrow{\delta_G}(T', T) \leq \delta$.*

Theorem 4. *Let $\delta > 0$ and $k > 0$ be an integer. There is an algorithm running in $O(kn^5)$ time that uses $O(kn^2)$ space for the leaf-restricted tree-tree simplification under $\overrightarrow{\delta_G}(T', T) \leq \delta$.*

6 Conclusion

In this paper, we studied the problem of approximating a graph with an alternative simpler graph with a minimum complexity preserving the Fréchet-like distances between them. To this end, we considered a Fréchet-like distance, known as Graph Distance, under different constraints in which the vertices of the simplified graph can be placed. While this was an initial work under such distances, we obtained a set of NP-hardness and algorithmic results depending on the problem variants. We believe some interesting variants can be left as open problems: (1) proposing any constant factor approximation algorithm for graph simplification under the graph distance, (2) proposing a fully polynomial-time algorithm for tree-tree simplification. Any further investigation on the aforementioned problems and/or obtaining new results for the simplified tree, selecting its vertices from anywhere in the ambient space, can be of interest.

Acknowledgment. Carola Wenk was supported by NSF Grant CCF-2107434, and Omrit Filtser by the Israel Science Foundation (Grant No. 2135/24). The authors thank the anonymous reviewers for their insightful comments, which improved the quality and presentation of the paper.

References

1. Abam, M.A., de Berg, M., Hachenberger, P., Zarei, A.: Streaming algorithms for line simplification. Discrete Comput. Geom. **43**(3), 497–515 (2010)
2. Agarwal, P.K., Har-Peled, S., Mustafa, N., Wang, Y.: Near linear time approximation algorithm for curve simplification. Algorithmica **42**(3–4), 203–219 (2005)
3. Akitaya, H., Buchin, M., Kilgus, B., Sijben, S., Wenk, C.: Distance measures for embedded graphs. Comput. Geom. Theory Appl. **95**(101743), 1–21 (2021)
4. Alt, H., Godau, M.: Computing the Fréchet distance between two polygonal curves. Int. J. Comput. Geom. Appl. **5**(1–2), 75–91 (1995)
5. Barequet, G., Chen, D.Z., Daescu, O., Goodrich, M.T., Snoeyink, J.: Efficiently approximating polygonal paths in three and higher dimensions. Algorithmica **33**(2), 150–167 (2002)
6. Bereg, S., Jiang, M., Wang, W., Yang, B., Zhu, B.: Simplifying 3D polygonal chains under the discrete Fréchet distance. In: Laber, E.S., Bornstein, C., Nogueira, L.T., Faria, L. (eds.) LATIN 2008. LNCS, vol. 4957, pp. 630–641. Springer, Heidelberg (2008). https://doi.org/10.1007/978-3-540-78773-0_54
7. de Berg, M., van Kreveld, M., Schirra, S.: Topologically correct subdivision simplification using the bandwidth criterion. Cartogr. Geograph. Inf. Syst. **25**(4), 243–257 (1998)
8. Bindewald, E., Shapiro, B.: RNA secondary structure prediction from sequence alignments using a network of k-nearest neighbor classifiers. RNA **12**(3), 342–352 (2006)
9. Bringmann, K., Chaudhury, B.R.: Polyline simplification has cubic complexity. J. Comput. Geom. **11**(2), 94–130 (2021). https://doi.org/10.20382/jocg.v11i2a5, https://doi.org/10.20382/jocg.v11i2a5. Special Issue of Selected Papers from SoCG 2019
10. Buzer, L.: Optimal simplification of polygonal chain for rendering. In: Proceedings of the 23rd Annual ACM Symposium on Computational Geometry (SoCG), pp. 168–174 (2007)
11. Chan, S., Chin, F.: Approximation of polygonal curves with minimum number of line segments or minimum error. Int. J. Comput. Geom. Appl. **6**(1), 59–77 (1996)
12. Chen, D.Z., Daescu, O., Hershberger, J., Kogge, P.M., Mi, N., Snoeyink, J.: Polygonal path simplification with angle constraints. Comput. Geom. **32**(3), 173–187 (2005)
13. Cheng, S.W., Huang, H.: Curve simplification and clustering under Fréchet distance. In: Proceedings of the ACM-SIAM Symposium on Discrete Algorithms (SODA), pp. 1414 – 1432 (2023)

14. Cheng, S.W., Huang, H.: Solving Fréchet distance problems by algebraic geometric methods. In: Proceedings of the ACM-SIAM Symposium on Discrete Algorithms (SODA), pp. 4502–4513 (2024)
15. Cheng, S.W., Huang, H., Jiang, L.: Simplification of trajectory streams (2025). https://arxiv.org/abs/2503.23025
16. Cignoni, P., Montani, C., Scopigno, R.: A comparison of mesh simplification algorithms. Comput. Graph. **22**(1), 37–54 (1998)
17. Douglas, D.H., Peucker, T.K.: Algorithms for the reduction of the number of points required to represent a digitized line or its caricature. Cartographica **10**(2), 112–122 (1973)
18. Estkowski, R., Mitchell, J.S.B.: Simplifying a polygonal subdivision while keeping it simple. In: Proceedings 17th Annual ACM Symposium on Computational Geometry, SCG 2001, pp. 40–49 (2001)
19. Filtser, O., Mirzanezhad, M., Wenk, C.: Minimum-complexity graph simplification under Fréchet-like distances (2021). https://arxiv.org/abs/2111.02591
20. Funke, S., Mendel, T., Miller, A., Storandt, S., Wiebe, M.: Map simplification with topology constraints: exactly and in practice. In: Proc. 19th Workshop on Algorithm Engineering and Experiments (ALENEX), pp. 185–196 (2017)
21. Godau, M.: A natural metric for curves—computing the distance for polygonal chains and approximation algorithms. In: Choffrut, C., Jantzen, M. (eds.) STACS 1991. LNCS, vol. 480, pp. 127–136. Springer, Heidelberg (1991). https://doi.org/10.1007/BFb0020793
22. Gudmundsson, J., Laube, P., Wolle, T.: Movement patterns in spatio-temporal data. In: Shekhar, S., Xiong, H., Zhou, X. (eds.) Encyclopedia of GIS, pp. 1362–1370. Springer, Cham (2007). https://doi.org/10.1007/978-3-319-17885-1_823
23. Guibas, L., Hershberger, J., Mitchell, J., Snoeyink, J.: Approximating polygons and subdivisions with minimum-link paths. Int. J. Comput. Geom. Appl. **3**(4), 383–415 (1993)
24. Imai, H., Iri, M.: An optimal algorithm for approximating a piecewise linear function. J. Inf. Process. **9**(3), 159–162 (1986)
25. Imai, H., Iri, M.: Polygonal approximations of a curve - formulations and algorithms. In: Toussaint, G.T. (ed.) Computational Morphology: A Computational Geometric Approach to the Analysis of Form. North-Holland, Amsterdam (1988)
26. van de Kerkhof, M., Kostitsyna, I., Löffler, M., Mirzanezhad, M., Wenk, C.: Global curve simplification. In: 27th Annual European Symposium on Algorithms (ESA 2019), Dagstuhl, Germany, vol. 144, pp. 1–14 (2019). http://drops.dagstuhl.de/opus/volltexte/2019/11188
27. van Kreveld, M., Löffler, M., Wiratma, L.: On optimal polyline simplification using the Hausdorff and Fréchet distance. In: Proceedings of the 34th International Symposium on Computational Geometry (SoCG), vol. 56, pp. 1–14 (2018)
28. Lee, Y.: Handwritten digit recognition using k nearest-neighbor, radial-basis function, and backpropagation neural networks. Neitrul Comput. **3**(3), 440–449 (1991)
29. Mendal, T.: Area-preserving subdivision simplification with topology constraints: exactly and in practice. In: Proceedings of the 20th Workshop on Algorithm Engineering and Experiments (ALENEX), pp. 117–128 (2018)
30. Meulemans, W.: Similarity measures and algorithms for cartographic schematization. Ph.D. thesis, Eindhoven University of Technology, Eindhoven University of Technology (2014)
31. Sester, M.: Cartographic generalization. J. Spat. Inf. Sci. (21), 5–11 (2020). https://doi.org/10.5311/JOSIS.2020.21.716. Invited Article 2

Drawing Reeb Graphs

Erin Chambers[1(✉)], Brittany Terese Fasy[2,3], Erfan Hosseini Sereshgi[3], and Maarten Löffler[3,4]

[1] University of Notre Dame, Notre Dame, IN, USA
echambe2@nd.edu
[2] Montana State University, Bozeman, MT, USA
brittany.fasy@montana.edu
[3] Tulane University, New Orleans, LA, USA
shosseinisereshgi@tulane.edu
[4] Utrecht University, Utrecht, The Netherlands
m.loffler@uu.nl

Abstract. Reeb graphs are simple topological descriptors with applications in topological data analysis and computational geometry. Despite their prevalence, visualization of Reeb graphs has received less attention. In this paper, we bridge an essential gap in the literature by exploring the complexity of drawing Reeb graphs. Specifically, we demonstrate that Reeb graph crossing number minimization is NP-hard, both for straight-lined and curved edges. On the other hand, we identify specific classes of Reeb graphs, namely paths and caterpillars, for which crossing-free drawings exist. We also give an optimal algorithm for drawing cycle-shaped Reeb graphs with the least number of crossings and provide initial observations on the complexities of drawing multi-cycle Reeb graphs. We hope that this work establishes the foundation for an understanding of the graph drawing challenges inherent in Reeb graph visualization and paves the way for future work in this area.

Keywords: Reeb Graphs · Graph Drawing · Minimum Crossing

1 Introduction

Reeb graphs have become an important tool in computational topology for visualizing continuous functions on complex spaces as a simplified discrete structure. Essentially, these graphs capture how the level sets of a function from a topological space to the real numbers evolve and connect. In a Reeb graph, the connected components of each level set become points, forming a graph, where the vertices are found at heights where topological changes occur and the edges represent the evolution of a single connected component; see Fig. 1. Originally introduced in [20], recent work on computing Reeb graph efficiently [16,19] has led to their increasing use in areas such as graphics, visualization, and shape analysis [3,22].

Despite how prevalent the use of Reeb graphs is, surprisingly little work has been done on generating nice drawings of these structures. The only work in this area we are aware of considers book embeddings of these graphs [18]; while

H. Fernau and B. Zhu (Eds.): IWOCA 2025, LNCS 15885, pp. 58–71, 2025.
https://doi.org/10.1007/978-3-031-98740-3_5

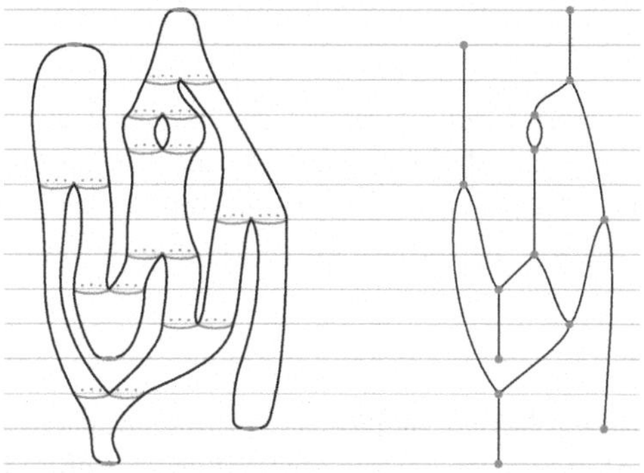

Fig. 1. A two-manifold and its Reeb graph.

of combinatorial interest, the algorithms seem less practical for easy viewing of larger Reeb graphs. So, this leads to a very natural question, both theoretical and practical: how difficult is it to draw a Reeb graph with minimal crossing?

Given the structural properties of Reeb graphs, there is an obvious connection to level drawings of graphs, as Reeb functions are simply real-valued labels on the vertices. *Layered graph drawing*, a.k.a. *hierarchical* or *Sugiyama-style* graph drawing is a graph drawing in which the vertices of a directed graph are drawn in horizontal rows, or *layers*, and in which the edges are y-monotone curves [1,8,21]; depending on the application, vertices may or may not be preassigned to levels. As with many drawing styles, layered drawings tend to be more readable when the number of crossings is low; a significant amount of effort has been directed towards crossing minimization. When a drawing without any crossings is possible, the graph is called *upward planar* (when the vertices are not preassigned to levels) or *level planar* (when they are). Other work characterizes forbidden structures in level planar graphs, concluding that there are infinitely many such forbidden minors in this class of level planar graphs [10]. Although it is possible to test whether a given directed acyclic graph with a single source and a single sink admits an upward planar drawing [14], the problem becomes NP-hard when there are multiple sources or sinks [15]. When the layers are preassigned, the problem is easier [17]. Recent work indicates that several algorithms for this are incorrect [11]; even in this setting, the details are tricky and the exact status is unclear. However, when a planar embedding is not possible, the problem of minimizing crossings is still of interest, and this problem is NP-hard, even when there are only two levels [12].

Building on our previous poster paper [4], this work initiates the study of crossing minimization in drawings of Reeb graphs, working toward practical solutions to the key problem of drawing these essential topological structures effectively. In particular, we demonstrate the inherent complexity of this task by proving that minimizing edge crossings is NP-hard. However, we make promising steps towards parameterization by cycle count, which is expected to be low in many applications. Specifically, we explore the structural properties and drawing algorithms for acyclic and single-cycle Reeb graphs. An extended version of the current conference paper, including all proofs, is available on ArXiv [5].

2 Preliminaries

We lay out the foundational definitions of general and generic Reeb graphs in Sect. 2.1, allowing us to define the problem of minimum crossing number for Reeb graphs formally in Sect. 2.2. Finally, we argue that we can focus our inquiry on the more specialized domain of refined Reeb graphs in Sect. 2.3.

2.1 Reeb Graphs

A Reeb graph, as first explained in [20], is a mathematical construct that represents the topological structure of a scalar function defined on a manifold. At a high level, the Reeb graph is a graph that encodes the connected components of level sets and the critical points of a function defined over a manifold.

To define the Reeb graph, let $\mathcal{M}$ be a compact, connected, and o-minimal manifold of dimension n, and let $h : \mathcal{M} \to \mathbb{R}$ be a smooth function. Define an equivalence relation on the points of $\mathcal{M}$ by setting $y \sim y'$ if y and y' are in the same path component of $h^{-1}(a)$ for some $a \in \mathbb{R}$. Assuming that the starting space and function are well-behaved (i.e., Morse functions on manifolds, or tame functions on constructible spaces [7,9]), the quotient space $G_h = \mathcal{M}/\sim$ is a finite graph G with the following structure: The vertices of G correspond to critical points of h, that is, the points $p \in \mathcal{M}$ whose differential is zero ($dh_p = 0$). The critical points of h partition $\mathcal{M}$ into regions called cells, where each cell is a locally path-connected compact space. Two vertices are connected by an edge in G if their corresponding critical points lie at the boundary of a common cell. A *Reeb graph* is the pair (G, h).

In graph theoretical terms, this means that we have a graph $G = (V, E)$, where the vertex set V comes with a *height* function $h : V \to \mathbb{R}$.

Let the degree $\deg(v)$ of a vertex be the number of edges in E with v as an endpoint. Furthermore, we define the *downward degree* $\deg_{\downarrow}(v)$ as the number of edges that connect v to a vertex of lower height: $\deg_{\downarrow}(v) = \{w \in V | (v, w) \in E, h(w) < h(v)\}$, and similarly the *upward degree* $\deg_{\uparrow}(v)$ as the number of edges that connect v to a vertex of higher height: $\deg_{\uparrow}(v) = \{w \in V | (v, w) \in E, h(w) > h(v)\}$. Note that $\deg_{\downarrow}(v) + \deg_{\uparrow}(v) = \deg(v)$ for all $v \in V$.

We define a *generic Reeb graph* as a Reeb graph (G, h) such that: every vertex has a unique height, so $h(v) \neq h(w)$ for any $v \neq w$; every vertex has $\deg(v) = 1$ or $\deg(v) = 3$; and furthermore $\deg_{\downarrow}(v) \leq 2$ and $\deg_{\uparrow}(v) \leq 2$.

2.2 Drawings of Reeb Graphs

A *drawing* of a Reeb graph is an immersion of G into $\mathbb{R}^2$. We have the following constraints on this immersion:

- Each vertex $v \in V$ is assigned to a unique point $(x, y) \in \mathbb{R}^2$, with the additional requirement that $y = h(v)$.
- Each edge $e = (v, w) \in E$ is assigned to a continuous y-monotone curve with v and w as endpoints.

This problem definition is somewhat related to the notions of *upwards* drawings and *level* drawings, given the height functions and edge constraint, although the concepts are not quite identical. In particular, in upwards planar, the exact height is not fixed for a vertex, but rather all (directed) edges must orient upwards. Level planar drawings generally fix integer-valued heights, and not real-valued, and algorithms in this setting generally seek drawings with no crossing, rather than attempting to minimize crossings as we do in this paper.

Minimizing edge crossings in a graph drawing is a desirable aesthetic criterion. This leads to the following computational problem:

Problem 1 (Reeb Graph Crossing Number). Given a Reeb graph, the Reeb graph crossing number (RGCN) is the minimum number of crossings among all drawings of that Reeb graph.

Ultimately, finding a drawing realizing the minimum number of crossings is the goal. Computing the crossing number is NP-hard in general [12], but to the best of our knowledge the Reeb crossing number has not been studied previously. Notably, the proof by Garey and Johnson [12] does not extend to this special case, as it uses high-degree vertices. Note that our proof does not rely on vertices having unique heights, but does ensure low-degree vertices.

2.3 Refined Reeb Graphs

Given a generic Reeb graph (G, h), the finite set of heights defines a set of 'levels' in the Reeb graph. By subdividing any longer edges so that all edges are between consecutive levels, we arrive at a refined Reeb graph:

Definition 1 (Refined Reeb Graph). *A* refined Reeb graph $(G = (V, E), h)$ *is a Reeb graph with the properties:*

- *every edge* $e = (v, w)$ *in* E *is between consecutive levels*
- $deg_{\downarrow}(v) \leq 2$ *and* $\deg_{\uparrow}(v) \leq 2$ *and* $\deg(v) \leq 3$ *(note: but no longer necessarily* $\deg(v) = 1$ *or* 3*, it can also be* 2*).*

The following lemma shows that a generic Reeb graph and its associated refined Reeb graph have the same crossing complexity. For a proof, see [5].

Lemma 1 (Constructing a Refined Reeb Graph). *For every generic Reeb graph, there is a refined Reeb graph, and their optimal drawings have the same number of crossings.*

3 Straight-Edge or Curved Drawings

A well-known result in graph theory is that a given graph has a plane embedding if and only if it has a straight-line plane embedding. We show here that the same is true for Reeb graphs without crossings. However, the same is not true for drawings with minimal crossing Reeb graphs: requiring the edges to be straight line segments may result in a higher crossing number.

Lemma 2. *Suppose that we are given a planar (zero-crossings) drawing of a refined, generic Reeb graph G, where every edge is an xy-monotone curve and every vertex is at its specified y-coordinate. Then, there exists another planar drawing of G, where the vertices are in the same order in the x-dimension, the vertices are at the same y-coordinates, and the edges are drawn as straight line segments.*

In [5], we prove this by adapting the "shift method" of [6], which is for planar triangulations, to this setting. However, when there are crossings present in the curved drawing, a similar statement does not hold, as we see in Fig. 2, the example used to prove the following lemma in [5].

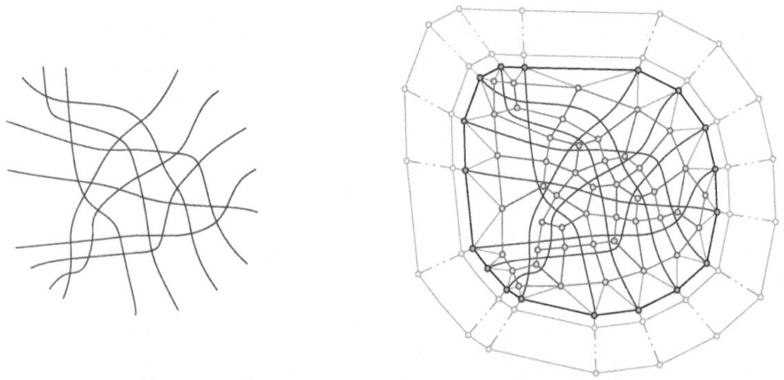

(a) Non-Stretchable Arrangement (b) Non-Stretchable Graph Drawing

Fig. 2. Building a non-stretchable graph drawing Lemma 3). (a) This arrangement of xy-monotone psuedolines is not stretchable, which means it cannot be drawn with straight lines. (b) This graph contains the pseudoline arrangement and its dual as sub-structures.

Lemma 3 (Non-Stretchable Graph Drawing). *There exists a Reeb graph* $(G = (V, E), h)$ *and a drawing* Γ *of* G *such that: every vertex* $v \in V$ *has y-coordinate* $h(v)$ *in* Γ*; every edge is an xy-monotone curve in* Γ*;* Γ *has k crossings; and there exists no straight-line drawing of G with* $\leq k$ *crossings.*

4 NP-Hardness

In this section, we prove the NP-hardness of drawing refined generic Reeb graphs.

Theorem 1 (RGCN is NP-Hard). *Given a generic Reeb graph* $(G = (V, E), h)$*, and an integer* k*, deciding if the crossing number is less than* k *is NP-hard.*

Our proof is inspired by the proof of Garey and Johnson [12]; however, we require several new ideas. Before we explore the NP-hardness proof, we explain the problem we use to prove NP-hardness and describe a crucial structure in the proof in detail. We transform a known NP-complete problem, optimal linear arrangement (OLA) [13], to the generic Reeb graph crossing number problem. We interpret this as "arranging" the vertices of the graph in a linear order, so that the total length of all edges of the graph is less than k. Formally, the OLA problem [13] is defined as follows: Given a graph $G = (V, E)$ and an integer k, is there a function $f :$ V $\rightarrow \{1, 2, ..., |V|\}$ such that $\sum_{\{v,w\} \in E} |f(v) - f(w)| \leq k$.

4.1 Triangular Hexagonal Grid

We construct a triangular hexagonal grid T_k using vertices and edges where k is the number of rows in the grid and each row i consists of i hexagons. A grid T_k consists of $k^2 + 4k + 1$ vertices and $\frac{3}{2}(k^2 + 3k)$ edges. The use of hexagonal grids maintains the characteristics of generic Reeb graphs, such as maximum vertex degree, while safeguarding the graph structure against changes caused by swapping vertices along the vertical axis. This precaution is essential as such alterations would result in a greater number of crossings per swaps; see Fig. 3. Therefore, we use it in Theorem 1 to build our gadget for the NP-hardness proof.

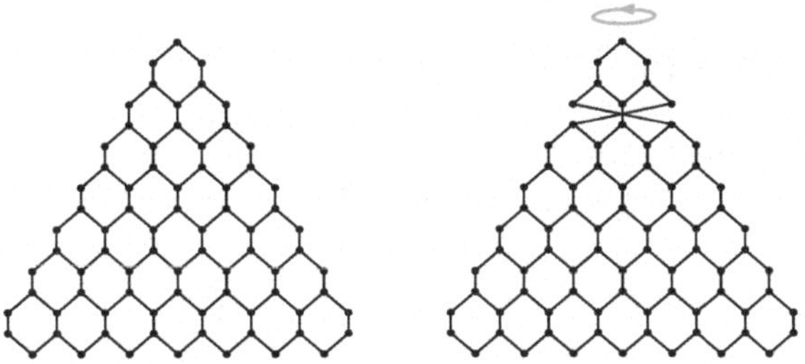

Fig. 3. A triangular hexagonal grid on the left and the same grid if we twist the structure vertically once over the second row.

4.2 Construction

Suppose that we have an instance of the OLA problem, given a graph $G = (V, E)$ and k; see Fig. 4. We construct a generic Reeb graph $H = (V_1 \cup V_2, E_1 \cup E_2 \cup E_3 \cup E_4)$ such that: We replicate every vertex in the set V to form the set V'. The vertices of V are positioned horizontally in an arbitrary order as $v_1, v_2, \ldots v_n$ and the same order is applied to the vertices of V'. Subsequently, all vertices in both V and V' are transformed into triangular hexagonal grids $T_{|E|}$, each consisting of $|E|^2+4|E|+1$ vertices with $|E|^2+1$ vertices at the bottom of the triangular structure. We then employ the middle vertex as a connector (analogous to an

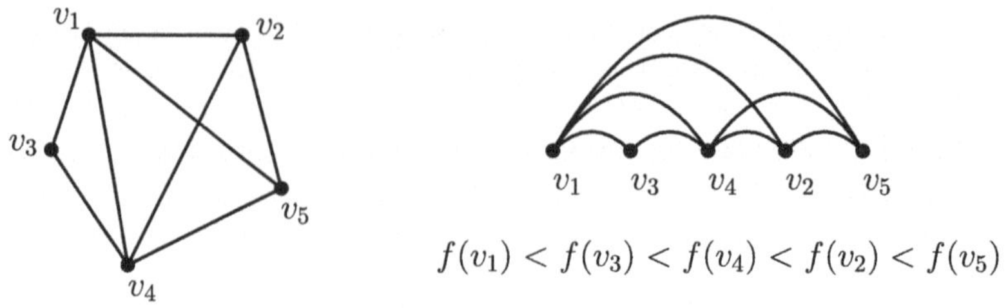

Fig. 4. Graph on the left and its image based on an optimal order defined by a function $f : V \rightarrow \{1, 2, ..., |V|\}$ on the right.

extension cord) to facilitate drawing of the edges of H. Furthermore, we establish edges between the corresponding $|E|^2$ vertices for each pair of hexagonal grids.

Figure 5 shows a construction based on an arbitrary order of the vertices.

- $V_1 = \{|E|^2 + 4|E| + 1 \text{ vertices}\}$
- $V_2 = \{2 \cdot \deg(v_i) \cdot |V| \text{ copies of } v_i \in V\}$
- $E_1 = \{(v_i, w_j')|v_i, w_j' \in V_2,\ (v_i, w_j) \in E,\ i > j\}$
- $E_2 = \{|E|^2 \text{ copies of } \{v_i, v_i'\}; v_i \in V, v_i' \in V'\}$
- $E_3 = \{(v_{i_p}, v_{i_{p+1}})|v_{i_p}, v_{i_{p+1}} \in V_2\}$
- $E_4 = \{\frac{3}{2}(|E|^2 + |E|) \text{ edges}\}$
- $k' = |E|^2(k - |E|) + (|E|^2 - 1)$

Note that constructing H and k' from G and k can be done in polynomial time. Additionally, H is a connected graph only if G is connected.

4.3 Correctness

Next, we argue that H has an embedding with at most k' crossings if and only if G has a linear arrangement of cost at most k through the following two lemmas whose proofs are found in [5].

Lemma 4. *A valid linear arrangement of G implies a valid embedding of H.*

Lemma 5. *A valid embedding of H implies a valid linear arrangement of G.*

Now, we are ready to prove Theorem 1 using the construction as stated above.

Proof (Proof of Theorem 1). Given an instance G of OLA, we construct a graph H as in Sect. 4.2, by Lemma 4, A valid linear arrangement of G implies a valid embedding of H, and by Lemma 5, the reverse is also true. Finally, note that H has polynomial size and the construction of H takes polynomial time. □

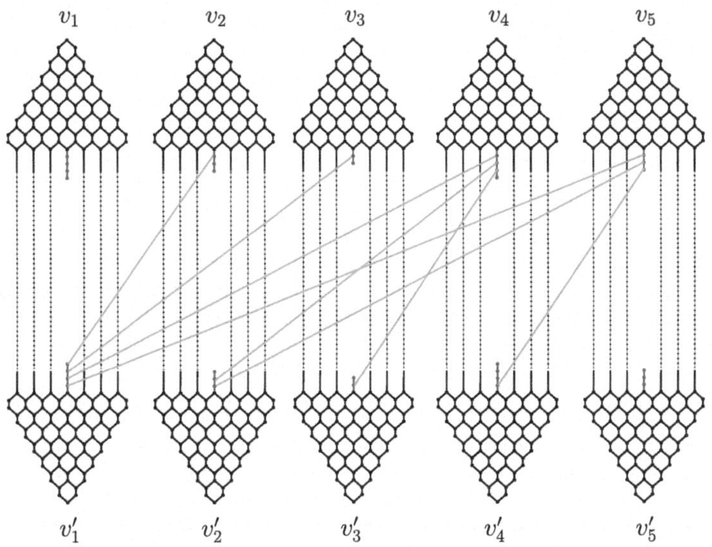

Fig. 5. The corresponding Reeb graph construction of Fig. 4 in our proof. Black vertices are V_1 and red vertices are V_2. E_1 edges are shown in orange, E_2 edges are dotted, E_3 are in red and E_4 are the edges of the triangular hexagonal grids represented in black.

5 Parameterized Complexity in the Number of Cycles

The number of cycles in a Reeb graph for oriented surfaces is connected to its genus. Therefore, in this section, we consider several special graph classes with zero or one cycle. Throughout this section, let (G, h) be a Reeb graph; we seek a drawing that minimizes the number of crossings.

5.1 No Cycles

Consider the case where G has no cycles (i.e., G is a tree). If we did not have the constraint that each vertex has a prescribed height, then G is a planar graph and hence can be drawn without crossings. If we instead consider the problem where the edges are pointed in a particular direction (upward or downward), it is still possible to draw G without crossings [2, Statement 6.1]. However, once we require the drawing to respect h, then it is possible that drawing G requires crossings.

We start with a basic observation, which has been used in the literature (e.g., [10]). If G has no branching points (i.e., is a single path), then drawing (G, h) without crossings is always possible:

Construction 1 (A Path). *Let (G, h) be a Reeb graph such that G is a single path. Then, label the vertices consecutively by the order they appear in the path: $v_1, v_2, \ldots, v_n$. We draw v_i at $(i, h(v_i))$, with straight-line edges connecting adjacent vertices on the path.*

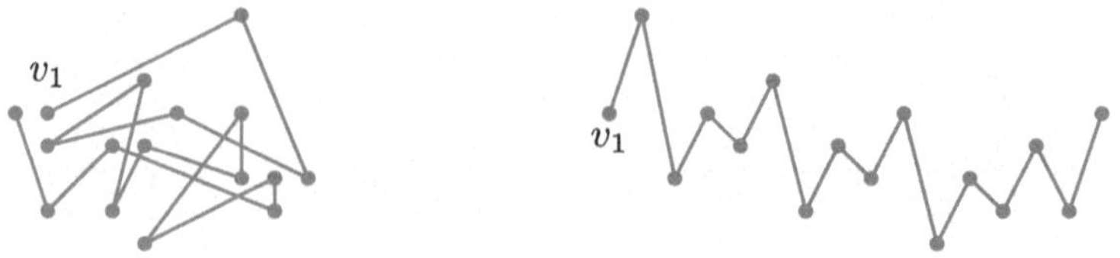

(a) Before (b) After

Fig. 6. Drawing the Reeb graph of a path. In (a), we see a drawing of the Reeb graph with seven crossings. We label one endpoint v_1, then draw consecutive vertices to the right to obtain the drawing without crossings shown in (b).

See Fig. 6 for an illustration of the Construction 1. Because no two edges share an x-coordinate, other than their shared endpoint, we see that this drawing has no crossings. In fact, the same result also applies to *caterpillars*, see Fig. 7.

Construction 2 (A Caterpillar). *Let* (G, h) *be a Reeb graph such that* G *is a caterpillar. We apply Construction 1 to the spine of* G*. Then, because each vertex* v *of* G *has* $\deg(v) \leq 3$*, we can simply draw the legs straight up or down.*

Fig. 7. Every caterpillar can be embedded without crossings.

Interestingly, because generic Reeb graphs have the property that all vertices are degree one or three, this means that all internal vertices of the spine have exactly one leg; furthermore, when the spine is bending up the leg points down and vice versa. For more complicated trees (e.g., lobsters) it is no longer true that they can always be drawn without crossings, and the problem of minimizing their crossings remains open.

5.2 One Cycle

We next consider drawing (G, h) when G is a cycle (and nothing more). Intuitively, we approach this type of graph by specifying the number of times that such a cycle must alternate up and down edges, as these are the potential crossings. We make this more precise as follows; see Fig. 8.

Definition 2 (Top-Down Iteration Number). *Let* (G, h) *be a Reeb graph and let* C *be a cycle in* G*. Let* $t = \max_{v \in C} h(v)$ *be the* top *level used by* C*, and* *let* $b = \min_{v \in C} h(v)$ *be the bottom level. Now, number the vertices of* C *in order of the cycle, starting with some vertex* v_1 *such that* $h(v_1) = t$. *Set* $t_1 = 1$, *and then set* $b_i = \min_{j > t_i}\{h(j) = b\}$ *and* $t_i = \min_{j > t_{i-1}}\{h(j) = t\}$ *be the first alternating indices where the cycle uses the top and bottom levels respectively. Let* k *be the total number of iterations. We say that* k *is the* top-down-iteration-number *of* C.

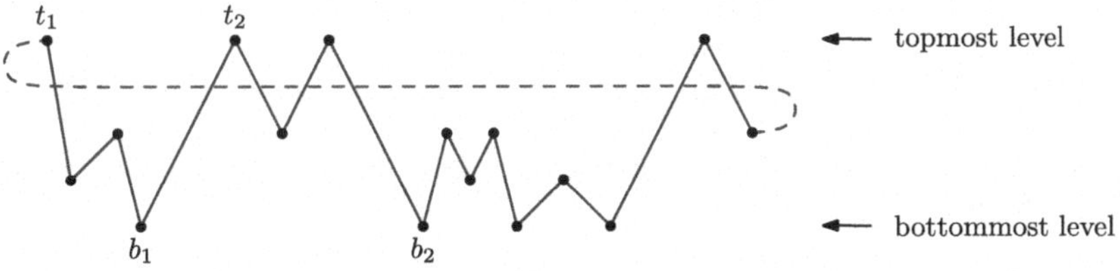

Fig. 8. An example of 0.8 cycle with top-down-iteration-number equal to 2.

We next state the main result we set out to prove in this section:

Theorem 2 (Cycle Drawing). *Let* G *be a cycle, and let* k *be the top-down iteration number of* G. *Then, there exists a drawing of* G *with* $k - 1$ *crossings.*

Before proving the theorem, we first build some intuition. When drawing a cycle, the main factor that influences the number of crossings is the number of vertices at the highest or lowest levels. We consider two special cases: where there is a single global maximum anvm, as in Fig. 9a, and where *every* vertex is a global maximum or minimum, as in Fig. 9b.

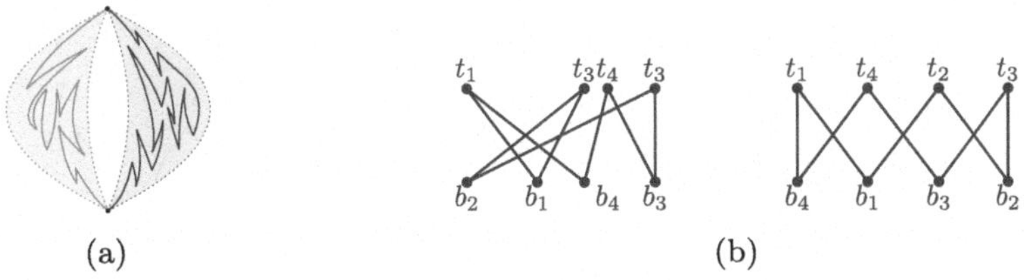

Fig. 9. (a) When there is a unique topmost and bottommost vertex, a cycle can always be drawn without crossings. (b) Drawing a cycle that alternates between two levels. (Left) A graph with the vertices labeled as described in Construction 3. (Right) The bowtie layout.

Lemma 6. *When* G *is a cycle and there is a unique topmost and a unique bottommost vertex, then* G *can be drawn without crossings.*

We next consider what we call a bowtie graph, where every vertex is a global maximum or minimum; see Fig. 9b for an illustration.

Construction 3 (Bowtie). *If G is a cycle with 2n vertices alternating between two heights, then the* bowtie construction *is created as follows:*

1. *Choose an arbitrary vertex on the top level to start; call it* $t_1 = v_1$.
2. *Label the next vertex (at the bottom level)* $b_1 = v_2$.
3. *Repeat, labeling* $t_2 = v_3, b_2 = v_4, t_3 = v_5, b_3 = v_6, \ldots, t_n = v_{2n-1}, b_n = v_2n$.
4. *For* $i = \{1, 2, \ldots n\}$, *draw* v_i *at* $(i, h(v_i))$.
5. *For* $i = \{n+1, n+2, \ldots 2n\}$, *draw* v_i *at* $(i - n, h(v_i))$.
6. *Connect* t_1 *and* b_n *with a vertical straight line.*
7. *For all* $i \in \{2, 3 \ldots 2n\}$, *connect* v_{i-1} *to* v_i *with a straight line.*

We next state two technical lemmas regarding the bowtie construction; proofs are found in [5].

Lemma 7 (Bowtie Construction is Optimal). *Let G be a graph that is a single cycle, with vertices alternating between two heights. Then, the bowtie construction provides a drawing that minimizes the number of crossings.*

Now, we sketch the global approach for the general case where G is a cycle. First, we identify critical vertices at the top and bottom level (following Definition 2). These critical vertices globally form a bowtie-like graph, which we draw using the bowtie technique. But then, the connections between critical vertices are not single edges but paths, which means we still need to draw them.

There are two types of subproblems; see Fig. 10. The subproblems of Type 1 are single paths that do not interfere with any other paths, so we can use Construction 1 to draw them without crossings. But the subproblems of Type 2 involve two crossing paths. We now show how to draw any such pair of crossing paths with only a single crossing.

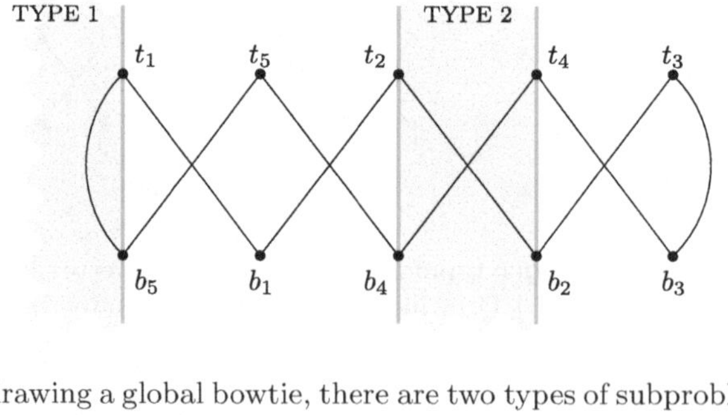

Fig. 10. When drawing a global bowtie, there are two types of subproblems to consider, as described in the proof of Theorem 2.

Lemma 8. *Every Type 2 subproblem has a solution with exactly one crossing.*

We now have all the ingredients to prove Theorem 2.

Proof (Proof of Theorem 2).

Given a cycle G and an associated map h, we first identify its top-down-iteration number k and the corresponding key vertices $t_1, \ldots, t_k$ and $b_1, \ldots, b_k$. Then, we first layout the subcycle of G consisting only of three key vertices using the bow-tie construction. Then we add the paths between the key vertices according to the two types of subproblems:

To solve subproblems of Type 1, we use Construction 1.

To solve subproblems of Type 2, we use Lemma 8 (Fig. 11).

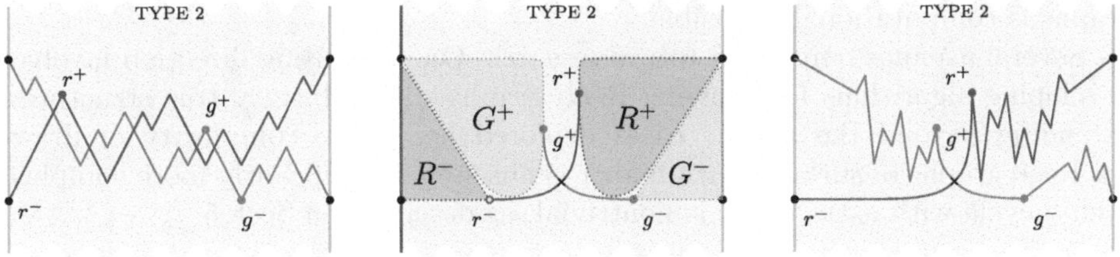

Fig. 11. A zoom in of the Type 2 subproblem from the proof of Lemma 8.

□

Remark 1 (One Cycle with Tassels). Going beyond drawing single-cycle Reeb graphs presents many challenges. Even when G has one cycle, but with *tassels*—bundles of edges that connect one vertex of the cycle to vertices of degree one—we were hopeful that a dynamic programming approach would work. The best hope would be if we could choose some leftmost edge (perhaps arbitrarily, perhaps try all $|E|$, then the remaining vertices can be split into an "upper portion" and a "lower portion" so that we have to calculate: (1) where the transition from the "upper" and "lower" portions is, of which there are $n - 1$ choices; (2) which x-order to place the vertices of the upper and lower portions, with the hope that a greedy approach works. Unfortunately, we have an example that would not be splittable into two such portions; see Fig. 12, where our tassels would be in place of the green edges.

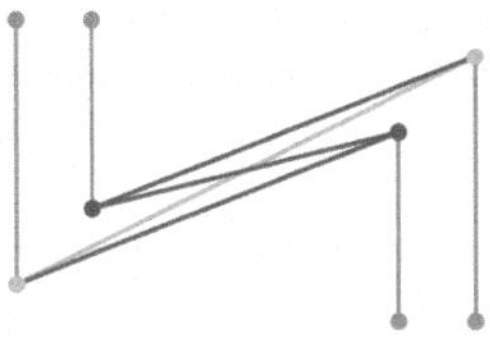

Fig. 12. An example that requires backtracking.

6 Conclusion and Discussion

We investigated the novel problem of drawing Reeb graphs. We established the computational complexity of minimizing crossing numbers, proving that the problem is NP-hard. Notably, we characterized classes of acyclic Reeb graphs, including paths and caterpillars, that can be drawn without crossings. For cyclic Reeb graphs, we derived conditions for crossing-free layouts, particularly for cycles with unique extrema. We also demonstrated the optimality of the *bowtie* configuration for cycles with alternating level vertices. Finally, we concluded by showing that minimizing crossing numbers for general cyclic Reeb graphs is computationally feasible.

Several avenues remain for future research. One promising direction involves developing algorithms for drawing Reeb graphs with arbitrary tree structures, extending beyond the acyclic cases explored here. The complexity of drawing Reeb graphs of surfaces with higher genus seems significantly more complex; even a cycle with extra edges is nontrivial, as described in Sect. 5.

Acknowledgments. We thank Sarah Percival for her participation in an earlier version of this work. This research began as a working group at the Annotated Signatures workshop in March 2023, funded by NSF CCF 2107434. Erin Chambers is partially supported by NSF CCF 2444309, Brittany Terese Fasy is partially supported by NSF CCF 2046730, and Erfan Hosseini Sereshgi is partially supported by NSF CCF 2107434.

Disclosure of Interests. The authors have no competing interests to disclose that are relevant to the content of this article.

References

1. Bastert, O., Matuszewski, C.: Layered drawings of digraphs. In: Kaufmann, M., Wagner, D. (eds.) Drawing Graphs. LNCS, vol. 2025, pp. 87–120. Springer, Heidelberg (2001). https://doi.org/10.1007/3-540-44969-8_5
2. Di Battista, G., Eades, P., Tamassia, R., Tollis, I.G.: Graph Drawing: Algorithms for the Visualization of Graphs, 1st edn. Prentice Hall PTR, USA (1998)
3. Biasotti, S., Giorgi, D., Spagnuolo, M., Falcidieno, B.: Reeb graphs for shape analysis and applications. Theor. Comput. Sci.: Comput. Algebraic Geom. Appl. **392**(13), 5–22 (2008). https://doi.org/10.1016/j.tcs.2007.10.018
4. Chambers, E., Fasy, B.T., Sereshgi, E.H., Löffler, M., Percival, S.: Drawing reeb graphs. In: 31st International Symposium on Graph Drawing and Network Visualization, Revised Selected Papers, Part II. Springer (2023)
5. Chambers, E., Fasy, B.T., Sereshgi, E.H., Löffler, M.: Drawing reeb graphs. arXiv:2504.21329 (2025)
6. De Fraysseix, H., Pach, J., Pollack, R.: How to draw a planar graph on a grid. Combinatorica **10**(1), 41–51 (1990). https://doi.org/10.1007/bf02122694
7. de Silva, V., Munch, E., Patel, A.: Categorified Reeb graphs. Discrete Comput. Geom. 1–53 (2016). https://doi.org/10.1007/s00454-016-9763-9
8. Di Battista, G., Eades, P., Tamassia, R., Tollis, I.G.: Layered Drawings of Digraphs, pp. 265–302. Prentice Hall (1998)

9. Edelsbrunner, H., Harer, J., Patel, A.K.: Reeb spaces of piecewise linear mappings. In: Proceedings of the Twenty-fourth Annual Symposium on Computational Geometry, SCG 2008, pp. 242–250. ACM, New York (2008). https://doi.org/10.1145/1377676.1377720
10. Estrella-Balderrama, A., Fowler, J.J., Kobourov, S.G.: On the characterization of level planar trees by minimal patterns. In: International Symposium Graph Drawing and Network Visualization (2009). https://api.semanticscholar.org/CorpusID:11606029
11. Fink, S.D., Pfretzschner, M., Rutter, I., Stumpf, P.: Level planarity is more difficult than we thought. In: 32nd International Symposium on Graph Drawing and Network Visualization (GD 2024), pp. 50–1. Schloss Dagstuhl–Leibniz-Zentrum für Informatik (2024)
12. Garey, M.R., Johnson, D.S.: Crossing number is NP-complete. SIAM J. Algebraic Discrete Methods **4**(3), 312–316 (1983). https://doi.org/10.1137/0604033
13. Garey, M.R., Johnson, D.S., Stockmeyer, L.: Some simplified NP-complete graph problems. Theor. Comput. Sci. **1**(3), 237–267 (1976). https://doi.org/10.1016/0304-3975(76)90059-1
14. Garg, A., Tamassia, R.: Upward planarity testing. Order **12**(2), 109–133 (1995). https://doi.org/10.1007/BF01108622
15. Garg, A., Tamassia, R.: On the computational complexity of upward and rectilinear planarity testing. SIAM J. Comput. **31**(2), 601–625 (2001). https://doi.org/10.1137/S0097539794277123
16. Harvey, W., Wang, Y., Wenger, R.: A randomized $O(m \log m)$ time algorithm for computing Reeb graphs of arbitrary simplicial complexes. In: Proceedings of the 2010 annual symposium on Computational geometry, SoCG 2010, pp. 267–276. ACM, New York (2010). https://doi.org/10.1145/1810959.1811005
17. Jünger, M., Leipert, S.: Level planar embedding in linear time. In: Kratochvíyl, J. (ed.) Graph Drawing, pp. 72–81. Springer, Heidelberg (1999)
18. Kurlin, V.: Book embeddings of Reeb graphs (2013). 10.48550/ARXIV.1312.1725
19. Parsa, S.: A deterministic $O(m \log m)$ time algorithm for the Reeb graph. In: Proceedings of the 28th Annual ACM Symposium on Computational Geometry, SoCG 2012. ACM (2012)
20. Reeb, G.: Sur les points singuliers d'une forme de pfaff completement integrable ou d'une fonction numerique [on the singular points of a completely integrable pfaff form or of a numerical function]. Comptes Rendus Acad. Sci. Paris **222**, 847–849 (1946). https://cir.nii.ac.jp/crid/1571417125676878592
21. Sugiyama, K., Tagawa, S., Toda, M.: Methods for visual understanding of hierarchical system structures. IEEE Trans. Syst. Man Cybern. SMC **11**(2), 109–125 (1981). https://doi.org/10.1109/TSMC.1981.4308636
22. Yan, L., et al.: Scalar field comparison with topological descriptors: properties and applications for scientific visualization. In: Computer Graphics Forum, vol. 40, pp. 599–633. Wiley Online Library (2021)

Monotone Partitions of Simple Polygons

Jaegun Lee[1(✉)], Hyojeong An[2], Hwi Kim[1], and Hee-Kap Ahn[3]

[1] Department of Computer Science and Engineering, Pohang University of Science and Technology, Pohang, Korea
{jagunlee,hwikim}@postech.ac.kr

[2] Graduate School of Artificial Intelligence, Pohang University of Science and Technology, Pohang, Korea
ahj8151@postech.ac.kr

[3] Graduate School of Artificial Intelligence, Department of Computer Science and Engineering, Pohang University of Science and Technology, Pohang, Korea
heekap@postech.ac.kr

Abstract. Given a simple polygon P with n vertices in the plane, we consider the problem of partitioning P into subpolygons using horizontal (parallel to the x-axis) line segments drawn inside P. We study three versions of the problem: we require (1) every subpolygon is x-monotone, (2) every subpolygon is y-monotone, and (3) every subpolygon is x- or y-monotone. The objective is to minimize the number of subpolygons in the partition. We give an $O(n)$-time algorithm for each version. The algorithm for version (1) improves upon the previously best $O(n \log n)$-time algorithm. We also show that version (3) is NP-complete if P contains holes.

1 Introduction

Partitioning a polygon into simpler components is a fundamental problem in computational geometry. A representative example is a triangulation of a simple polygon [2]. Various other geometric objects such as trapezoids, convex polygons, and star-shaped polygons have also been considered [1,3,7,14].

A simple polygon is *monotone for a line* ℓ if, for any line ℓ' perpendicular to ℓ, the intersection of the polygon with ℓ' is connected. Many computationally hard geometric problems become more tractable when a given polygon is monotone [5,13]. Thus, a possible strategy to solve some geometric problems on a polygon is to decompose the polygon into monotone pieces, and then to apply the known algorithms that work on a monotone polygon. This decomposition is called a *monotone partition*. There is a fair amount of work on the monotone partition problem [6,8,9,12,15].

In this paper, we study the problem of partitioning a simple polygon into x-monotone or y-monotone subpolygons, using horizontal line segments (open line segments lying in the interior of the polygon) each connecting two boundary points of the polygon, which we call *diagonals*.

H. Fernau and B. Zhu (Eds.): IWOCA 2025, LNCS 15885, pp. 72–85, 2025.
https://doi.org/10.1007/978-3-031-98740-3_6

The objective is to minimize the *number* of subpolygons in the partition. We call this the *minimum monotone partition* problem. We study three versions of the problem: (1) every subpolygon is x-monotone, (2) every subpolygon is y-monotone, and (3) every subpolygon is x- or y-monotone. See Fig. 1. Given a polygon, we refer to a solution of the minimum monotone partition problem as its *optimal partition*.

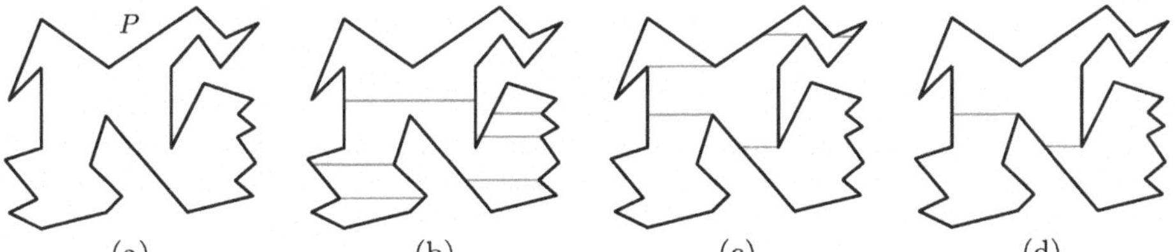

Fig. 1. (a) A simple polygon P. (b) Optimal x-monotone partition. (c) Optimal y-monotone partition. (d) Optimal x- or y-monotone partition.

There are two interesting aspects of our problem. First, the diagonals we use are all *horizontal*, that is, parallel to the x-axis. This may reflect the situation where we can cut a given shape only in a certain direction (e.g., with a guillotine blade). Second, we allow subpolygons in our partition to be monotone for the cutting direction or its orthogonal direction. This is a relaxation to the previous case [11]. The number of subpolygons can be reduced substantially by allowing two monotone directions. See Fig. 1.

For the sake of completeness, we formally define a *partition* of a polygon: it is a set of interior-disjoint subpolygons whose union equals the polygon. We say that a partition is x-monotone (y-monotone) if every subpolygon in the partition is x-monotone (y-monotone).

Previous Work. There has been a fair amount of research on monotone partitioning of a simple polygon, using line segments connecting two distinct vertices of the polygon, not necessarily horizontal. Lee and Preparata [9] proposed an $O(n \log n)$-time algorithm to partition a polygon into subpolygons, each being monotone for some line, where n is the complexity of the polygon. Later, Keil [6] considered the same problem while minimizing the size of the partition. He developed an $O(n^4 R)$-time algorithm, where R is the number of reflex vertices of the polygon.

Liu and Ntafos [12] presented an algorithm to partition a simple polygon into minimum number of *uniformly* monotone subpolygons, that is, subpolygons all of which are monotone for a certain line. Their algorithm runs in $O(nR^2 \log n + nR^3 + R^5)$ time. Recently, Kim et al. [8] made an improvement by presenting an $O(nR \log n + R^5)$-time algorithm.

The work by Liu [11] is comparable to our research. Given a rectilinear polygon with rectilinear holes with n vertices in total, his algorithm finds a partition of the polygon into minimum number of x-monotone subpolygons using horizontal lines, and runs in $O(n \log n)$ time. He did not mention whether a better algorithm exists when there is no hole.

Our Results. We give a linear time algorithm for all three versions of the minimum monotone partition problem. The algorithm for version (1) improves upon the previously best $O(n \log n)$-time algorithm by Liu [11]. We also show that version (3) is NP-complete if holes are allowed. The NP-completeness result remains true even when the input polygon is rectilinear. This is a stark contrast to the $O(n \log n)$-time algorithm for version (1) for any rectilinear polygon.

For the problem of finding an orientation θ that minimizes the size of an optimal partition of the polygon, we present two algorithms. The first algorithm runs in $O(n + R^2 \log R)$ time using $O(n + R^2)$ space, while the second runs in $O(n + R^2 \log n)$ time using $O(n)$ space.

Sketches of Our Algorithms. We first show that there is a set S of $O(n)$ diagonals such that the diagonal set of an optimal monotone partition is contained in S. Here, $S = S_1 \cup S_2$, where S_1 is the set of diagonals, each incident to a reflex vertex, and S_2 is the set of horizontal diagonals that pass through the midpoint between the topmost and the bottommost vertex of each subpolygon in the partition induced by S_1. We call the partition induced by S the *horizontal partition*.

Since the polygon is simple, we can represent the dual graph [2] of the horizontal partition in a rooted tree T. Each node of T corresponds to a convex polygon in the horizontal partition. Our algorithm is a dynamic programming running on T. For a node v of T, let $Q(v)$ be the union of the convex polygons corresponding to the nodes contained in the subtree rooted at v. A subproblem in our algorithm consists of a node v and a set z of states, and its solution is an optimal partition of $Q(v)$ in state z. In each subproblem, we have optimal partitions for every pair (v_i, z_i) of a child node v_i and a state z_i, and we determine how to solve the subproblem for v and z from the partitions of the child nodes.

The monotonicity of the partition, however, is not guaranteed even if each partition corresponding to a child node is monotone. This is because a reflex portion may appear in the merge. It takes too much time to test every possible set of merged partitions. We show that after some appropriate preprocessing, we can solve a subproblem for a node v in time linear to the number of child nodes. Thus, our algorithm computes an optimal partition using $O(n)$ time and space.

For the problem of finding an optimal orientation, we show that there are only a small number of events at which the size of an optimal partition changes. Each event is defined by a reflex vertex of P (its incident edges) or two such vertices (the line segment connecting them). We compute and sort the events according to the orientation of the events. Our algorithm computes the dual tree only once at the first event. It never computes the whole tree again, but updates some necessary parts, and computes an optimal partition for the current event. We achieve the linear space complexity by computing the next event on the fly.

Finally, we show that version (3) of the problem is NP-complete when the polygon has holes. We give a reduction from the planar 3-SAT problem. The reduction constructs a rectilinear polygon from a given planar 3-SAT formula. We show that there is a monotone partition of the polygon of size at most k, for a given integer k if and only if the formula is satisfiable.

2 Partitioning Simple Polygons

We give algorithms for the minimum monotone partition problem when the input polygon is simple. In Sect. 2.1, we construct a rooted tree which is a dual to the horizontal partition of P. We present dynamic programming algorithms running on the tree in Sect. 2.2. Our algorithms run in linear time. Using our algorithms, we solve the problem of finding an orientation minimizing the size of an optimal partition in Sect. 2.3.

There is a polygon whose optimal partition has size $\Omega(n)$. See Fig. 2(a). Thus, our algorithm is optimal even without considering the time for reading the input.

2.1 Dual Rooted Tree

Throughout this paper, we denote by P the input simple polygon with n vertices. The algorithm begins with computing the dual rooted tree corresponding to the *horizontal partition* of P.

We first show that we can compute a set S of $O(R)$ horizontal diagonals such that there is an optimal partition whose diagonal set is contained in S. Let S_1 be the set of diagonals, each incident to a reflex vertex of P. Let S_2 be the set of diagonals, each passing through the midpoint between the uppermost and the lowermost vertex of a subpolygon in the partition induced by S_1.

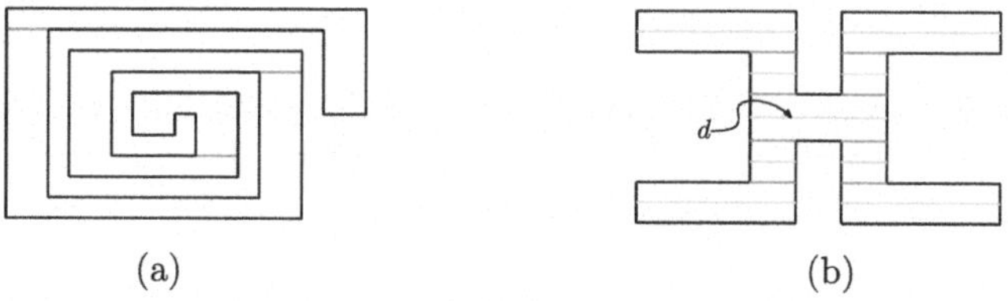

Fig. 2. (a) An n-gon whose optimal x- or y-monotone partition has size $\Omega(n)$. (b) S_1 consists of the red segments, and S_2 consists of the blue segments. The optimal x- or y-monotone partition is induced by the single diagonal d. (Color figure online)

There is an optimal partition such that each diagonal in the partition is contained in $S_1 \cup S_2$. Suppose not and let d be such a diagonal. Because $d \notin S_1$, no endpoint of d is a reflex vertex of P. Then we can move d vertically until it meets any diagonal in S_2 while maintaining the optimality of the partition.

Observe that there is a simple polygon none of whose optimal solution is induced by diagonals only in S_1. See Fig. 2(b).

We define the *horizontal partition* of P as the partition of P induced by $S_1 \cup S_2$. Consider the dual graph T corresponding to the partition. Each node in T corresponds to a convex polygon $R(v)$ in the partition, and there is an edge between two nodes v_1 and v_2 if and only if $R(v_1)$ and $R(v_2)$ share the boundary. Since P is simple, T is a tree.

We select the root v_r of T such that the convex polygon $R(v_r)$ contains the lexicographically first vertex v_1 of P based on the y-coordinate. That is, $y(v_1) \leqslant y(v_2)$, or $y(v_1) = y(v_2)$ and $x(v_1) \leqslant x(v_2)$ for any other vertex v_2 of P, where $x(\cdot)$ and $y(\cdot)$ denote the x- and the y-coordinate of a point, respectively.

To construct the dual rooted tree, we use Chazelle's triangulation algorithm [2]. Given a simple polygon, it computes the partition induced by the horizontal extensions. Then, it constructs the dual tree of the partition. By duplicating each node of the tree and connecting them by an edge, we can make the tree the dual of the horizontal partition. See Fig. 3. Finally, we make the tree rooted. All of this can be done in linear time. Thus, we have the following lemma.

Lemma 1. *We can construct the dual rooted tree corresponding to the horizontal partition of P in linear time.*

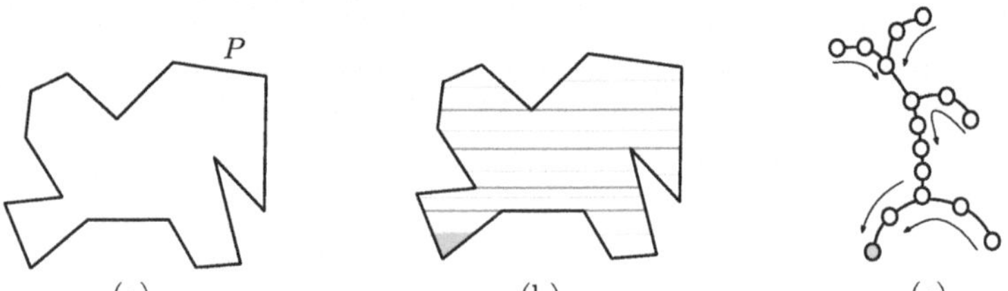

(a) (b) (c)

Fig. 3. (a) Simple polygon P. (b) A horizontal partition of P. The gray convex polygon corresponds to the root of the dual rooted tree. (c) Dual rooted tree of the horizontal partition of P.

2.2 Dynamic Programming

In this subsection, we present dynamic programming algorithms running on the dual rooted tree T. For ease of description, we assume that P is *rectilinear*. Later, we will remove the assumption so that the algorithm works for any simple polygon. A rectilinear polygon is a polygon all whose edges are axis-aligned. The substructures determining its monotonicity are *reflex edges* - edges whose both end vertices are reflex (we assume that the polygon does not have three consecutive and collinear vertices). We simply call them *dents*. It is easy to see that a rectilinear polygon is x-monotone if and only if there is no vertical dent, and y-monotone if and only if there is no horizontal dent. A vertical dent of P is a *left dent* (*right dent*) if the interior of P lies locally to the right (left) of it.

Every Subpolygon is x-Monotone. We first consider version (1). For a node v of T, let $Q(v)$ be the union of $R(v')$'s for all nodes v' contained in the subtree of T rooted at v. Because P is rectilinear, $R(v)$ is a rectangle for every node v of T. For the simplest algorithm, one may define the subproblem for v as an optimal partition of $Q(v)$. In other words, only one optimal partition is stored for each node. In the following, we will show that this algorithm may fail to compute an optimal partition of P correctly, which makes it necessary to elaborate the subproblem definition. We describe this algorithm since it may enhance the understanding of the problem.

The algorithm runs in a bottom-up manner. The base case corresponds to a leaf node of T. For any such leaf node v, $Q(v)$ is a rectangle $R(v)$, and its unique optimal partition is $\{R(v)\}$. The inductive step corresponds to an internal node v of T. Among the diagonals of P contained in the boundary $\partial R(v)$, the algorithm determines which ones to put in $Q(v)$.

Let $v_1, ..., v_m$ be the child nodes of v. Let $\mathsf{P}(v_i)$ be the optimal partition of $Q(v_i)$ stored by the algorithm, for $i = 1, ..., m$. The goal is to compute an optimal partition $\mathsf{P}(v)$ of $Q(v)$ given $\mathsf{P}(v_i)$'s. Observe that $\mathsf{P}(v)$ is x-monotone if and only if the *base polygon* $B(\mathsf{P}(v))$ of $\mathsf{P}(v)$, which is the subpolygon in $\mathsf{P}(v)$ containing $R(v)$, is x-monotone. All other subpolygons in $\mathsf{P}(v)$ are x-monotone by the induction hypothesis. We have two choices for each v_i: either to place the diagonal $R(v_i) \cap R(v)$ in $\mathsf{P}(v)$ or not to place it. In the latter case, $B(\mathsf{P}(v_i))$ is *merged* into $R(v)$. We simply say that the child node v_i is merged into v.

A difficulty here is that even if the base polygon of every child node merged into v is x-monotone, the base polygon of v may not be x-monotone. This happens when there is a vertical dent in $R(v) \cup B(\mathsf{P}(v_i))$, while there is no such vertical dent in $B(\mathsf{P}(v_i))$. Moreover, if we add a diagonal to remove such a dent, the diagonal in S_2 may not be taken into account, resulting in a non-optimal partition. To resolve this issue, for each edge (v_i, v) of T, we store information on whether the diagonal $R(v_i) \cap R(v)$ corresponding to the edge is incident to any dent as follows. We denote such a diagonal by $\overline{v_i}$.

Edge Information. Let ℓ and r be the left and the right endpoints of $\overline{v}$, respectively. If ℓ does not lie on a left dent of P, we set $\mathsf{left}(\overline{v}) =$ None. Otherwise, we set $\mathsf{left}(\overline{v})$ according to the position of ℓ on the left dent, say e, as follows. We set $\mathsf{left}(\overline{v}) =$ middle if ℓ is not an endpoint of e. If ℓ is an endpoint of e, there are two cases. We set $\mathsf{left}(\overline{v}) =$ start if the left side of $R(p(v))$, where $p(v)$ denotes the parent node of v, is contained in e, and set $\mathsf{left}(\overline{v}) =$ end otherwise. In a similar way, we set $\mathsf{right}(\overline{v})$ according to the location of r. See Fig. 4.

Lemma 2. *We can compute* $\mathsf{left}(\overline{v})$ *and* $\mathsf{right}(\overline{v})$ *for all nodes* v *in* $O(n)$ *time.*

Subproblem Definition. With the edge information, we give our subproblem formulation by defining the *state* of a partition. It tells us whether there is a vertical

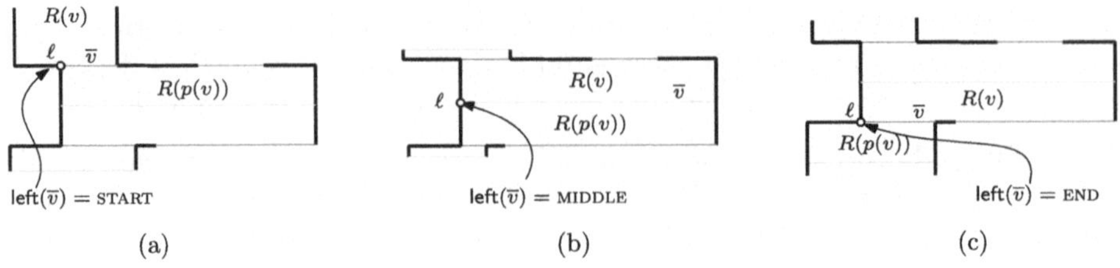

Fig. 4. The left endpoint ℓ of $\overline{v}$ lies on a dent. (a) $\mathsf{left}(\overline{v}) = \text{START}$ since ℓ is an endpoint of the dent containing the left side of $R(p(v))$. (b) $\mathsf{left}(\overline{v}) = \text{MIDDLE}$ since ℓ is not an endpoint of the dent. (c) $\mathsf{left}(\overline{v}) = \text{END}$ since ℓ is an endpoint of the dent not containing the left side of $R(p(v))$.

dent in $K(\mathsf{P}(v)) := (P \setminus Q(v)) \cup B(\mathsf{P}(v))$ not yet removed. For an x-monotone partition $\mathsf{P}(v)$ of $Q(v)$, we say that $\mathsf{P}(v)$ is in state $z = (\mathsf{l}, \mathsf{r})$ for $\mathsf{l}, \mathsf{r} \in \{\mathsf{True}, \mathsf{False}\}$, by the following rules (See Fig. 5).

- $z.\mathsf{l} = \mathsf{True}$ if and only if the left endpoint of $\overline{v}$ lies on a left dent of $K(\mathsf{P}(v))$.
- $z.\mathsf{r} = \mathsf{True}$ if and only if the right endpoint of $\overline{v}$ lies on a right dent of $K(\mathsf{P}(v))$.

We define our subproblem as follows.

$$\mathsf{OPT}(v, z) := (\text{optimal partition of } Q(v) \text{ in state } z)$$

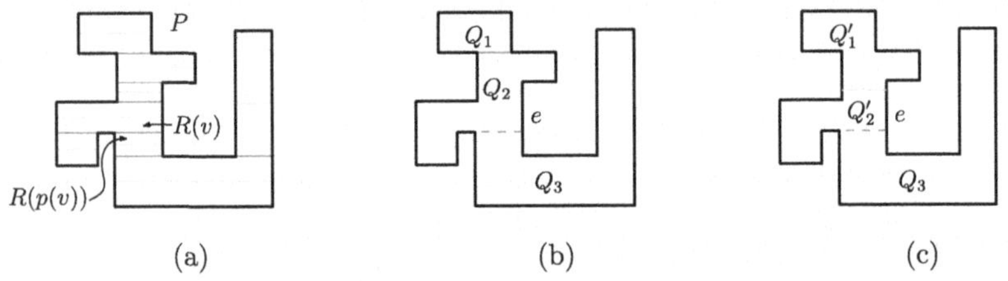

Fig. 5. $\mathsf{P}(v)$ for a rectilinear polygon P. (a) Horizontal partition of P. (b) If $\mathsf{P}(v) = \{Q_1, Q_2\}$, $\overline{v}$ (red dashed line) is incident to a right dent e in $K(\mathsf{P}(v)) = Q_2 \cup Q_3$. Thus, $z.\mathsf{r} = \mathsf{True}$. (c) If $\mathsf{P}(v) = \{Q'_1, Q'_2\}$, e is not a dent in $K(\mathsf{P}(v)) = Q'_2 \cup Q_3$. Thus, $z.\mathsf{r} = \mathsf{False}$. (Color figure online)

Recall that a base case corresponds to a leaf node of T. If v is a leaf node, we have $\mathsf{OPT}(v, (\mathsf{False}, \mathsf{False})) = \{R(v)\}$ with its size 1. We have $\mathsf{OPT}(v, z) = \varnothing$ with its size ∞, for every other state z. An inductive case corresponds to an internal node v of T: solve $\mathsf{OPT}(v, z)$ for v and a fixed state z, given $\mathsf{OPT}(v_i, z_i)$ for every pair of a child node v_i and a state z_i of a partition of $Q(v_i)$, for the child nodes $v_1, \ldots, v_m$ of v.

We define a pair (c_i, z_i) such that c_i is true if and only if $B(\mathsf{OPT}(v_i, z_i))$ is merged into $R(v)$. Our goal is to compute $\mathcal{S} = \{(c_1, z_1), \ldots, (c_m, z_m)\}$ for (v, z)

such that the resulting partition is of minimum size among all partitions of $Q(v)$ in state z. Before describing how our algorithm computes such $\mathcal{S}$, we verify that this subproblem definition is valid.

Lemma 3. *For any node v of T and any state z, there always is a set of pairs $\{(c_1, z_1), \ldots, (c_m, z_m)\}$ corresponding to an optimal partition.*

For a set $\mathcal{S} = \{(c_1, z_1), (c_2, z_2), \ldots, (c_m, z_m)\}$, the size of the partition $\mathsf{P}(v)$ of $Q(v)$ determined by $\mathcal{S}$ is as follows. Given a state z, our goal is to find a set $\mathcal{S}^*$ among all such sets $\mathcal{S}$, minimizing the size of the partition of state z.

$$|\mathsf{P}(v)| = 1 + \sum_{\{i \mid c_i = \mathsf{True}\}} \Big(|\mathsf{OPT}(v_i, z_i)| - 1\Big) + \sum_{\{i \mid c_i = \mathsf{False}\}} |\mathsf{OPT}(v_i, z_i)|$$

If $c_i = \mathsf{False}$ (v_i is not merged into v), the state z_i affects neither the monotonicity of $B(\mathsf{P}(v))$ nor the state z, so it is optimal to choose a state that minimizes the partition size. Now suppose that $c_i = \mathsf{True}$. We consider $\mathsf{left}(\overline{v_i})$ and $\mathsf{right}(\overline{v_i})$. If $\mathsf{left}(\overline{v_i}) = \mathsf{right}(\overline{v_i}) = \textsc{None}$, then we have $z_i.\mathsf{l} = z_i.\mathsf{r} = \mathsf{False}$ for any state z_i, and thus merging v_i into v does not generate any vertical dent. The optimal choice is again to choose a state that minimizes the partition size. The remaining cases of $\mathsf{left}(\overline{v_i})$ are as follows, which also hold for $\mathsf{right}(\overline{v_i})$ analogously.

Case 1. $\mathsf{left}(\overline{v_i}) = \textsc{end}$. Then $z_i.\mathsf{l} = \mathsf{False}$, because otherwise there is a vertical dent in $B(\mathsf{P}(v))$. See Fig. 6(b).

Case 2. $\mathsf{left}(\overline{v_i}) = \textsc{start}$ or $\mathsf{left}(\overline{v_i}) = \textsc{middle}$. Then the left dent e, on which the left endpoint of $\overline{v_i}$ lies, has one endpoint in $Q(v_i)$ and another endpoint, say p, in $P \setminus Q(v_i)$.

Subcase 2.1. $p \in Q(v)$ and $p \notin \overline{v}$. Then there is a child node $v_j \neq v_i$ such that $p \in Q(v_j)$. If all of c_i, c_j, $z_i.\mathsf{l}$ and $z_j.\mathsf{l}$ are true, then $e \in B(\mathsf{P}(v))$. Thus, we choose the pairs (c_i, z_i) and (c_j, z_j) minimizing the size such that any one of c_i, c_j, $z_i.\mathsf{l}$ and $z_j.\mathsf{l}$ is false. See Fig. 6(c).

Subcase 2.2. $p \notin Q(v)$ or $p \in \overline{v}$. Then the left endpoint of $\overline{v}$ lies on e. We have $z.\mathsf{l} = \mathsf{True}$ if and only if $z_i.\mathsf{l} = \mathsf{True}$. See Fig. 6(d).

We analyze the time complexity of the algorithm. For each subproblem $\mathsf{OPT}(v, z)$, we can find a pair (c_i, z_i) for every child node v_i leading to an optimal partition of $Q(v)$ in $O(1)$ time, in every case above, except Subcase 2.1. In that case, we need to consider all pairs (c_i, z_i) and (c_j, z_j) for two child nodes v_i and v_j, which can be enumerated in $O(1)$ time. Thus, for each node v of T, we can solve the subproblem $\mathsf{OPT}(v, z)$ for any state z in time linear in the number of child nodes of v. Our algorithm hence runs in linear time.

Every Subpolygon is y-Monotone. We now consider version (2) of the problem. Here we have to remove every horizontal dent in P. Since we use horizontal diagonals only, we add a diagonal at an *endpoint* of each horizontal dent.

A horizontal dent may appear if two child nodes v_i, v_j of a node v are merged such that $R(v) \cap R(v_i)$ and $R(v) \cap R(v_j)$ are on the same (top or bottom) side

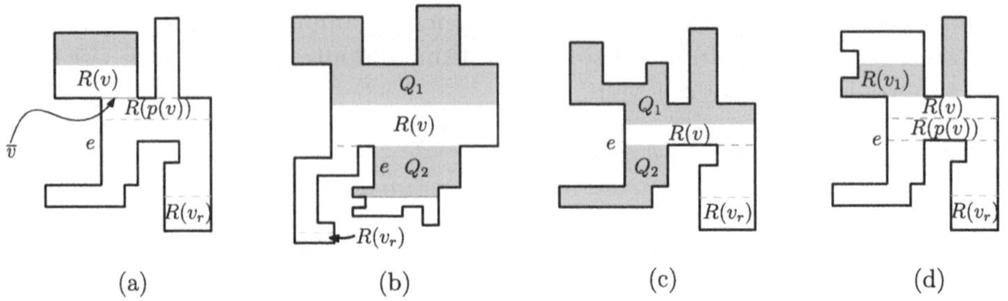

Fig. 6. The base polygon of the partitions for child nodes of v (gray regions) are merged into $R(v)$. (a) We have $\mathsf{left}(\overline{v})$ = START. Accordingly, $z.\mathsf{l}$ = True regardless of which child nodes are merged. (b) The top endpoint of the dent e is contained in $R(v) \cap Q_2$, and thus $B(\mathsf{P}(v))$ is not x-monotone. (c) The top endpoint of e is contained in Q_1, and the bottom endpoint is contained in Q_2. Thus, $B(\mathsf{P}(v))$ is not x-monotone. (d) $\mathsf{left}(\overline{v})$ = MIDDLE, $\mathsf{left}(\overline{v_1})$ = START, and $z_1.\mathsf{l}$ = True. As a result, $z.\mathsf{l}$ = True.

of $R(v)$. It may also appear by a single child node. Consider a child node v_i of v such that $R(v_i) \cap R(v)$ and $R(v) \cap R(p(v))$ are on the same side of $R(v)$. If v_i is merged into v, then merging v into $p(v)$ generates a horizontal dent.

To handle such a case with a single node, we define the state as $z = (\mathsf{y})$. It consists of only one boolean flag: this seems redundant but it helps better understand the state definition for version (3). For a partition $\mathsf{P}(v)$ of v, we have $z.\mathsf{y}$ = True if and only if $\overline{v}$ has an endpoint incident to a horizontal dent e of $K(\mathsf{P}(v))$ such that e is contained in $R(v)$. See Fig. 7.

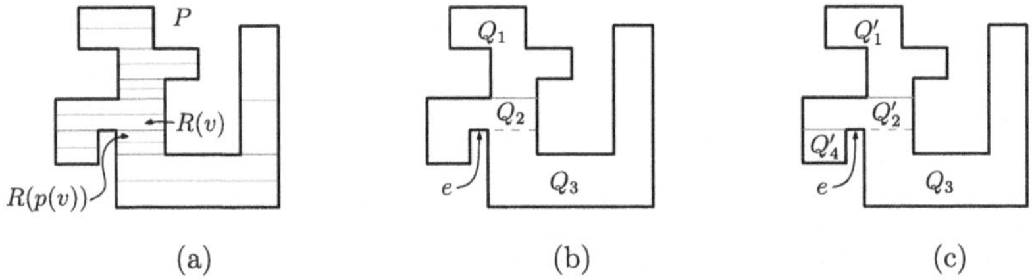

Fig. 7. Rectilinear polygon P and its partitions $\mathsf{P}(v)$. (a) Horizontal partition of P. (b) If $\mathsf{P}(v) = \{Q_1, Q_2\}$, $\overline{v}$ (red dashed) is incident to a horizontal dent e in $K(\mathsf{P}(v)) = Q_2 \cup Q_3$, and e is in $B(\mathsf{P}(v)) = Q_2$. Thus, $z.\mathsf{y}$ = True. (c) If $\mathsf{P}(v) = \{Q'_1, Q'_2, Q'_4\}$, e is not a dent in $K(\mathsf{P}(v)) = Q'_2 \cup Q'_3$. Thus, $z.\mathsf{y}$ = False.

We solve the subproblem $\mathsf{OPT}(v, z)$ as follows. Among the child nodes of v, there is at most one child node v' such that $\overline{v'}$ is a diagonal in S_2. For the other child nodes $v_1, \ldots, v_m$, the diagonals $\overline{v_1}, \ldots, \overline{v_m}$ lie on the same side of $R(v)$. Thus, at most one among them can be merged into v. In other words, we have $z.\mathsf{y}$ = True if and only if $R(p(v))$ lies on the same side of $R(v)$ with $v_1, \ldots, v_m$ for the parent node $p(v)$, and one of the child nodes is merged.

As in version (1), if $c_i = \mathsf{False}$, it is optimal to choose a state z_i minimizing the size of the partition. Thus, consider when only one of $\{c_1, \ldots, c_m\}$, say c_i, is true. Here the partition size equals the partition size for the case when all of $\{c_1, \ldots, c_m\}$ are false, minus $\min_z |\mathsf{OPT}(v_i, z)| - |\mathsf{OPT}(v_i, (\mathsf{False}))| + 1$. We can find an index i maximizing it in time linear in the number of child nodes. Thus, our algorithm runs in linear time.

Every Subpolygon is x- or y-Monotone. We give a sketch of our algorithm for version (3). We define the state of a partition with four variables: l and r from version (1), y from version (2), and the variable α for the monotonicity of the base polygon.

- $z.\alpha$ is xy if and only if $B(\mathsf{P}(v))$ is x-monotone and y-monotone.
- $z.\alpha$ is x if and only if $B(\mathsf{P}(v))$ is x-monotone but not y-monotone.
- $z.\alpha$ is y if and only if $B(\mathsf{P}(v))$ is y-monotone but not x-monotone.

For the subproblems $\mathsf{OPT}(v, z)$ with $z.\alpha = x$, we use the approaches for versions (1). The only difference is that we merge the child nodes whose base polygon of the partition is x- or xy-monotone. For subproblems with $z.\alpha = y$, we merge child nodes with partitions of y- or xy-monotone base polygon. Finally, for subproblems with $z.\alpha = xy$, we merge child nodes only with xy-monotone base polygons. Considering all constraints from the x-monotone case (version (1)) and the y-monotone case (version (2)) does not increase the time complexity, and we can obtain an optimal partition of P in linear time.

Removing the Assumption that P is Rectilinear. For a simple polygon P that is not necessarily rectilinear, the monotonicity of a polygon in a partition of P can be broken not only by reflex edges, but also by reflex vertices of the polygon. Such a reflex vertex is a *x-peak* (*y-peak*) if both its neighboring vertices lie on the same half-plane defined by the vertical (horizontal) line through the vertex. Considering each x- and y-peak as a degenerate reflex edge (of zero length), we can compute an optimal partition in the same way as above algorithms.

If we explicitly compute the partitions for each pair (v, z) of a node v and a state z, we need $O(nR)$ time and space, because a partition of $Q(v)$ is of size $O(n)$ and there are $O(R)$ nodes. To spare time and space, we only store the size of an optimal partition. Still, we can implicitly store all such optimal partitions by keeping at each subproblem (v, z) the optimal set of pairs $\{c_i, z_i\}$ for child nodes. After finding the size of an optimal partition of P, we compute the partition explicitly by traversing T in a top-down manner: for each node v starting from the root, we add diagonals corresponding to $c_i = \mathsf{False}$, and recursively proceed towards the bottom. This takes $O(n)$ time. Thus, we have our main result.

Theorem 1. *For a simple polygon P with n vertices, we can compute a partition of P into minimum number of x-monotone/y-monotone/x- or y-monotone polygons, using only horizontal diagonals, in $O(n)$ time using $O(n)$ space.*

2.3 Finding Optimal Orientations

Suppose we allow rotating the polygon P. Let P_θ be the polygon obtained by rotating P by θ in clockwise direction. The size of an optimal x-monotone/y-monotone/x- or y-monotone partition may differ at distinct values of $\theta \in [0, \pi)$. Thus, it is natural to consider the problem of finding an orientation θ^* minimizing the size of a monotone partition. Such an orientation is *optimal.*

We present two algorithms for the problem. Our algorithms compute *events*, which are angles at which the size of an optimal partition of P_θ could change. There are three types of events: (1) a peak or a reflex edge appears, (2) a peak or a reflex edge disappears, and (3) the diagonal between two peaks becomes parallel to the x-axis. The size of an optimal partition changes only at these events. Events of types (1) and (2) occur only when some edge of P is parallel to the x- or y-axis, and thus there are $O(R)$ such events, while events of type (3) occur $O(R^2)$ times. If we use our linear-time algorithm for all $O(R^2)$ events, we get an $O(nR^2)$-time, $O(n)$-space algorithm. We can improve it.

First Algorithm. We compute all events of types (1) and (2) in $O(R)$ time, and all events of type (3) in $O(R^2)$ time. The orientations are sorted in ascending order, which can be done in $O(R^2 \log R)$ time and $O(R^2)$ space. We compute the dual rooted tree T at $\theta = 0$ and make it a balanced binary tree T' in $O(n)$ time. The complexity of T' is still $O(R)$.

Our algorithm updates T' as θ increases from 0 to π. For events of types (1) and (2), we only need to update the edge information stored in T'. For an event of type (3), let $\overline{v_i}$ and $\overline{v_j}$ be the extensions of the nodes corresponding to the two peaks that create the event. We have to update the location of these nodes in T'. Since the height difference between v_i and v_j is $O(\log R)$, it takes $O(\log R)$ time to handle an event. Therefore, the optimal orientation can be found in $O(n + R^2 \log R)$ time using $O(n + R^2)$ space.

Second Algorithm. Our second algorithm is slower than the first one, but it uses less space. It also computes events, but it maintains the space complexity as linear by not computing them all at once. It stores only one event for each peak, and uses ray shooting queries to compute the next event. We start from the initial orientation $\theta = 0$ and maintain a priority queue that stores the next event for each reflex vertex of P. Whenever an orientation θ is popped from the priority queue, we compute the next event for the reflex vertex corresponding to θ and push it into the priority queue.

We can compute all events of types (1) and (2) in $O(R)$ time. To compute events of type (3), we traverse P in counterclockwise order from the opposite reflex vertex where the last event of type (3) occurred and determine the next reflex vertex visible to p, which takes $O(\log n)$ time using ray shooting queries [4]. Since we perform $O(R)$ ray shooting queries per reflex vertex, we use $O(R^2 \log n)$ additional time. Thus, the optimal orientation can be found in $O(n + R^2 \log n)$ time using $O(n)$ space.

Theorem 2. *Given a simple polygon P with n vertices, including R reflex vertices, we can find an orientation $\theta^* \in [0, \pi)$ such that the optimal x-monotone/y-monotone/x- or y-monotone partition of P_θ is minimized in $O(n+R^2 \log R)$ time using $O(n + R^2)$ space, or in $O(n + R^2 \log n)$ time using $O(n)$ space.*

3 Partitioning Polygon with Holes

We show that the following decision version of the minimum monotone partition problem for version (3) is NP-complete if P has holes. Our construction works even when we restrict P to be rectilinear.

Problem 1. Given a rectilinear polygon P with rectilinear holes and a positive integer k, decide whether there is a partition P of P into x- or y-monotone subpolygons with $|\mathsf{P}| \leqslant k$.

For a given partition P of P, we can check whether $|\mathsf{P}| \leqslant k$ and all subpolygons in P are x- or y-monotone in linear time. Thus, Problem 1 is in NP.

For an NP-hardness proof, we use the reduction from the planar 3-satisfiability (P3SAT) problem. In P3SAT, a 3-SAT formula F with a variable set $X = \{x_1, ..., x_n\}$ and a clause set $C = \{c_1, ..., c_m\}$ is given such that the graph $G(F) = (V, E)$ with $V = X \cup C$ and $E = \{(x_i, c_j) \mid x_i \text{ or } \overline{x_i} \text{ is a literal in } c_j\}$ is planar. Here, each variable in X has value either 1 (True) or 0 (False), and each clause in C consists of exactly three variables in X. It is well-known that P3SAT is NP-complete [10].

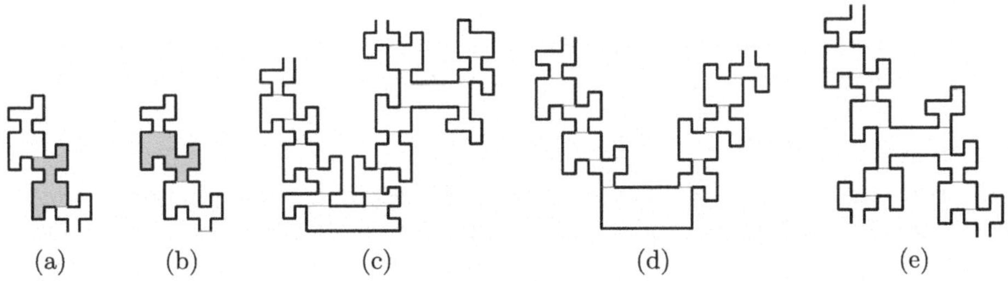

Fig. 8. (a) y-monotone partition of a variable gadget corresponding to True. (b) x-monotone partition of a variable gadget corresponding to False. (c) Negation gadget. Blue diagonals correspond to True, and red diagonals correspond to False. (d) Turn gadget. (e) Split gadget. (Color figure online)

Our reduction algorithm generates a rectilinear polygon for a given P3SAT formula, so that there is a partition of the polygon of size at most k if and only if the formula is satisfiable. The rectilinear polygon consists of subpolygons which we call *gadgets*. There are five types of gadgets: variable, clause, negation, turn and split. The variable, clause and negation gadgets work in the same way as their counterparts in the P3SAT formula. The two remaining types of gadgets are to correctly connect the variable, clause, and negation gadgets.

In a *variable gadget*, horizontal dents and vertical dents appear on its boundary in an alternating manner. There are two ways to get an optimal partition: each diagonal removes two horizontal dents (y-monotone partition), or it removes two vertical dents (x-monotone partition). See Fig. 8(a, b). The former corresponds to True, and the latter corresponds to False.

A *negation gadget* (Fig. 8(c)) is connected to a variable gadget if is used as a negative literal. We use *turn gadgets* (Fig. 8(d)) and *split gadgets* (Fig. 8(e)) to make connections between the variable gadgets and the clause gadgets.

A *clause gadget* is a single rectangle connected to three variable gadgets. See Fig. 9. All three variable gadgets are connected to the top side of the rectangle. For each variable gadget, a diagonal is placed on the top side if and only if the variable is false. If any of the three variables is true, we can merge the lowermost part (subpolygon) of the corresponding variable gadget into the rectangle, reducing the size of the partition by one. The merged subpolygon is y-monotone but not x-monotone, see Fig. 9(a). However, even if we have more than one True variable, we can merge only one such variable gadget into the rectangle as in Fig. 9(b).

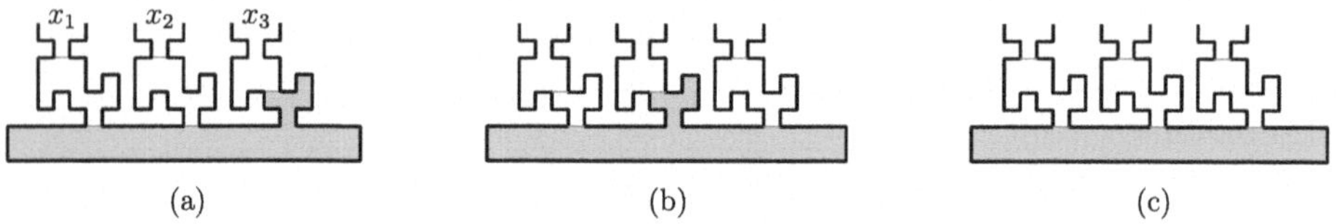

Fig. 9. A clause gadget with variables x_1, x_2, and x_3. (a) x_1 = False, x_2 = False, x_3 = True. (b) x_1 = True, x_2 = True, x_3 = False. (c) x_1 = False, x_2 = False, x_3 = False.

Theorem 3. *Given a rectilinear polygon P with holes and a positive integer k, deciding whether there is a monotone partition of P whose size is at most k is NP-complete.*

References

1. de Berg, M., Cheong, O., van Kreveld, M., Overmars, M.: Computational Geometry: Algorithms and Applications, 3rd edn. Springer, Heidelberg (2008)
2. Chazelle, B.: Triangulating a simple polygon in linear time. Discrete Comput. Geom. **6**(3), 485–524 (1991). https://doi.org/10.1007/BF02574703
3. Chazelle, B., Dobkin, D.P.: Optimal convex decompositions. In :Computational Geometry. Machine Intelligence and Pattern Recognition, vol. 2, pp. 63–133. North-Holland (1985)
4. Chazelle, B., et al.: Ray shooting in polygons using geodesic triangulations. Algorithmica **12**(1), 54–68 (1994)

5. Garey, M.R., Johnson, D.S., Preparata, F.P., Tarjan, R.E.: Triangulating a simple polygon. Inf. Process. Lett. **7**(4), 175–179 (1978)
6. Keil, J.M.: Decomposing a polygon into simpler components. SIAM J. Comput. **14**(4), 799–817 (1985)
7. Kim, H., Lee, J., Ahn, H.-K.: Rectangular partitions of a rectilinear polygon. Comput. Geom. **110**, 101965 (2023)
8. Kim, H., Lee, J., Ahn, H.-K.: Uniformly monotone partitioning of polygons. Theoret. Comput. Sci. **1012**, 114724 (2024)
9. Lee, D.T., Preparata, F.P.: Location of a point in a planar subdivision and its applications. SIAM J. Comput. **6**, 594–606 (1977)
10. Lichtenstein, D.: Planar formulae and their uses. SIAM J. Comput. **11**(2), 329–343 (1982)
11. Liu, R.: Partitioning rectilinear polygons into simpler components. Ph.D. thesis, The University of Texas at Dallas (1985)
12. Liu, R., Ntafos, S.: On decomposing polygons into uniformly monotone parts. Inf. Process. Lett. **27**, 85–89 (1988)
13. Oks, E., Sharir, M.: Minkowski sums of monotone and general simple polygons. Discrete Comput. Geom. **35**(2), 223–240 (2006)
14. Preparata, F.P., Shamos, M.I.: Computational Geometry: An Introduction. Springer, Heidelberg (1985)
15. Wei, X., Joneja, A., Mount, D.M.: Optimal uniformly monotone partitioning of polygons with holes. Comput. Aided Des. **44**, 1235–1252 (2012)

Graph Algorithms

A Linear Delay Algorithm of Enumerating Strongly-Connected Induced Subgraphs Based on SSD Set System

Kan Shota(✉) and Kazuya Haraguchi

Graduate School of Informatics, Kyoto University, Kyoto, Japan
{shota.kan,haraguchi}@amp.i.kyoto-u.ac.jp

Abstract. In this paper, we propose a linear delay algorithm that enumerates all strongly-connected induced subgraphs of a given digraph. We consider the problem in a general framework based on what we call Superset-Subset-Disjoint (SSD) set system. Introducing basic properties of this novel set system, we show that, given a graph, the family of vertex subsets that induce k-edge-connected subgraphs is an SSD system. We then develop an algorithm that enumerates all subsets in an SSD system and show that the delay is linear with respect to the ground set size and the running time of three oracles that implicitly describe the SSD system. Finally, for the purpose of enumerating all strongly-connected induced subgraphs, we present linear-time implementation of the three oracles.

Keywords: SSD set system · Edge-connectivity · Dominator tree · Subgraph enumeration

1 Introduction

Let $G = (V, E)$ be a directed graph (digraph) or an undirected graph with a vertex set V and an edge set E. We denote $n := |V|$ and $m := |E|$. For $S \subseteq V$, we denote by $G[S]$ the *subgraph induced by* S. An undirected graph G is *k-edge-connected* if at least k edges must be removed to disconnect G. A directed graph G is *k-edge-connected* if at least k edges must be removed to make G not strongly-connected. Hence, a 1-edge-connected digraph is *strongly-connected.* Being a fundamental concept of graph theory, edge-connectivity has attracted much attention of researchers from algorithm theory and network analysis for many years [3,4,7,18,23].

An *enumeration problem* in general asks to output all required solutions without duplication and has many applications in such fields as data mining [1,2,5,15] and bioinformatics [14,17,19]. *Subgraph enumeration* is among the most important research topics, and efficient enumeration algorithms have

This work is partially supported by JSPS KAKENHI Grant Number 25K14993. The preprint appears as [20].

H. Fernau and B. Zhu (Eds.): IWOCA 2025, LNCS 15885, pp. 89–102, 2025.
https://doi.org/10.1007/978-3-031-98740-3_7

been developed for various classes of subgraphs; e.g., minimal strongly-connected subgraphs [13], maximal cliques [5,6,16], and pseudo-cliques [22].

Delay of an enumeration algorithm refers to any computation time in the following three cases: (i) computation time between the start of the algorithm and the output of the first solution; (ii) computation time between any two consecutive outputs; and (iii) computation time between the last output and the halt of the algorithm. An enumeration algorithm achieves *polynomial delay* (resp., *linear delay*) if the delay is bounded by a polynomial (resp., a linear function) with respect to the input size, where polynomial delay (and also linear delay) is among the strongest classes of enumeration algorithms [12].

The following theorem is the main result of this paper.

Theorem 1. *Given a directed graph $G = (V, E)$, we can enumerate all strongly-connected induced subgraphs in $O(n + m)$ delay.*

Theorem 1 improves the delay bound $O(n^5 m)$ in Theorem 5(i) of [9], which is the best known result for the enumeration of strongly-connected induced subgraphs.

In order to solve this subgraph enumeration problem, we introduce a general model based on set systems. Let U be an element set and $\mathcal{S} \subseteq 2^U$ be a family of subsets of elements. We call $(U, \mathcal{S})$ or $\mathcal{S}$ a *set system*. Given $(U, \mathcal{S})$, we call each $S \in \mathcal{S}$ a *solution*. Let $C \subseteq U$. If a solution S is a proper subset of C, then we call S a *proper subset solution* (*PSS* for short) of C. We call an inclusion-wise maximal PSS of C a *MaxPSS* of C. We call a nonempty subset Y of C an *removable set* (*RS*) of C if $C \setminus Y$ is a solution. We call an inclusion-wise minimal RS of C a *MinRS* of C. We denote by $\textsc{MaxPSS}_{\mathcal{S}}(C)$ the family of all MaxPSSs of C and by $\textsc{MinRS}_{\mathcal{S}}(C)$ the family of all MinRSs of C. Observe that $X \in \textsc{MaxPSS}_{\mathcal{S}}(C)$ implies $C \setminus X \in \textsc{MinRS}_{\mathcal{S}}(C)$ and $Y \in \textsc{MinRS}_{\mathcal{S}}(C)$ implies $C \setminus Y \in \textsc{MaxPSS}_{\mathcal{S}}(C)$. For subsets $I \subseteq C \subseteq U$, we denote $\mathcal{S}(C) := \{S \in \mathcal{S} \mid S \subseteq C\}$; $\mathcal{S}(C, I) := \{S \in \mathcal{S} \mid I \subseteq S \subseteq C\}$; and $\textsc{MinRS}_{\mathcal{S}}(C, I) := \{Y \in \textsc{MinRS}_{\mathcal{S}}(C) \mid Y \cap I = \emptyset\}$. We say that a subset C is *MaxPSS-disjoint* if all MaxPSSs of C are pairwise disjoint and *MinRS-disjoint* if all MinRSs of C are pairwise disjoint.

For $k \in \mathbb{Z}_+$ and a graph $G = (V, E)$, we consider a set system $(V, \mathcal{S}_{G,k})$ such that $\mathcal{S}_{G,k} \triangleq \{S \subseteq V \mid G[S] \text{ is } k\text{-edge-connected}\}$. We call $\mathcal{S}_{G,k}$ a *k-edge-connected system*, and in particular, we call it a *strongly-connected system* if G is directed and $k = 1$. If G is clear from the context, then we abbreviate $\mathcal{S}_{G,k}$ into $\mathcal{S}_k$. Furthermore, for $C \subseteq V$, we abbreviate $\textsc{MaxPSS}_{\mathcal{S}_k}(C)$ and $\textsc{MinRS}_{\mathcal{S}_k}(C)$ into $\textsc{MaxPSS}_k(C)$ and $\textsc{MinRS}_k(C)$, respectively. We say that G is MaxPSS-disjoint (resp., MinRS-disjoint) if V is MaxPSS-disjoint (resp., MinRS-disjoint).

A system $(U, \mathcal{S})$ is a *Subset-Disjoint* (*SD*) *system* [21] if, for any two solutions $S, S' \in \mathcal{S}$ such that $S \supsetneq S'$, every MinRS Y of S is either a subset of S' (i.e., $Y \subseteq S'$) or disjoint with S' (i.e., $Y \cap S' = \emptyset$). Assuming an oracle that computes a MinRS of a query solution, the authors in [21] show that all solutions of a given SD system $(U, \mathcal{S})$ can be enumerated in linear delay with respect to $|U|$ and the oracle running time. They also show that the 2-edge/vertex-connected system for an undirected graph is an SD system, yielding a linear-delay algorithm that enumerates all solutions in the system, which improves the delay bound $O(n^3m)$ in Corollary 2 of [11].

However, it is hard to apply their framework to other subgraph enumeration problems. For example, we can easily see that the strongly-connected system for a directed graph is not an SD system.

In this paper, to design a similar general enumeration framework, we propose what we call SSD system, an extension of SD system.

Definition 1. *A set system* $(U, \mathcal{S})$ *is a* Superset-Subset-Disjoint (SSD) system *if, for any two solutions* $S, S' \in \mathcal{S}$ *such that* $S \supsetneq S'$*, every MinRS* Y *of* S *satisfies at least one of the following:* Y *is a superset of* S' $(Y \supseteq S')$*;* Y *is a subset of* S' $(Y \subseteq S')$*; and* Y *is disjoint with* S' $(Y \cap S' = \emptyset)$.

Every SD system is an SSD system, but the converse is not true. We will show that $\mathcal{S}_{G,k}$ is an SSD system (Lemma 4) to utilize the properties of SSD systems for the enumeration problem in strongly-connected systems.

We assume that an SSD system $(U, \mathcal{S})$ is given implicitly by means of the following three oracles.

- $\Lambda_{\text{id}}(S)$ (identification): For a solution $S \in \mathcal{S}$, it returns TRUE if S is MaxPSS-disjoint and FALSE otherwise.
- $\Lambda_{\text{MaxPSS}}(S, I)$ (generation of MaxPSSs): For any MaxPSS-disjoint solution $S \in \mathcal{S}$ and a subset $I \subseteq S$, it returns all MaxPSSs X of S such that $X \supseteq I$.
- $\Lambda_{\text{MinRS}}(S, I)$ (generation of MinRSs): For any MinRS-disjoint solution $S \in \mathcal{S}$ and a subset $I \subseteq S$, it returns all MinRSs Y of S such that $Y \cap I = \emptyset$.

Let τ_{id}, τ_{MaxPSS} and τ_{MinRS} denote upper bounds on the computation time of oracles Λ_{id}, Λ_{MaxPSS} and Λ_{MinRS}, respectively. Let σ_{id}, σ_{MaxPSS} and σ_{MinRS} denote upper bounds on the computation space used by oracles Λ_{id}, Λ_{MaxPSS} and Λ_{MinRS}, respectively.

The following Theorems 2 and 3 play essential roles in proving Theorem 1.

Theorem 2. *Let* $(U, \mathcal{S})$ *be an SSD system that is implicitly given by oracles* Λ_{id}, Λ_{MaxPSS} *and* Λ_{MinRS}*. For a solution* $S \in \mathcal{S}$ *and a nonempty subset* $I \subseteq S$*, we can enumerate all solutions in* $\mathcal{S}(S, I)$ *in* $O(|S| + \tau_{\text{id}} + \tau_{\text{MaxPSS}} + \tau_{\text{MinRS}})$ *delay and in* $O(|S| + \sigma_{\text{id}} + \sigma_{\text{MaxPSS}} + \sigma_{\text{MinRS}})$ *space.*

Theorem 3. *Let* $G = (V, E)$ *be a strongly-connected digraph. We can perform the following tasks in* $O(n + m)$ *time:*

(i) *determining whether* G *is MaxPSS-disjoint or not;*
(ii) *generating all MaxPSSs of* G *if* G *is MaxPSS-disjoint; and*
(iii) *generating all MinRSs of* G *if* G *is MinRS-disjoint.*

Theorem 3 indicates that, for a strongly-connected system, we can implement the three oracles so that they run in linear time. The remainder of this paper is mainly devoted to the proofs of Theorems 1 to 3.

The paper is organized as follows. We provide terminology and notation in Sect. 2. We show properties of general SSD systems and k-edge-connected systems in Sect. 3. We then provide proof for Theorem 2 in Sect. 4 and those for Theorems 3 and 1 in Sect. 5. Finally, we give concluding remarks in Sect. 6. Some proofs and details are omitted due to space limitations. See the preprint [20].

2 Preliminaries

We denote by $\mathbb{Z}$ and $\mathbb{Z}_+$ the sets of integers and nonnegative integers, respectively. For $p, q \in \mathbb{Z}$ $(p \leq q)$, let $[p, q] := \{p, p+1, \ldots, q\}$.

Set Systems. For two sets A, B, if $A \cap B = \emptyset$, then we may write their union $A \cup B$ as $A \sqcup B$ (i.e., disjoint union) in order to emphasize that they are disjoint. We say that A and B *intersect* if $A \cap B \neq \emptyset$, $A \setminus B \neq \emptyset$ and $B \setminus A \neq \emptyset$.

A solution S in a set system $(U, \mathcal{S})$ can be MaxPSS-disjoint and MinRS-disjoint at the same time. It is trivial that this happens when $|\textsc{MinRS}_{\mathcal{S}}(S)| \in \{0, 1\}$. The following lemma provides a necessary and sufficient condition that this happens when $|\textsc{MinRS}_{\mathcal{S}}(S)| \geq 2$.

Lemma 1. *Let $(U, \mathcal{S})$ be any set system and $S \in \mathcal{S}$ be a solution such that $|\textsc{MinRS}_{\mathcal{S}}(S)| \geq 2$. Then S is MaxPSS-disjoint and MinRS-disjoint if and only if $|\textsc{MinRS}_{\mathcal{S}}(S)| = 2$ holds and $\textsc{MinRS}_{\mathcal{S}}(S)$ is a partition of S.*

A set system $(U, \mathcal{S})$ is *confluent* if, for any three solutions $S, S', S'' \in \mathcal{S}$, $S'' \subseteq S \cap S'$ implies $S \cup S' \in \mathcal{S}$. A confluent system is not necessarily an SD system or an SSD system.

A set system $(U, \mathcal{S})$ is *laminar* if no two solutions $S, S' \in \mathcal{S}$ intersect. Any laminar system $(U, \mathcal{L})$ is an SSD system and the converse is not true.

Graphs. Throughout the paper, we assume a graph to be simple and directed except Sect. 3, where a graph may be undirected or may contain multiedges. We do not deal with self-loops in any part of the paper.

Let $G = (V, E)$ be a digraph. For a vertex subset $S \subseteq V$, we denote by $G - S$ the induced subgraph $G[V \setminus S]$. When $S = \{v\}$, we write $G - \{v\}$ as $G - v$ for simplicity. For an edge subset $F \subseteq E$, we denote by $G - F$ the subgraph $(V, E \setminus F)$. When $F = \{e\}$, we write $G - \{e\}$ as $G - e$.

A *path* in $G = (V, E)$ is represented as $v_1 \to v_2 \to \cdots \to v_{\ell+1}$, such that $(v_i, v_{i+1}) \in E$ for $i \in [1, \ell]$. A path is *Hamiltonian* if it visits every vertex in V exactly once. For convenience, we may represent a simple path $v_1 \to v_2 \to \cdots \to v_{\ell+1}$ by a vertex set $\{v_1, v_2, \ldots, v_{\ell+1}\}$ unless no confusion arises.

An undirected graph (resp., a digraph) $G = (V, E)$ is *connected* (resp., *strongly-connected*) if there is an s, t-path for every distinct vertices $s, t \in V$. A vertex subset $S \subseteq V$, or $G[S]$, is a *connected component* (resp., *strongly-connected component*) if S is inclusion-wise maximal such that $G[S]$ is connected (resp., strongly-connected). An *articulation point* (resp., a *strong articulation point*) of G is a vertex $v \in V$ such that the number of connected components (resp., strongly-connected components) of $G - v$ is greater than that of G.

Let $G = (V, E)$ be a digraph. For a vertex $v \in V$, we denote by $\Gamma_G^+(v)$ and $\Gamma_G^-(v)$ the out-neighborhood and in-neighborhood of v in G, respectively. For a subset $S \subseteq V$, let $\Gamma_G^+(S) := (\bigcup_{v \in S} \Gamma_G^+(v)) \setminus S$ and $\Gamma_G^-(S) := (\bigcup_{v \in S} \Gamma_G^-(v)) \setminus S$. The *converse of* G is represented as $G^\top = (V, E^\top)$.

Dominator Trees. A *flow graph* with a start vertex s, denoted by G_s, is a directed graph such that every vertex is reachable from s. A strongly-connected graph can be regarded as a flow graph by taking any vertex as the start vertex.

For a strongly-connected digraph $G = (V, E)$, let us designate a start vertex $s \in V$. There is an s, v-path and v, s-path for every $v \in V$. We say that a vertex u *dominates* v, or equivalently, that u is a *dominator of* v if every s, v-path visits u. We denote by $\mathrm{Dom}(v)$ the set of dominators of v, where $s, v \in \mathrm{Dom}(v)$ holds for all $v \in V$. Let us define a binary relation R on V such that $u\mathrm{R}v \Leftrightarrow u \in \mathrm{Dom}(v)$. The binary relation R is reflexive and transitive. The transitive reduction of R forms a tree rooted at s, which is called the *dominator tree (of* G_s*)* [10], where, for $u, v \in V$, u is the parent of v if and only if $\mathrm{Dom}(v) = \mathrm{Dom}(u) \cup \{v\}$. We denote the dominator tree by DT_s. For $v \in V$, we denote by $\pi_s(v)$ the parent of v in DT_s, where we define $\pi_s(s) := \textsc{Nil}$. We denote by $Ch_s(v)$ the set of children of v in DT_s. We will also consider the dominator tree of the converse $G_s^\top$ and denote it by $DT_s^\top$. For $v \in V$, we denote the parent of v in $DT_s^\top$ by $\pi_s^\top(v)$, where we define $\pi_s^\top(s) := \textsc{Nil}$, and the set of children of v by $Ch_s^\top(v)$.

3 SSD Set Systems

In this section, we provide important properties of a general SSD set system and a k-edge-connected system $\mathcal{S}_{G,k}$ for given graph G and integer $k \in \mathbb{Z}_+$.

Lemma 2. *For an SSD system* $(U, \mathcal{S})$*, let* $S \in \mathcal{S}$ *be a solution such that* $|\textsc{MaxPSS}_\mathcal{S}(S)| \geq 2$*. For* $X, X' \in \textsc{MaxPSS}_\mathcal{S}(S)$ $(X \neq X')$*,*

(i) *if* $X \cap X' \neq \emptyset$*, then the union of any two sets in* $\textsc{MaxPSS}_\mathcal{S}(S)$ *is* S*; and*
(ii) *if* $X \cap X' = \emptyset$*, then* S *is MaxPSS-disjoint.*

The following Lemma 3 is the counterpart of Lemma 2 concerning MinRSs.

Lemma 3. *For an SSD system* $(U, \mathcal{S})$*, let* $S \in \mathcal{S}$ *be a solution such that* $|\textsc{MinRS}_\mathcal{S}(S)| \geq 2$*. For* $Y, Y' \in \textsc{MinRS}_\mathcal{S}(S)$ $(Y \neq Y')$*,*

(i) *if* $Y \cap Y' \neq \emptyset$*, then the union of any two sets in* $\textsc{MinRS}_\mathcal{S}(S)$ *is* S*; and*
(ii) *if* $Y \cap Y' = \emptyset$*, then* S *is MinRS-disjoint.*

An interesting implication of these two lemmas is that whether a solution S of an SSD system is MaxPSS-disjoint (resp., MinRS-disjoint) or not can be decided by checking whether any two MaxPSSs (resp., MinRSs) of S are disjoint or not, as stated in Lemma 2(ii) (resp., Lemma 3(ii)).

Lemma 2 is enough for proving the following Theorem 4.

Theorem 4. *Let* $(U, \mathcal{S})$ *be an SSD system. Any solution* $S \in \mathcal{S}$ *is MaxPSS-disjoint and/or* S *is MinRS-disjoint.*

k-Edge-Connected Systems. For $k \in \mathbb{Z}_+$ and a graph $G = (V, E)$, which is either undirected or directed and may contain multiedges, the system $(V, \mathcal{S}_k)$ is confluent and SSD.

Lemma 4. *For any $k \in \mathbb{Z}_+$ and graph $G = (V, E)$, the system $(V, \mathcal{S}_k)$ is confluent and an SSD system.*

Using Lemma 4, we can show the following Theorem 5 that describes a decomposition of G based on k-edge-connectivity.

Theorem 5. *Let $k \in \mathbb{Z}_+$ and $G = (V, E)$ be a graph with $|V| \geq 2$. Then G is MaxPSS-disjoint and/or MinRS-disjoint (with respect to $\mathcal{S}_{G,k}$), and if G is MaxPSS-disjoint, then* $\textsc{MaxPSS}_k(V)$ *is a partition of V.*

Theorem 5 appears to be trivial. When G is not k-edge-connected (i.e., $V \notin \mathcal{S}_k$), it is easy to show that $\textsc{MaxPSS}_k(V) := \{X_1, X_2, \ldots, X_q\}$ is a partition of V. Furthermore, $X_1, X_2, \ldots, X_q$ can be generated by existing algorithms. For example, for $k = 1$, textbook algorithms [7] generate them in $O(n+m)$ time by decomposition into connected components (resp., strongly-connected components) when G is undirected (resp., directed).

The point of Theorem 5 is that we admit G itself to be k-edge-connected (i.e., $V \in \mathcal{S}_k$). In this case, if V is given as input, the existing algorithms will output V as the unique solution, which is not a MaxPSS of V.

4 Enumerating Solutions in SSD System

In this section, we show a linear-delay algorithm that enumerates all solutions in a given SSD system, as a proof for Theorem 2.

Our strategy is binary partition, a fundamental strategy of enumeration. For any set system $(U = \{u_1, u_2, \ldots, u_p\}, \mathcal{S})$ (i.e., not necessarily SSD), we have a partition of the solution set as follows;

$$\mathcal{S} = \mathcal{S}(U, \emptyset) = \mathcal{S}(\emptyset, \emptyset) \sqcup \Big(\bigsqcup_{i=1}^{p} \mathcal{S}(U \setminus \{u_1, \ldots, u_{i-1}\}, \{u_i\}) \Big). \tag{1}$$

Based on this partition, we enumerate solutions in each $\mathcal{S}(U \setminus \{u_1, \ldots, u_{i-1}\}, \{u_i\})$, $i \in [1, p]$ independently. We do not need to take into account the first term $\mathcal{S}(\emptyset, \emptyset)$ since it is empty. We partition each $\mathcal{S}(U \setminus \{u_1, \ldots, u_{i-1}\}, \{u_i\})$ further by using a nice structure that is peculiar to SSD system. The following Lemmas 5 and 6 provide the partition of $\mathcal{S}(U \setminus \{u_1, \ldots, u_{i-1}\}, \{u_i\})$, which are analogous to Lemmas 3 and 4 in [21] for SD system.

Lemma 5. *For an SSD system $(U, \mathcal{S})$, let $S \in \mathcal{S}$ be a solution. It holds that $\mathcal{S}(S, I) = \mathcal{S}(S \setminus Y, I) \sqcup \mathcal{S}(S, I \cup Y)$ for any nonempty subset $I \subseteq S$ and $Y \in$* $\textsc{MinRS}_{\mathcal{S}}(S, I)$.

Lemma 6. *For an SSD system $(U, \mathcal{S})$, let $S \in \mathcal{S}$ be a solution. If S is not MaxPSS-disjoint, then for any nonempty subset $I \subseteq S$, it holds that $\mathcal{S}(S, I) = \{S\} \sqcup (\bigsqcup_{i=1}^{q} \mathcal{S}(S \setminus Y_i, I \sqcup Y_1 \sqcup \cdots \sqcup Y_{i-1}))$, where* $\textsc{MinRS}_{\mathcal{S}}(S, I) := \{Y_1, Y_2, \ldots, Y_q\}$.

Algorithm 1. An algorithm to enumerate all solutions in $\mathcal{S}(S, I)$ for a solution S in an SSD system and its nonempty subset $I \subseteq S$

Input: An SSD system $(U, \mathcal{S})$ that is implicitly given by means of oracles Λ_{id}, $\Lambda_{\mathrm{MaxPSS}}$ and Λ_{MinRS}, a solution $S \in \mathcal{S}$ and a nonempty subset $I \subseteq S$
Output: All solutions in $\mathcal{S}(S, I)$

procedure $\textsc{EnumSSD}(S, I, d)$
  Output S **if** d is even;
  if $\Lambda_{\mathrm{id}}(S)$ returns $\textsc{True}$ **then**
    $\mathcal{X} := \Lambda_{\mathrm{MaxPSS}}(S, I)$;
    for each $X \in \mathcal{X}$ **do**
      Execute $\textsc{EnumSSD}(X, I, d+1)$
    end for
  else
    $\mathcal{Y} := \Lambda_{\mathrm{MinRS}}(S, I)$; $J := I \cup \left(\bigcup_{Y \in \mathcal{Y}} Y\right)$;
    for each $Y \in \mathcal{Y}$ **do**
      $J := J \setminus Y$; Execute $\textsc{EnumSSD}(S \setminus Y, J, d+1)$
    end for
  end if;
  Output S **if** d is odd
end procedure

For an SSD system $(U, \mathcal{S})$ and a nonempty subset $I \subseteq U$, let $S \in \mathcal{S}$ be a solution such that $S \supseteq I$. If S is MaxPSS-disjoint, then there is an immediate partition of $\mathcal{S}(S, I)$ such that

$$\mathcal{S}(S, I) = \{S\} \sqcup \Big(\bigsqcup_{X \in \textsc{MaxPSS}_{\mathcal{S}}(S):\ I \subseteq X} \mathcal{S}(X, I) \Big), \tag{2}$$

where there is at most one MaxPSS X of S with $I \subseteq X$ since S is MaxPSS-disjoint and I is nonempty. Otherwise (i.e., if S is MinRS-disjoint), Lemma 6 provides a partition of $\mathcal{S}(S, I)$ such that

$$\mathcal{S}(S, I) = \{S\} \sqcup \Big(\bigsqcup_{i=1}^{q} \mathcal{S}(S_i, I_i) \Big), \tag{3}$$

where $\textsc{MinRS}_{\mathcal{S}}(S, I) = \{Y_1, Y_2, \dots, Y_q\}$, $S_i = S \setminus Y_i$ and $I_i = I \sqcup Y_1 \sqcup \dots \sqcup Y_{i-1}$, $i \in [1, q]$.

We can enumerate all solutions in $\mathcal{S}(S, I)$ as follows; if S is MaxPSS-disjoint, then we output solutions in $\mathcal{S}(X, I)$ for $X \in \textsc{MaxPSS}_{\mathcal{S}}(S)$ with $X \supseteq I$ recursively, and otherwise, we output solutions in $\mathcal{S}(S_i, I_i)$, $i \in [1, q]$ recursively, where S can be output before or after the above solutions. This procedure is summarized in Algorithm 1. To enumerate all solutions in $\mathcal{S}(S, I)$, it suffices to invoke $\textsc{EnumSSD}(S, I, 0)$.

Proof for Theorem 2. See Algorithm 1. The procedure $\textsc{EnumSSD}(S, I, d)$ outputs S precisely in its execution. If S is MaxPSS-disjoint, which is identified in line 3, then all solutions in the second term of the right hand in (2) are output recursively (line 6), and otherwise, all solutions in the second term of the right hand in (3) are output recursively (line 11). These show the correctness. □

5 Linear-Time Generation of MaxPSSs in Strongly-Connected Systems

Assume that a strongly-connected graph $G = (V, E)$ is given such that $n = |V| \geq 2$. In this section, we present a linear-time algorithm to generate all MaxPSSs of V with respect to $\mathcal{S}_{G,1}$. For simplicity, we may employ abuse notations in this section such that MaxPSS of V is represented as a "MaxPSS of G", an MinRS of V for $G^\top$ is represented as a "MinRS of $G^\top$", and so on.

A naive approach to generating MaxPSSs of G is that we generate strongly-connected components of $G - v$ for all $v \in V$, and then remove duplicates and non-maximal solutions somehow. The time complexity of this method is $\Omega(n(n+m)+N)$, where N denotes the sum of $|X|$ for all $X \in \textsc{MaxPSS}_{\mathcal{S}}(V)$. Note that N can be $\Omega(n^2)$; e.g., a complete digraph. Using Corollary 6.10 of [8], we can generate MaxPSSs of G in $O(n^2)$ time.

Our purpose is to generate MaxPSSs of G in linear time with respect to n, m and N. Recall that, if Y is a MinRS of G, then $V \setminus Y$ is a MaxPSS of G. We derive several properties of MinRSs of G in Sect. 5.1. In particular, for any vertex $s \in V$, we show that every MinRS Y of G with $s \notin Y$ exists as a path in the dominator trees DT_s and $DT_s^\top$ such that the end vertex is a leaf. This enables us to generate all MinRSs Y with $s \notin Y$ by traversing DT_s and $DT_s^\top$. By Theorem 5, G is MaxPSS-disjoint and/or MinRS-disjoint. In Sect. 5.2, we show how to decide whether G is MinRS-disjoint or not and explain algorithms to generate all MaxPSSs of G in each case. This section concludes with the proofs for Theorems 3 and 1.

5.1 Properties of MinRSs

We can enlarge any solution $S \in \mathcal{S}_1$ by "attaching" a directed path to S.

Lemma 7. *Let $G = (V, E)$ be a digraph and $S \in \mathcal{S}_1$ be a solution. For any directed path $P = \{v_1, v_2, \ldots, v_q\}$ in $G - S$, if $v_1 \in \Gamma_G^+(S)$ and $v_q \in \Gamma_G^-(S)$, then $S \cup P \in \mathcal{S}_1$.*

Lemma 8 states that, for any MinRS Y of a solution S, there is a Hamiltonian path in $G[Y]$ for which no "shortcut" exists. It is also shown that the starting vertex of the Hamiltonian path is the only vertex in Y that has incoming edges from $X := S \setminus Y$; and that the end vertex is the only vertex in Y that has outgoing edges to X.

Lemma 8. *For a strongly-connected graph $G = (V, E)$, let $Y \in \textsc{MinRS}_1(V)$ and $X := V \setminus Y \in \textsc{MaxPSS}_1(V)$.*

(i) *In* $G[Y]$, *a path exists from a vertex in* $\Gamma_G^+(X)$ *to a vertex in* $\Gamma_G^-(X)$ *and such a path is Hamiltonian.*
(ii) *Let us denote a Hamiltonian path in* $G[Y]$ *by* $v_1 \to v_2 \to \cdots \to v_q$. *For* $i, j \in [1, q]$, *if* $j - i \geq 2$, *then* $(v_i, v_j) \notin E$.
(iii) *It holds that* $|\Gamma_G^+(X)| = |\Gamma_G^-(X)| = 1$.

Proof. (i) Observe that neither $\Gamma_G^+(X)$ nor $\Gamma_G^-(X)$ is empty since G is strongly-connected. Let $P := \{u_1, u_2, \ldots, u_\ell\} \subseteq Y$ be a simple path such that $u_1 \in \Gamma_G^+(X)$ and $u_\ell \in \Gamma_G^-(X)$. Such P exists since there is a directed path in G from u_1 to any vertex in X by the strong-connectivity of G and the path must visit a vertex in $\Gamma_G^-(X)$. If $P \subsetneq Y$, then by Lemma 7, $G[(V \setminus Y) \cup P] = G[V \setminus (Y \setminus P)]$ is strongly-connected. The nonempty subset $Y \setminus P$ would be an RS of V, which contradicts the minimality of Y. We see that $P = Y$ holds and it is a Hamiltonian path in $G[Y]$.
(ii) Suppose $(v_i, v_j) \in E$. Then $Y' := \{v_1, \ldots, v_i, v_j, \ldots, v_q\} \subsetneq Y$ would be a simple non-Hamiltonian path in $G[Y]$ such that $v_1 \in \Gamma_G^+(X)$ and $v_q \in \Gamma_G^-(X)$, contradicting (i).
(iii) If $|\Gamma_G^+(X)| > 1$, there is $i \in [2, q]$ such that $v_i \in \Gamma_G^+(X)$. Then $\{v_i, v_{i+1}, \ldots, v_q\} \subsetneq Y$ would be a simple non-Hamiltonian path in $G[Y]$ such that $v_i \in \Gamma_G^+(X)$ and $v_q \in \Gamma_G^-(X)$, contradicting (i). It is analogous that $|\Gamma_G^-(X)| = 1$ holds. □

MinRSs and Dominator Trees. For a strongly-connected digraph $G = (V, E)$ and $s \in V$, we consider where MinRSs of G exist in the dominator tree DT_s. Let us confirm fundamental properties on converse and dominator tree.

Lemma 9. *For a strongly-connected graph* $G = (V, E)$, *let* $Y \subseteq V$, $X := V \setminus Y$ *and* $s, t \in X$ $(s \neq t)$.

(i) Y *is an RS of* G *if and only if* Y *is an RS of* $G^\top$.
(ii) Y *is a MinRS of* G *if and only if* Y *is a MinRS of* $G^\top$.
(iii) *If* Y *is an RS of* G, *then* t *is reachable from* s *in both* $DT_s - Y$ *and* $DT_s^\top - Y$.

Lemma 10 provides a necessary and sufficient condition that a subset $Y \subseteq V$ with $s \notin Y$ is a MinRS of G.

Lemma 10. *For a strongly-connected graph* $G = (V, E)$ *and* $s \in V$, *let* $Y \subsetneq V$ *be a nonempty proper subset such that* $s \notin Y$. *Then* Y *is a MinRS of* G *if and only if the following* (i) *to* (iii) *hold.*

(i) *For each* $u \in Y$, *it holds that* $Ch_s(u) \subseteq Y$ *and* $Ch_s^\top(u) \subseteq Y$.
(ii) $DT_s[Y]$ *is a directed path.*
(iii) $DT_s^\top[Y]$ *is a directed path whose orientation is the reverse of* $DT_s[Y]$.

Proof. Let $Y = \{v_1, v_2, \ldots, v_q\}$.
($\Longrightarrow$) By Lemma 8(i), suppose that $v_1 \to v_2 \to \cdots \to v_q$ is the Hamiltonian path in $G[Y]$. (i) If not, then for some $u \in Y$, there would be a child $t \in Ch_s(u)$ (or $t \in Ch_s^\top(u)$) that cannot be reached from s in $DT_s - Y$ (or $DT_s^\top - Y$),

Algorithm 2. An algorithm to generate all MinRSs of V that do not contain a given vertex for a strongly-connected graph $G = (V, E)$

Input: A strongly-connected graph $G = (V, E)$ such that $|V| \geq 2$ and a vertex $s \in V$
Output: All MinRSs of V that do not contain s
1: **procedure** GEN-MINRSS-WITHOUT-ROOT($G = (V, E), s$)
2: Construct the dominator trees DT_s and $DT_s^\top$;
3: **for each** leaf v of DT_s such that $|Ch_s^\top(v)| \leq 1$ **do**
4: $u := v$; $Y := \{u\}$;
5: **while** u is not a leaf of $DT_s^\top$, $|Ch_s(\pi_s(u))| = 1$, $|Ch_s^\top(\pi_s(u))| \leq 1$ and $\pi_s^\top(\pi_s(u)) = u$ **do**
6: $u := \pi_s(u)$; $Y := Y \cup \{u\}$
7: **end while**;
8: Output Y **if** u is a leaf of $DT_s^\top$
9: **end for**
10: **end procedure**

contradicting Lemma 9(iii). (ii) In G_s, every s, v_i-path, $i \in [1, q]$ visits v_1 by Lemma 8(iii), and by Lemma 8(ii), the only v_1, v_i-path is $v_1 \to v_2 \to \cdots \to v_i$. We have that $\mathrm{Dom}(v_i) = \mathrm{Dom}(v_{i-1}) \cup \{v_i\}$ for any $i \in [2, q]$, meaning that $v_1 \to v_2 \to \cdots \to v_q$ is a directed path in DT_s. (iii) $G^\top$ is strongly-connected and Y is a MinRS of $G^\top$. By Lemma 8(i), there is a Hamiltonian path in $G^\top[Y]$, where we see that it is $v_q \to v_{q-1} \to \cdots \to v_1$. Applying the same argument as (ii) to $G_s^\top$, $v_q \to v_{q-1} \to \cdots \to v_1$ is a directed path in $DT_s^\top$.

($\Longleftarrow$) By (ii), we represent $DT_s[Y]$ as a directed path $v_1 \to v_2 \to \cdots \to v_q$, where $DT_s^\top[Y]$ is represented as $v_q \to v_{q-1} \to \cdots \to v_1$ by (iii). By (i), every vertex $t \in V \setminus Y$ is reachable from s in both $DT_s - Y$ and $DT_s^\top - Y$, indicating that there are s, t-path and t, s-path in $G - Y$. We see that $G - Y$ is strongly-connected by transitivity of strong-connectivity and that Y is an RS of G.

We show the minimality of Y. It is trivial if $|Y| = 1$. Let $|Y| \geq 2$. Suppose any MinRS Y' of G such that $Y' \subseteq Y$. If $v_1 \in Y \setminus Y'$, then any vertex in Y' would dominate v_1 in $G_s^\top$ and v_1 is not reachable from s in $DT_s^\top - Y'$, which contradicts Lemma 9(iii). Hence we have $v_1 \in Y'$. If there is $i \in [2, q]$ such that $v_i \in Y \setminus Y'$, then v_1 would dominate v_i in G_s, and v_i is not reachable from s in $DT_s - Y'$, which contradicts Lemma 9(iii). Thus, $Y' = Y$ holds. □

By (i) and (ii) of Lemma 10, any MinRS Y of G with $s \notin Y$ contains a leaf in DT_s. Note that there may be a leaf of DT_s that belongs to no MinRS of G.

Now we are ready to describe an algorithm to generate all MinRSs of G that does not contain the root s. The algorithm is shown in Algorithm 2.

Lemma 11. *For a strongly-connected graph $G = (V, E)$ such that $|V| \geq 2$ and $s \in V$, Algorithm 2 generates all MinRSs Y of G such that $s \notin Y$ in $O(n + m)$ time and space.*

Proof. We provide a proof only for the correctness here. The while-loop from line 5 to line 7 begins with $u := v$ for a leaf v of DT_s and repeats $u := \pi_s(u)$

(i.e., we move to the parent of u) and $Y := Y \cup \{u\}$ as long as the conditions in line 5 are satisfied.

It holds that $s \notin Y$ for any output Y; If u is a child of s, then the last condition $\pi_s^\top(\pi_s(u)) = u$ in line 5 is not satisfied since $\pi_s^\top(\pi_s(u)) = \pi_s^\top(s) = \text{NIL} \neq u$.

We claim that each leaf in DT_s is contained in at most one MinRS Z such that $s \notin Z$. By Lemma 10(ii), $DT_s[Z]$ is a directed path. Also by (i), it is a path from a vertex a to a leaf b, where it is possible that $a = b$. If there are two MinRSs that contain b, it would contradict that each vertex except the root has its unique parent.

We show that, for each leaf v of DT_s, the algorithm outputs the MinRS Z such that $s \notin Z$ and $v \in Z$ if exists. When we exit the while-loop, we already observed that $s \notin Y$ holds. Y is the set of vertices between a leaf v and its ancestor u in DT_s and hence $DT_s[Y]$ is a directed path and satisfies Lemma 10(ii). We also see that $DT_s^\top[Y]$ is a directed path and hence satisfies Lemma 10(iii); by the last condition of line 5, Y is the set of vertices between an inner node v and its descendant u in $DT_s^\top$. For Lemma 10(i), $|Ch_s(a)| \leq 1$ and $|Ch_s^\top(a)| \leq 1$ hold for every $a \in Y$. While $Ch_s(a) \subseteq Y$ is satisfied for every $a \in Y$, $Ch_s^\top(u) \subseteq Y$ is not satisfied if u, the end vertex of the directed path $DT_s^\top[Y]$, is not a leaf of $DT_s^\top$. However, such Y is not output by line 8. We have shown that the output Y satisfies Lemma 10(i) to (iii) and that $Y = Z$ holds. □

5.2 Generating MaxPSSs

How we generate MaxPSSs of G depends on whether G is MinRS-disjoint or not. We can decide whether G is MinRS-disjoint or not as follows. Choosing a vertex $s \in V$, we first generate all MinRSs $Y \in \text{MINRS}_1(V)$ such that $s \notin Y$ by Algorithm 2. The MinRSs generated by Algorithm 2 are pairwise disjoint. If two or more MinRSs are generated, then G is MinRS-disjoint. Otherwise, pick up a vertex in the generated MinRS, say t. We again run Algorithm 2 to generate all MinRSs $Y \in \text{MINRS}_1(V)$ such that $t \notin Y$. If there are two or more MinRSs are generated, then G is MinRS-disjoint. Otherwise, if the two MinRSs generated by the two runs are disjoint, then G is MinRS-disjoint.

Lemma 12. *For a strongly-connected graph $G = (V, E)$ such that $|V| \geq 2$, we can determine whether G is MinRS-disjoint or not in $O(n+m)$ time and space.*

We can decide whether G is MaxPSS-disjoint or not in $O(n)$ time as follows. If G is not MinRS-disjoint, then we can conclude that G is MaxPSS-disjoint by Theorem 4. Otherwise, it is possible that G is MinRS-disjoint as well as MaxPSS-disjoint, which occurs only in the case described in Lemma 1.

Now we are ready to present algorithms to generate all MaxPSSs of MinRS-disjoint graphs (Lemma 13) and MaxPSS-disjoint graphs (Lemma 14). For a MinRS-disjoint graph, we generate all MinRSs that do not contain a given vertex s using Algorithm 2, and output the corresponding MaxPSSs. Then, we select a vertex t from one of these MinRSs, and generate all MinRSs that do not contain t, outputting corresponding MaxPSSs only for those containing s. For a

MaxPSS-disjoint graph, we generate the unique MinRS Y that does not contain a given vertex using Algorithm 2, output its corresponding MaxPSS, and list all strongly-connected components of $G[Y]$ as MaxPSSs.

Lemma 13. *For a strongly-connected MinRS-disjoint graph $G = (V, E)$, we can generate all MaxPSSs of G in $O(n + m + N)$ time and $O(n + m)$ space.*

Lemma 14. *For a strongly-connected MaxPSS-disjoint graph $G = (V, E)$, we can generate all MaxPSSs of G in $O(n + m)$ time and space.*

Proof for Theorem 3. We can determine whether G is MinRS-disjoint in $O(n + m)$ time by Lemma 12. We can also determine whether G is MaxPSS-disjoint or not in additional $O(n)$ time computation. If G is MinRS-disjoint, then we can generate MinRSs of G in $O(n+m)$ time by Lemma 13, where we output Y_s and Y_t instead of $V \setminus Y_s$ and $V \setminus Y_t$. Otherwise, we can generate them in $O(n + m)$ time by Lemma 14 since G is MaxPSS-disjoint in this case. □

Proof for Theorem 1. For a digraph $G = (V, E)$, the strongly-connected system $(V, \mathcal{S}_1)$ is an SSD system by Lemma 4. Let $V = \{v_1, v_2, \ldots, v_n\}$. To generate all solutions in $\mathcal{S}_1$, we partition $\mathcal{S}_1$ into $\mathcal{S}_1(V \setminus \{v_1, v_2, \ldots, v_{i-1}\}, \{v_i\})$, $i \in [1, n]$, based on (1). The solution set $\mathcal{S}_1(V \setminus \{v_1, v_2, \ldots, v_{i-1}\}, \{v_i\})$ is not empty since $\{v_i\}$ belongs to this set, and the only inclusion-maximal solution in $\mathcal{S}_1(V \setminus \{v_1, v_2, \ldots, v_{i-1}\}, \{v_i\})$ is the strongly-connected component that contains the vertex v_i in the subgraph $G - \{v_1, v_2, \ldots, v_{i-1}\}$. Denoting by S_i this maximal solution, we can enumerate all solutions in $\mathcal{S}_1(V \setminus \{v_1, v_2, \ldots, v_{i-1}\}, \{v_i\})$ by executing $\textsc{EnumSSD}(S_i, \{v_i\}, 0)$ in Algorithm 1 in linear delay with respect to $|V|$ and oracle running times by Theorem 2.

We can generate each S_i, $i \in [1, n]$ in $O(n + m)$ time. By Theorem 3, the three oracles for $\mathcal{S}_1$ can be implemented so that the time complexity is $O(n + m)$. □

6 Concluding Remarks

In this paper, given a digraph, we proposed an algorithm that enumerates all strongly-connected induced subgraphs in linear delay (Theorem 1). For this, we introduced a novel set system, SSD (Superset-Subset-Disjoint) system, and developed an algorithm for enumerating all solutions in a given SSD system (Theorem 2) and provided an efficient implementation of the oracles (Theorem 3).

We have considered strongly-connected system (i.e., $\mathcal{S}_{G,k}$ for a digraph G and $k = 1$). Extensions to $k \geq 2$ or undirected graphs remains as future work. The graph decomposition based on MaxPSS is interesting by itself. We leave as future work precise comparisons with existing decomposition techniques and exploration of their applications to other graph problems.

References

1. Agrawal, R., Imieliński, T., Swami, A.: Mining association rules between sets of items in large databases. In: Proceedings of the 1993 ACM SIGMOD International Conference on Management of Data, SIGMOD 1993, pp. 207–216 (1993). https://doi.org/10.1145/170035.170072
2. Agrawal, R., Srikant, R.: Fast algorithms for mining association rules in large databases. In: Proceedings of the 20th International Conference on Very Large Data Bases, VLDB 1994, pp. 487–499 (1994)
3. Ahuja, R.K., Magnanti, T.L., Orlin, J.B.: Network Flows: Theory, Algorithms, and Applications. Prentice Hall Inc., USA (1993)
4. Bang-Jensen, J., Gutin, G.Z.: Digraphs: Theory, Algorithms and Applications, 2nd edn. Springer (2008). https://doi.org/10.1007/978-1-84800-998-1
5. Chang, L., Yu, J.X., Qin, L.: Fast maximal cliques enumeration in sparse graphs. Algorithmica **66**, 173–186 (2012). https://doi.org/10.1007/s00453-012-9632-8
6. Conte, A., Grossi, R., Marino, A., Versari, L.: Sublinear-space and bounded-delay algorithms for maximal clique enumeration in graphs. Algorithmica **82**(6), 1547–1573 (2019). https://doi.org/10.1007/s00453-019-00656-8
7. Cormen, T.H., Leiserson, C.E., Rivest, R.L., Stein, C.: Introduction to Algorithms, 4th edn. MIT Press (2022)
8. Georgiadis, L., Italiano, G.F., Parotsidis, N.: Strong connectivity in directed graphs under failures, with applications. SIAM J. Comput. **49**(5), 865–926 (2020). https://doi.org/10.1137/19M1258530
9. Haraguchi, K., Nagamochi, H.: Enumeration of support-closed subsets in confluent systems. Algorithmica **84**(5), 1279–1315 (2022). https://doi.org/10.1007/s00453-022-00927-x
10. Italiano, G.F., Laura, L., Santaroni, F.: Finding strong bridges and strong articulation points in linear time. Theoret. Comput. Sci. **447**, 74–84 (2012). https://doi.org/10.1016/j.tcs.2011.11.011
11. Ito, T., Sano, Y., Yamanaka, K., Hirayama, T.: A polynomial delay algorithm for enumerating 2-edge-connected induced subgraphs. IEICE Trans. Inf. Syst. **E105.D**(3), 466–473 (2022). https://doi.org/10.1587/transinf.2021FCP0005
12. Johnson, D.S., Yannakakis, M., Papadimitriou, C.H.: On generating all maximal independent sets. Inf. Process. Lett. **27**(3), 119–123 (1988). https://doi.org/10.1016/0020-0190(88)90065-8
13. Khachiyan, L., Boros, E., Elbassioni, K., Gurvich, V.: On enumerating minimal dicuts and strongly connected subgraphs. Algorithmica **50**, 159–172 (2008). https://doi.org/10.1007/s00453-007-9074-x
14. Khiste, N., Ilie, L.: E-MEM: efficient computation of maximal exact matches for very large genomes. Bioinformatics **31**(4), 509–514 (2014). https://doi.org/10.1093/bioinformatics/btu687
15. Kuramochi, M., Karypis, G.: Finding frequent patterns in a large sparse graph. Data Min. Knowl. Disc. **11**(3), 243–271 (2005). https://doi.org/10.1007/s10618-005-0003-9
16. Makino, K., Uno, T.: New algorithms for enumerating all maximal cliques. In: Hagerup, T., Katajainen, J. (eds.) SWAT 2004. LNCS, vol. 3111, pp. 260–272. Springer, Heidelberg (2004). https://doi.org/10.1007/978-3-540-27810-8_23
17. Roux de Bézieux, H., Lima, L., Perraudeau, F., Mary, A., Dudoit, S., Jacob, L.: CALDERA: finding all significant de Bruijn subgraphs for bacterial GWAS. Bioinformatics **38**(Supplement_1), i36–i44 (2022). https://doi.org/10.1093/bioinformatics/btac238

18. Schrijver, A.: Combinatorial Optimization: Polyhedra and Efficiency, Algorithms and Combinatorics, vol. 24. Springer, Heidelberg (2003)
19. Sese, J., Seki, M., Fukuzaki, M.: Mining networks with shared items. In: Proceedings of 19th ACM International Conference on Information and Knowledge Management (CIKM 2010), pp. 1681–1684 (2010). https://doi.org/10.1145/1871437.1871703
20. Shota, K., Haraguchi, K.: SSD set system, graph decomposition and hamiltonian cycle (2024). https://doi.org/10.48550/arXiv.2408.04615
21. Tada, T., Haraguchi, K.: A linear delay algorithm in SD set system and its application to subgraph enumeration. J. Comput. Syst. Sci. (2025). https://doi.org/10.1016/j.jcss.2025.103637
22. Uno, T.: An efficient algorithm for solving pseudo clique enumeration problem. Algorithmica **56**(1), 3–16 (2008). https://doi.org/10.1007/s00453-008-9238-3
23. Williamson, D.P.: Network Flow Algorithms. Cambridge University Press (2019). https://doi.org/10.1017/9781316888568

Exact Learning of Weighted Graphs Using Composite Queries

Michael T. Goodrich, Songyu Liu(✉), and Ioannis Panageas

University of California, Irvine, Irvine, USA
{goodrich,songyul4,ipanagea}@uci.edu
https://ics.uci.edu/~goodrich/, https://ics.uci.edu/~theory/,
https://panageas.github.io/

Abstract. In this paper, we study the ***exact learning*** problem for weighted graphs, where we are given the vertex set, V, of a weighted graph, $G = (V, E, w)$, but we are not given E. The problem, which is also known as ***graph reconstruction***, is to determine all the edges of E, including their weights, by asking queries about G from an oracle. As we observe, using simple shortest-path length queries is not sufficient, in general, to learn a weighted graph. So we study a number of scenarios where it is possible to learn G using a subquadratic number of composite queries, which combine two or three simple queries.

Keywords: graph theory · machine learning · exact learning · graph reconstruction

1 Introduction

The problem of exactly learning the edges of a graph from queries, which is also known as "graph reconstruction," is well-studied and well-motivated for both practical and theoretical reasons; e.g., see [1–7,12–15,17,19–21,24–27,29,30]. For instance, from a practical perspective, there are many contexts in which we know the vertices of a graph but must issue queries to learn its edges. Examples include social network science, where we know the members of a social network, but must issue queries to learn their relationships, such as friendship, collaboration, or even romantic partners; see, e.g., [8,28,31]. Other practical examples include learning phylogenetic relationships through genetic tests [3], learning faults in circuit boards [12,15,27], learning road networks from GPS traces [5], and learning connections in a network, such as a peer-to-peer network or the Internet via routing queries [4].

Furthermore, from the perspective of theoretical computer science, there are many interesting complexity theory and information theory questions that arise in graph reconstruction (e.g., see [2–6,13,19,21,24,25,29,30]) including the following:

– What types of queries are sufficient to exactly learn all the edges in a graph and which ones are not?

H. Fernau and B. Zhu (Eds.): IWOCA 2025, LNCS 15885, pp. 103–117, 2025.
https://doi.org/10.1007/978-3-031-98740-3_8

- What kinds of queries can learn the edges of a graph using a subquadratic number of queries?
- What are the best algorithms for solving the graph reconstruction problem for a given set of queries?
- Are there non-trivial lower bounds for the number of queries needed to solve the graph reconstruction problem?
- What types of graphs are amenable to efficient exact learning algorithms?

Thus, there is ample practical and theoretical motivation for studying the graph reconstruction problem, where the vertices of the graph are known and we can only learn information about the edges through queries to an oracle with full knowledge of the graph. The goal is to find all the edges while minimizing the number of queries to the oracle.

Most of the existing literature on the graph reconstruction problem focuses on the case where the unknown graph is connected and unweighted, with notable results including work by Kannan, Mathieu, and Zhou [19,24]. In this paper, we study the consequences of dropping these two assumptions, which leads to some interesting avenues of algorithmic exploration. For example, as we observe, previous types of queries that are sufficient to learn unweighted connected graphs turn out to be insufficient to learn all disconnected weighted graphs. Thus, we need new types of queries to solve the graph reconstruction problem for disconnected weighted graphs. Indeed, taking inspiration from how combinations of drugs can be used to treat diseases [16], such as cancer and AIDS, the methods we explore in this paper involve solving the graph reconstruction problem through composite queries, which combine different types of queries in concert to efficiently learn the edges and their weights in a graph.

1.1 Queries

Let us therefore consider different types of queries that might be used to exactly learn a weighted graph efficiently. Assume that a given weighted graph, G, has n known vertices, m unknown edges, and let D denote the maximum degree of any vertex in the graph.

Wang and Honorio study the problem of reconstructing weighted directed trees using what the authors call "additive queries", which return the sum of the edge weights on the directed path between two given vertices and 0 otherwise [30]. Of course, unlike in trees, there can be multiple paths between two vertices in a general graph; hence, let us define the following analogous query for a possibly disconnected weighted graph:

- **Distance query:** define the query, q_d, to return the sum of the edge weights on a shortest weighted path between two given vertices and $+\infty$ if there is no path.

Note that we use $+\infty$ instead of 0 so that the distance returned by the query satisfies the triangle inequality. Kannan *et al.* [19] show that a connected unweighted graph can be learned using an oracle that returns an entire shortest path between a given pair of vertices, or an oracle that returns the distance of a shortest path between them, where distance is measured by the number of edges, which is also known as the "hop count." Unfortunately, as Fig. 1 shows, the weighted shortest path may not correspond to the unweighted shortest path; hence, a solution to the weighted case cannot easily be derived from existing solutions to the unweighted case. Moreover, the distance query, q_d, only returns a single number, not an entire path in the graph.

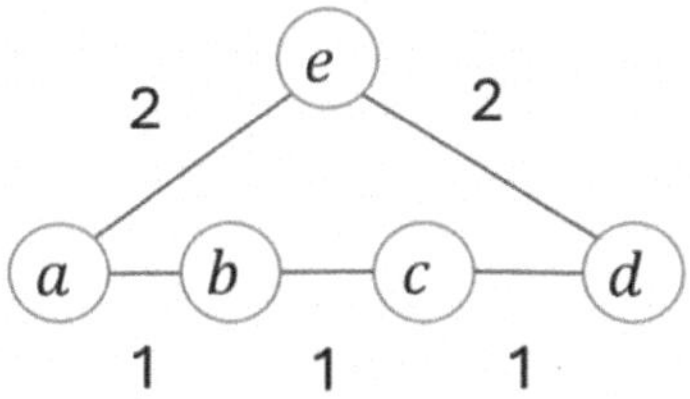

Fig. 1. The weighted shortest path and the unweighted shortest path from a to d are different. In a weighted graph, the bottom path is the shortest. However, if we ignore the weights and treat it as an unweighted graph, then the top path will be the shortest with only two hops.

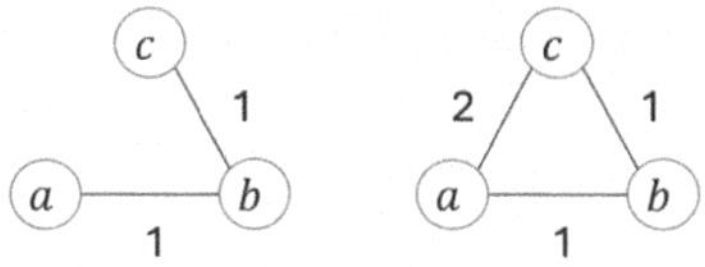

Fig. 2. Using query q_d alone, the two graphs will lead to the same query results for all pairs of vertices. The existence of the edge ac cannot be confirmed or denied. ac is called a transitive edge (Definition 1) in the graph on the right.

Indeed, there are weighted graphs that cannot be reconstructed using q_d queries alone. See Fig. 2. Related to this issue, the notion of transitive edges was introduced to describe similar issues by Wang and Honorio [30] and we redefine it here in our context.

Definition 1. *Let* $G = (V, E, w)$ *be an undirected and weighted graph where* $w(u, v)$ *returns the weight of an edge,* uv*, and 0 if there is no edge. An edge* uv *is* ***transitive*** *if a path exists between* u *and* v *whose weighted length* $l \leq w(u, v)$.

Transitive edges cannot be recovered by q_d alone. It might be possible to exclude such edges by imposing specific constraints on the graph. However, we want our algorithm to work on a more general class of graphs, so we need an additional type of query. In particular, we introduce the following additional type of query:

- **Edge-weight query:** the query, q_w, takes two vertices, u and v, and returns $w(u,v)$.

Using q_w alone, it is at least possible to reconstruct any weighted graph, but it is necessary to query every pair of vertices, leading to an $\Theta(n^2)$ brute-force solution. We refer to this algorithm as "EXHAUSTIVE-QUERY". To accommodate disconnected graphs and obtain better query complexities, therefore, we define yet another type of query and we consider layering the input graph based on weight thresholds. In network mapping, tools such as `traceroute` provide delay information, which can be seen as distances between vertices [4,19]. Weights above a certain threshold can indicate high delay in the network.

Dividing a weighted graph into layers based on the weights of the edges revealed interesting within-layer and across-layer connections between topology and edge weights in real-world datasets [11]. This motivates us to layer the input graph. Following the definition in [11], layer W_{thr} of a graph G is defined as $G[w \geq W_{\text{thr}}]$, the subgraph of G including all vertices of G, but only edges with weight $w \geq W_{\text{thr}}$. W_{thr} is a weight threshold that will change in our algorithm. This subgraph may be disconnected. If $w_1 < w_2, G[w \geq w_1] \supseteq G[w \geq w_2]$. All the queries can take an additional parameter, W_{thr}, and give us information within this subgraph. This ability to query within a subgraph is essential because the distances between vertices in this subgraph may differ significantly from their distances in the original graph. There is no straightforward relationship between these two distances under our general assumptions on edge weights that we will explain in the next section.

Let $q_d(u, v, W_{\text{thr}})$ return the sum of the edge weights on a shortest weighted path between vertices u and v in $G[w \geq W_{\text{thr}}]$, and $+\infty$ if there is no such path. Similarly, let $q_w(u, v, W_{\text{thr}})$ return $w(u, v)$ in $G[w \geq W_{\text{thr}}]$ and 0 when there is no edge. Let EXHAUSTIVE-QUERY(V, W_{thr}) call $q_w(u, v, W_{\text{thr}})$ for every pair of vertices in V. To work with disconnected graphs, we need one more type of query, which can detect connected components efficiently. Relying solely on q_d for this purpose would result in prohibitively high query complexity. Suppose a graph has n vertices and k connected components. [23] showed a lower bound of $kn - \binom{k+1}{2}$ queries and k can be as large as $\Theta(n)$ in our case. This necessitates another type of query. Liu and Mukherjee [23] proposed the following oracle to learn k connected components using $\Theta(n \log k)$ queries: for any vertex u and set $S \subseteq V$ not containing u,

$$\alpha_m(u, S) = \begin{cases} 1 & \text{if for some } v \in S,\ u \text{ and } v \text{ belong to the same component,} \\ 0 & \text{otherwise.} \end{cases}$$

Incorporating the weight threshold, we have:

- **Component query:** let $q_c(u, S, W_{\text{thr}})$ return $\alpha_m(u, S)$ in $G[w \geq W_{\text{thr}}]$.

Our composite query q therefore, is a mixture of $q_w(u, v, W_{\text{thr}}), q_d(u, v, W_{\text{thr}})$, and $q_c(u, S, W_{\text{thr}})$. Every time we perform a query, we can choose the query type that is most important. In addition, we assume that edges in the original input graph have weight at least 1, and allow composite queries that fix $W_{\text{thr}} = 1$, in which case these queries return the corresponding quantities in the original input graph.

1.2 Related Work

As mentioned above, existing literature on the graph reconstruction problem tends to focus on connected unweighted graphs and the following query oracles defined for such graphs, with distance determined by hop count, e.g., see [4,19]:

- An unweighted ***distance*** oracle in an unweighted graph returns the hop-count distance between a given pair of vertices.
- A ***shortest-path*** oracle returns an ordered list of the vertices of a shortest path between a given pair of vertices.
- A **k*th-hop*** oracle returns the kth vertex on a shortest path between a given pair of vertices.

For example, Kannan, Mathieu, and Zhou [19] present an algorithm for reconstructing an unweighted, connected, bounded-degree graph using $\tilde{O}(n^{3/2})$ distance queries,[1] where one randomly selects centers and then partitions the graph into slightly overlapping Voronoi-cell subgraphs based on the chosen centers until the graph clusters were all small enough to perform an exhaustive query on each cluster. Afshar *et al.* [5] proposed a similar approach with different Voronoi cell sizes and a query complexity that depended on the dual graph connecting neighboring cells. Mathieu and Zhou [24] focused on random regular graphs, giving an algorithm that uses only $\tilde{O}(n)$ distance queries in this special case.

If we view the graph as an unknown non-negative $\binom{n}{2}$ dimensional vector, x_G, with $\text{supp}\,(x_G)$ denoting the non-zero coordinates, the following queries can be defined [7]:

- ***Linear queries:*** Given any non-negative $\binom{n}{2}$ dimension vector a_G, what is $a_G \cdot x_G$?
- ***OR queries:*** Given any subset S of the $\binom{n}{2}$ dimensions, is $\text{supp}\,(x_G) \cap S$ empty?

Linear queries are more general than cross-additive queries, which in weighted graphs return the sum of the weights of the edges crossing between two given sets of vertices, and Choi [13] presents one of the few existing results for reconstructing graphs with real weights using cross-additive queries.

A closely related, and more powerful query, is the general ***additive*** query. For any given set of vertices, the additive query returns the sum of the weights of the edges with both endpoints in the set.[2] Using this additive query, Mazzawi [25]

[1] As is common, we use $\tilde{O}(*)$ to denote asymptotic bounds that ignore polylogarithmic factors.

[2] Note that this is not the same as the additive query defined by Wang and Honorio in the context of weighted trees [30].

gives a reconstruction algorithm that matches the information-theoretic lower bound for graphs with positive integer weights. Alternatively, when the weights are bounded real numbers that are all positive (or negative), Kim [20] reduces the graph reconstruction problem using additive queries to a coin-weighing problem.

1.3 Our Results

In this paper, we study weighted graph reconstruction problems for graphs that are not necessarily connected and for connected graphs, and in both cases weights are real numbers in the range, $[1, W_{\max}]$. Table 1 shows our results. The first row where the weighted graph need not be connected is our main contribution. The second row is a natural generalization of the result presented in [19] and we provide the detailed analysis for this case in Appendix of the full version of this paper. The main body of this paper will focus on the first case. The idea of our approach is to divide the graph into layers and use a Voronoi-cell approach to reconstruct each connected component in each layer. This way we will discover edges with weight in consecutive intervals, $[1,2), [2,4), [4,8)$, and so on. Although an unweighted tree reconstruction algorithm based on layers was proposed in [9], it is not comparable to our approach. In [9], layers were defined to partition vertices based on their distances to the root of the tree. This is more natural in their context because trees admit a balanced vertex separator of size 1 [9]. However, it is more convenient to define layers based on edges in our case.

We assume the input graph has m edges and maximum degree $D \gg 1$. We further assume that $m/D = \omega(1)^3$. This condition of $m/D = \omega(1)$ is relatively mild. For instance, if $D = o(n)$ and the original input graph is connected, then $m \geq n - 1$ and the condition will be satisfied. For reconstructing graphs of unbounded degree using distance queries, Kannan *et al.* established an $\Omega(n^2)$ lower bound, demonstrating that bounded degree $D = o(n)$ is indeed a necessary assumption [19]. We can further simplify the results when $D = O(\text{polylog}(n))$, an assumption made by Kannan *et al.* [19]. However, we highlight the dependency on D in our query complexity results. If $W_{\max} = 1$, then the input graph is effectively unweighted, which implies a query complexity of $\tilde{O}(D^3 n^{3/2})$ when the input graph is connected, matching the complexity of the algorithm by Kannan *et al.*

Table 1. Our results for weighted graphs with maximum degree D. α is a positive constant to be defined in the next section.

Connected?	Queries	Query complexity
No	$q_w(u, v, W_{\text{thr}}), q_d(u, v, W_{\text{thr}}), q_c(u, S, W_{\text{thr}})$	$\tilde{O}\left(\left(1 + \frac{1}{\alpha}\log D\right) D^3 n^{3/2}\right)$
Yes	$q_w(u, v, 1), q_d(u, v, 1)$	$\tilde{O}(D^{3W_{\max}} n^{3/2})$

[3] $\omega(1)$ denotes a function that tends to infinity as the underlying parameter n tends to infinity.

Our algorithms introduce a number of interesting techniques for solving the graph reconstruction problem for weighted and possibly disconnected graphs. For example, the approach of using queries with weight thresholds allows us to design efficient query algorithms even with edge weights that can vary significantly. Typically, when adapting algorithms from unweighted graphs to weighted graphs, the runtime becomes dependent on the aspect ratio, defined as the maximum weight divided by the minimum weight [10]. Therefore, it is desirable to have a graph with a small aspect ratio. Transforming a graph into another one with a smaller aspect ratio while preserving its shortest path structure is extremely difficult [10]. In contrast, our approach goes beyond adapting and makes it possible to get rid of this dependency. This stems from our probabilistic model where edge weights are randomly drawn from a distribution. Such a model may represent scenarios like road networks, where vertices correspond to intersections, edges to roads, and edge weights to traffic conditions. While the graph structure remains fixed, the traffic follows a probability distribution. This probabilistic model is a fundamental distinction between our work and that of [19]. Building upon a specific edge weight distribution, we are able to eliminate the $\log_2 W_{\max}$ factor that appears in our algorithm, which is a key contribution of this paper. Furthermore, existing literature only uses one type of query to reconstruct graphs. A single type of query may have limitations but at the same time may excel at certain tasks. The constituent parts of our composite query either fail to reconstruct the graph or do so inefficiently. However, we can design a powerful algorithm by combining them.

2 Preliminaries

We largely follow the notation introduced by Kannan *et al.* [19], for the sake of consistency. Let $G = (V, E, w)$ be an undirected and weighted graph, where V is the vertex set, E is the edge set, and w is the weight function. When the context is clear, we also use w to denote the weight of a single edge. Let $d(u, v)$ denote the weighted shortest path distance of the underlying graph. The underlying graph can change since we work with different layers of the graph and different connected components. In most cases, the distance is the one returned by the query q_d and should be clear from the context. For a subset of vertices $S \subseteq V$ and a vertex $v \in V$, define $d(S, v) = \min_{s \in S} d(s, v)$. Define δ as the hop distance, i.e. the unweighted shortest path distance. For $v \in V$, define the neighborhood of v as $N(v) = \{u \in V : \delta(u, v) \leq 1\}$, and define the neighborhood of v within 2 hops as $N_2(v) = \{u \in V : \delta(u, v) \leq 2\}$.

For a pair of distinct vertices $u, v \in V$, we say that uv is an edge of G if and only if $uv \in E$. For a subset of vertices $S \subseteq V$, let $G[S]$ be the subgraph induced by S. A closely related notation is $G[w \geq W_{\text{thr}}]$, the edge induced subgraph of G including only edges with weight $w \geq W_{\text{thr}}$ where W_{thr} is a weight threshold.

When we pass a vertex $s \in V$ and a subset $T \subseteq V$ to a query, we perform the query for every pair in their Cartesian product. When we pass two subsets of vertices, we also perform the query for every pair in their Cartesian product.

Recall that edges have real weights of at least 1. We explore two theoretical models regarding the weights, a general one and a specific one. When the input is a weighted graph or tree, existing literature assumes that the weights are fixed numbers [13,30]. In contrast, we study a model where the structure of the input graph is fixed and the edge weights are sampled from a continuous distribution whose support is $[1, +\infty)$. We use $F(w)$ to denote its cumulative distribution function (CDF). Suppose the graph has m edges and we have m independent and identically distributed (i.i.d.) samples from this distribution. The maximum weight of the graph $W_{\max}$ is the largest order statistic and its CDF is $F(w)^m$. We assume that the observed value of this random variable in the input graph is known to the algorithms and we also denote it by $W_{\max}$. The distinction will be clear from the context.

In Appendix of the full version of this paper, we make no further assumptions about the distribution of edge weights, only that they fall within the range $[1, W_{\max}]$. Section 3 maintains this range constraint but introduces a specific distribution for the weights. Experiments on real-world graphs suggest that edge weights follow a power law distribution (Pareto distribution) [22]. Specifically, we use a Pareto distribution with lower threshold 1 and parameter $\alpha > 0$. Therefore, the edge weights are i.i.d. random variables with CDF

$$F(w) = 1 - w^{-\alpha}, \quad \text{for } w \in [1, +\infty).$$

If we want to make sure the variance exists, we can assume $\alpha > 2$. However, as long as it is a positive constant, our results will hold.

3 Queries with a Threshold

In this section, we design a layer-by-layer reconstruction (LBL-R) algorithm (Algorithm 1) that can reconstruct weighted graphs that are not necessarily connected using composite queries. We then prove its correctness and show that its query complexity is $\tilde{O}\left(\left(1 + \frac{1}{\alpha} \log D\right) D^3 n^{3/2}\right)$.

Algorithm 1. Layer-by-layer reconstruction (LBL-R)

function LBL-R(V)
  $E \leftarrow \{\}$
  for $j \leftarrow 0, 1, 2, \ldots, \lfloor \log_2 W_{max} \rfloor$ **do**
    $W_{\text{thr}} \leftarrow 2^j$
    $C \leftarrow$ FIND-CONNECTED-COMPONENTS(V, W_{thr})
    for $c \in C$ **do**
      if $|c| \leq n^{1/4}$ **then** ▷ A small connected component.
        $E_{jc} \leftarrow$ EXHAUSTIVE-QUERY(c, W_{thr})
      else
        $E_{jc} \leftarrow$ RECONSTRUCT(c, W_{thr})
      $E \leftarrow E \cup E_{jc}$
    if $\max_{c \in C} |c| \leq n^{1/4}$ **then** ▷ All the connected components are small.
      break
  return E

3.1 Description of LBL-R

First, we provide a high-level description of our algorithm. This algorithm takes as input the set V, which is all the vertices in the given graph, and recovers all the edges of the graph. It works iteratively. During iteration j, we set $W_{\text{thr}} = 2^j$ and find the connected components of $G[w \geq W_{\text{thr}}]$. If a connected component is small, then we can afford an exhaustive search. This will discover all of its edges with weight $w \geq W_{\text{thr}}$. If a connected component is large, we will call RECONSTRUCT, which has a lower query complexity than an exhaustive search, but is not guaranteed to find the same set of edges. The missing edges will be found in subsequent iterations. If all the connected components are small, we will exhaustively search all of them and this is the time to break out of the outer loop.

Remark 1. In fact, once EXHAUSTIVE-QUERY is called on a small connected component in iteration j, we do not need additional queries involving its vertices. It is possible to exclude these vertices after this iteration. The asymptotic query complexity will be the same, so we do not include this change for simplicity (Fig. 3).

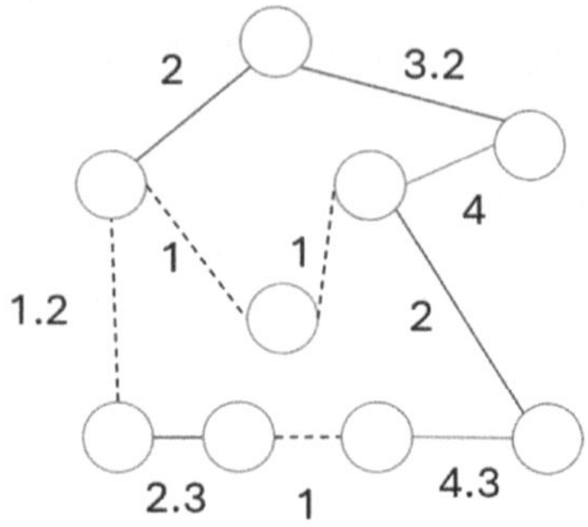

Fig. 3. Iteration 1 of LBL-R (Algorithm 1). $W_{\text{thr}} = 2$. Edges with weight $w < W_{\text{thr}}$ have been discovered and are represented by dashed lines. $G[w \geq W_{\text{thr}}]$ has three connected components, one of which only has one vertex. Edges with weight $w \in [W_{\text{thr}}, 2W_{\text{thr}})$ are guaranteed to be discovered and are represented by solid blue lines. Edges with weight $w \geq 2W_{\text{thr}}$ will be discovered in future iterations and are represented by solid orange lines.

Next, we explain the subroutines used in our main algorithm. Due to space limit, we defer their pseudocode to Appendix of the full version of this paper. As a direct result of our component query, FIND-CONNECTED-COMPONENTS finds connected components of $G[w \geq W_{\text{thr}}]$ with query complexity $\tilde{O}(n)$ [23]. The other subroutine RECONSTRUCT can be treated as a black box algorithm for our purpose. RECONSTRUCT and its subroutines are adaptations of the algorithms presented in [19], using our specific queries. While our LBL-R algorithm builds upon this subroutine used for the unweighted, connected case in [19], we have introduced significant modifications to reconstruct a different class of graphs. When the parameter W_{thr} is set to 1, our RECONSTRUCT subroutine includes the original one in [19] as a special case. Now we briefly review this original version for completeness.

For a connected graph G_c with bounded degree, it first finds a set of vertices A and each vertex $a \in A$ will become the center of an extended Voronoi cell $D_a \subseteq V$. These cells are overlapping. Then an exhaustive search is performed within each cell and the union of vertex-induced subgraphs $\bigcup_{a\in A} G_c[D_a]$ will cover all the edges. Finally, because this whole procedure is randomized, its query complexity may differ. If the number of queries issued is too large, it will stop the ongoing procedure and retry.

We now consider the query complexity of RECONSTRUCT. The query complexity proofs in [19] can be directly extended to accommodate any metric between vertices of a connected graph with bounded degree. We present the following lemma without proof.

Lemma 1 (Extension of Sect. 2.2.2 in [19]). *For connected graphs with n vertices and maximum degree D, the randomized algorithm RECONSTRUCT has expected query complexity $\tilde{O}(D^3 n^{3/2})$.*

The query complexity result assumes that n is large, so we do not apply RECONSTRUCT to all connected components. Note that the expectation is with respect to (wrt) the random choices in the algorithm and does not depend on the random weights of our input graph. Also, this lemma only states the query complexity and not the correctness. In fact, RECONSTRUCT by itself may not correctly reconstruct our input graph and a more detailed analysis is in the subsequent section.

3.2 Correctness of LBL-R

To show the correctness of our algorithm LBL-R, we first analyze what happens in one iteration for one connected component. Our analysis focuses on RECONSTRUCT-SUB, which is the core procedure repeatedly executed by the RECONSTRUCT algorithm, and the result will directly extend to RECONSTRUCT. The detailed pseudocode for RECONSTRUCT-SUB and its subroutines are in Appendix of the full version of this paper. We can show that RECONSTRUCT will discover edges with weights $w \in [W_{\text{thr}}, 2W_{\text{thr}})$ within a single connected component during one iteration of our main algorithm. When the threshold in the queries is W_{thr}, the algorithm might also discover edges with weight $w \geq 2W_{\text{thr}}$ but there is no guarantee.

Now we can show the correctness of LBL-R (Algorithm 1). During any iteration, for a connected component, if the EXHAUSTIVE-QUERY case is executed, we will discover its edges with weight $w \geq W_{\text{thr}}$, which is a superset of $[W_{\text{thr}}, 2W_{\text{thr}})$. Otherwise, we will discover its edges with weight $w \in [W_{\text{thr}}, 2W_{\text{thr}})$. The union of these intervals will cover all the possible weights when the outer loop finishes. If the break statement is triggered, EXHAUSTIVE-QUERY will discover all edges with weight $w \geq W_{\text{thr}}$. The correctness of LBL-R (Algorithm 1) thus follows.

Theorem 1 (Main correctness). *The output of LBL-R is all edges of G.*

3.3 Query Complexity of LBL-R

As previously indicated, the central challenge of our analysis lies in eliminating the $\log_2 W_{\max}$ factor in LBL-R (Algorithm 1). The lemmas and proofs that accomplish this constitute the primary technical contribution of our work and distinguish us from the prior work by [19]. We first consider the query complexity for one iteration of our main algorithm. The query complexity to find all the connected components is much smaller than the other steps since we have component queries at our disposal. Then, we compute the query complexity to process all connected components. Finally, we analyze when the break statement will be triggered.

Let us focus on the inner for loop in LBL-R (Algorithm 1). For some connected components, we will perform an exhaustive search, and for the rest, we will call RECONSTRUCT. The sizes of these connected components are carefully chosen so that for each group of components, the query complexity has a small upper bound.

Lemma 2 (Query complexity for all connected components). *The inner for loop in LBL-R has expected query complexity $\tilde{O}(D^3 n^{3/2})$. The expectation is wrt the random choices in the RECONSTRUCT algorithm and this result holds for all input graphs.*

If we directly incorporate the outer loop, the overall query complexity will depend on $\lfloor \log_2 W_{max} \rfloor$. This is undesirable since W_{max} can be quite large. However, we can leverage our edge weight distribution to eliminate this term and significantly improve the query complexity. Recall that throughout this section, edge weights are sampled from a Pareto distribution. For a particular input graph, all its weights are fixed. The distribution specifies all the weight configurations that this graph may have. The following lemma shows that under our assumptions, the maximum weight is $\Omega(D^{\frac{1}{\alpha}})$ asymptotically almost surely (a.a.s.).[4] While a stronger result could be established, this lemma is sufficient for our subsequent analysis and effectively demonstrates that the maximum weight is large. Recall that $m/D = \omega(1)$, the weights are m i.i.d. samples, and $W_{\max}$ is their maximum. Since the Pareto distribution has a heavy tail, when we draw sufficiently many samples from it, the maximum will be arbitrarily larger than $D^{\frac{1}{\alpha}}$ a.a.s.

Lemma 3 (Maximum weight). *Let $w^* = D^{\frac{1}{\alpha}} > 1$, then for any constant $c > 1$, $\lim_{n \to +\infty} \Pr(W_{max} > cw^*) = 1$.*

At any iteration j, the weight of an edge may or may not be below the threshold. For any edge, the probability that it is in $G[w \geq W_{\text{thr}}]$ is $p :- \Pr(w \geq W_{\text{thr}}) :- T(W_{\text{thr}})$. We notice the similarity between this and the binomial random graph $G(n,p)$ [18]. In $G(n,p)$, there are n vertices, and any vertex can connect to all other vertices with probability p. The size of connected components in $G(n,p)$ has been well studied. In our model, we can show an analogous result which allows us to analyze at which iteration the break statement will be triggered.

Lemma 4 (Part of Theorem 5.4 in [18]). *In a binomial random graph $G(n,p)$, if $np = c \in (0,1)$, then a.a.s. the largest connected component has $O(\log(n))$ vertices.*

Lemma 5 (Largest connected component). *In our model, if $Dp = c \in (0,1)$, then a.a.s. the largest connected component has $O(\log(n))$ vertices.*

Now we can prove our main lemma in this section. The result suggests that in LBL-R (Algorithm 1) the outer loop is in fact superficial. We do not need to finish the outer loop to fully recover the underlying graph. This will allow us to get rid of the query complexity's dependency on the maximum weight and obtain an efficient algorithm. The proof outline is as follows: W_{thr} increases with each iteration, and since weights follow a Pareto distribution, $p = \Pr(w \geq W_{\text{thr}}) =$

[4] a.a.s. means the statement holds true with probability tending to one as the number of vertices tends to infinity.

$T(W_{\text{thr}})$ will decrease. Once Dp falls below 1, we can apply the lemma above. Consequently, the break statement in LBL-R (Algorithm 1) will be triggered no later than this point and we can calculate when this occurs.

Lemma 6 (Early termination). *Under the following assumptions:*

1. $m/D = \omega(1)$.
2. *Edge weights are i.i.d. random variables with CDF*

$$F(w) = 1 - w^{-\alpha} \quad \text{for } w \in [1, \infty), \text{where } \alpha > 0 \text{ is a constant,}$$

and the maximum of these random variables is W_{max}.

We have the following result: a.a.s. the break statement in LBL-R will terminate the outer loop at least i iterations early for any natural number i.

Theorem 2 (Query complexity). *For weighted graph reconstruction using composite queries, there is a randomized algorithm with query complexity* $\tilde{O}\left(\left(1 + \frac{1}{\alpha}\log D\right) D^3 n^{3/2}\right)$.

4 Conclusion

In this paper, we demonstrate that for weighted and potentially disconnected graphs with bounded degree whose weights follow a Pareto distribution, combining distance queries with edge-weight and component queries enables efficient graph reconstruction. An interesting direction for future work would be to establish lower bound results when only two of the three query types are allowed for interesting subclasses of graphs. For instance, if a graph has no transitive edges, can edge-weight queries be omitted? Similarly, is it possible to reconstruct subclasses of graphs without component queries? As previously noted, $\Omega(n^2)$ distance queries are necessary to identify all connected components in a general graph. However, stronger assumptions on the graph structure and edge weights might lead to better lower bounds.

Disclosure of Interests. The authors have no competing interests to declare that are relevant to the content of this article.

References

1. Abrahao, B., Chierichetti, F., Kleinberg, R., Panconesi, A.: Trace complexity of network inference. In: Proceedings of the 19th ACM SIGKDD International Conference on Knowledge Discovery and Data Mining, KDD 2013, pp. 491–499. ACM, Chicago (2013). https://doi.org/10.1145/2487575.2487664, https://dl.acm.org/doi/10.1145/2487575.2487664
2. Afshar, R., Goodrich, M.T.: Exact learning of multitrees and almost-trees using path queries. In: Latin American Symposium on Theoretical Informatics, pp. 293–311 (Aug 2022). https://doi.org/10.48550/arXiv.2208.04216, https://www.semanticscholar.org/paper/a11e550eff669a4056d3957693fceb60add0bd47, arXiv:2208.04216 [cs]

3. Afshar, R., Goodrich, M.T., Matias, P., Osegueda, M.C.: Reconstructing biological and digital phylogenetic trees in parallel. Drops-idn/v2/document/10.4230/lipics.esa.2020.3 (2020). https://doi.org/10.4230/LIPIcs.ESA.2020.3, https://drops.dagstuhl.de/entities/document/10.4230/LIPIcs.ESA.2020.3
4. Afshar, R., Goodrich, M.T., Matias, P., Osegueda, M.C.: Mapping networks via parallel kth-hop traceroute queries. Drops-idn/v2/document/10.4230/lipics.stacs.2022.4 (2022). https://doi.org/10.4230/LIPIcs.STACS.2022.4, https://drops.dagstuhl.de/entities/document/10.4230/LIPIcs.STACS.2022.4
5. Afshar, R., Goodrich, M.T., Ozel, E.: Efficient exact learning algorithms for road networks and other graphs with bounded clustering degrees. Drops-idn/v2/document/10.4230/lipics.sea.2022.9 (2022). https://doi.org/10.4230/LIPIcs.SEA.2022.9, https://drops.dagstuhl.de/entities/document/10.4230/LIPIcs.SEA.2022.9
6. Arunachaleswaran, E.R., De, A., Kannan, S.: Reconstructing ultrametric trees from noisy experiments. In: Proceedings of the 34th International Conference on Algorithmic Learning Theory, pp. 82–114. PMLR (2023). https://proceedings.mlr.press/v201/arunachaleswaran23a.html. iSSN: 2640-3498 shortConferenceName: ALT
7. Assadi, S., Chakrabarty, D., Khanna, S.: Graph connectivity and single element recovery via linear and OR queries (2021). http://arxiv.org/abs/2007.06098, arXiv:2007.06098 [cs]
8. Backstrom, L., Kleinberg, J.: Romantic partnerships and the dispersion of social ties: a network analysis of relationship status on Facebook. In: 17th ACM Conference on Computer Supported Cooperative Work & Social Computing, pp. 831–841. CSCW (2014). https://doi.org/10.1145/2531602.2531642
9. Bastide, P., Groenland, C.: Optimal distance query reconstruction for graphs without long induced cycles (2024). https://doi.org/10.48550/arXiv.2306.05979, arXiv:2306.05979 [cs]
10. Bernstein, A., Bodwin, G., Wein, N.: Are there graphs whose shortest path structure requires large edge weights? Drops-idn/v2/document/10.4230/lipics.itcs.2024.12 (2024). https://doi.org/10.4230/LIPIcs.ITCS.2024.12, https://drops.dagstuhl.de/entities/document/10.4230/LIPIcs.ITCS.2024.12
11. Bu, F., Kang, S., Shin, K.: Interplay between topology and edge weights in real-world graphs: concepts, patterns, and an algorithm. Data Min. Knowl. Discov. **37**(6), 2139–2191 (2023). https://doi.org/10.1007/s10618-023-00940-w
12. Chen, C., Hwang, F.: Detecting and locating electrical shorts using group testing. IEEE Trans. Circuits Syst. **36**(8), 1113–1116 (1989). https://doi.org/10.1109/31.192423, http://ieeexplore.ieee.org/document/192423/
13. Choi, S.S.: Polynomial time optimal query algorithms for finding graphs with arbitrary real weights. In: Proceedings of the 26th Annual Conference on Learning Theory, pp. 797–818. PMLR (2013). https://proceedings.mlr.press/v30/Choi13.html. iSSN 1938-7228
14. Culberson, J.C., Rudnicki, P.: A fast algorithm for constructing trees from distance matrices. Inf. Process. Lett. **30**(4), 215–220 (1989). https://doi.org/10.1016/0020-0190(89)90216-0, https://linkinghub.elsevier.com/retrieve/pii/0020019089902160
15. Fang, S.C.: Detecting electrical shorts on printed circuit boards. Int. J. Prod. Res. **28**(6), 1031–1037 (1990). https://doi.org/10.1080/00207549008942773, https://www.tandfonline.com/doi/full/10.1080/00207549008942773
16. Foucquier, J., Guedj, M.: Analysis of drug combinations: current methodological landscape. Pharmacol. Res. Perspect. **3**(3), e00149 (2015). https://doi.org/10.1002/prp2.149, https://bpspubs.onlinelibrary.wiley.com/doi/abs/10.1002/prp2.149

17. Jagadish, M., Sen, A.: Learning a bounded-degree tree using separator queries. In: Jain, S., Munos, R., Stephan, F., Zeugmann, T. (eds.) Algorithmic Learning Theory, pp. 188–202. Springer, Heidelberg (2013)
18. Janson, S., Luczak, T., Rucinski, A.: Random Graphs. Wiley (2011). Google-Books-ID: RjnqVoB4VmUC
19. Kannan, S., Mathieu, C., Zhou, H.: Graph reconstruction and verification. ACM Trans. Algorithms **14**(4), 1–30 (2018). https://doi.org/10.1145/3199606, https://dl.acm.org/doi/10.1145/3199606
20. Kim, J.H.: Finding weighted graphs by combinatorial search. Arxiv abs/1201.3793 (2012). https://www.semanticscholar.org/paper/6ff61e772de7d2a649c5979ea239bd26b375ec01, arXiv:1201.3793 [cs, math]
21. Krivelevich, M., Zhukovskii, M.: Reconstructing random graphs from distance queries (2024). https://doi.org/10.48550/arXiv.2404.18318, arXiv:2404.18318 [math]
22. Kumar, R., Liu, P., Charikar, M., Benson, A.R.: Retrieving top weighted triangles in graphs. In: Proceedings of the 13th International Conference on Web Search and Data Mining, WSDM 2020, pp. 295–303. ACM, Houston (2020). https://doi.org/10.1145/3336191.3371823, https://dl.acm.org/doi/10.1145/3336191.3371823
23. Liu, X., Mukherjee, S.: Tight query complexity bounds for learning graph partitions. In: Proceedings of Thirty Fifth Conference on Learning Theory, pp. 167–181. PMLR (2022). https://proceedings.mlr.press/v178/liu22a.html. iSSN 2640-3498
24. Mathieu, C., Zhou, H.: A simple algorithm for graph reconstruction. Drops-idn/v2/document/10.4230/lipics.esa.2021.68 (2021). https://doi.org/10.4230/LIPIcs.ESA.2021.68, https://drops.dagstuhl.de/entities/document/10.4230/LIPIcs.ESA.2021.68
25. Mazzawi, H.: Optimally reconstructing weighted graphs using queries: (extended abstract). In: Proceedings of the Twenty-first Annual ACM-SIAM Symposium on Discrete Algorithms, pp. 608–615. Proceedings, Society for Industrial and Applied Mathematics (2010). https://doi.org/10.1137/1.9781611973075.51, https://epubs.siam.org/doi/10.1137/1.9781611973075.51
26. Shi, W., West, D.B.: Diagnosis of wiring networks: an optimal randomized algorithm for finding connected components of unknown graphs. SIAM J. Comput. **28**(5), 1541–1551 (1999). https://doi.org/10.1137/S0097539795288118, http://epubs.siam.org/doi/10.1137/S0097539795288118
27. Shi, W., West, D.B.: Structural diagnosis of wiring networks: finding connected components of unknown subgraphs. SIAM J. Discrete Math. **14**(4), 510–523 (2001). https://doi.org/10.1137/S0895480100371286, http://epubs.siam.org/doi/10.1137/S0895480100371286
28. Stadtfeld, C., Pentland, A.S.: Partnership ties shape friendship networks: a dynamic social network study. Soc. Forces **94**(1), 453–477 (2015). https://doi.org/10.1093/sf/sov079
29. Stegehuis, C., Weedage, L.: Reconstruction of geometric random graphs with the simple algorithm. Arxiv abs/2407.18591 (2024). https://doi.org/10.48550/arXiv.2407.18591, https://www.semanticscholar.org/paper/0487c10a7522b6cc483b7e7c6e2821b31bf581bd, arXiv:2407.18591 [cs, math]
30. Wang, Z., Honorio, J.: Reconstructing a bounded-degree directed tree using path queries (2017). https://arxiv.org/abs/1606.05183
31. Yang, S.H., Smola, A.J., Long, B., Zha, H., Chang, Y.: Friend or frenemy? Predicting signed ties in social networks. In: 35th International ACM SIGIR Conference on Research and Development in Information Retrieval, pp. 555–564. SIGIR (2012). https://doi.org/10.1145/2348283.2348359

Monotone Classes, Even Graphs and the Hamiltonian Cycle Problem

Vadim Lozin(✉)

Mathematics Institute, University of Warwick, Coventry CV4 7AL, UK
V.Lozin@warwick.ac.uk

Abstract. It is known that the Hamiltonian cycle problem can be solved in a minor-closed class X in polynomial time if and only if X excludes a planar graph, or equivalently, if and only if the tree-width is bounded in X (unless $P = NP$). The family of monotone classes of graphs extends the family of minor-classes, and the polynomial-time solvability of the problem in this family extends far beyond graphs of bounded tree-width. However, the complexity of the problem in the family of monotone classes remains a terra incognita even in the case of a single forbidden subgraph. In the present paper, we identify several crucial landmarks in this territory, introduce the notion of even graphs that plays the fundamental role in the study of the Hamiltonian cycle problem, and prove a number of breakthrough results of both positive and negative nature.

Keywords: Hamiltonian cycle · Polynomial algorithm · Monotone class

1 Introduction

A Hamiltonian cycle in a graph is a cycle containing all vertices of the graph. Determining whether a graph has a Hamiltonian cycle is one of the first problems that has been shown to be NP-complete. Moreover, the problem remains NP-complete under substantial restrictions, for instance, for planar graphs [4], graphs of vertex degree at most 3 or $K_{1,5}$-free split graphs [9]. On the other hand, in some classes it admits a polynomial-time solution, for instance, for graphs of bounded tree- [1] or clique-width [3]. From these examples we conclude that there is a simple criterion for polynomial-time solvability of the problem in the family of minor-closed classes of graphs: unless $P = NP$, the problem admits a polynomial-time solution in a minor-closed class X if and only if X excludes (does not contain) at least one planar graph. This conclusion follows from the above discussion and the result of Robertson and Seymour stating that graphs excluding a planar graph as a minor have bounded tree-width [10].

Taking into account the intimate relationship between the Hamiltonian cycle and Travelling Salesman problems, the closedness under vertex and edge deletion in the study of the Hamiltonian cycle problem is well justified, but the closedness under edge contraction is not. By omitting the latter requirement from the definition of minor-closed classes, we obtain the more general family of monotone

H. Fernau and B. Zhu (Eds.): IWOCA 2025, LNCS 15885, pp. 118–131, 2025.
https://doi.org/10.1007/978-3-031-98740-3_9

classes. With this extension the situation becomes substantially more complex in the sense that within the family of monotone classes the Hamiltonian cycle problem admits polynomial-time solutions far beyond classes of bounded tree-width. In particular, no criterion for polynomial-time solvability of the problem on monotone classes is available even in the case of a single excluded subgraph. A preliminary discussion on this topic is presented in Sect. 2. Then in Sect. 3 we introduce the notion of even graphs that plays a fundamental role in the study of the Hamiltonian cycle problem on monotone classes. We prove the NP-completeness of the problem for even graphs and derive a number of important consequences from this conclusion. In Sect. 4, we prove several results related to polynomial-time solvability of the problem on monotone classes. In the rest of the present section, we introduce basic terminology and notation.

We consider only simple (undirected, without loops and multiple edges) graphs. The *neighbourhood* of a vertex v is the set of vertices adjacent to v, and the *degree* of v is the number of its neighbours. A graph is *sub-cubic* if all its vertices have degree at most 3. Vertices of degree exactly 3 will be called *cubic*.

A class of graphs is *monotone* if it is closed under taking subgraphs and is *hereditary* if it is closed under taking induced subgraphs. If a graph G does not belong to a class X, then we say that X *excludes* G, or that G is forbidden for X. Every hereditary class X can be characterized by a set M of minimal forbidden induced subgraphs, i.e. minimal (with respect to the induced subgraph relation) graphs that do not belong to X, in which case we say that graphs in X are M-free. If M is a finite set, then we call X a *finitely-defined* class.

Similarly, every monotone class X can be characterized by a set M of minimal forbidden subgraphs, i.e. minimal (with respect to the subgraph relation) graphs that do not belong to X, in which case we say that graphs in X are M-subgraph-free. Clearly, every monotone class is hereditary, and hence, a monotone class admits a characterization both in terms of minimal forbidden subgraphs and in terms of minimal forbidden induced subgraphs. The two sets of forbidden graphs for a monotone class may coincide or may be different, but what is important is that either both of them are finite or both are infinite, which is not difficult to see. Therefore, we can apply the term 'finitely-defined' to a monotone class without specifying which of the two forbidden sets we refer to.

2 Preliminary Discussion

A helpful criterion for the NP-completeness of an algorithmic problem Π in a finitely-defined hereditary class X is the notion of boundary classes: Π is NP-complete in X if and only if X contains at least one boundary class for Π [5].

The first two boundary classes for the Hamiltonian cycle problem have identified in [5]. One of them is the class Φ of graphs every connected component of which is a caterpillar with hairs of arbitrary length: these are sub-cubic trees in which all cubic vertices belong to the same path (see an example in Fig. 1). Therefore, the Hamiltonian cycle problem is polynomial-time solvable in a finitely-defined class X *only if* X excludes a graph from Φ (unless $P = NP$). Moreover,

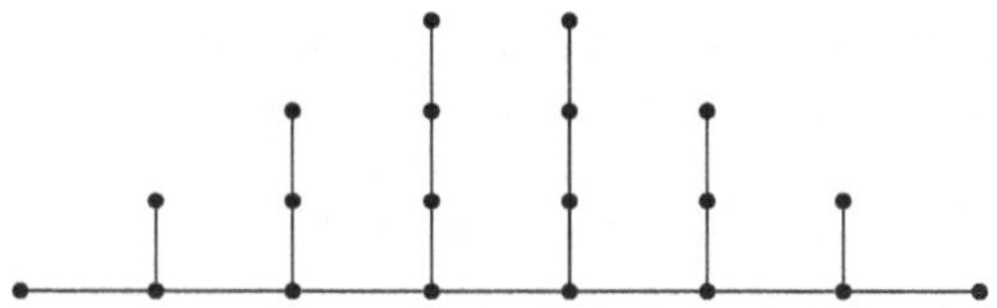

Fig. 1. A caterpillar with hairs of arbitrary length

as we show in [7], in the study of the Hamiltonian cycle problem on monotone classes we can be restricted to *connected* forbidden graphs from Φ.

One more boundary class identified in [5] is the class Φ^{Δ} of graphs that can be obtained from graphs in Φ by replacing each cubic vertex with a triangle and all their induced subgraphs. Together the two boundary classes imply the NP-completeness of the problem for F-subgraph-free graphs for infinitely many graphs $F \in \Phi$, including the two graphs in Fig. 2 [6].

Fig. 2. The graphs Ξ_1 and Ξ_2

To reveal one more boundary class, let us observe that in each graph in Φ every cubic vertex has a non-cubic neighbour. It is not difficult to check that if we apply the transformation represented in Fig. 3 to a cubic vertex for which x is non-cubic, then it transforms G into a graph, which has a Hamiltonian cycle if and only if G has.

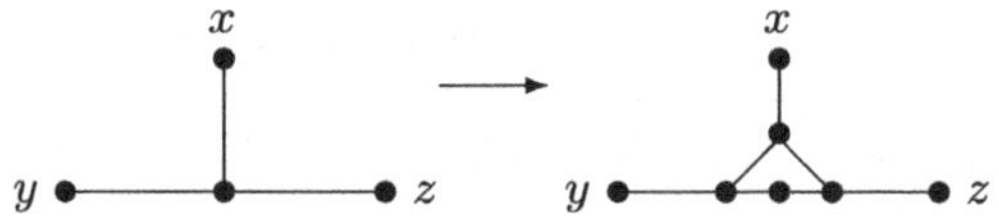

Fig. 3. Replacement of a cubic vertex with a square

We denote by $\Phi^{\square}$ the class of all graphs obtained from graphs in Φ by replacing each cubic vertex with a square (as shown in Fig. 3) and all their induced subgraphs. From the above discussion we conclude that if F is a graph such that none of the graphs containing F as a spanning subgraph belongs to $\Phi^{\square}$, then the Hamiltonian cycle problem is NP-complete in the class of F-subgraph-free graphs. This implies the NP-completeness of the problem in F-subgraph-free graphs for a variety of graphs $F \in \Phi$. Some of them are listed in the following theorem.

Theorem 1. *The Hamiltonian cycle problem is NP-complete for Ξ_3-subgraph-free graphs and for Ξ_4-subgraph-free graphs (see Fig. 4).*

Fig. 4. The graphs Ξ_3 and Ξ_4 in Theorem 1

3 Even Graphs and the Hamiltonian Cycle Problem

Let us call vertices of degree 2 in a graph G *white* and vertices of degree at least 3 *black*. In the study of the Hamiltonian cycle problem we may assume that G has no vertices of degree 0 or 1, since otherwise G has no Hamiltonian cycle.

The *black graph* is a subgraph of G induced by black vertices and a *black component* is a connected component in the black graph. A *white path* is a path all vertices of which are white, except for the endpoints, which are black.

Clearly, every edge incident to a white vertex belongs to any Hamiltonian cycle in G, if G has any. Therefore, we may assume that every black vertex b of G is adjacent to at most two white vertices, because otherwise G has no Hamiltonian cycle. In the study of the problem on monotone classes, we may further assume that b is incident to at most one white vertex. Indeed, if b has exactly two white neighbours, then all other edges incident to b can be deleted, in which case b becomes white.

Definition 1. *An even graph is a connected graph in which every black component is an induced cycle of even length, and every black vertex has exactly one white neighbour.*

The importance of even graphs in the study of the Hamiltonian cycle problem will become clearer in due course. For the time being, let us observe that if an even graph has a Hamiltonian cycle C, then C contains exactly half of the edges in each black component, and this half creates a perfect matching in the black component. Clearly, every even cycle contains exactly two perfect matchings, and therefore, the Hamiltonian cycle problem reduces for an even graph to choosing in each black component one of the two perfect matchings for the inclusion in the solution.

Since the number of black components in even graphs can be arbitrarily large, it is not surprising that the problem is NP-complete for these graphs, which we formally prove in Theorem 2. However, the complexity of the problem depends not only on the number of black components, but also on their sizes. To reveal this dependency, let us denote by Σ_p the set of even graphs in which every black component has at most p vertices.

Theorem 2. *The Hamiltonian cycle problem is NP-complete in the class* Σ_8.

Theorem 2 implies some important corollaries for the Hamiltonian cycle problem on monotone classes. To state one of them, let us denote by H_k the graph represented in Fig. 5. Clearly, H_k belongs to Φ for any $k \geq 1$.

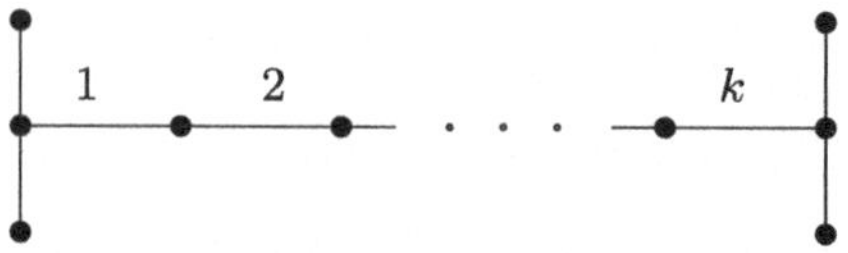

Fig. 5. Graph H_k

Corollary 1. *The Hamiltonian cycle problem is NP-complete in the class of* H_k*-subgraph-free graphs for all* $k \geq 6$.

Proof. Let Σ_8^k be the set of even graphs in Σ_8 in which all white paths have length more than k. From Theorem 2, we conclude that the problem is NP-complete in Σ_8^k for each fixed k. Indeed, every graph in Σ_8 can be transformed into a graph in Σ_8^k by introducing more vertices on short white paths, and this operation does not affect the Hamiltonicity of the input graph.

It is not difficult to see that every graph in Σ_8^k is H_k-subgraph-free for all $k \geq 6$, and hence the problem is NP-complete in the class of H_k-subgraph-free graphs for all $k \geq 6$. □

Together with the results in Sect. 2 this corollary leads to the following interesting conclusion.

Corollary 2. *The Hamiltonian cycle problem is NP-complete in the class of* F*-subgraph-free graphs for any connected graph* $F \in \Phi$ *with at least* 5 *hairs.*

To determine the complexity of the problem in H_k-subgraph-free graphs for $k < 6$, we need to solve the following problem.

Σ_6**-problem.** Determine the complexity of the Hamiltonian cycle problem in Σ_6.

If the Σ_6-problem is NP-complete, then the Hamiltonian cycle problem is NP-complete in H_k-subgraph-free graphs for $k = 4, 5$ by analogy with Corollary 1. We claim that the converse also is true, i.e. the Hamiltonian cycle problem in H_k-subgraph-free graphs for $k = 4, 5$ can be reduced in polynomial time to graphs in Σ_6. This reduction is long and is beyond the scope of this paper.

In Σ_4, the Hamiltonian cycle problem can be solved in polynomial time, as we shall see in the next section, and this conclusion is an important ingredient in various polynomial-time results for the problem in monotone classes.

4 Polynomial-Time Results

In the preceding sections, we presented a collection of negative results, i.e. the results that deal with the NP-complete cases of the problem. Knowing what is negative without knowing what is positive is just half-knowledge. Therefore, in this section we turn to polynomial-time results.

Let F be a connected graph in Φ and let X be the class of F-subgraph-free graphs. Assuming that the Σ_6-problem is NP-complete and that X is of unbounded tree-width, we conclude, with the help of the results in the previous sections, that the Hamiltonian cycle problem can be solved in polynomial time in X only if F is one of the following four graphs H_1, H_2, H_3, T (Fig. 6) or a graph obtained from H_1, H_2, H_3, T by subdividing their pendant edges.

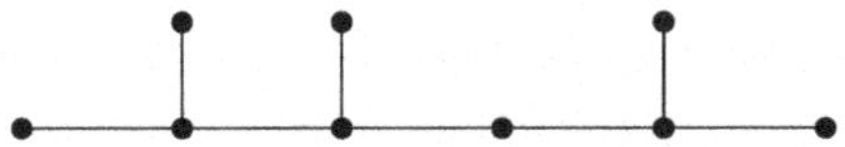

Fig. 6. The graph T

For the T-subgraph-free graphs, the Hamiltonian cycle problem was solved in polynomial time in [8]. Observe that T contains all of H_1, H_2, H_3 as (induced) subgraphs. Therefore, in what follows we analyze subdivisions of H_1, H_2, H_3, T.

Let us denote by F^k the graph obtained from F by subdividing each of its pendant edges by k new vertices. We note that in the class of T^1-subgraph-free graphs the problem is NP-complete by Theorem 1, as T^1 contains Ξ_3.

In the class of H_1^k-subgraph-free graphs, the problem can be solved in polynomial time for all values of k. To outline the argument, let us introduce the following terminology, which will be used throughout this section.

Definition 2. *Assume the input graph G contains a Hamiltonian cycle C. An edge of G that does not belong to C will be called a chord. A chord splits C into two paths. The length of a shortest of them will be called the length of the chord.*

It is not difficult to see that if an H_1^k-subgraph-free graph has a Hamiltonian cycle, then every chord in this cycle has length less than $2k$. This observation together with the following theorem proved in [6] solve the problem for H_1^k-subgraph-free graphs for each k.

Theorem 3. *If every chord in a Hamiltonian cycle (or path) has length at most k, then the graph has tree-width at most $2k$.*

In [6], the problem was solved for H_2^1-subgraph-free graphs by showing that for graphs in this class containing a Hamiltonian cycle,

($\star$) there is a threshold k such that long chords (of length more than k) do not cross (when drawn inside the cycle) short chords (of length at most k).

This property allows to separate graphs with Hamiltonian cycles where all chords are short from graphs where all chords are long. For graphs of the first type (short chords), the problem can be solved according to Theorem 3. For H_2^1-subgraph-free graphs with long chords, the Hamiltonian cycle problem effectively reduces to graphs in Σ_4.

Unfortunately Property ($\star$) destroys for H_2^k-subgraph-free graphs for $k > 1$ and for H_3^k-subgraph-free graphs for $k \geq 1$. Nonetheless, as we show in Sect. 4.2, the problem admits a polynomial-time solution for H_3^1-subgraph-free graphs. Observe that by adding to H_3^1 four more vertices, we obtain the graph Ξ_4 (Fig. 4), in which case the Hamiltonian cycle problem becomes NP-complete.

Similarly to the solutions for H_2^1-subgraph-free graphs, the solution for H_3^1-subgraph-free graphs includes two main ingredients. One of them deals with graphs of bounded tree-width and the other can be viewed as an extension of graphs in Σ_4. Together they form a general framework, which is of independent interest and which is described in Sect. 4.1. The reduction from general H_3^1-subgraph-free graphs to graphs that satisfy the framework is presented in Sect. 4.2.

Throughout this section we use the following terminology. A *grey component* is an extension of a black component B obtained by adding to it all white paths that connect a vertex in B to a vertex in B.

Assume the input graph G contains a Hamiltonian cycle C. A maximal section of consecutive vertices of C that belong to the same grey component will be called a *grey interval*, or simply an *interval*. According to our assumption, every black vertex is adjacent to at most one white vertex, and hence every grey interval consists of at least 2 vertices.

Let B be a grey component of G. We will call the vertices of B that have neighbours outside of B the *contact vertices*. Note that B may consists of one or more intervals. Each interval gives rise to exactly two contact vertices, and hence the number of contact vertices in B must be even, since otherwise the graph has no Hamiltonian cycle.

4.1 Hamiltonian Cycles in Σ_4 and More General Classes

In this section, we discuss the problem for graphs in which every grey component has 2 or 4 contact vertices. Obviously, every graph in Σ_4 possesses this property.

Assume first that a grey component B has exactly 2 contact vertices. Then we apply the following procedure.

(α) If B has exactly two contact vertices, then check if B has a Hamiltonian path connecting the contact vertices. If such a path does not exist, then stop: the input graph has no Hamiltonian cycle. Otherwise, choose arbitrarily a Hamiltonian path connecting the contact vertices, include the edges of the path in the solution and delete all other edges from B.

Procedure (α) destroys grey components with two contact vertices, i.e. after its implementation all vertices in such components become white. Observe that this may affect the structure of other grey components by reducing the number of their contact vertices. However, for an easier description, we *do not update* the structure of other grey components.

Suppose now that a grey component B has four contact vertices v_1, v_2, v_3, v_4. If G has a Hamiltonian cycle, then the vertices of B can be partitioned into two parts each of which forms a path connecting a pair of contact vertices. We will call such a partition a *pairing* (of contact vertices) and will refer to a pairing as the set of edges in the two paths. Also, we will say that two pairings are of the same type, if they pair the contact vertices in the same way. Clearly, if all possible pairings in B have the same type, then it is irrelevant which one to choose, since non-contact vertices of B have no neighbours outside of B. Note that a pairing may disconnect the graph after the removal of edges not in the pairing. However, if B admits pairings of at least two different types, then at least one of them necessarily keeps the graph connected.

Lemma 1. *If B is a grey component that admits pairings of at least two different types, then at least one of them does not disconnect the graph.*

Proof. Let B be a grey component with contact vertices v_1, v_2, v_3, v_4 and let L and R be two pairings of different types, say L pairs v_1 with v_2 and v_3 with v_4, while R pairs v_1 with v_3 and v_2 with v_4. Assume to the contrary that

(1) the deletion of all edges of B except for the edges of L disconnects the graph into two components C_{12} (containing vertices v_1, v_2) and C_{34} (containing vertices v_3, v_4), and
(2) the deletion of all edges of B except for the edges of R disconnects the graph into two components C_{13} (containing vertices v_1, v_3) and C_{24} (containing vertices v_2, v_4).

Note that (1) separates v_1 from v_3 and v_4, while (2) separates v_1 from v_4. Therefore, after the deletion of *all* edges of B vertex v_1 is separated from all other contact vertices. In other words, after the deletion of all edges of B, vertices v_1, v_2, v_3, v_4 belong to pairwise different connected components, say V_1, V_2, V_3, V_4, respectively.

We observe that in each connected component V_i vertex v_i has degree 1 (it is adjacent to a white vertex only). Any other vertex of odd degree in V_i (if there is any) is black, i.e. appears in some grey component B'. In the graph $G[B']$ induced by B' the number of odd vertices is even (by the Handshake lemma). Attaching to $G[B']$ four white neighbours changes the parity of exactly four vertices of B' and hence leaves the number of vertices of B' with odd degrees in the graph G even. Since all vertices of B' belong to only one of the components V_i, we conclude that in each component V_i the number of vertices of odd degree is odd. This is not possible by the Handshake lemma and hence either (1) or (2) is not valid, i.e. we can keep one of the pairings and delete all other edges of B without disconnecting G. □

Lemma 1 justifies the following procedure.

(β) If B has exactly four contact vertices, then
   (β_1) check if B admits any pairing. If not, then stop: the input graph has no Hamiltonian cycle.
   (β_2) If all pairings in B have the same type, then choose any of them, include the edges of the pairing in the solution and delete all other edges from B. If this operation disconnects the graph, then stop: the input graph has no Hamiltonian cycle.
   (β_3) If B admits pairings of at least two different types, then choose one of them that does not disconnect the graph, include the edges of the pairing in the solution and delete all other edges from B.

Procedure (β), similarly to (α), destroys grey components to which it applies, i.e. all vertices of the component become white, and again we *do not update* the structure of grey components affected by this transformation. However, it is important to emphasize that β_3 must be applied to black components one by one, since the choice of a pairing in one of such components may affect the choice of pairings in the others. The above discussion leads to the following conclusion.

Theorem 4. *Let G be a graph in which every grey component has* 2 *or* 4 *contact vertices. If Procedures (α) and (β) can be implemented in polynomial time, then the Hamiltonian cycle problem can be solved in polynomial time for G.*

It is not difficult to see that for graphs in Σ_4, every grey component with two contact vertices necessarily has a Hamiltonian path connecting them, and such a path can be easily detected, and that every grey component with four contact vertices admits two pairings of different types, and a pairing that does not disconnect the graph can be easily identified. As a result, we make the following conclusion.

Corollary 3. *Every graph in Σ_4 has a Hamiltonian cycle and such a cycle can be found in polynomial time.*

4.2 A Polynomial Algorithm for H_3^1-Subgraph-Free Graphs

In this section, we develop a polynomial-time algorithm for the Hamiltonian cycle problem in the class of H_3^1-subgraph-free graphs. For ease of notation, throughout the section we denote H_3^1 by R. In the proof, we will refer to any copy of the forbidden subgraph as a sequence $(v_1, v_2, \ldots, v_{12})$ of vertices ordered according to Fig. 7.

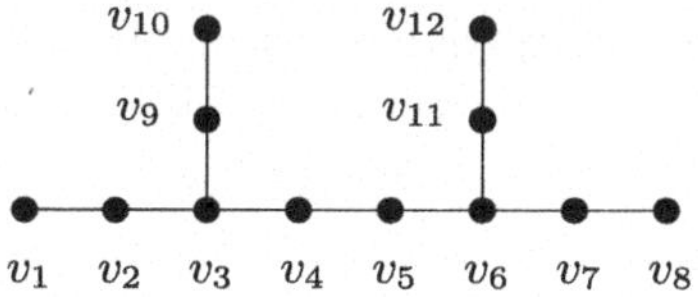

Fig. 7. The graph $R = H_3^1$

We start with some helpful rules to solve the problem and a set of reductions simplifying the input graph. We emphasize that by deleting an edge or a vertex from an R-subgraph-free graph, we obtain an R-subgraph-free graph again. The following six rules and reductions are obvious.

(R1) If the graph has vertices of degree 0 or 1, then stop: G has no Hamiltonian cycle.
(R2) If the graph is disconnected, then stop: G has no Hamiltonian cycle.
(R3) If the graph contains a vertex adjacent to more than two white vertices, then stop: G has no Hamiltonian cycle.
(R4) If the graph contains a vertex x adjacent to exactly two white vertices, then delete the edges connecting x to all other its neighbours (if there are any).
(R5) If two adjacent black vertices x and y are connected by a white path, then delete the edge xy.
(R6) If the graph contains a white path with at least 4 white vertices, then replace it with a path containing 3 white vertices (it is not difficult to see that this operation does not create the forbidden graph and does not change the Hamiltonicity of the graph).

Now we introduce a reduction applicable to a graph G containing an induced subgraph shown on the left in Fig. 8, in which vertices a, b, c have degree 3 in G. The reduction depends on the degree of x. If the degree of x also is 3, the reduction consists in deleting the edges ab and xc. Otherwise, it transforms the graph as shown in Fig. 8. We will refer to this reduction as the *diamond reduction* and will denote it by (R7).

Fig. 8. The diamond reduction.

Claim 1. Let G' be a graph obtained from G by the diamond reduction. Then G has a Hamiltonian cycle if and only if G has. Moreover, if G is R-subgraph-free, then so is G'.

One more reduction is illustrated in Fig. 9. It is applicable to a graph G containing an induced subgraph shown on the left, in which vertices a, b, c have degree 3 in G, and moreover, a and b have white neighbours. We will refer to this reduction as the *butterfly reduction* and will denote it by (R8).

Fig. 9. The butterfly reduction.

Claim 2. Let G' be a graph obtained from G by the butterfly reduction. Then G has a Hamiltonian cycle if and only if G has.

In our algorithm to solve the problem we implement the above rules and reductions whenever they are applicable. In what follows, we show that if G contains a Hamiltonian cycle, then every grey component

(A) either is small (contains at most 18 vertices) and has exactly 4 contact vertices
(B) or has bounded tree-width (at most 28) and at most 2 contact vertices.

To this end, we prove a number of results under the assumption that G has a Hamiltonian cycle C. The *C-distance* between two vertices of G is the distance between them in the cycle C. To prove (A), we will show that every grey component B containing either a chord of length at least 15 or a chord connecting two distinct intervals of B is small (contains at most 18 vertices) and has exactly 4 contact vertices. Then (B) will follow from (A) and Theorem 3. The proof will be given through a sequence of technical claims. In the next two claims we denote the vertices of the graph listed along the cycle C by $x_0, x_2, \ldots, x_{n-1}$.

Claim 3. Let x_0x_k be a chord of length at least 8. Then

x_1 is adjacent to x_j only if $j \in \{n-1, 3, k-4, k-2, k-1, k, k+1, k+2, k+4\}$,
x_2 is adjacent to x_j only if $j \in \{0, 4, k-2, k-1, k, k+1, k+2\}$,
x_3 is adjacent to x_j only if $j \in \{0, 1, 5, k\}$.

Remark. In this claim, we required that both parts of the cycle C defined by the chord x_0x_k are sufficiently large (have length at least 8). This requirement is sufficient but not necessary and in most cases the claim remains valid even if one of the parts (or both) is smaller than 8. For instance, the first sentence of the proof remains valid even if the path $(x_k, x_{k+1}, \ldots x_{n-1}, x_0)$ has length 3. In what follows we will apply Claim 3 taking this remark into account.

Claim 4. If x_0x_k is a chord of length at least 11, then x_1 is not adjacent to x_{k-4}, x_{k+4}.

Claims 3 and 4 show that in the presence of a sufficiently long chord there are severe restrictions on the neighbourhoods of the vertices located near the endpoints of the chord. The restrictions become even more severe if we have two long chords close to each other, as in the four subgraphs shown in Fig. 10. These four subgraphs will play a special role in our solution, so we will refer to them as *special subgraphs.* The importance of these structures comes from the fact that grey components containing them are small. This will be proved in several steps (claims). In the proofs, we denote the vertices of the graph by following the patterns shown in Fig. 10.

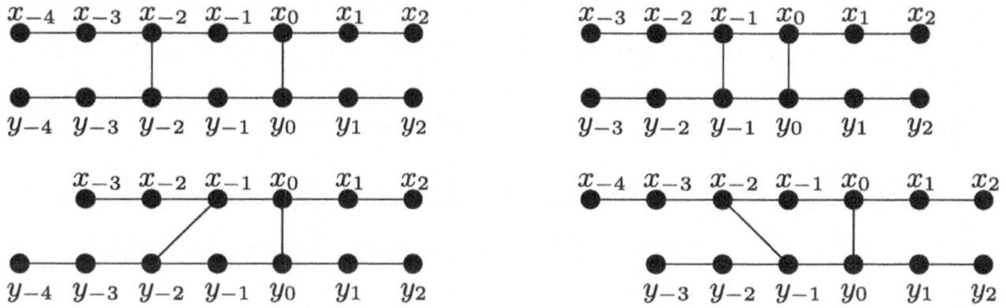

Fig. 10. Four special subgraphs: the horizontal edges belong to the Hamiltonian cycle C and the remaining two edges are chords.

Claim 5. Let G contain a special subgraph represented in Fig. 10. If the length of the chord x_0y_0 is at least 11, then vertex x_3 can be adjacent only to x_1 (in addition to x_2 and x_4).

Claim 6. Let G contain a special subgraph represented in Fig. 10. If the length of the chord x_0y_0 is at least 11, then at least one of x_3 and x_4 is white.

Claim 7. If x_0y_0 is a chord of length at least 13, then G contains a special subgraph containing x_0 and y_0.

Claim 8. If x_0y_0 is a chord of length at least 15, then the grey component B containing x_0 and y_0 consists of at most 18 vertices and has 4 contact vertices.

Claim 9. If a grey component has at least 4 contact vertices, then it consists of at most 18 vertices and has exactly 4 contact vertices.

Corollary 4. *If B is a grey component with more than* 18 *vertices, then the tree-width of of the subgraph of G induced by B is at most* 28 *and B has at most* 2 *contact vertices.*

Proof. If B contains more than 18 vertices, then it has at most 2 contact vertices by Claim 9 Also, by Claim 8 every chord in B has length at most 14 and hence the tree-width of $G[B]$ is at most 28 by Theorem 3. □

Theorem 5. *The Hamiltonian cycle problem can be solved for R-subgraph-free graphs in polynomial time.*

Proof. Let G be an R-subgraph-free graph. To avoid trivialities, we assume that the number of vertices in G is sufficiently large, say at least 20. Otherwise, the problem can be solved by brute-force.

To determine if G has a Hamiltonian cycle, we first apply to it rules and reductions (R1)–(R8) as long as they are applicable. If the algorithm does not stop at this stage, we partition the graph into grey components and white paths connecting them.

Assume first that all vertices of G belong to the same grey component. Then the graph has a Hamiltonian cycle *only if* all chords have length at most 14 (by Claim 8 and hence the tree-width of G is at most 28 (by Theorem 3). In linear time we can check if the tree-width of G is at most 28 (see [2]), and if not, then we conclude that G has no Hamiltonian cycle, and if yes, we solve the problem by an algorithm for graphs of bounded tree-width [1].

Now suppose that G has at least two grey components. If at least one of them has an odd number of contact vertices, or has more than 4 contact vertices, or contains more than 18 vertices and has more than 2 contact vertices, or contains more than 18 vertices and has tree-width more than 28, then we conclude that G has no Hamiltonian cycle (by Claims 8 and 9 and Corollary 4). Otherwise, the problem can be solved in polynomial time by Theorem 4, because in this case Procedures (α) and (β) can be implemented in polynomial time. In particular, a Hamiltonian path in a grey component of bounded tree-width can be detected in polynomial time [1]. □

5 Concluding Remarks and Open Questions

In this paper, we proved a number of negative and positive results related to the Hamiltonian cycle problem in monotone classes of graphs. However, a variety of questions in this area remains open. Perhaps the most intriguing one is the complexity of the problem in the class Σ_6. Assuming that the problem is NP-complete in this class, the results of this paper together with the solution for H_2^1-subgraph-free graphs [6] lead to the following conditional dichotomy.

Proposition 1. *The Hamiltonian cycle problem in the class of H_k^1-subgraph-free graphs can be solved in polynomial time if and only if $k \leq 3$ (assuming $P \neq NP$ and the Σ_6-problem is NP-complete).*

Whether the solutions for H_2^1-subgraph-free and H_3^1-subgraph-free graphs can be extended to H_2^k-subgraph-free and H_3^k-subgraph-free graphs for larger values of k is another open question.

The last potential point for improvement (under the assumption that the Σ_6-problem is NP-complete) is the solution for T-subgraph-free graphs (see Fig. 6). Theorem 1 shows that this solution cannot be extended too much (compare Ξ_3 and T). To what extent can the solution for T-subgraph-free graphs be extended is one more challenging question.

References

1. Arnborg, S., Proskurowski, A.: Linear time algorithms for NP-hard problems restricted to partial k-trees. Discrete Appl. Math. **23**, 11–24 (1989)
2. Bodlaender, H.L.: A linear time algorithm for finding tree-decompositions of small treewidth. SIAM J. Comput. **25**, 1305–1317 (1996)
3. Borie, R.B., Parker, R.G., Tovey, C.A.: Solving problems on recursively constructed graphs. ACM Comput. Surv. **41**, 4 (2008)
4. Garey, M.R., Johnson, D.S., Tarjan, R.E.: The planar Hamiltonian circuit problem is NP-complete. SIAM J. Comput. **5**, 704–714 (1976)
5. Korpelainen, N., Lozin, V., Malyshev, D., Tiskin, A.: Boundary properties of graphs for algorithmic graph problems. Theoret. Comput. Sci. **412**, 3545–3554 (2011)
6. Lozin, V.: The Hamiltonian cycle problem and monotone classes. In: Lecture Notes in Computer Science, vol. 14764, pp. 460–471 (2024)
7. Lozin, V.: Graph problems and monotone classes. Discrete Appl. Math. (submitted)
8. Vadim Lozin, B.M., Pandey, S., Paulusma, D., Siggers, M., Smith, S., van Leeuwen, E.J.: Complexity framework for forbidden subgraphs II: edge subdivision and the "H"-graphs (2022). https://doi.org/10.4230/LIPIcs.ISAAC.2024.47
9. Renjith, P., Sadagopan, N.: Hamiltonian cycle in $K_{1,r}$-free split graphs - a dichotomy. Internat. J. Found. Comput. Sci. **33**, 1–32 (2022)
10. Robertson, N., Seymour, P.D.: Graph minors. V. Excluding a planar graph. J. Combin. Theory Ser. B **41**, 92–114 (1986)

Covering Vertices by 4^+-Paths: A Simpler Local Search Coupled with a More Delicate Amortization

Mingyang Gong[1], Guangting Chen[2](✉), Guohui Lin[1](✉), Eiji Miyano[3], and Abbinash Ranjitkar[1]

[1] Department of Computing Science, University of Alberta, Edmonton, Canada
{mgong4,guohui,ranjitka}@ualberta.ca

[2] Zhejiang University of Water Resources and Electric Power, Hangzhou, China
gtchen@hdu.edu.cn

[3] Kyushu Institute of Technology, Iizuka, Japan
miyano@ai.kyutech.ac.jp

Abstract. We consider the problem to cover the maximum number of vertices in a graph using a collection of vertex-disjoint long paths, where by "long" each path has at least four vertices. We propose a simple local search operation to examine the neighborhoods of up to four extra vertices for every pair of existing paths to seek for improvement, resulting in an $O(|V|^7)$-time $\frac{5}{3}$-approximation algorithm. The performance analysis is done via a delicate amortization scheme, in which the vertices in the computed solution are partitioned into four groups in order to receive tokens from the optimal solution. The novelty in the amortization scheme is to allow a vertex with a larger group index to receive more tokens than a vertex with a smaller group index, so that for each path in the computed solution, its total received tokens are well balanced and are shown to be no more than $\frac{5}{3}$ times its order.

Keywords: Vertex covering · path cover · local search · amortized analysis · approximation algorithm

1 Introduction

Given a simple graph $G = (V, E)$, a *path cover* is a collection of vertex-disjoint paths such that each vertex of G is on one of the paths. The Path Cover problem [6] aiming to minimize the number of paths has been well-studied and the readers can refer to [1,2,7,17,18] for major milestones and those on the related problems. The *Hamiltonian path* problem is an extreme special case of Path Cover, which asks for the existence of a single path that contains all the vertices. An interesting intermediate variant between these two problems, referred to as "*path cover by long paths*", seeks for covering the maximum number of vertices in the graph, where by "long" it means every path contains at least a certain number of vertices [10,11].

H. Fernau and B. Zhu (Eds.): IWOCA 2025, LNCS 15885, pp. 132–145, 2025.
https://doi.org/10.1007/978-3-031-98740-3_10

Given an integer $k \geq 1$, a k-path (k^+-path, resp.) contains exactly k (at least k, resp.) vertices. We denote the variant as MPC^{k+}, which formally claims every k^+-path to be long and seeks for covering the maximum number of vertices by a collection of vertex-disjoint long paths. One sees that when $k = 1$, the problem is trivial; but when k is part of the input, the decision version of the problem is the *longest path* problem, which includes the Hamiltonian path problem and is NP-complete. It is then interesting to explore the case where k is a fixed constant greater than 1.

The problem MPC^{2+} is equivalent to finding a path cover to minimize the number of 1-paths, which can be computed in $O(|V||E|)$ time [4]. MPC^{3+} is shown polynomial-time solvable too [3]. In fact, when $k = 3$, any vertex on a 3^+-path has degree at most 2 and at least one of them has degree exactly 2; that is, a 3^+-path is exactly a 2-*piece* defined in [12]. Hartvigsen at al. [12] have presented a polynomial-time algorithm for the 2-piece packing problem, to cover the maximum number of vertices. When $k \geq 4$, MPC^{k+} is NP-hard [14].

As observed by Gong et al. [10,11], MPC^{k+} can be seen as a special case of the maximum weighted $(2k-1)$-set packing problem [6]. The state-of-the-art $(k - \frac{1}{63,700,992} + \epsilon)$-approximation algorithm for the latter problem [13,16] thus is applicable for MPC^{k+}. In [10,11], Gong et al. present the first $O(|V|^{k+1})$-time $(0.4394k + 0.6576)$-approximation algorithm directly for MPC^{k+}, which is 2.4- and $\frac{19}{7}(\approx 2.714)$-approximation when $k = 4$ and 5, respectively. The algorithm applies three local operations to increase the number of covered vertices, by examining some special local neighborhoods containing up to k extra vertices of an existing long path. Additionally, Gong et al. [10,11] propose two more local operations for $k = 4$, to either increase the number of covered vertices or keep it unchanged but increase the number of 4-paths in the solution. This results in an $O(|V|^8)$-time 2-approximation algorithm for MPC^{4+}.

In [8], Gong et al. propose a completely different approach for MPC^{4+} that results in an improved $O(\min\{|V|^5, |E|^2|V|^2\})$-time 1.874-approximation algorithm. Their novel observation is that the number of end-vertices in a maximum matching (i.e., twice the size of the maximum matching), which can be computed in $O(\sqrt{|V|}|E|)$ time [15], is comparable to the optimum of MPC^{4+}. Briefly, the algorithm greedily connects the edges of a maximum matching M to the vertices not in $V(M)$ to form as many 5-paths or *bad components* as possible, where a bad component contains exactly one edge of M but cannot form a 4^+-path by itself. It then computes a *path-cycle cover* to interconnect the maximum number of bad components and to form 4^+-paths locally. If the obtained solution fails to involve a sufficiently large portion of the end-vertices in M, then the algorithm recursively finds an additional collection of 4^+-paths in a smaller remainder subgraph of G. These design and analysis techniques were non-trivially extended by Gong et al. [9] to an $O(|V|^{2.5}|E|^2)$-time 2.511-approximation algorithm for MPC^{5+}, in which the bad components involve *bi-stars* (a subgraph containing exactly two vertices of degree 2 or greater).

1.1 Our Contribution

We fix a graph $G = (V, E)$ for discussion. In this paper, we switch back to local search and propose a simpler local search algorithm for MPC^{4+} that has only one operation. Our algorithm starts with an empty solution; the operation examines the neighborhoods containing up to four extra vertices of a combination of at most two existing paths in the current solution and seeks to cover more vertices in such a neighborhood. This way, we contribute an $O(|V|^7)$-time $\frac{5}{3}$-approximation algorithm for MPC^{4+}.

Our operation includes most operations proposed in the literature, such as the ones in [10,11], as special cases, but it is easier to present, to be detailed in Sect. 3. The approximation ratio of $\frac{5}{3}$, however, is proved by a more delicate amortization scheme. Let $\mathcal{P}$ denote the solution produced by our algorithm and $\mathcal{Q}$ denote a fixed optimal solution. The novel design idea in the amortization scheme is to partition the vertices of (i.e., *covered by*) $\mathcal{P}$ into four groups, so that a vertex in a group of a larger index can be assigned with more tokens. For the token assignment, each vertex $v \in V(\mathcal{Q})$ has exactly one token, and the token is distributed to some vertices of $V(\mathcal{P})$ that are "close" to v, in the way that a larger fraction of the token is assigned to the one with a larger group index. This way, we are able to balance all the cases of token distribution and show that each path $P \in \mathcal{P}$ receives at most $\frac{5}{3}|V(P)|$ tokens. The amortization scheme is presented in Sect. 4, and the performance analysis of the algorithm is in Sects. 5 and 6. We conclude the paper in Sect. 7.

Due to the space limit, we choose to elaborate our algorithm and the design ideas in details, but omit the proofs of the technical lemmas.

2 Basic Definitions

We first introduce some basic definitions. A graph G' is a subgraph of $G = (V, E)$ if its vertex and edge sets are subsets of V and E, respectively. Given a subgraph G', we use $V(G')$ and $E(G')$ to denote the vertex and edge sets of G', respectively, where $|V(G')|$ is called the *order* of G'.For any subset $V' \subseteq V$, let $E' = \{u\text{-}v \in E \mid u, v \in V'\}$ and the subgraph $G[V'] = (V', E')$ is *induced by* V'.

A *path* is a subgraph of G such that the vertices can be ordered as a sequence and the edge set consists of the edges connecting two consecutive vertices. Given a positive integer k, a k-path has order exactly k and a k^+-path has order at least k. In the following, a (2ℓ)-path is presented as $u_0\text{-}\ldots\text{-}u_{\ell-1}\text{-}v_{\ell-1}\text{-}\ldots\text{-}v_0$ and a $(2\ell+1)$-path is presented as $u_0\text{-}\ldots\text{-}u_\ell\text{-}v_{\ell-1}\text{-}\ldots\text{-}v_0$. The vertices u_0, v_0 are the endpoints of the path and one sees that the vertex subscripts indicate the distance from the closest endpoint (defined as the number of edges in between). The vertex groups are defined using their indices, as will be seen in Definition 1. Given a path P and an external vertex $u \in V \setminus V(P)$, if u is adjacent to an endpoint of P in G, then the notation $u\text{-}P$ or $P\text{-}u$ denotes the path extending P to u, depending on to which endpoint u is adjacent. Similarly, if u, v are adjacent to the two endpoints of P, respectively, then $u\text{-}P\text{-}v$ denotes the path extending P from one end to u and from the other end to v.

3 The Algorithm

We introduce the local search operation first, followed by the algorithm.

We use $\mathcal{P}$ to denote the current solution during the algorithm, which is initialized empty and becomes the output at termination. Since an 8^+-path can be cut into a 4-path and another vertex-disjoint path, we always assume the order of every path in $\mathcal{P}$ is in between 4 and 7. In the operation, we select up to 4 vertices in $V \setminus V(\mathcal{P})$ and up to two paths P, P' in $\mathcal{P}$, and construct the subgraph G_1 induced by these vertices. One sees that the order of G_1 is at most $4 + 2 \cdot 7 = 18$. We then enumerate all possible feasible solutions for G_1 in $O(1)$ time to find an optimal solution. If the optimal solution is better than $\{P, P'\}$ (i.e., covers more vertices in G_1), then the operation replaces the two paths P, P' by the optimal solution, thus to update $\mathcal{P}$.

Operation 1. *Given a collection $\mathcal{P}$ of vertex-disjoint 4^+-paths, select up to 4 vertices in $V \setminus V(\mathcal{P})$ and at most two paths P, P' in $\mathcal{P}$. Let G_1 be the subgraph induced by these selected vertices. The operation replaces $\{P, P'\}$ by an optimal solution (which is a collection of vertex-disjoint 4^+-paths of order up to 7) for G_1 if it covers more vertices than $\{P, P'\}$.*

We remark that Operation 1 includes previously proposed four local operations "Add", "Rep", "DoubleRep" and "Look-ahead" in [10, 11] as special cases. The "Re-cover" operation in [10, 11] is proposed to increase the number of 4-paths in the solution. Operation 1 doesn't aim at this side-objective, and one will see from the later analysis that this sub-objective does not affect the approximation ratio of $\frac{5}{3}$.

Our local search algorithm is denoted as Approx1, which iteratively applies Operation 1 to improve $\mathcal{P}$ and terminates when the operation is not applicable.

Lemma 1. *The algorithm* Approx1 *runs in $O(|V|^7)$ time.*

Proof. One sees that there are $O(|V|^6)$ choices of up to 4 vertices in $V \setminus V(\mathcal{P})$ and at most two paths in $\mathcal{P}$, i.e., each iteration takes $O(|V|^6)$ time as computing an optimal solution for the induced subgraph resulted by each choice is done in $O(1)$ time. It is clear that there are $O(|V|)$ iterations, and therefore the algorithm Approx1 terminates in $O(|V|^7)$ time. □

For ease of presentation, we use the following set of notations.

Notation 1. *In the sequel performance analysis, $\mathcal{P}$ denotes the solution returned by the algorithm* Approx1 *and $\mathcal{Q}$ denotes a fixed optimal solution for the input graph G, which is assumed without loss of generality to be $G = (V(\mathcal{P}) \cup V(\mathcal{Q}), E(\mathcal{P}) \cup E(\mathcal{Q}))$. The graph $\mathcal{Q} - \mathcal{P}$ is the one obtained from $\mathcal{Q}$ after removing all the vertices of $V(\mathcal{P})$ and their incident edges in $E(\mathcal{Q})$. As a result, each connected component of $\mathcal{Q} - \mathcal{P}$ is a path, which is a sub-path of some path in $\mathcal{Q}$.*

4 The Amortization Scheme

In the amortization scheme, each vertex of $V(\mathcal{Q})$ has exactly one token and we present how to distribute the tokens of $V(\mathcal{Q})$ to $V(\mathcal{P})$. Later in the performance analysis, we estimate the total tokens received by each path in $\mathcal{P}$, which are the upper bounds on the approximation ratio. In more details, we first partition the vertices of $V(\mathcal{P})$ into four groups in Definition 1, guided by the idea that a vertex in a group of a larger index can later receive more token. The following lemma discusses the vertices u_2, v_2 on a 6-path in $\mathcal{P}$, as an example to inspire the partitioning into different groups in Definition 1.

Lemma 2. *Suppose u-v is an edge on a path $P \in \mathcal{P}$. Then u, v cannot be adjacent to a common vertex of $V \setminus V(\mathcal{P})$. Moreover, if P is a 6-path, then at most one of u_2, v_2 can be adjacent to a vertex of $V \setminus V(\mathcal{P})$.*

By Lemma 2, if u_2, v_2 are on a 6-path in $\mathcal{P}$, then at most one of u_2, v_2 can be adjacent to a vertex of $V \setminus V(\mathcal{P})$. According to this property and the similar ones, we group the vertices of $V(\mathcal{P})$ as follows.

Definition 1. *(Grouping) We partition the vertices of $V(\mathcal{P})$ into the following four groups:*

1. *The group H_0 contains u_0, v_0 of all paths and u_2, v_2 of 7-paths;*
2. *the group H_1 contains u_1, v_1 of all paths;*
3. *the group H_2 contains u_2 of 5-paths;*
4. *the group H_3 contains u_3 of 7-paths.*
5. *For the vertices u_2, v_2 on a 6-path, at most one of them is adjacent to a vertex of $V \setminus V(\mathcal{P})$ in $\mathcal{Q}$ (by Lemma 2). If u_2 (v_2, resp.) is such a vertex, then we assign u_2 and v_2 (v_2 and u_2, resp.) to the group H_2 and H_0, respectively; if none of u_2, v_2 is such a vertex, then we assign both of u_2, v_2 to H_2.*

Those u_2, v_2 on a 6-path or 7-path that are assigned to the group H_0 are defined as special.

Note that in item 5 of Definition 1, when none of u_2, v_2 is adjacent to any vertex of $V \setminus V(\mathcal{P})$ in $\mathcal{Q}$, both of them are assigned to the group H_2. The reason behind is that, in this case, u_2, v_2 do not receive tokens from $V \setminus V(\mathcal{P})$ and therefore they are set up to take over some tokens for the vertices in $H_0 \cup H_1$.

Lemma 3. *For each path $Q \in \mathcal{Q}$, $V(Q) \cap V(\mathcal{P}) \neq \emptyset$; for any path $R \in \mathcal{Q} - \mathcal{P}$, $|V(R)| \leq 3$.*

Lemma 4. *Suppose u-R is a path in G where u is on a path $P \in \mathcal{P}$ and $R \in \mathcal{Q} - \mathcal{P}$. If $u \in H_i$, then $|V(R)| \leq i$.*

In our amortization scheme, a vertex $v \in V(\mathcal{P})$ receives tokens from its "close" vertices in $V(\mathcal{Q})$. The concept of "close" is defined below.

Definition 2. *(Close segment and critical vertex)*

- *A path is a* segment *if it is a sub-path of a path in $\mathcal{Q}$.*
- *For a vertex $v \in V(\mathcal{P}) \cap V(\mathcal{Q})$, a segment $S = R$-w is a* close segment *of v if v-S is a segment too, where if R is non-empty then $R \in \mathcal{Q} - \mathcal{P}$ and if w is non-empty then $w \in V(\mathcal{P})$.*
- *In the last item, when w exists, w is a* close *vertex of v (and the other way around, v-R is a close segment of w, and v is a close vertex of w); furthermore, w is the* critical *vertex of v if w is the only close vertex of v, or the one with a larger or equal group index.*
- *A close segment S of v is* maximal *if there is no other close segment of v containing all the vertices of S.*

Remark 1. Suppose R_1-v-R_2 is a segment where $v \in V(\mathcal{P})$ and R_1, R_2 are two (*longest*, skipped in the sequel) paths in $\mathcal{Q} - \mathcal{P}$ incident to v. Since each path of $\mathcal{Q}$ has order at least 4, if $|V(R_1)| \le 1$ and $|V(R_2)| \le 1$, then v has a close vertex and thus a critical vertex. Specifically, if $v \in H_0 \cup H_1$ and $v \in V(\mathcal{Q})$, then by Lemma 4, v has a close and critical vertex.

We are now ready to present our amortization scheme.

Definition 3. *(Amortization scheme) First consider a path $R \in \mathcal{Q} - \mathcal{P}$ and let v-R-w be a segment where $v, w \in V(\mathcal{P})$ or void if they do not exist. By Lemma 3, at least one of v, w exists. Without loss of generality, assume v exists and the assignment of the tokens of $V(R)$ is as follows:*

1. *If w is void, then we assign the $|V(R)|$ tokens to v.*
2. *Suppose w exists.*
   2.1. *(S1) If $v \in H_1$ and $w \in H_3$, then by Lemma 4, $|V(R)| = 1$ and we assign the whole token to w;*
   2.2. *Except the special case S1, we assign $\frac{1}{2}|V(R)|$ tokens to each of v, w.*

Secondly, consider a vertex $v \in V(\mathcal{P}) \cap V(\mathcal{Q})$ and let w be the critical vertex of v when $v \in H_0 \cup H_1$ (by Remark 1). The assignment of the one token of v is done as follows:

3. *If $v \in H_2 \cup H_3$, then we assign the one token to v itself.*
4. *Suppose $v \in H_1$.*
   4.1. *If $w \in H_2 \cup H_3$, then we assign $\frac{2}{3}$ and $\frac{1}{3}$ token to v and w, respectively;*
   4.2. *(S2) If $w \in H_0$ and there is a segment v-w-x where $x \in H_3$, then we assign $\frac{2}{3}$ and $\frac{1}{3}$ token to v and x, respectively;*
   4.3. *Except the special case S2, if $w \in H_0 \cup H_1$, then we assign the one token to v.*
5. *Suppose $v \in H_0$.*
   5.1. *If $w \in H_3$, then we assign the one token to w;*
   5.2. *If $w \in H_0$, then we assign the one token to v;*
   5.3. *(S3) If $w \in H_1$ and there is a segment v-w-x where $x \in H_3$, then we assign the one token to x;*

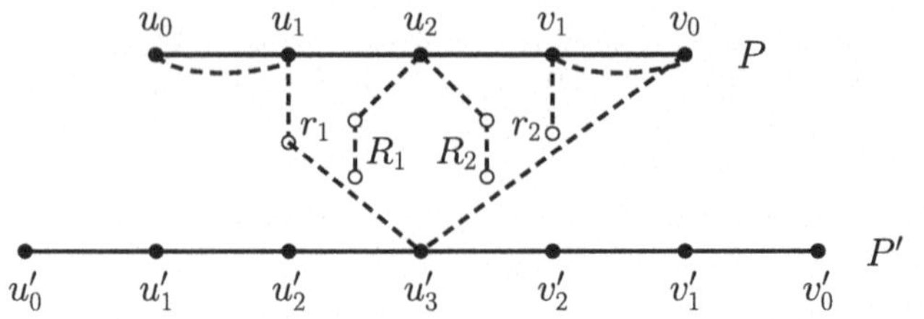

Fig. 1. The solid and dashed edges are in $E(\mathcal{P})$ and $E(\mathcal{Q})$, respectively. The empty vertices are in $V(\mathcal{Q} - \mathcal{P})$. The paths u_0-...-v_0 (P) and u'_0-...-v'_0 (P') are in $\mathcal{P}$ while the paths u_0-u_1-r_1-u'_3-v_0-v_1-r_2 and R_1-u_2-R_2 are in $\mathcal{Q}$. The path P gets $\frac{25}{3}$ tokens in total and the vertex u'_3 gets $\frac{11}{3}$ tokens.

5.4. Except the special case S3, if $w \in H_1 \cup H_2$, then we assign $\frac{1}{2}$ token to each of v, w.

In Fig. 1, we use an instance to illustrate the amortization scheme including the special cases S1 (in Step 2.1) and S2 (in Step 4.2). In the figure, the solid edges are in $E(\mathcal{P})$, and $\mathcal{P}$ contains a 5-path $P = u_0$-...-v_0 and a 7-path $P' = u'_0$-...-v'_0. The optimal solution $\mathcal{Q}$ contains a 7-path u_0-u_1-r_1-u'_3-v_0-v_1-r_2 and a 5-path R_1-u_2-R_2. We calculate below the tokens received by the path P and the vertex u'_3. By the grouping in Definition 1, $u_0, v_0 \in H_0$, $u_1, v_1 \in H_1$, $u_2 \in H_2$ and $u'_3 \in H_3$. Since u_1 is the critical vertex of u_0, u_0 receives $\frac{1}{2}$ token from itself. Note that S1 holds for r_1 and r_1 gives the one token to u'_3. The vertex u'_3 is the critical vertex of u_1 and therefore, u_1 receives $\frac{1}{2}$ and $\frac{2}{3}$ token from u_0 and itself, respectively. That is, u_1 receives $\frac{7}{6}$ tokens in total. For the vertex u_2, it receives 5 tokens from R_1, R_2 and itself. For the vertex v_1, the one token of r_2 is assigned to v_1. Moreover, v_1-v_0-u'_3 is a segment and the special case S2 holds. It follows that v_1 receives $\frac{2}{3}$ token from itself and u'_3 receive $\frac{1}{3}$ and one token from v_1 and v_0, respectively. In conclusion, P gets $\frac{25}{3}$ tokens and u'_3 gets $\frac{11}{3}$ tokens. In the main theorem, we prove that our algorithm is a $\frac{5}{3}$-approximation and therefore, the 5-path P in Fig. 1 shows a tight configuration for the algorithm.

Remark 2. Each vertex $v \in V(\mathcal{P}) \cap V(\mathcal{Q})$ can receive tokens from itself and its two maximal close segments. If $x \in H_3$ and the special case S2 or S3 happens, then x can receive tokens from the vertex v mentioned in S2 and S3.

5 The Structures of $\mathcal{P}$

In this section, we show the structures of the paths in $\mathcal{P}$. The proofs of the lemmas in this section heavily rely on Operation 1, by use of the termination condition. That is, if a lemma does not hold, then Operation 1 would be applicable, contradicting Lemma 1. Due to space limit, we only show full details for the proof of Lemma 5, but use an example to present the "informal" proofs to the other technical lemmas.

The next two lemmas show that a special vertex (see Definition 1) cannot be adjacent to a vertex in $H_0 \cup H_1$ if some specific condition is satisfied.

Lemma 5. *Suppose v_2 is a special vertex on a 6-path $P \in \mathcal{P}$ and v_2-$u' \in E$ where u' is on another path $P' \neq P$ in $\mathcal{P}$. Then the following three statements hold:*

1. *$u' \notin H_0$;*
2. *if $|V(P')| = 5$ and u'_2 is adjacent to a vertex of $V \setminus V(\mathcal{P})$, then $u' \notin H_1$;*
3. *if $|V(P')| \geq 6$, then $u' \notin H_1$.*

Proof. Since v_2 is a special vertex, by Definition 1, $v_2 \in H_0$ and u_2 is adjacent to a vertex $r \in V \setminus V(\mathcal{P})$ in $\mathcal{Q}$. We next prove the three statements as follows.

Statement 1: We prove the statement by contradiction to assume $u' \in H_0$. By Definition 1 again, u' is a special vertex or u' is an endpoint of P'. We distinguish three cases.

Case 1. u' is a special vertex on a 6-path. Without loss of generality, we assume $u' = v'_2$. Then by Definition 1, there exists a vertex $r' \in V \setminus V(\mathcal{P})$ adjacent to u'_2 in $\mathcal{Q}$. If $r = r'$, then when Operation 1 chooses the paths P, P' and the vertex r, we can construct two 4^+-paths u_0-u_1-u_2-r-u'_2-u'_1-u'_0 and v_0-v_1-v_2-v'_2-v'_1-v'_0, which cover one more vertex than P, P'. Otherwise, $r \neq r'$. When Operation 1 chooses the paths P, P' and two vertices r, r', we can construct three 4^+-paths v_0-v_1-v_2-v'_2-v'_1-v'_0, u_0-u_1-u_2-r and u'_0-u'_1-u'_2-r', which cover two more vertices than P, P'.

Case 2. u' is a special vertex on a 7-path. Without loss of generality, we assume $u' = v'_2$. When Operation 1 chooses the paths P, P' and the vertex r, we can construct three paths v_0-v_1-v_2-v'_2-v'_1-v'_0, u_0-u_1-u_2-r and u'_0-u'_1-u'_2-u'_3, which cover one more vertex than P, P'.

Case 3. u' is an endpoint of P'. Without loss of generality, we assume $u' = u'_0$. When Operation 1 chooses the two paths P, P' and the vertex r, we can construct two 4^+-paths r-u_2-u_1-u_0 and v_0-v_1-v_2-u'_0-...-v'_0, which cover one more vertex than P, P'. Therefore, we can conclude that Operation 1 is always applicable, a contradiction.

Statement 2: We prove the statement by contradiction to assume $u' \in H_1$. Without loss of generality, by Definition 1, we further assume $u' = u'_1$. Let r' be a vertex of $V \setminus V(\mathcal{P})$ that is adjacent to u'_2. If $r' = r$, then when Operation 1 chooses the paths P, P' and the vertex r, we can construct two 4^+-paths u_0-u_1-u_2-r-u'_2-v'_1-v'_0 and v_0-v_1-v_2-u'_1-u'_0, which cover one more vertex than P, P'. Otherwise, $r' \neq r$. When Operation 1 chooses the paths P, P' and the vertices r, r', we can construct three 4^+-paths r-u_2-u_1-u_0, r'-u'_2-v'_1-v'_0 and v_0-v_1-v_2-u'_1-u'_0, which cover two more vertices than P, P'. In conclusion, Operation 1 is applicable, a contradiction.

Statement 3: Similarly, we assume $u' = u'_1$. Since $|V(P')| \geq 6$, when Operation 1 chooses the paths P, P' and the vertex r, we can construct three 4^+-paths r-u_2-u_1-u_0, v_0-v_1-v_2-u'_1-u'_0 and u'_2-...-v'_0, which cover one more vertex than P, P'. Therefore, Operation 1 is applicable, a contradiction. Then the lemma is proved. □

Lemma 6. *Suppose P is a 5^+-path of $\mathcal{P}$ and there exists an index $t \in \{2,3\}$ such that u_t-R is a segment where $R \in \mathcal{Q} - \mathcal{P}$. Moreover, u-v_2' is an edge of E where $u \in V(P)$ and v_2' is a special vertex on a 7-path $P' \neq P$ in $\mathcal{P}$. Then the following three statements hold:*

1. *$u \notin H_0$;*
2. *if $|V(P)| = 5$, then $u \notin H_1$;*
3. *if $|V(R)| \geq 2$, then $u \notin H_1$.*

Proof. See Fig. 1 for an illustration, where u_2-R_1 is a segment and therefore, none of u_0, u_1, v_1, v_0 is adjacent to v_2' (Statements 1 and 2). □

The following lemma shows that for a path $P \in \mathcal{P}$, if a certain condition is satisfied, then the vertex u_i cannot be adjacent to v_j in the graph G except for some specific values of i, j.

Lemma 7. *Suppose P is a 5^+-path of $\mathcal{P}$ and there exists an index $t \in \{2,3\}$ such that u_t-R is a segment where $R \in \mathcal{Q} - \mathcal{P}$. Moreover, u_i, v_j are two vertices of $V(P)$ with $i < t$. Then the following three statements hold:*

1. *If $|V(R)| > \frac{1}{2}(|V(P)| - 3)$, then u_i-$v_j \notin E$;*
2. *if $|V(P)| = 5$ and $|V(R)| = 1$, then u_i-$v_j \notin E$ unless $(i,j) = (1,1)$;*
3. *if $|V(P)| = 7$ and $|V(R)| = 2$, then u_i-$v_j \notin E$ unless $(i,j) \in \{(1,1),(2,2)\}$.*

Proof. Let us see the proof of Statement 2 on a special case where u_0 is adjacent to v_1. In this case, when Operation 1 chooses P and the vertex of R, we can construct a 6-path R-u_2-u_1-u_0-v_1-v_0, which contradicts Operation 1 is not applicable. □

The next lemma is vital for us to prove that the two vertices u_0, u_1 of a path can receive at most 3 tokens in total in Lemma 15.

Lemma 8. *Suppose u_1 in a path P in $\mathcal{P}$ is adjacent to a vertex of $V \setminus V(\mathcal{P})$. If $|V(P)| \geq 5$, then u_0-$u_2 \notin E$. Moreover, u_0 is not adjacent to a vertex of H_0.*

We next consider a 5^+-path $P \in \mathcal{P}$ and there exists an index $t \in \{2,3\}$ such that u_t-R is a segment where $R \in \mathcal{Q} - \mathcal{P}$. In the next five lemmas, we show that if $|V(R)|$ is relatively large, then the vertices of P except u_t are in $H_0 \cup H_1$ and their critical vertices are in $H_2 \cup H_3$ (see Remark 1). In other words, once u_t receives a large number of tokens, then the number of tokens received by the other vertices of P will be relatively smaller.

Lemma 9. *Suppose P is a 5^+-path of $\mathcal{P}$ and there exists an index $t \in \{2,3\}$ such that u_t-R is a segment where $R \in \mathcal{Q} - \mathcal{P}$. Moreover, u_1-r is an edge in E where $r \in V \setminus V(\mathcal{P})$. If $|V(R)| \geq 2$ or $|V(P)| = 5$, then $r \notin V(R)$ and r is not adjacent to any vertex of $V(P) \setminus \{u_1, u_3\}$.*

Lemma 10. *Suppose u_2 is on a 5- or 6-path $P \in \mathcal{P}$ and; u_0 or u_1 is adjacent to a vertex u' in G where u' is on a path P' in $\mathcal{P}$. Moreover, u_2-R is a segment where $R \in \mathcal{Q} - \mathcal{P}$. Then the following statements hold:*

1. *If* $|V(R)| = 2$, *then* $u' \notin (H_0 \cup H_1) \setminus \{u_0, u_1\}$;
2. *if* $|V(P)| = 5$, *then* $u' \notin H_0 \setminus u_0$.

Proof. One can see the red edge u_1-u_1' in Fig. 2 for an illustration. Basically, if the edge exists, then Operation 1 is applicable when P, P' and the vertices of R are chosen, resulting in a 6-path R-u_2-v_2-v_1-v_0 and a 7-path u_0-u_1-u_1'-...-v_0'. □

Lemma 11. *Suppose* u_2 *is on a* 5- *or* 6*-path* $P \in \mathcal{P}$ *and;* u_1-r *is an edge of* E *where* $r \in V \setminus V(\mathcal{P})$. *Moreover,* u_0 *or* r *is adjacent to a vertex* u' *in* G *where* u' *is on a path* P' *in* $\mathcal{P}$. *If* u_2-R *is a segment where* $R \in \mathcal{Q} - \mathcal{P}$, *then the following statements hold:*

1. *If* $|V(R)| = 2$, *then* $u' \notin (H_0 \cup H_1 \cup H_2) \setminus u_1$;
2. *if* $|V(P)| = 5$, *then* $u' \notin (H_0 \cup H_1) \setminus u_1$.

Proof. One can see the red edges u_0-u_2' and r-u_2' in Fig. 2 for an illustration. Basically, if one of the two edges, say u_0-u_2' exists, then Operation 1 is applicable when P, P' and the vertices of R, r are chosen, resulting in a 6-path R-u_2-v_2-v_1-v_0 and a 7-path r-u_1-u_0-u_2'-...-v_0'. □

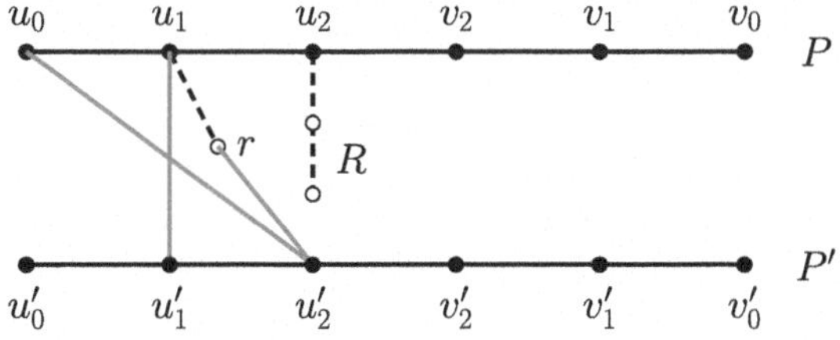

Fig. 2. The paths u_0-...-v_0 and u_0'-...-v_0' are in $\mathcal{P}$ while the paths u_1-r and u_2-R are two segments. By the first statement of Lemma 10, the red edge u_1-u_1' does not present in E. Similarly, by the first statement of Lemma 11, none of the red edges u_0-u_2' and r-u_2' would exist in E. (Color figure online)

Lemma 12. *Suppose* P *is a* 6- *or* 7*-path in* $\mathcal{P}$ *and there exists an index* $t \in \{2, 3\}$ *such that* R-u_t *is a segment where* $R \in \mathcal{Q} - \mathcal{P}$. *Moreover,* v_1-u' *is an edge in* E *where* u' *is on a path* P' *in* $\mathcal{P}$. *Then the following statements hold:*

1. *If* $|V(R)| = 2$, *then* $u' \notin H_0 \setminus \{v_0, v_2\}$;
2. *if* $|V(R)| = 3$, *then* $u' \notin (H_0 \cup H_1) \setminus \{v_0, v_2\}$.

Proof. One can see the red edge v_1-u_0' in Fig. 3 for an illustration. Basically, if the edge exists, then Operation 1 is applicable when P, P' and R are chosen, resulting in a 5-path R-u_2-u_1-u_0 and an 8-path v_0'-...-u_0'-v_1-v_0 (which can be broken down into two 4-paths). □

Lemma 13. *Suppose P is a 6- or 7-path in $\mathcal{P}$ and there exists an index $t \in \{2,3\}$ such that R-u_t is a segment where $R \in \mathcal{Q} - \mathcal{P}$. Moreover, v_1-x-u' is a segment where $x \in \{v_0, v_2\}$ or $x \in V \setminus V(\mathcal{P})$ and; u' is on a path P' in $\mathcal{P}$. Then the following two statements hold:*

1. *If $|V(R)| = 2$, then $u' \notin (H_0 \cup H_1) \setminus \{v_0, v_1, v_2\}$ unless $|V(P)| = 7$ and $(x, u') = (v_2, u_2)$;*
2. *If $|V(R)| = 3$, then $u' \notin (H_0 \cup H_1 \cup H_2) \setminus \{v_0, v_1, v_2\}$.*

Proof. One can see the red edges v_2-u'_1, v_0-u'_1 and r-u'_1 in Fig. 3 for an illustration. Basically, if one of them, say v_2-u'_1, exists, then Operation 1 is applicable when P, P' and R are chosen, resulting in a 5-path R-u_2-u_1-u_0 and an 8-path v_0-v_1-v_2-u'_1-...-v'_0 (which can be broken down into two 4-paths). □

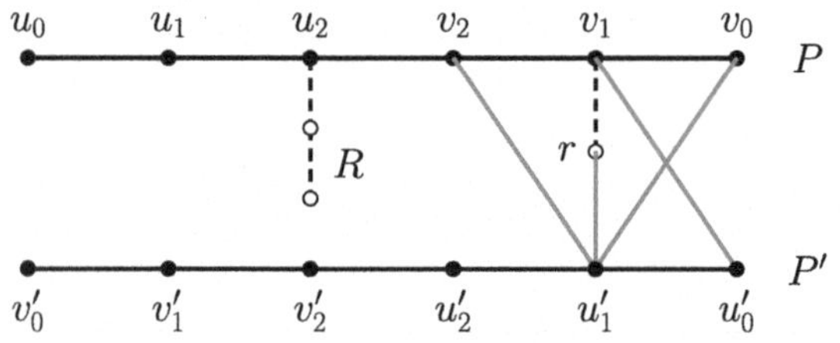

Fig. 3. The paths u_0-...-v_0 and u'_0-...-v'_0 are in $\mathcal{P}$ while the paths u_1-r and u_2-R are two segments. By the first statement of Lemma 12, the red edge v_1-u'_0 does not present in E. Similarly, by the first statement of Lemma 13, none of the red edges v_0-u'_1, r-u'_1 and v_2-u'_1 would exist in E. (Color figure online)

Lemma 14. *Suppose P is a 7-path in $\mathcal{P}$ and R-u_3 is a segment where R is a 2-path of $\mathcal{Q} - \mathcal{P}$. If u_2-$v_2 \in E(\mathcal{Q})$ and u_3-$w \in E(\mathcal{Q})$ where $w \notin V(R)$, then $w \in \{u_2, v_2\}$ or $w \in H_2 \cup H_3$.*

6 The Performance Analysis

In this section, we analyze the total tokens received by each path in $\mathcal{P}$. For each vertex $u \in V(\mathcal{P})$, we use the notation $\tau(u)$ to denote the number of tokens received by u. Similarly, for each path $P \in \mathcal{P}$, we use $\tau(P)$ to denote the total number of tokens received by the vertices of P.

Lemma 15. *The following two statements hold:*

1. *Each vertex of H_0 can receive at most one token;*
2. $\tau(u_0) + \tau(u_1) \leq 3$.

We next define two important kinds of edges: balanced edge and nice edge. These two definitions help us identify the common structures used in the later technical Lemmas 19 and 20 and their proofs. For example, in Lemma 19, we actually prove that the edge u_0-u_1 is balanced or nice.

Definition 4. *(Balanced edge) Suppose u-v is an edge of a path P in $\mathcal{P}$ where $u \in H_0$ and $v \in H_1$. The edge u-v is* balanced *if v is not adjacent to any vertex of $V \setminus V(\mathcal{P})$ in $\mathcal{Q}$ and one of the following two conditions is satisfied:*

B1. None of u, v is adjacent to any vertex of $H_0 \setminus u$ in $\mathcal{Q}$;
B2. none of u, v is adjacent to any vertex of $(H_0 \cup H_1) \setminus \{u, v\}$ in $\mathcal{Q}$.

We remark that B2 is stronger than B1.

Lemma 16. *Given a balanced edge u-v, $\tau(u) + \tau(v) \le 2$ and if B2 holds, then $\tau(u) + \tau(v) \le \frac{5}{3}$.*

Definition 5. *(Nice edge) Suppose u-v is an edge of a path P in $\mathcal{P}$ where $u \in H_0$ and $v \in H_1$. The edge u-v is* nice *if one of the following two conditions is satisfied:*

N1. v is not adjacent to any vertex of $H_0 \setminus u$ in $\mathcal{Q}$. Moreover, if v-x-w is a segment where $w \in V(\mathcal{P})$ and, $x = u$ or $x \in V \setminus V(\mathcal{P})$, then $w \in H_2 \cup H_3$;
N2. v is not adjacent to any vertex of $(H_0 \cup H_1) \setminus u$ in $\mathcal{Q}$. Moreover, if v-x-w is a segment where $w \in V(\mathcal{P})$ and, $x = u$ or $x \in V \setminus V(\mathcal{P})$, then $w \in H_3$.

We remark that the condition N2 is stronger than N1. When v is not adjacent to any vertex of $V \setminus V(\mathcal{P})$ in $\mathcal{Q}$, then u-v becomes a balanced edge and Nj can imply Bj, for $j = 1, 2$. In Fig. 1, the edges u_0-u_1, v_0-v_1 are both nice and N2 holds.

Lemma 17. *Given a nice edge u-v, $\tau(u) + \tau(v) \le \frac{8}{3}$ and if N2 holds, then $\tau(u) + \tau(v) \le \frac{5}{3}$.*

Lemma 18. *Consider the subpath v_2-v_1-v_0 of a 6- or 7-path $P \in \mathcal{P}$. Suppose $v_2 \in H_0$ and the edge v_2-v_0 is in $E(\mathcal{Q})$. The following two statements hold:*

1. *If v_1 is not adjacent to any vertex of $H_0 \setminus \{v_0, v_2\}$ in $\mathcal{Q}$, then $\tau(v_2) + \tau(v_1) + \tau(v_0) \le 3$;*
2. *if none of v_2, v_1, v_0 is adjacent to any vertex of $(H_0 \cup H_1 \cup H_2) \setminus \{v_2, v_1, v_0\}$ in $\mathcal{Q}$, then $\tau(v_2) + \tau(v_1) + \tau(v_0) \le 2$.*

Lemma 19. *Suppose P is a 5- or 6-path in $\mathcal{P}$ and u_2-R is a segment where R is in $\mathcal{Q} - \mathcal{P}$. Then the following two statements hold:*

1. *If $|V(P)| = 5$, then $\tau(u_0) + \tau(u_1) \le \frac{8}{3}$;*
2. *if $|V(R)| = 2$, then $\tau(u_0) + \tau(u_1) \le \frac{5}{3}$.*

Lemma 20. *Suppose P is a 6- or 7-path in $\mathcal{P}$ and there exists an index $t \in \{2, 3\}$ such that R-u_t is a segment where R is a 2^+-path of $\mathcal{Q} - \mathcal{P}$. Then the following three statements hold:*

1. *If* $|V(P)| = 6$, *then* $\tau(v_2) + \tau(v_1) + \tau(v_0) \leq \frac{19}{6}$;
2. *if* $|V(P)| = 7$, $|V(R)| = 2$ *and* u_2-$v_2 \notin E(\mathcal{Q})$, *then* $\tau(v_2)+\tau(v_1)+\tau(v_0) \leq \frac{19}{6}$;
3. *if* $|V(R)| = 3$, *then* $\tau(v_2) + \tau(v_1) + \tau(v_0) \leq 2$.

Recall that $\tau(P)$ is the total tokens received by the path $P \in \mathcal{P}$. We have the following upper bounds on $\tau(P)$, distinguished by the order of P.

Lemma 21. *If P is a 5-path in $\mathcal{P}$, then* $\tau(P) \leq \frac{25}{3}$.

Lemma 22. *If P is a 6-path in $\mathcal{P}$, then* $\tau(P) \leq 10$.

Lemma 23. *If P is a 7-path in $\mathcal{P}$, then* $\tau(P) \leq \frac{35}{3}$.

Theorem 1. Approx1 *is an* $O(|V|^7)$*-time* $\frac{5}{3}$*-approximation algorithm for* MPC^{4+}.

Proof. By Lemma 1, it is sufficient to show $\tau(P) \leq \frac{5}{3}|V(P)|$ for each path $P \in \mathcal{P}$. Recall that $|V(P)| \in \{4, 5, 6, 7\}$. If $|V(P)| = 4$, then by the second statement of Lemma 15, $\tau(u_0) + \tau(u_1) \leq 3$ and by symmetry, $\tau(v_0) + \tau(v_1) \leq 3$. Therefore, $\tau(P) \leq 6 < \frac{5}{3}|V(P)|$. Together with Lemmas 21, 22 and 23, the theorem is proved. □

7 Conclusion

We presented a $\frac{5}{3}$-approximation algorithm that runs in $O(|V|^7)$ time for MPC^{4+}. In the algorithm, we propose a simple local search operation, which is more general than previously proposed local search operations such as those in [10,11]. For the performance analysis, we designed a more delicate amortization scheme, in which the vertices covered in the computed solution are partitioned into four groups, with a larger group index allowing a vertex to receive more tokens from the optimal solution. Such a weighting strategy balances all the possible cases, that the total tokens received by every path P are no more than $\frac{5}{3}|V(P)|$.

Both the local search operation and the amortization scheme can possibly be extended for larger values of k or general k, which is one of our future research. On the other hand, we are also interested in non-trivial lower bounds on the approximation ratio, as we believe MPC^{k+} is APX-hard through a reduction from maximum 2-SAT [6] similarly as in [5].

Acknowledgments. The research is supported by the NSERC Canada, the Alberta Graduate Excellence Scholarship and the Alberta Innovate Graduate Scholarship, the NNSF China (Grants 12371316, 12471301), the MST China (Grant G2023016016L), and the KAKENHI Japan (Grant JP24K02902) and JST CREST Japan (Grant JPMJR1402).

References

1. Asdre, K., Nikolopoulos, S.D.: A linear-time algorithm for the k-fixed-endpoint path cover problem on cographs. Networks **50**, 231–240 (2007)
2. Asdre, K., Nikolopoulos, S.D.: A polynomial solution to the k-fixed-endpoint path cover problem on proper interval graphs. Theor. Comput. Sci. **411**, 967–975 (2010)
3. Chen, Y., et al.: Path cover with minimum nontrivial paths and its application in two-machine flow-shop scheduling with a conflict graph. J. Comb. Optim. **43**, 571–588 (2022)
4. Chen, Y., Goebel, R., Lin, G., Su, B., Xu, Y., Zhang, A.: An improved approximation algorithm for the minimum 3-path partition problem. J. Comb. Optim. **38**, 150–164 (2019)
5. Eto, H., Kawaharada, S., Lin, G., Miyano, E., Ozdemir, T.: Directed path partition problem on directed acyclic graphs. In: Proceedings of IWOCA 2024. LNCS, vol. 14764, pp. 314–326 (2024)
6. Garey, M.R., Johnson, D.S.: Computers and Intractability: A Guide to the Theory of NP-Completeness. W. H. Freeman and Company, San Francisco (1979)
7. Gómez, R., Wakabayashi, Y.: Nontrivial path covers of graphs: existence, minimization and maximization. J. Comb. Optim. **39**, 437–456 (2020)
8. Gong, M., Chen, Z.-Z., Lin, G., Wang, L.: An approximation algorithm for covering vertices by 4^+-paths. In: Proceedings of COCOA 2023. LNCS, vol. 14461, pp. 459–470 (2023)
9. Gong, M., Chen, Z.-Z., Lin, G., Wang, L.: Approximately covering vertices by order-5 or longer paths. In: Proceedings of COCOON 2024 (2024)
10. Gong, M., Edgar, B., Fan, J., Lin, G., Miyano, E.: Approximation algorithms for covering vertices by long paths. Algorithmica **86**, 2625–2651 (2024)
11. Gong, M., Fan, J., Lin, G., Miyano, E.: Approximation algorithms for covering vertices by long paths. In: Proceedings of MFCS 2022, pp. 53:1–53:14 (2022)
12. Hartvigsen, D., Hell, P., Szabó, J.: The k-piece packing problem. J. Graph Theory **52**, 267–293 (2006)
13. Hochbaum, D.S.: Efficient bounds for the stable set, vertex cover and set packing problems. Discret. Appl. Math. **6**, 243–254 (1983)
14. Kobayashi, K., et al.: Path cover problems with length cost. Algorithmica **85**, 3348–3375 (2023)
15. Micali, S., Vazirani, V.V.: An $O(\sqrt{|V|}|E|)$ algorithm for finding maximum matching in general graphs. In: Proceedings of FOCS, vol. 1980, pp. 17–27 (1980)
16. Neuwohner, M.: An improved approximation algorithm for the maximum weight independent set problem in d-claw free graphs. In: Proceedings of STACS 2021. LIPIcs, vol. 187, pp. 53:1–53:20 (2021)
17. Pao, L.L., Hong, C.H.: The two-equal-disjoint path cover problem of matching composition network. Inf. Process. Lett. **107**, 18–23 (2008)
18. Rizzi, R., Tomescu, A.I., Mäkinen, V.: On the complexity of minimum path cover with subpath constraints for multi-assembly. BMC Bioinform. **15**, S5 (2014)

Bicluster Editing with Overlaps: A Vertex Splitting Approach

Faisal N. Abu-Khzam(✉), Lucas Isenmann, and Zeina Merchad

Department of Computer Science and Mathematics, Lebanese American University, Beirut, Lebanon
faisal.abukhzam@lau.edu.lb

Abstract. The BiCluster Editing problem aims at editing a given bipartite graph into a disjoint union of bicliques via a minimum number of edge deletion or addition operations. As a graph-based model for data clustering, the problem aims at a partition of the input dataset, which cannot always obtain meaningful clusters when some data elements are expected to belong to more than one cluster each. To address this limitation, we introduce the Bicluster Editing with Vertex Splitting problem (BCEVS) which consists of finding a minimum sequence of edge editions and vertex splittings such that the result graph is a disjoint union of bicliques. The vertex splitting operation consists of replacing a vertex v with two vertices whose union of neighborhoods is the neighborhood of v. We also introduce the problem of Bicluster Editing with One-Sided Vertex Splitting (BCEOVS) where we restrict the splitting operations to the first set of the bipartition (often corresponding to data elements in the input raw data). We prove the two problems are NP-complete even when restricted to bipartite planar graphs of maximum degree three. Moreover, assuming the Exponential Time Hypothesis holds, there is no $2^{o(n)}n^{O(1)}$-time (resp. $2^{o(\sqrt{n})}n^{O(1)}$-time) algorithm for BCEVS and BCEOVS on bipartite (resp. planar) graphs with maximum degree three where n is the number of vertices of the graph. Furthermore we prove both problems are APX-hard and solvable in polynomial time on trees. On the other hand, we prove that BCEOVS is fixed parameter tractable with respect to solution size and admits a polynomial size kernel.

Keywords: Biclustering · Bicluster Editing · Vertex Splitting

1 Introduction

Cluster Editing is a classical problem with numerous applications across fields, most importantly in computational biology and gene expressions [19,21]. The problem, as originally proposed, involves determining whether a graph G can be

This research project was supported by the Lebanese American University under the President's Intramural Research Fund PIRF0056.

H. Fernau and B. Zhu (Eds.): IWOCA 2025, LNCS 15885, pp. 146–159, 2025.
https://doi.org/10.1007/978-3-031-98740-3_11

transformed into a graph consisting of a disjoint union of cliques through the addition or deletion of at most k edges. Cluster Editing was shown to be NP-hard [16,20], and multiple parameterized and approximation algorithms have been developed to address it [5,6,8–12].

When the input and output graphs are restricted to bipartite graphs, the problem is referred to as BICLUSTER EDITING. The main question is whether it is possible to modify (by deleting or adding) at most k edges in the input bipartite graph so that the resulting graph becomes a disjoint union of bicliques. This problem is greatly used in the field of computational biology and in the analysis of gene expressions [17], among other things. BICLUSTER EDITING has been shown to be NP-complete [4], and several parameterized and approximation algorithms have been proposed. The simplest fixed-parameter algorithm of Bicluster Editing runs in $O(4^k + |E|)$ (simply by exhaustively trying all possible editing operations when an induced path of length three is found) [18]. Guo *et al.* [13] improved the running time to $O(3.24^k + |E|)$. More recently, Xiao and Kou improved the running time bound to $O^*(2.9312^k)$ [24] while Tsur proposed a branching algorithm with a running time of $O^*(2.636^k)$ [22] and further improved it to run in $O^*(2.22^k)$ [23].

Cluster Editing with Overlapping Communities extends the traditional Cluster Editing problem by allowing vertices to belong to more than one cluster, or *to split* among them. This particular vertex splitting operation permits a vertex v to be replaced by two vertices whose combined neighborhoods is the neighborhood of v. Consequently, the data element represented by v can simultaneously belong to more than one cluster, which is an obvious practical objective. Vertex splitting has been introduced and studied in a number of recent articles [2,3]. In this paper, we use this operation for the first time in the realm of biclustering and study the complexity of the corresponding problems.

Biclustering is often used when processing raw data given in some tabular form, where rows correspond to data elements and columns correspond to features or some other form of attributes. Typically, clustering algorithms aim at grouping the data elements (the rows). By introducing vertex splitting as another (optional) editing operation, we consider the case where splitting is allowed on both sides of the bipartite graph, as well as the case where it is allowed only on one side (the data elements only). The two corresponding problems are BICLUSTER EDITING WITH VERTEX SPLITTING (BCEVS) and BICLUSTER EDITING WITH ONE SIDED VERTEX SPLITTING (BCEOVS). In general, this approach can uncover significant relationships that conventional biclustering techniques might overlook.

Our Contribution. We prove that both BCEVS and BCEOVS are NP-complete, even when restricted to planar graphs of maximum degree three. In addition, and assuming the EXPONENTIAL TIME HYPOTHESIS holds, we prove that there is no $2^{o(n)}n^{O(1)}$-time algorithm for BCEVS and BCEOVS on bipartite graphs with maximum degree three. We also show that there is no $2^{o(\sqrt{n})}n^{O(1)}$-time algorithm when the input is restricted to planar graphs (again, modulo the ETH). Furthermore we prove that both problems are APX-hard. On the positive side, we show that BCEOVS is fixed-parameter tractable and admits a polynomial size kernel, and we also show that the two problems are solvable in polynomial time on trees.

2 Preliminaries

We adopt the following common graph-theoretic terminology. A graph $G = (V, E)$ is said to be *bipartite* if its vertex set V can be divided into two disjoint sets A and B such that every edge $e \in E$ connects a vertex in A to a vertex in B. That is, there are no edges between vertices within the same subset. If G is a bipartite graph consisting of subsets A and B, then a biclique in G is defined by two subsets $A' \subseteq A$ and $B' \subseteq B$ such that every vertex in A' is connected to every vertex in B'. The *open neighborhood* $N(v)$ of a vertex v is the set of vertices adjacent to it. The degree of v is the number of edges incident on v, which is $|N(v)|$ since we only consider simple graphs. A *geodesic path*, or shortest path, between two vertices in a graph is a path that has the smallest number of edges among all paths connecting these vertices. An *induced subgraph* $G[X]$ where $X \subseteq V$, is a subgraph formed from a graph G by taking a subset X of its vertices and all of the edges in G that have both endpoints in X.

We consider the following operations on a bipartite graph $G = (A, B, E)$: edge addition, edge deletion and vertex splitting. Vertex splitting is an operation that replaces a vertex v by two copies v_1 and v_2 such that $N(v) = N(v_1) \cup N(v_2)$. The operation results in a new bipartite graph. An *exclusive* vertex split requires that $N(v_1) \cap N(v_2) = \emptyset$. In this paper, we do not assume a split is exclusive, but our proofs apply to this restricted version. We define the following new problems.

Bicluster Editing with Vertex Splitting
Given: A bipartite graph $G = (A, B, E)$, along with positive integer k.
Question: Can we transform G into a disjoint union of bicliques by performing at most k edits including vertex splitting operations?

Bicluster Editing with One-Sided Vertex Splitting
Given: A graph $G = (A, B, E)$, along with positive integer k.
Question: Can we transform G into a disjoint union of bicliques by performing at most k edits where only vertices from the set B are subjected to vertex splitting operations?

A geodesic of length 3 in a bipartite graph consists of 4 vertices that are not part of a biclique (simply because the endpoints are non-adjacent). Such a geodesic is treated as a forbidden structure, or conflict, that is to be *resolved* in order to obtain a disjoint union of bicliques. To explicate, we say that an operation resolves a geodesic of length 3 in a bipartite graph G if, after applying this operation, the two end vertices of the geodesic are at distance different from 3. We say that two geodesics of length 3 are *independent* if there does not exist an operation which resolves simultaneously both of them.

Observation 1 *There are six types of operations that can resolve a geodesic (a, b, c, d): deleting one of the edges ab, bc, cd, or adding the edge ad or splitting the vertex b (resp. c) such that one copy is adjacent to a (resp. b) and not to c (resp. d) and the other copy is adjacent to c (resp. d) and not to a (resp. b). Furthermore, there are several ways to split vertex b or c to resolve the geodesic if these vertices are incident to other vertices.*

3 Complexity of BCEVS and BCEOVS

The objective is to exhibit a reduction to Bicluster Editing with Vertex Splitting and Bicluster Editing with One-sided Vertex Splitting from a variant of 3SAT. For a 3-CNF formula F, we denote by m the number of clauses of F.

Construction 1 *Consider a 3-CNF formula F. For every variable v, we denote by $d(v)$ the number of clauses where v appears and we denote by $c(v)_1, \ldots, c(v)_{d(v)}$ the clauses where v appears.*

For every clause c where a variable v appears, let j be the index of c in the list, defined above, of the clauses where v appears, we define v_c as $v_{6(j-1)+1}$ (resp. $v_{6(j-1)+3}$) if v appears positively (resp. negatively). We create a graph G_F as follows:

- *For each variable v, we create a cycle $v_1, \ldots, v_{6d(v)}$. We consider the indices modulo $6d(v)$ (so for example $v_{6d(v)+1} = v_1$).*
- *For each clause c, we create a vertex c (we identify a clause and its vertex).*
- *For each clause c, containing the variables u, v, w, we add the edges cu_c, cv_c and cw_c.*

The following Lemma follows immediately from the above construction, while the proofs of the other two lemmas are omitted due to space constraints.

Lemma 1. *Given a 3-CNF formula F with clauses set C, the graph G_F has $19m$ vertices where m is the number of clauses of F. We define A as the vertices of the variables which have an even index. We define B as the vertices of the variables which have an odd index and the vertices of the clauses. Then the obtained graph G_F is bipartite with bipartition (A, B). Furthermore, the graph has maximum degree 3.*

Lemma 2. *A cycle of length $6k$, with $k \geq 2$, requires at least $2k$ operations to be turned into a disjoint union of bicliques. The solution sequences of operations of length $2k$ are the three following ones: delete the edges $v_{1+3i}v_{2+3i}$ for every i, delete the edges $v_{2+3i}v_{3+3i}$ for every i and delete the edges $v_{3+3i}v_{4+3i}$ for every i.*

Lemma 3. *Let a cycle $v_1, \ldots, v_{6k}$ of length $6k$ and a strictly increasing sequence $i_1, \ldots, i_b$ of integers in $[1, 6k]$ such that the edges $v_{i_{j-1}}v_{i_j}$ and $v_{i_j}v_{i_{j+1}}$ are deleted for every $j \in [1, b]$. Suppose that for every $j \in [1, b]$, $i_{j+1} - i_j$ equals 0 or 2 modulo 6. Then the remaining paths require at least $2k - b$ operations to be turned into a disjoint union of bicliques.*

Theorem 1. *BCEOVS and BCEVS are NP-complete even when restricted to bipartite graphs of maximum degree 3.*

Proof. These problems are clearly in NP. Let F be a 3-CNF, we denote the set of clauses by C and the set of variables by V. Let G be the bipartite graph obtained by Construction 1. We set $k = 8m$ where m is the number of clauses. Let us prove that the graph G has a sequence of at most k operations such that it turns G into a disjoint union of bicliques if and only if F is satisfiable.

Assume that F is satisfiable. For every true (resp. false) variable v we delete the edges $v_{1+3i}v_{2+3i}$ (resp. $v_{2+3i}v_{3+3i}$) for every i. For every clause c, there exists a variable v appearing in c which satisfies c. Let u and w be the two other variables appearing in c. We delete the edges vu_c and vw_c.

The resulting graph is a union of bicliques because the variable cycles have been turned into disjoint paths of length 2 and for every clause, the remaining edge is connected to the middle of a path of length 2. Therefore the connected components are stars with 4 vertices and paths of length 2. We have done $6d(v)/3 = 2d(v)$ deletions for every variable v. We have done two edge deletions for every clause. In total, we have done $2m + \sum_{v \in V} 2d(v) = (2+2\cdot 3)m = 8m = k$ operations (Fig. 1).

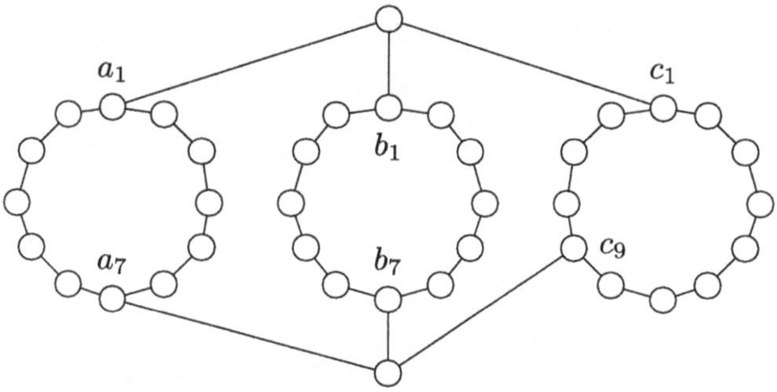

Fig. 1. The graph constructed by Construction 1 for the 3-CNF $(a \vee b \vee c) \wedge (a \vee b \vee \overline{c})$.

Assume that G can be turned into a disjoint union of bicliques with a sequence of at most k operations. For every variable v, we denote by $Op(v)$ the set of edge editions and vertex splittings done between vertices of the cycle of v. For every clause c, we denote by $Op(c)$ the set of the splits of c and edge deletions which are incident to c and edge additions between c and the vertices $u_c, u_{c-1}, u_{c+1}, v_c, v_{c-1}, v_{c+1}, w_c, w_{c-1}, w_{c+1}$ where u, v and w are the variables in c. These sets are pair-wise disjoint.

For each variable v, according to Lemma 2, $Op(v)$ is of size at least $2d(v)$. We denote by a_0 (resp. a_1) the number of clause c such that $Op(c)$ is of size 0 (resp. 1). We denote by a_2 the number of clauses such that $Op(c)$ is of size at least 2. Thus $m = a_0 + a_1 + a_2$. Let us prove that $a_0 = 0$ and $a_1 = 0$.

Let c be a clause such that $Op(c)$ is of size 0. In the BCEOVS problem the edges $u_{c-1}u_c$ and u_cu_{c+1} must be deleted for each variable u appearing in c. Suppose there exists u such that it is not the case, then there still exist the geodesic $u_{c\pm1}, u_c, c, v_c$ where v is another variable appearing in c. In the BCEVS problem, it is also possible to solve the problems of the geodesics $u_{c\pm1}, u_c, c, v_c$ by splitting the vertex v. So there are three cases for each variable u appearing in c. Either both edges $u_{c-1}u_c$ and u_cu_{c+1} are deleted, either one edge among $u_{c-1}u_c$ and u_cu_{c+1} is deleted and u is split, either only u is split.

Let c be a clause such that $Op(c)$ is of size 1. In the BCEOVS problem there exists two variables u and v appearing in c such that the edges $u_{c-1}u_c$, u_cu_{c+1} and $v_{c-1}v_c$, v_cv_{c+1} are deleted. Indeed, if c is split or if an edge incident to c is deleted, then there remains two variables u and v appearing in c such that the edges cu_c and cv_c are still present. We deduce that the four previous edges must be deleted. Otherwise there is an edge addition in $Op(c)$ which occurs between vertices of u and of v. For geodesics where the third variable w appears in c, the edges $u_{c-1}u_c$, u_cu_{c+1}, as well as $v_{c-1}v_c$ and v_cv_{c+1}, should be removed. In the BCEVS problem, there exists two vertices appearing in c such that either both edges $u_{c-1}u_c$ and u_cu_{c+1} are deleted, either one edge among $u_{c-1}u_c$ and u_cu_{c+1} is deleted and u is split, either only u is split.

For every variable v, we denote by $B(v)$ the clauses c such that v is appearing in c and such that $|Op(c)| \leq 1$ and such that the two edges $v_{c-1}v_c$ and v_cv_{c+1} are deleted. We denote by $b(v)$ the size of $B(v)$ and by $r(v)$ the number of clauses c such that v is appearing in c and such that $|Op(c)| \leq 1$ and such that v_c is split and such that the two edges $v_{c-1}v_c$ and v_cv_{c+1} are deleted. Thus $3a_0 + 2a_1 = \sum_{v \in V} b(v) + r(v)$ by double counting.

Let v be a variable. For every $c \in B(v)$, the edges $v_{c-1}v_c$ and v_cv_{c+1} are deleted. All the indices of $B(v)$ are 0, 2 modulo 6 by definition of the graph. In reference to Lemma 3, $Op(v)$ is of size at least $2d(v) + b(v)$. The total number of operations is at least

$$\geq \sum_{c \in C} |Op(c)| + \sum_{v \in V} |Op(v)|$$

$$\geq 2a_2 + a_1 + \sum_{v \in V}(2d(v) + b(v) + r(v)) \geq 2a_2 + a_1 + \sum_{v \in V} 2d(v) + \sum_{v \in V}(b(v) + r(v))$$

$$\geq 2a_2 + a_1 + 6m + (3a_0 + 2a_1) \geq 2m + a_0 + a_1 + \sum_{v \in V} 2d(v) \geq k + a_0 + a_1 .$$

As the sequence is of length at most k, we deduce that $a_0 = a_1 = 0$. Thus $|Op(c)| \geq 2$ for every clause c. We conclude that $|Op(c)| = 2$ for every clause c and $|Op(v)| = 2d(v)$ for every variable v. In reference to Lemma 2, either we delete the edges $v_{1+3k}v_{2+3k}$ for every k, either we delete the edges $v_{2+3k}v_{3+3k}$ for every k, either we delete the edges $v_{3+3k}v_{4+3k}$ for every k. In the second case, we assign v to positive. Otherwise we assign v to negative (Fig. 2).

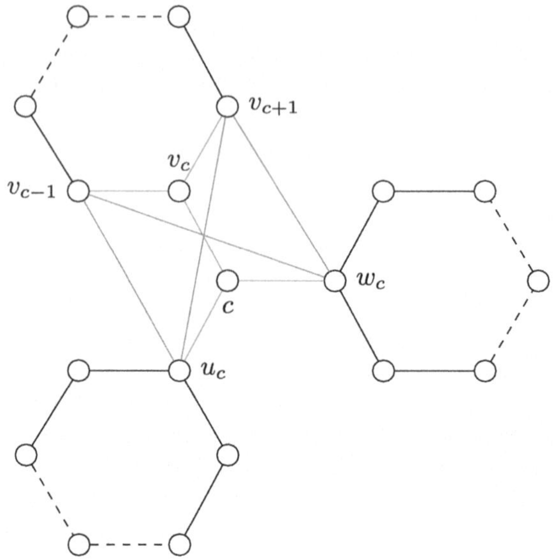

Fig. 2. Example of a clause c where the three green edges incident to c are not deleted and the three green edges incident to v_c are not deleted. The edges in gray should be added or the vertex v_c should be split. (Color figure online)

Let us prove that this assignment satisfies all the clauses. Let c be a clause. If no variable appearing in c is positive, then all these variables are either negative or undefined. For every variable v appearing in c, v_c is connected to a path of length 2 in the variable cycle. Then we need at least 3 operations to solve the 3 conflicts c, v_c, v_{c+1}, v_{c+2} (or c, v_c, v_{c-1}, v_{c-2}) for each variable v appearing in c. This contradicts the fact that $Op(c)$ is of size at most 2. We deduce that there exists a variable v appearing in c which has a positive assignment. Therefore, this assignment satisfies all the clauses. We conclude that the problems are NP-complete.

Corollary 1. *BCEVS and BCEOVS remain NP-Complete on bipartite planar graphs with maximum degree three.*

Since the previous construction is linear in the number of vertices and (resp. Planar) 3-SAT does not admit a $2^{o(n)}n^{O(1)}$ (resp. $2^{o(\sqrt{n})}n^{O(1)}$) time algorithm [7]; unless the Exponential Time Hypothesis (ETH) fails, we conclude the following:

Corollary 2. *Assuming the ETH holds, there is no $2^{o(n)}n^{O(1)}$ (resp. $2^{o(\sqrt{n})}n^{O(1)}$) time algorithm for BCEVS and BCEOVS on bipartite (resp. planar) graphs with maximum degree three where n is the number of vertices of the graph.*

The proofs of the following theorems are omitted due to space constraints.

Theorem 2. *BCEVS and BCEOVS are APX-hard.*

Theorem 3. *BCEVS and BCEOVS are solvable in polynomial time on trees.*

Theorem 4. *Bicluster Editing with One-sided Vertex Splitting can be solved in $O^*(5^{3k}k^{2k})$, where k is the solution size. Therefore, it is fixed parameter tractable.*

4 A Polynomial-Size Kernel for Bicluster Editing with One-Sided Vertex Splitting

We utilize an equivalent formulation of the Bicluster Editing with One-sided Vertex Splitting problem, where only the vertices of the set B are allowed to split. The objective is to find a cover of the vertices of a bipartite graph $G = (A, B, E)$ such that the restriction to A is a partition.

Definition 1. *An A-partitioning cover C of a bipartite graph $G = (A, B, E)$ is a set of subsets of $A \cup B$ covering the vertices of G such that the restrictions of the subsets of C to A is a partition of A. We define the cost of C as follows:*

- *For every vertex $a \in A$, there exists a unique subset X of C such that $a \in X$ such that the cost of a is defined as $|(B \cap X) \setminus N(a)| + |\overline{X} \cap N(a)|$ (where $\overline{X}$ denotes the complementary of X in $A \cup B$).*
- *For every vertex $b \in B$, the cost of b is defined as $j-1$ where j is the number of subsets of C containing b.*

Finally, the cost of C is defined as: $cost(C) = \sum_{a \in A} cost(a) + \sum_{b \in B} cost(b)$.

The cost of a vertex $a \in A$ will correspond to the number of edited edges incident to a and the cost of a vertex $b \in B$ will correspond to the number of times a vertex b is split. We will say that the edge ab, where $a \in$ will be *deleted* if the unique subset X of C containing a does not contain b (Fig. 3).

Lemma 4. *A bipartite graph has an A-partitioning cover of cost at most k if and only if there exists a sequence of length at most k of edge editions and vertex splittings on B.*

Definition 2. *Let G be a graph. A twin class is a maximal subset of the vertices of G which have the same neighborhood.*

Observe that the twin classes of a graph partition the vertices of this graph.

Lemma 5 (Twin adapted A-partitioning cover). *Let $G = (A, B, E)$ be a bipartite graph. There exists an A-partitioning cover C of G of minimum cost such that for every twin class T and every subset X of C, then either $T \cap X = \emptyset$ or $T \subseteq X$.*

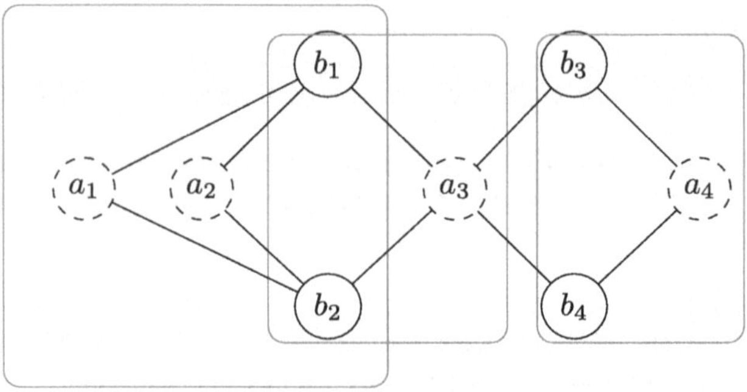

Fig. 3. Example of an A-partitioning cover of cost 4 where $A = \{a_1, a_2, a_3, a_4\}$. $cost(b_1) = cost(b_2) = 1$ because both vertices are in 2 subsets and $cost(a_3) = 2$ because the endvertices of the edges a_3b_3 and a_3b_4 are not in the same subset.

Proof. Let C be a minimum A-partitioning cover of G. Consider a vertex $a \in A$ having the minimum cost among its twins. Let P the subset of C containing a. Let a' be a twin of a which is not in P. We move a' to P. The new cost of a' is $cost(a)$. By minimality of a, the cost of X decreases. We can now assume that all the twins of a are in the same subset. We repeat this operation for every vertex of the set A.

Let b be a vertex of B, whose subsets containing b are noted $P_1, \ldots, P_r$, minimizing the quantity $r - 1 + \sum_{i=1}^{r} P_i \cap \overline{N(b)} + \overline{\cup_{i=1}^{r} P_i} \cap N(b)$. Let b' be a twin of b. Then move b' so that it is contained in the same subsets as b. After this operation the cost of X has decreased because the difference between the new cost of X and the original cost of X is:

$$r - 1 + \sum_{i=1}^{r} P_i \cap \overline{N(b)} + \overline{\cup_{i=1}^{r} P_i} \cap N(b) - (r' - 1 + \sum_{i=1}^{r'} P_i' \cap \overline{N(b')} + \overline{\cup_{i=1}^{r'} P_i'} \cap N(b'))$$

which is at most 0 by minimality of b. We repeat this operation for every twin of b. By repeating these operations for every vertex of B, we have found an A-partitioning cover of minimum cost which satisfies the property.

Lemma 6. *Let $G = (A, B, E)$ be a connected bipartite graph with k twin classes in A. Then the cost of any A-partitioning cover is at least $\sqrt{k} - 1$.*

Proof. By Lemma 5, consider an A-partitioning cover C of G of minimum cost which is adapted to the twin classes of G . We denote by c the number of subsets of C.

Construct a graph G' with one vertex for every subset of C and one vertex for every vertex of B. Thus G' has $n' = c + |B|$ vertices. Connect in G' the vertex of a subset X of C to a vertex b if $b \in N[X]$. As G is connected, then G' is also connected. Thus the number m' of edges of G' is at least $n' - 1 = c + |B| - 1$.

Let us now prove that $cost(C) \geq \sum_{b \in B}(d(b) - 1)$ where $d(b)$ denotes the degree of b in G'. For any vertex b of B, we denote by $del(b)$ the number of edges incident to b that will be deleted: it is the number of vertices $a \in A$ such that $ab \in E$ and such that the unique subset X of C containing a does not contain b.

Let b be a B vertex. We denote by s the cost of b. Then b is in $s+1$ subsets of C. Let $X_1, \dots, X_d$ be the subsets of C such that $b \in \cap_{i=1}^{d} N[X_i]$. Remark that $s+1 \leq d$. If $s = d-1$, then $cost(b) + del(b) \geq d-1$. Otherwise $s \leq d-2$. There exists at least $d-(s+1)$ subsets of $X_1, \dots, X_d$ in which b is not contained. Therefore, for each subset $X \in C$ such that $b \notin X$ but $b \in N[X]$, there exists a vertex $a \in X$ such that $b \in N(a)$. Thus $cost(b) + del(b) \geq s + (d-s-1) = d-1$.

Remark that $cost(C) = \sum_{b \in B} cost(b) + \sum_{a \in A} cost(a) \geq \sum_{b \in B}(cost(b) + del(b))$. Because of the previous paragraph, we conclude that we have the inequality $cost(C) \geq \sum_{b \in B}(d(b) - 1)$.

As G' is bipartite, the number m' of edges of G' equals $\sum_{b \in B} d(b)$. Thus, $cost(C) \geq m' - |B|$. As $m' \geq c + |B| - 1$, then the cost of C is at least $c - 1$.

Let us now make a disjunction of cases. If $c \geq \sqrt{k}$, then $cost(C) \geq \sqrt{k} - 1$ because of the previous inequality.

Otherwise, assume $c \leq \sqrt{k}$. As C is a twin adapted A-partitioning cover of G, $k = \sum_{i=1}^{c} ATC(X_i)$ where $ATC(X_i)$ is the number of A twin classes contained in X_i. Thus there exists a subset X of C containing at least $\sqrt{k}$ A twin classes because otherwise $ATC(X_i) < \sqrt{k}$ for every i and thus $\sum_{i=1}^{c} ATC(X_i) < c\sqrt{k} \leq k$, a contradiction.

Let x be the number of A twin classes contained in X. Let us show that $\sum_{a \in A \cap X} cost(a) \geq x - 1$.

If every A vertex of X has a non zero cost, then the sum of the costs of the A vertices of X is at least $|X| \geq x$. Otherwise, suppose that there exists a vertex $a \in A$ in X of cost 0. Then $N(a)$ must be in X. For every $a' \in A$ such that a' is a vertex of X of a different twin class than a, $N(a') \neq N(a)$. Let $b \in N(a') \Delta N(a)$. If $b \in N(a') \setminus N(a)$ then $b \notin X$, because otherwise the cost of a would be non zero. Therefore the edge $a'b$ will be deleted, and thus the cost of a' is at least 1. If $b \in N(a) \setminus N(a')$. As $b \in X$, the edge ab will be added and then the cost of a' is at least 1. We deduce that the sum of the costs of the A vertices in X is at least $x - 1$, because except for the twin class of a, every other vertex is of cost at least 1.

Hence, in any case, the sum of the costs of the A vertices in X is at least $x - 1$. As $x \geq \sqrt{k}$, we deduce that the cost of C is at least $\sqrt{k} - 1$. By minimality of C, we conclude that every A partitioning cover has a cost of at least $\sqrt{k} - 1$.

Lemma 7 (Reduction). *Let G be a bipartite graph given along with an integer k. Consider the twin classes $T_1, \dots, T_p$ of G. For every $i \in [p]$, we consider a subset T_i' of T_i of size $k+1$ if $|T_i| \geq k+1$, otherwise we set T_i' to T_i. Then G has an A-partitioning cover of cost at most k if and only $G[\cup T_i']$ has an A-partitioning cover of cost at most k.*

Proof. We denote by G' the graph $G[\cup T_i']$.

Assume that there exists an A-partitioning cover C of the vertices of G of cost at most k. We consider the cover C' induced by the cover C on the vertices of G' (which is an induced subgraph of G). The costs of the vertices can only be decreased by deleting vertices. Therefore the cover of C' must have a smaller cost than C.

Assume that G' has an A-partitioning cover of cost at most k. Thus there exists an A-partitioning cover C' of G' of cost at most k which is adapted to the twin classes. We define an A-partitioning cover C of G by extending the subsets of C' to the twin classes of G.

We shall prove that C and C' have the same cost. Let x be a vertex of a twin class which has been reduced. Then x must have k twins. As the cover is adapted to the twin classes, then the twins of x are contained in the same subsets. Therefore they must all have the same cost. It is therefore impossible for the cost of x to be non zero, otherwise the cost of x and all of its twins would be larger than $k+1$. This implies that $cost(x) = 0$ in C'. Therefore the cost of x is also 0 in C. Hence C and C' have the same cost.

Lemma 8. *Let $G = (A, B, E)$ be a connected bipartite graph such that all twin classes are of size at most k and having a twin adapted A-partitioning cover C of cost at most k. Then there are at most $4k^4$ twin classes in B.*

Proof. Consider a twin class T of A, and a vertex $a \in T$. Then $|T| \leq k$. Let t be the number of twin classes that have a vertex in $N(T)$. Let us prove that $t \leq 2k+1$.

Assume that $t \geq 2k+2$. Then there exists B vertices $y_1, y_2, \ldots, y_{2k+2}$ in $N(T)$ of different twin classes. Consider two different indices i and j. Let us prove that there exists a vertex $a' \in A$ such that the edge $a'y_i$ or $a'y_j$ is either deleted or added.

As y_i and y_j have a different neighborhood, there exists $a' \in A$ such that a' is adjacent to y_i but not to y_j (we can swap i and j if a' is not adjacent to y_i but to y_j). The vertex a' is not in T (otherwise a' would be adjacent to y_i and y_j). Thus $a' \neq a$.

If a and a' are in a same subset Y of C, and y_j is also in this set, then the edge $a'y_j$ will be added. Otherwise $y_j \notin Y$ and in this case the edge ay_j will be deleted. Otherwise a and a' are in different subsets X and X' of C. If y_i is not in both X and X', then the edge ay_i or ay_j is deleted. Otherwise y_i is in both X and X'. Thus the cost of y_i is at least 1 (Fig. 4).

We deduce that in every case, either an edge adjacent to y_i or y_j is deleted or added, or y_i has a cost of at least 1. By summing up all these cases for every pair $y_{2i-1}y_{2i}$ for every $i \in \{1, \ldots, k+1\}$, we conclude that the cost of C is at least $k+1$. A contradiction. Hence $t \leq 2k+1$.

By our hypothesis, the twin classes are of size at most k. Therefore $N(T)$ is of size at most $k(2k+1) \leq 2k^2$. Because of Lemma 6, there are at most $(k+2)^2 - 1$ twin classes in A (otherwise the cost of C would be at least $k+1$). As the graph is connected, each twin class in B is connected to at least one twin class in A. Thus there are at most $((k+2)^2 - 1) \cdot 2k^2 \leq 4k^4$ twin classes in B.

Theorem 5. *Bicluster Editing with One-sided Vertex Splitting has an $O(k^5)$ kernel.*

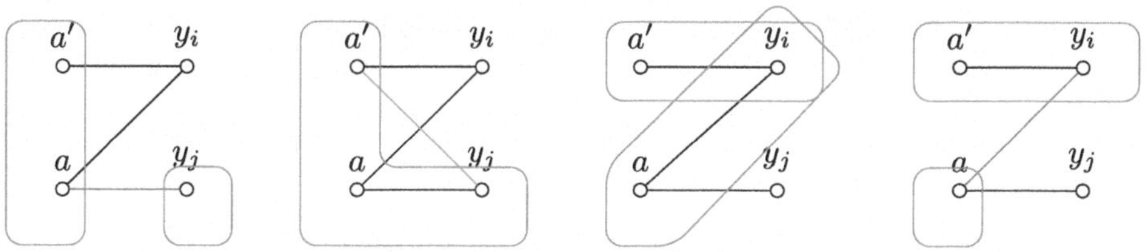

Fig. 4. On the left, a and a' are in the same set of the cover and y_j is not in the same set. In the second case, a, a' and y_j are in the same set, therefore $a'y_j$ will be added. In the third figure, a and a' are in different sets both containing y_i; therefore y_i is split at least once. In the fourth figure, a and a' are not in the same set and y_i is not in the set of a, therefore the edge ay_i is added.

Proof. Computing the twin classes can be done in $O(n^2)$. If there are more than $(k+2)^2 - 1$ twin classes in A or more than $4(k+1)^4$ twin classes in B, then we set G' to a path of length $4(k+1)$. This graph needs at least $k+1$ operations to be turned into a disjoint union of bicliques because there are $k+1$ disjoint geodesics of length 3 each. Now because of Lemmas 6, 7 and 8, we must have a no-instance.

Otherwise there are at most $(k+2)^2 - 1$ twin classes in A and $4(k+1)^4$ twin classes in B. We set G' to be the graph constructed by Lemma 7. Then, in this case, G' has at most $2k^2 + 4(k+1)^4$ twin classes which are all of size at most $k+1$. Therefore G' is of size at most $7k^5$, and G' has a cover of cost at most k if and only if G has a cover of cost at most k.

We conclude that, in all cases, we have constructed a reduced problem instance (the graph G') of size at most $7k^5$, in polynomial time, such that G' has a cover of cost at most k if and only if G has a cover of cost at most k.

5 Concluding Remarks

This paper introduces the Bicluster Editing with Vertex Splitting problem (BCEVS) and the Bicluster Editing with One-Sided Vertex Splitting problem (BCEOVS). Both BCEVS and BCEOVS have been shown to be NP-complete even when restricted to bipartite planar graphs of degree at most three. We also proved the two problems are APX-hard. On the positive side, a fixed-parameter algorithm was presented for BCEOVS and the two problems are proved to be solvable in polynomial-time on trees. This latter result might seem to be of limited importance, but it suggests that the problems might be fixed-parameter tractable when parameterized by the treewidth of the input graphs, which is hereby posed as an open problem.

Future work may focus on proving the NP-completeness of (either of) the two problems on other classes of bipartite graphs. The APX-hardness of the problems leads to the question whether finding a polynomial time constant-factor approximation is possible. We have further shown that BCEOVS is FPTwith respect to the number k of operations. In addition, we have shown an $O(k^5)$

kernel for the BCEOVS problem. It would be interesting to look for a better kernel bound and to extend these results to BCEVS. Another interesting future work would be to consider other auxiliary parameters such as twin-width, as well as local parameters (as in [1,14,15]) such as a bound on the number of times a vertex can be split. This latter bound is motivated by real applications where a data element cannot belong to an arbitrary large number of clusters.

References

1. Abu-Khzam, F.N.: On the complexity of multi-parameterized cluster editing. J. Discret. Algorithms **45**, 26–34 (2017)
2. Abu-Khzam, F.N., et al.: Cluster editing with vertex splitting. Discret. Appl. Math. **371**, 185–195 (2025)
3. Abu-Khzam, F.N., Egan, J., Gaspers, S., Shaw, A., Shaw, P.: Cluster editing with vertex splitting. In: Lee, J., Rinaldi, G., Mahjoub, A.R. (eds.) ISCO 2018. LNCS, vol. 10856, pp. 1–13. Springer, Cham (2018). https://doi.org/10.1007/978-3-319-96151-4_1
4. Amit, N.: The bicluster graph editing problem. Ph.D. thesis, Tel Aviv University, Bicluster NP-hard (2004)
5. Böcker, S.: A golden ratio parameterized algorithm for cluster editing. J. Discret. Algorithms **16**, 79–89 (2012)
6. Chen, J., Meng, J.: A 2k kernel for the cluster editing problem. J. Comput. Syst. Sci. **78**(1), 211–220 (2012)
7. Cygan, M., et al.: Lower Bounds Based on the Exponential-Time Hypothesis, pp. 467–521. Springer, Cham (2015)
8. D'Addario, M., Kopczynski, D., Baumbach, J., Rahmann, S.: A modular computational framework for automated peak extraction from ion mobility spectra. BMC Bioinform. **15**(1) (2014)
9. Fadiel, A., et al.: Computational analysis of mass spectrometry data using novel combinatorial methods. AICCSA **6**, 8–11 (2006)
10. Fellows, M., Langston, M., Rosamond, F., Shaw, P.: Efficient parameterized preprocessing for cluster editing. In: Csuhaj-Varjú, E., Ésik, Z. (eds.) FCT 2007. LNCS, vol. 4639, pp. 312–321. Springer, Heidelberg (2007). https://doi.org/10.1007/978-3-540-74240-1_27
11. Gramm, J., Guo, J., Hüffner, F., Niedermeier, R.: Graph-modeled data clustering: exact algorithms for clique generation. Theory Comput. Syst. **38**(4), 373–392 (2005)
12. Guo, J.: A more effective linear kernelization for cluster editing. Theor. Comput. Sci. **410**(8–10), 718–726 (2009)
13. Guo, J., Wang, J., Li, M., Pan, Y.: A novel algorithm for finding biclusters with coherent values in gene expression data. In: Proceedings of the 2007 IEEE Symposium on Computational Intelligence in Bioinformatics and Computational Biology, pp. 253–258 (2007)
14. Heggernes, P., Lokshtanov, D., Nederlof, J., Paul, C., Telle, J.A.: Generalized graph clustering: recognizing (p, q)-cluster graphs. In: Thilikos, D.M. (ed.) WG 2010. LNCS, vol. 6410, pp. 171–183. Springer, Heidelberg (2010). https://doi.org/10.1007/978-3-642-16926-7_17
15. Komusiewicz, C., Uhlmann, J.: Cluster editing with locally bounded modifications. Discret. Appl. Math. **160**(15), 2259–2270 (2012)

16. Kriváanek, M., Morávek, J.: NP-hard problems in hierarchical-tree clustering. Acta Informatica **23**(3), 311–323 (1986)
17. Madeira, S.C., Oliveira, A.L.: Biclustering algorithms for biological data analysis: a survey. IEEE/ACM Trans. Comput. Biol. Bioinf. **1**(1), 24–45 (2004)
18. Protti, F., da Silva, M.D., Szwarcfiter, J.L.: Applying modular decomposition to parameterized cluster editing problems. Theory Comput. Syst. **44**(1), 91–104 (2009)
19. Rahmann, S., Wittkop, T., Baumbach, J., Martin, M., Truß, A., Böcker, S.: Exact and heuristic algorithms for weighted cluster editing. In: Computational Systems Bioinformatics, pp. 391–401. World Scientific (2007)
20. Shamir, R., Sharan, R., Tsur, D.: Cluster graph modification problems. Discret. Appl. Math. **144**(1–2), 173–182 (2004)
21. Tan, Y., et al.: Gene editing: an instrument for practical application of gene biology to plant breeding. J. Zhejiang Univ.-Sci. B **21**(6), 460–473 (2020)
22. Tsur, D.: Faster algorithms for finding the maximum bicluster in bipartite graphs. Theor. Comput. Sci. **852**, 1–9 (2021)
23. Tsur, D.: Faster parameterized algorithms for bicluster editing and flip consensus tree. Theor. Comput. Sci. **953**, 113796 (2023)
24. Xiao, X., Zhang, Y., Zhang, Y., Wang, Y., Wang, Y.: A simple and efficient biclustering algorithm for large-scale gene expression data. IEEE/ACM Trans. Comput. Biol. Bioinf. **19**(1), 29–40 (2022)

Vector Spaces of Graphs Closed Under Isomorphism

Vadim Lozin[1(✉)] and D. V. Zakharova[2]

[1] Mathematics Institute, University of Warwick, Coventry CV4 7AL, UK
V.Lozin@warwick.ac.uk
[2] Nizhny Novgorod University, Nizhny Novgorod, Russia
Darya.Zakharova@itmm.unn.ru

Abstract. The cycle space of a graph G is the set of all spanning Eulerian subgraphs of G, i.e. spanning subgraphs in which all vertices have even degrees. This is a vector space over the finite field of two elements with the operation of symmetric difference of the edge sets. We observe that this space is closed under isomorphism, i.e. if a subgraph H of G belongs to it, then any subgraph of G isomorphic to H also belongs to this space. Obviously, the same is true for the space of all spanning subgraphs of G. How many vector spaces closed under isomorphism can a graph have? It is known that if G is the complete graph K_n, then it has at most 14 vector spaces closed under isomorphism regardless of the value of n. In the present paper, we reveal the relationships between these spaces and prove that all of them can be obtained from the above two by means of four operations. We also reveal some relationships between vector spaces of a complete bipartite graph closed under isomorphism.

Keywords: Cycle space · Cut space · Complete graph

1 Introduction

Let G be a simple graph, i.e. a finite undirected graph without loops and multiple edges. We denote the vertex set of G by $V(G)$ and the edge set by $E(G)$. A graph H is a *spanning subgraph* of G if $V(H) = V(G)$ and $E(H) \subseteq E(G)$. The set of all spanning subgraphs of G forms a vector space $\mathcal{E}(G)$ over the finite field of two elements with the operation of symmetric difference of the edge sets. Alternatively, this a binary vector space of dimension $m = |E(G)|$, i.e. the set of all binary vectors of length (the number of bits) m, with the operation of addition mod 2. Each binary vector in this space represents a subset of $E(G)$, and we will use the terminology of subsets and vectors interchangeably. We will also use the same sign $\oplus$ for the operation of symmetric difference of sets and addition mod 2 of binary vectors.

The space $\mathcal{E}(G)$ has a prominent subspace consisting of the Eulerian spanning subgraphs of G, i.e. spanning subgraphs in which all vertices have even degrees, also known as the *cycle space* of G [5]. There are many other subspaces of $\mathcal{E}(G)$.

H. Fernau and B. Zhu (Eds.): IWOCA 2025, LNCS 15885, pp. 160–172, 2025.
https://doi.org/10.1007/978-3-031-98740-3_12

In particular, the set of all linear combinations of any collection of vectors from $\mathcal{E}(G)$ creates a subspace, and the number of such subspaces can be arbitrarily large for sufficiently large $n = |V(G)|$. However, the cycle space and $\mathcal{E}(G)$ have a peculiar property: they are closed under an isomorphism, i.e. if a subgraph H of G belongs to one of these spaces, then any subgraph of G isomorphic to H also belongs to the same space. In [7], it was shown that if G is the complete graph K_n, then the number of vector spaces closed under isomorphism is at most 14 regardless of the value of n. In the present paper, we reveal the relationships between these spaces and prove that all of them can be obtained from $\mathcal{E}(G)$ and the cycle space by means of four operations.

The organization of the paper is as follows. In the rest of the present section we introduce basic terminology and notation used in the paper. In Sect. 2 we present the 14 spaces of K_n and in Sect. 3 we describe the four operations over vector spaces. In Sect. 4 we show how the 14 spaces of K_n can be obtained from $\mathcal{E}(G)$ and the cycle space by means of the four operations described in Sect. 3, while in Sect. 5 we reveal some relationships between vector spaces of a complete bipartite graph closed under isomorphism. In Sect. 6 we discuss some open problems related to the topic of the paper.

For a graph G and a vertex $v \in V(G)$, the *degree* of v is the number of edges incident to v. If the degree of v is 0, we call v an *isolated vertex* and say that v creates a trivial connected component of G. By P_n, C_n and $K_{n,m}$ we denote a chordless path with n vertices, a chordless cycle with n vertices, and a complete bipartite graph with parts of size n and m, respectively. Also, if G is a graph, then nG stands for the disjoint union of n copies of G. Throughout the paper the term "subgraph" means a "spanning subgraph".

2 The 14 Vector Spaces of a Complete Graph

The family of 14 vector spaces of a complete graph K_n that are closed under isomorphism has been identified in [7]. It consists of two subfamilies $B_1, \ldots, B_7$ and $L_1, \ldots, L_7$ defined as follows.

- B_1 all subgraphs,
- B_2 subgraphs with an even number of edges,
- B_3 subgraphs in which the degrees of all vertices have the same parity,
- B_4 subgraphs in which the degrees of all vertices have the same parity that coincides with the parity of the number of edges,
- B_5 subgraphs in which all vertices have even degrees,
- B_6 $= B_2 \cap B_3$,
- B_7 $= B_2 \cap B_5$.

- L_1 the empty (edgeless) subgraph,
- L_2 the empty subgraph and the complete graph,
- L_3 complete bipartite subgraphs in which all vertices have even degrees,
- L_4 complete bipartite subgraphs and their complements in which all vertices have even degrees,

L_5 complete bipartite subgraphs,
L_6 complete bipartite subgraphs in which all vertices have even degrees, and the complements of complete bipartite subgraphs in which all vertices have odd degrees,
L_7 complete bipartite subgraphs and their complements.

In this notation, $B_1 = \mathcal{E}(K_n)$ and B_5 is the cycle space of K_n. The space L_5 is known in the literature as the *cut space* [2], and for K_n it consists of complete bipartite subgraphs. In the terminology of vector spaces, L_5 is the orthogonal complement of B_5. In Sect. 4, we show that B_i and L_i are orthogonal to each other for all $i = 1, 2, \ldots, 7$.

The fact that each of the sets $B_1, \ldots, B_7$ and $L_1, \ldots, L_7$ is a space (i.e. closed under $\oplus$) can be checked by direct inspection. We note that for an odd n, there exist at most 8 spaces, because in this case $B_3 = B_5$, $B_4 = B_6 = B_7$, and $L_3 = L_4 = L_6 = \emptyset$.

3 Operations on Vector Spaces

In this section, we analyze various operations on vector spaces from a purely algebraic point of view, i.e. we consider the set $\mathcal{C}$ of all binary vectors of length m, which forms an m-dimensional binary vector space, and analyze operations on subspaces of this set. We refer to any subspace of $\mathcal{C}$ simply as a space.

For two binary vectors $X = (x_1, x_2, \ldots, x_m)$ and $Y = (y_1, y_2, \ldots, y_m)$ we denote $X \oplus Y = (x_1 \oplus y_1, x_2 \oplus y_2, \ldots, x_m \oplus y_m)$.

By K we denote the complete vector, i.e. the vector of all 1s, and for a binary vector X, we denote by $\overline{X}$ the complement of X, i.e. $\overline{X} = X \oplus K$. Also, for a set A of vectors, we denote $\overline{A} = \{X \ : \ \overline{X} \in A\}$.

The *Hamming weight* of a binary vector X is the number of non-zero entries in X. It is well-known (and not difficult to see) that the set $\mathcal{C}_{even}$ of all vectors of even Hamming weight in $\mathcal{C}$ forms a subspace of $\mathcal{C}$, i.e. is closed under $\oplus$.

3.1 Orthogonal Complementation

Two binary vectors $X = (x_1, x_2, \ldots, x_m)$ and $Y = (y_1, y_2, \ldots, y_m)$ are said to be *orthogonal* to each other if $x_1y_1 \oplus x_2y_2 \oplus \ldots \oplus x_my_m = 0$. In other words, X and Y are orthogonal if they have an even number of non-zero entries in the intersection.

For a vector space A, let us denote by $A^{\perp}$ the set of all vectors each of which is orthogonal to all vectors in A. The set $A^{\perp}$ is called the *orthogonal complement* of A and is known to be a vector space itself. This is true for general vector spaces, not necessarily binary. For binary spaces we additionally have $A = (A^{\perp})^{\perp}$. In Sect. 4, we will show that B_i is the orthogonal complement of L_i, and vice versa, for each $i = 1, \ldots, 7$.

3.2 Even Restriction and Complement Expansion

Now we introduce two more operations on vector spaces. One of them is denoted and defined as $\lfloor A \rfloor = \mathcal{C}_{even} \cap A$ and is called the *even restriction* of A. Since the intersection of two vector spaces is a vector space, $\lfloor A \rfloor$ is a space for any space A. The second operation is denoted and defined as $\lceil A \rceil = A \cup \overline{A}$ and is called the *complement expansion* of A.

Theorem 1. *For any space A, the set $\lceil A \rceil$ is a vector space. Also, $\lceil A \rceil = A$ if and only if A contains the complete vector.*

Proof. If A contains the complete vector K, then for any vector $X \in A$, we have $\overline{X} = X \oplus K \in A$ and $\lceil A \rceil = A$. If $K \notin A$, then for any $X \in A$ we have $\overline{X} \notin A$ (else $K = X \oplus X \oplus K = X \oplus \overline{X} \in A$) and hence $\lceil A \rceil \neq A$.

To show the closedness of $\lceil A \rceil$, consider two vectors $X, Y \in \lceil A \rceil$. If both of them belong to A, then $X \oplus Y \in A$, since A is a space. If $X \in A$ and $Y \in \overline{A}$, i.e. $\overline{Y} \in A$, then $X \oplus \overline{Y} \in A$ and hence $X \oplus Y = X \oplus \overline{Y} \oplus K = \overline{X \oplus \overline{Y}} \in \overline{A} \subseteq \lceil A \rceil$. Finally, if $X \notin A$ and $Y \notin A$, then $\overline{X} \in A$ and $\overline{Y} \in A$, implying $\overline{X} \oplus \overline{Y} \in A$, and hence $X \oplus Y = \overline{X} \oplus K \oplus \overline{Y} \oplus K = \overline{X} \oplus \overline{Y} \in A \subseteq \lceil A \rceil$.

□

Now we prove a key result connecting orthogonal complement, even restriction and complement expansion.

Theorem 2. *For any space A, $\lfloor A \rfloor^{\perp} = \lceil A^{\perp} \rceil$ and $\lceil A \rceil^{\perp} = \lfloor A^{\perp} \rfloor$.*

Proof. To prove the first equality, we consider an arbitrary vector $X \in \lfloor A \rfloor^{\perp}$ and assume by contradiction that X does not belong to $\lceil A^{\perp} \rceil$. In particular, X does not belong to $A^{\perp}$ and hence there is a vector Y in A, which is not orthogonal to X. On the other hand, since $X \in \lfloor A \rfloor^{\perp}$, X is orthogonal to every vector in $\lfloor A \rfloor \subseteq A$. Thus Y belongs to $A - \lfloor A \rfloor$, i.e. the Hamming weight of Y is odd. We also observe that X is not orthogonal to any other vector Y' of odd Hamming weight in A, because $Y \oplus Y'$ belongs to $\lfloor A \rfloor$ and X is orthogonal to all vectors in $\lfloor A \rfloor$.

Clearly, the complement expansion of any vector space contains the complete vector K. Therefore, since X does not belong to $\lceil A^{\perp} \rceil$, and $\lceil A^{\perp} \rceil$ contains the complete vector, we conclude that $\overline{X}$ also does not belong to $\lceil A^{\perp} \rceil$. On the other hand, since every vector in $\lfloor A \rfloor$ has an even Hamming weight and X is orthogonal to each of them, we conclude that $\overline{X}$ also is orthogonal to each vector in $\lfloor A \rfloor$, i.e. $\overline{X} \in \lfloor A \rfloor^{\perp}$. Now, by analogy with X, we conclude that $\overline{X}$ is not orthogonal to any vector of odd Hamming weight in A. In particular, $\overline{X}$ is not orthogonal to Y. But now we have a contradiction. Indeed, since the Hamming weight of Y is odd, exactly one of X and $\overline{X}$ must be orthogonal to Y. This contradiction shows that $\lfloor A \rfloor^{\perp} \subseteq \lceil A^{\perp} \rceil$.

To prove the inverse inclusion $\lfloor A \rfloor^{\perp} \supseteq \lceil A^{\perp} \rceil$, consider an arbitrary vector X in $\lceil A^{\perp} \rceil$ and assume by contradiction that X does not belong to $\lfloor A \rfloor^{\perp}$. Therefore, there is a vector $Y \in \lfloor A \rfloor$ which is not orthogonal to X. Since $\lceil A^{\perp} \rceil$ contains the complete vector K, $\overline{X}$ also belongs to $\lceil A^{\perp} \rceil$, and since the Hamming weight of Y is even, $\overline{X}$ also is not orthogonal to Y.

Since X and $\overline{X}$ belong to $\lceil A^{\perp} \rceil$, one of them belongs to $A^{\perp}$ and hence is orthogonal to each vector in A. On the other hand, neither X nor $\overline{X}$ is orthogonal to the vector Y in $\lfloor A \rfloor \subset A$, which is a contradiction that completes the proof of the first equality. The second one can be proved by analogy. □

3.3 Core Spaces and Odd Complementation

Let us say that a vector space A is a *core space* if $A \neq \lfloor A \rfloor$ and $A \neq \lceil A \rceil$. We denote by A_0 the set of vectors of even Hamming weight and by A_1 the set of vectors of odd Hamming weight in A. We call the family of all vector spaces containing $\lfloor A \rfloor$ and contained in $\lceil A \rceil$ the *A-block*.

Theorem 3. *For any core space A, the A-block consists of the following five vector spaces:*

$$\begin{aligned} A &= A_0 \cup A_1, \\ \lceil A \rceil &= A_0 \cup A_1 \cup \overline{A}_0 \cup \overline{A}_1 \\ \lfloor A \rfloor &= A_0 \\ A_* &= A_0 \cup \overline{A}_0, \\ A^* &= A_0 \cup \overline{A}_1. \end{aligned}$$

The five spaces are pairwise distinct.

Proof. Since A is a space, $\lceil A \rceil$ and $\lfloor A \rfloor$ are spaces. Also, if the dimension of A is k, then the dimension of $\lceil A \rceil$ is $k+1$ and the dimension of $\lfloor A \rfloor$ is $k-1$. Therefore, any other space in the A-block has dimension k. We also observe that the four sets $A_0, A_1, \overline{A}_0, \overline{A}_1$ are pairwise distinct and disjoint, since otherwise their union contains fewer than 2^{k+1} vectors.

Next, we note that A_* is a space, because it is the complement expansion of $\lfloor A \rfloor$. Also, if the length of the vectors in A is odd, then A^* is the even restriction of $\lceil A \rceil$. However, this is not the case if the vectors in A have an even length. So, we provide a complete proof showing that A^* is a space, which is valid regardless of the length of the vectors in A.

To show that A^* is a vector space, consider two vectors X and Y in A^*. If both of them belong to A_0, then $X \oplus Y$ belongs to A_0, since A_0 is a vector space. Assume $X \in A_0$ and $Y \in \overline{A}_1$. Then $\overline{Y} \in A_1$, $X \oplus \overline{Y} \in A_1$ and $\overline{X \oplus \overline{Y}} \in \overline{A}_1$. Since $\overline{X \oplus \overline{Y}} = X \oplus \overline{Y} \oplus K = X \oplus Y \oplus K \oplus K = X \oplus Y$, we conclude that $X \oplus Y \in \overline{A}_1$.

Finally, let $X \in \overline{A}_1$ and $Y \in \overline{A}_1$. Then $\overline{X} \in A_1$, $\overline{Y} \in A_1$, $\overline{X} \oplus \overline{Y} \in A_0$ and hence $X \oplus Y = X \oplus K \oplus Y \oplus K = \overline{X} \oplus \overline{Y} \in A_0$.

We have proved that each of A, $\lceil A \rceil$, $\lfloor A \rfloor$, A_*, A^* is a space, and since $A_0, A_1, \overline{A}_0, \overline{A}_1$ are pairwise distinct, we conclude that the five spaces are pairwise distinct.

It remains to show that there are no other spaces in the A-block. To this end, we observe the following:

($\star$) For any space B and two vectors $X \notin B$ and $Y \in B$, we have $X \oplus Y \notin B$, since otherwise $X = (X \oplus Y) \oplus Y \in B$. For a vector $X \notin B$, we denote $B \oplus X = \{X \oplus Y : Y \in B\}$ and observe that $B \cup (B \oplus X)$ is a vector space with $Z_1 \oplus Z_2 \in B \oplus X$ for any two vectors $Z_1, Z_2 \in B \oplus X$.

Since $A_0 \cup A_1$, $A_0 \cup \overline{A}_0$, $A_0 \cup \overline{A}_1$ are spaces and the four sets $A_0, A_1, \overline{A}_0, \overline{A}_1$ are pairwise disjoint, we conclude with the help of ($\star$) that

if $X \in A_1$, then $A_0 \oplus X = A_1$,
if $X \in \overline{A}_1$, then $A_0 \oplus X = \overline{A}_1$,
if $X \in \overline{A}_0$, then $A_0 \oplus X = \overline{A}_0$.

Therefore, any space containing $\lfloor A \rfloor$ and contained in $\lceil A \rceil$ coincides with one of the five spaces described in the statement of the lemma. □

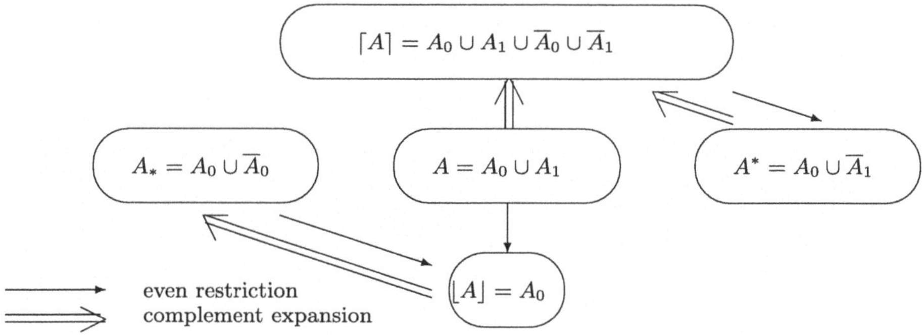

Fig. 1. The A-block for an odd m

For an odd value of m (the length of the vectors), the vectors in $\overline{A}_1$ have an even Hamming weight and the vectors in $\overline{A}_0$ have an odd Hamming weight. In this case, A^* is the even restriction of $\lceil A \rceil$, while $\lfloor A \rfloor$ is the even restriction of A_*. Therefore, all spaces in the A-block can be obtained from A by even restriction and complement expansion as shown in Fig. 1. The same applies to the five spaces orthogonal to the spaces in the A-block.

If m is even, then the vectors in $\overline{A}_1$ have an odd Hamming weight and the vectors in $\overline{A}_0$ have an even Hamming weight, in which case A^* cannot be obtained from A by even restriction and complement expansion. To overcome this difficulty, we introduce the operation of *odd complementation* that transforms A to A^*. In other words, the odd complementation transforms each vector of an odd Hamming weight in A to its complement. In the next theorem we show that this operation behaves nicely on the orthogonal spaces in the case of an even m.

Theorem 4. *If the length of the vectors in A is even, then $(A^*)^\perp = (A^\perp)^*$.*

Proof. To prove the theorem, we observe that

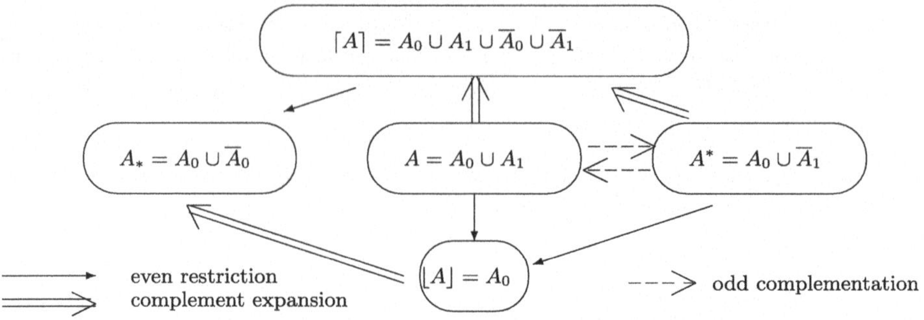

Fig. 2. The A-block for an even m

(1) if X is a vector of an even Hamming weight, then X is orthogonal to a vector Y if and only if X is orthogonal to $\overline{Y}$,
(2) if X and Y both have odd Hamming weights and an even length, then X is orthogonal to Y if and only if $\overline{X}$ is orthogonal to $\overline{Y}$,

which follows from the definition of orthogonality.

Now let X be a vector in $(A^*)^{\perp}$, and assume first that the Hamming weight of X is odd. Since X is orthogonal to all vectors in A^*, $\overline{X}$ is orthogonal to all vectors in A by (2). Therefore, $\overline{X} \in A^{\perp}$ and hence $X \in (A^{\perp})^*$ by definition of the *-operation. Suppose next that the Hamming weight of X is even. Since X is orthogonal to all vectors in A^*, X is orthogonal to all vectors in A by (1). Therefore, $X \in A^{\perp}$ and hence $X \in (A^{\perp})^*$ by definition of the *-operation. This proves that $(A^*)^{\perp} \subseteq (A^{\perp})^*$.

Conversely, let X be a vector in $(A^{\perp})^*$, and assume first that the Hamming weight of X is odd. Then $\overline{X}$ belongs to $A^{\perp}$. Since $\overline{X}$ is orthogonal to all vectors in A, X is orthogonal to all vectors in A^* by (2). Therefore, $X \in (A^*)^{\perp}$. Suppose next that the Hamming weight of X is even. Then X belongs to $A^{\perp}$. Since X is orthogonal to all vectors in A, X is orthogonal to all vectors in A^* by (1). Therefore, $X \in (A^*)^{\perp}$. This proves that $(A^*)^{\perp} \supseteq (A^{\perp})^*$, and completes the proof of the theorem. □

The relationship between the five spaces of the A-block for an even m is shown in Fig. 2.

4 The Relationships Between the 14 Spaces of K_n

The results of the previous section reveal the relationships between the 14 spaces of K_n shown in Figs. 3, 4, 5, 6. We formally prove this in Theorem 5. In addition, in Theorem 5 we combine the results of the previous section together with Lemmas 1 and 2 (below) to develop an alternative proof of the fact that there are no other spaces of K_n closed under isomorphism.

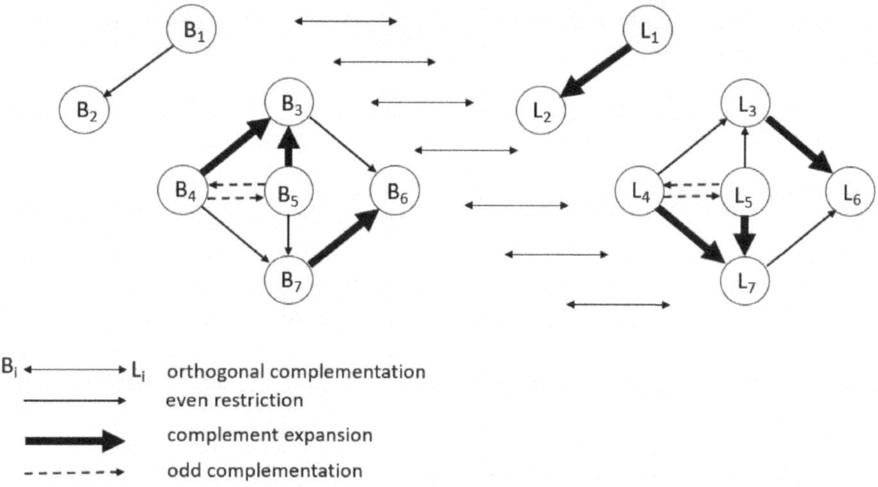

Fig. 3. Relationships between spaces for $n = 0 \bmod 4$

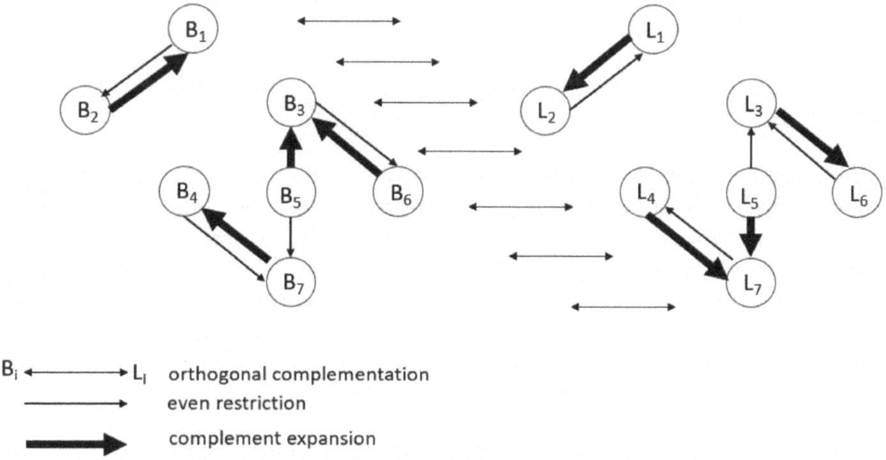

Fig. 4. Relationships between spaces for $n = 2 \bmod 4$

A proof of the following two lemmas can be found in [7]. We provide an alternative (shorter) proof for Lemma 2 and a slightly modified (also a bit shorter) proof for Lemma 1 to make the paper self-contained. In the proofs, we use the following conventions and notations.

- To simplify the notation, we ignore isolated vertices in the description of spanning subgraphs of K_n.
- For a spanning subgraph G of K_n, we denote by $[G]$ the minimal vector space of K_n closed under isomorphism that contains G.

- If G is a labeled graph with vertices a and b, then $G^{(a,b)}$ is the graph obtained from G by relabeling a to b, and b to a.

Lemma 1. $B_2 = [P_3]$ *and* $B_7 = [C_4]$.

Proof. Obviously, every graph in $[P_3]$ has an even number of edges, and hence $[P_3] \subseteq B_2$. To show that $B_2 \subseteq [P_3]$, assume the contrary and let $G \in B_2 - [P_3]$ be a minimum (in terms of $|E(G)|$) counterexample. If G contains a vertex v of degree at least 2, then $G \oplus P_3$ (with v being the center of the P_3 and $E(P_3) \subseteq E(G)$) is a smaller counterexample. Otherwise, each non-trivial connected component of G is a K_2 and the number of such components is even. But $2K_2$ is the symmetric difference of two copies of P_3, and hence $G \in [P_3]$, contradicting the assumption.

Obviously, $[C_4] \subseteq B_7$. To show that $B_7 \subseteq [C_4]$, assume the contrary and let $G \in B_7 - [C_4]$ be a minimum counterexample. If $E(P_4) \subseteq E(G)$, then $G \oplus C_4$ (with $E(P_4) \subset E(C_4)$) is a smaller counterexample. Otherwise, each non-trivial connected component of G is a K_3 and the number of such components is even. But $2K_3$ is the symmetric difference of three copies of C_4, and hence $G \in [C_4]$, contradicting the assumption. □

Lemma 2. *Let X be a vector space of K_n closed under isomorphism. If $X \not\subseteq L_7$, then $C_4 \in X$. If additionally $X \not\subseteq B_3$, then $P_3 \in X$.*

Proof. Consider a spanning subgraph G of K_n and two vertices a, b in G. We denote by P the set of vertices adjacent to exactly one of a and b, and observe that the graph $H = G \oplus G^{(a,b)}$ is isomorphic to $K_{2,p}$, where $p = |P|$ (possibly $p = 0$). It is easy to see that if $G \notin L_7$, then vertices a and b can be chosen so that $p \geq 1$ and $2 + p < n$, i.e. H contains an isolated vertex. Let c be a vertex adjacent to a, b, and let d be an isolated vertex in H. Then the graph

$$H \oplus H^{(a,d)} \oplus H^{(c,b)} \oplus H^{(a,d)(c,b)}$$

is isomorphic to C_4, which can be checked by direct inspection.

If $G \notin B_3$ and the degrees of a and b have different parities, then p is odd. If $p = 1$, then $H = P_3$. Otherwise, $H \oplus C_4 = K_{2,p-2}$ (with $E(C_4) \subseteq E(H)$) and thus P_3 can be obtained by repeated applications of this operations. □

Theorem 5. *Each vector space of the complete graph K_n closed under isomorphism coincides with one of $B_1, \ldots, B_7$, $L_1, \ldots, L_7$. All of them can be obtained from B_1 and B_5 by means of orthogonal complementation, even restriction, complement expansion and odd complementation as shown in Figs. 3, 4, 5, 6. In particular, B_i is orthogonal to L_i, and vice versa, for all $i = 1, 2, \ldots, 7$.*

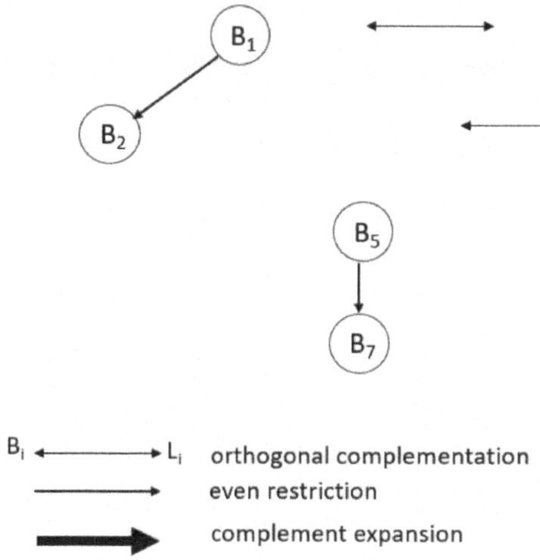

Fig. 5. Relationships between spaces for $n = 1 \bmod 4$

Proof. We denote by m the number of edges of K_n and assume that $n > 4$ to avoid trivialities. Suppose first that n is even. Then B_5 is a core space, because $K_n \in \lceil B_5 \rceil - B_5$ and $C_{n-1} \in B_5 - \lfloor B_5 \rfloor$. Therefore, by Theorem 3, the B_5-block consists of the following five pairwise disjoint spaces:

B_5,
$B_3 = \lceil B_5 \rceil$
$B_7 = \lfloor B_5 \rfloor$
$B_4 = (B_5)^*$ for an even m, and $B_4 = (B_5)_*$ for an odd m,
$B_6 = (B_5)_*$ for an even m, and $B_6 = (B_5)^*$ for an odd m.

Any other space X which is not a subset of L_7 contains B_2 (by Lemmas 1 and 2) and hence coincides with either B_2 or B_1, because the dimension of B_2 is $m - 1$ and the dimension of B_1 is m.

If X is a subset of L_7, then $X^\perp$ is not. Indeed, C_4 does not belong to L_7 for $n > 4$. On the other hand, it is not difficult to see that C_4 is orthogonal to each graph in L_7 and hence belongs to $X^\perp$. Therefore, all spaces that are subsets of L_7 are orthogonal to the spaces that are not subsets of L_7. Obviously, B_1 is orthogonal to L_1, and B_2 is orthogonal to L_2. Also, it is a well-known fact that B_5 is orthogonal to L_5. Together with Theorems 2 and 4 this proves the relationships between the spaces shown in Fig. 3 for an even m, ($n = 0 \bmod 4$), and in Fig. 4 for an odd m ($n = 2 \bmod 4$).

If n is odd, then B_3 coincides with B_5, because the number of vertices of odd degree must be even. Therefore, the family of spaces that are not subsets of L_7 reduces to B_1, B_2, B_5 and B_7. As before, each space which is a subset of L_7 is orthogonal to a space which is not. In this case, the relationships between the spaces is shown in Fig. 5 for an even m, ($n = 1 \bmod 4$), and in Fig. 6 for an odd m ($n = 3 \bmod 4$). □

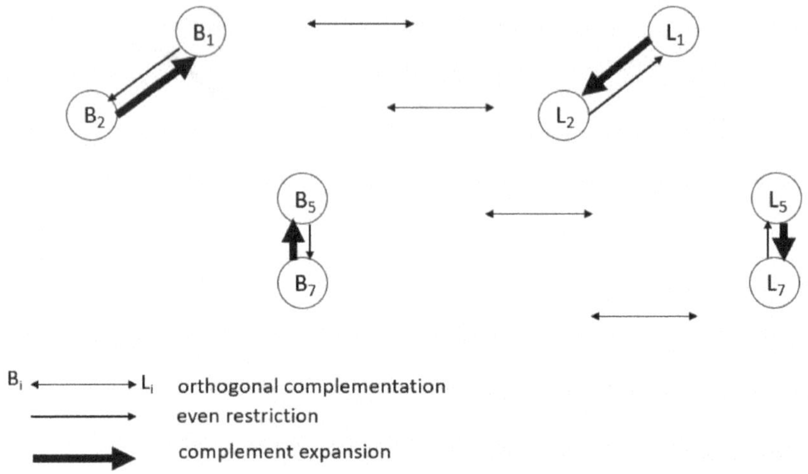

Fig. 6. Relationships between spaces for $n = 3 \bmod 4$

5 Vector Spaces of a Complete Bipartite Graph

By a bipartite graph we mean a graph given together with a bipartition of its vertex set into two color classes (independent sets) A and B. Bipartite isomorphism is an isomorphism between two bipartite graphs that respects the colors. The vector spaces of a complete bipartite graph closed under bipartite isomorphism have been identified in [1]. The list of these spaces is different from the spaces of a complete graph, but it is also finite. In the present section we analyze the relationship between five of the vector spaces of a complete bipartite graph and reveal a striking similarity between the structure of this relationship and the structure of the B_5-block (for $n = 0 \bmod 4$).

We describe a spanning subgraph of a complete bipartite graph $K_{a,b}$ with $a = |A|$ and $b = |B|$ by a binary $a \times b$ matrix. For $\alpha, \beta \in \{0, 1\}$, we denote by $D(\alpha, \beta)$ the set of spanning subgraphs of $K_{a,b}$ in which

all vertices in A have even degrees if $\alpha = 0$ and odd degrees if $\alpha = 1$,
all vertices in B have even degrees if $\beta = 0$ and odd degrees if $\beta = 1$.

The five vector spaces of our interest are represented in Fig. 7. To describe the relationship between these spaces, we adapt the definitions of the three operations introduced in Sect. 3 to the bipartite case as follows:

- *even restriction* of a space D is the operation of restricting to matrices in D in which all rows have even Hamming weights,
- *complement expansion* of a space D is the operation of adding to D all matrices obtained from the matrices in D by complementing the first row,
- *odd complementation* of a space D is the operation of complementing the first row in each matrix of D in which all rows have odd Hamming weights.

Observe that the updated definitions coincide with the original ones, when restricted to vectors, i.e. matrices with one row. It is now routine to check that the relationship between the five spaces with an even b can be described as shown in Fig. 7. It is similar to the structure of the B_5-block (for $n = 0 \bmod 4$).

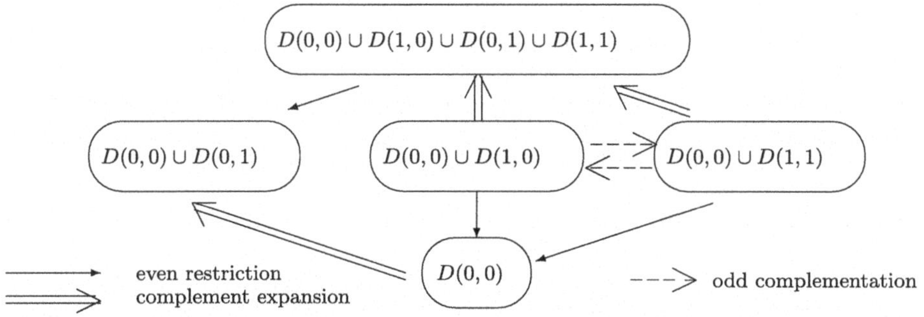

Fig. 7. The relationship between five spaces

The similarity becomes even more intriguing, if we assign an alternative meaning to the notation $D(\alpha, \beta)$ as follows: $D(\alpha, \beta)$ is the set of spanning subgraphs of K_n in which

the number of edges is even if $\alpha = 0$ and is odd if $\alpha = 1$,
all vertices have even degrees if $\beta = 0$ and odd degrees if $\beta = 1$.

In this case, Fig. 7 describes both the relationship between the five spaces of a complete bipartite graph (in the original definition of $D(\alpha, \beta)$) and the relationship between the five spaces of the B_5-block (in the alternative definition of $D(\alpha, \beta)$). We leave the problem of describing the relationship between all spaces of a complete bipartite graph open and postpone it for future research. Some other open problems are mentioned in the concluding section.

6 Concluding Remarks and Open Problems

In this paper, we revealed the relationships between vector spaces of a complete graph closed under isomorphism. Our results show that the cycle space plays one of the central roles in this family, emphasizing the importance of this space. In particular, we showed that most of the other spaces can be obtained from the cycle space by a number of operations described in Sect. 3. The results of Sect. 3 are of independent interest and they complement various results on binary vector spaces obtained in [3,4,6].

Characterizing vector spaces closed under isomorphism for other graphs remains an interesting open problem. One more challenging question is whether the family of vector spaces closed under isomorphism always admits a finite characterization within any class of graphs.

References

1. Alekseev, V.E., Zakharova, D.V.: On linear spaces of bipartite graphs. Zhurnal Srednevolzhskogo matematicheskogo obshchestva **26**(1), 11–19 (2024). https://doi.org/10.15507/2079-6900.26.202401.11-19. (in Russian)
2. Biggs, N.: Algebraic potential theory on graphs. Bull. Lond. Math. Soc. **29**(6), 641–682 (1997)
3. Brouwer, A.E.: An inequality in binary vector spaces. Discret. Math. **59**(3), 315–317 (1986)
4. Cohen, G., Frankl, P.: On tilings of the binary vector space. Discret. Math. **31**(3), 271–277 (1980)
5. Hartman, I.B.-A.: Long cycles generate the cycle space of a graph. Eur. J. Combin. **4**(3), 237–246 (1983)
6. Shank, H.: Some parity results on binary vector spaces. Ars Combin. **8**, 107–108 (1979)
7. Zakharova, D.V.: Symmetric linear spaces of graphs. Diskret. Mat. **23**(2), 103–107 (2011). (in Russian). Translation in Discrete Math. Appl. **21**(3), 303–308 (2011)

Average Sensitivity of Breadth-First Search Algorithms on Grids

Ryan Assari and Qian-Ping Gu(✉)

School of Computing Science, Simon Fraser University, Burnaby, Canada
{maa154,qgu}@sfu.ca

Abstract. The average sensitivity of a graph algorithm is a measure for the stability of the algorithm when the input graph G is perturbed by changes in edges/nodes of G. The measure is quantified by the earth mover's distance between the algorithm's outputs on the original and perturbed graphs. Stable algorithms (of small average sensitivity) are highly desirable. We analyze the average sensitivity of the breadth-first search (BFS) algorithm with BFS trees as outputs when the perturbation removes one edge from the input graph. It is known that for an arbitrary graph G, the average sensitivity of the BFS algorithm is $\Theta(|V(G)|)$. We show that for an $m \times n$ 2-dimensional grid graph, there is a deterministic BFS algorithm with average sensitivity at most 2, and there is a randomized BFS algorithm with average sensitivity at most 2 for $m = n$. We prove that for an $n_1 \times n_2 \times n_3$ 3-dimensional grid, there is a deterministic BFS algorithm with average sensitivity at most $\frac{4}{3} + \frac{2}{n}$ and there is a randomized BFS algorithm with average sensitivity at most $\frac{2}{3} + \frac{4}{n}$ for $n = n_1 = n_2 = n_3$. These show that BFS algorithms have much better average sensitivity on grid graphs than $\Theta(|V(G)|)$ on arbitrary graphs.

Keywords: Graph algorithms · average sensitivity · breadth-first search · grid graphs

1 Introduction

The average sensitivity of a graph algorithm is a measure for the stability of the algorithm when there is a perturbation (represented by some changes in edges/nodes) in the input graph G. Let G' be the graph obtained from G by the perturbation. The stability of the algorithm can be measured by calculating how close the outputs of the algorithm are when G and G' are given as inputs. The average sensitivity is defined in several areas. For example, the average sensitivity of Boolean functions is introduced in complexity theory, learning theory and study of noise sensitivity [4,6,18]. In data science, the average sensitivity of clustering algorithms on missing edges/subgraphs in the input graph is important on error measurement [12,16]; and PageRank algorithms are expected to produce stable rankings when there is a small perturbation in the linkage pattern

H. Fernau and B. Zhu (Eds.): IWOCA 2025, LNCS 15885, pp. 173–187, 2025.
https://doi.org/10.1007/978-3-031-98740-3_13

[14]. Stable algorithms received much attention in learning theory and numerical analysis as well [2,5,17,23]. A dedicated study on the average sensitivity of graph algorithms appears in [21].

For a graph algorithm which computes a set of edges (or nodes) of a graph as output, the average sensitivity of the algorithm is based on some distance between the output set for the original graph G and the output set for the perturbed graph G'. The earth mover's distance (introduced in [19]) between the two output sets is commonly used. Since perturbations on different edges/nodes of the input graph may create different outputs, the average sensitivity of an algorithm is defined as the average distance over all affected edges/nodes. The average sensitivity analysis of algorithms is important because in many applications the input graph is dynamic, and it is often impossible or costly to always keep track of the input graph. Algorithms of small average sensitivity are desirable in applications where the stability is crucial, as the algorithms produce solutions on available input instance that remain close to those on the original instance when perturbations occur.

Varma and Yoshida introduced the notion of the k-average sensitivity for perturbations of removing k edges from G and use the term *average sensitivity* for 1-average sensitivity [21]. They analyze the average sensitivity of algorithms for several fundamental graph problems, including the single source shortest paths, minimum spanning tree, global minimum cut, minimum s-t cut, maximum matching, minimum vertex cover and 2-coloring problems [21]. Average sensitivity of graph algorithms for many other important problems have also been studied, for example, dynamic programming algorithms for the maximum weight chain problem [9], algorithms for Euclidean (k, l)-clustering problem [24], and decision tree learning algorithms [7]. Average sensitivity of approximation algorithms for the maximum matching problem and knapsack problem can be found in [10,25].

Breadth-first search (BFS) is a fundamental graph algorithm. A key output of the BFS algorithm is a BFS tree which has numerous applications. The average sensitivity of the BFS algorithm can provide valuable insights into the robustness and reliability of the BFS trees generated by the algorithm. It is shown that for an arbitrary graph G, the average sensitivity of the single source shortest paths algorithms is $\Theta(|V(G)|)$, implying that the average sensitivity of the BFS algorithm with BFS trees as outputs is $\Theta(|V(G)|)$ [21]. The average sensitivity of the BFS algorithm on special classes of graphs is yet to be studied.

Grid graphs are an important class of graphs that are commonly used to model many networks and structures like roads, city blocks, traffic management etc. The BFS algorithm on grid graphs has wide applications, including robotics and path planning [13,15], disaster management and evacuation planning [3, 11], network and communication [1], healthcare and epidemic modeling [13,26], urban planning and traffic management [8], agriculture and irrigation systems [22], and games and puzzles [20]. The average sensitivity of the BFS algorithm on grids is of great importance in designing more stable algorithms, protocols and disease intervention strategies and resource allocation in these applications.

Following the notion of the average sensitivity in [21] for the perturbation of removing one edge from the original graph G, we analyze the average sensitivity of deterministic and randomized BFS algorithms on $m \times n$ 2-dimensional (2D) and $n_1 \times n_2 \times n_3$ 3-dimensional (3D) grid graphs to provide a guideline for applying stable BFS algorithms.

Given a graph G and a source node s, the BFS algorithm first marks s as partially explored; while there is a partially explored node do the following: select in the first-in-first-out order a partially explored node u and check every edge $\{u, v\}$ incident to u; if v is not partially/fully explored then mark v partially explored and include $\{u, v\}$ in the BFS tree; after every edge $\{u, v\}$ is checked, mark u fully explored. When there are multiple such edges $\{u, v\}$, these edges can be checked in an arbitrary order. Different checking orders may create different BFS trees. For a node u not on the perimeter of a 2D grid, there are four edges $\{u, v\}$ incident to u and there are $4! = 24$ checking orders for u. When a deterministic order is used to check the edges $\{u, v\}$, the BFS algorithm is called a *deterministic BFS (DBFS) algorithm*, while using a randomized order is called a *randomized BFS (RBFS) algorithm.*

We analyze the average sensitivities for DBFS and RBFS algorithms on 2D and 3D grids. Our results for $m \times n$ 2D grids is summarized as follows:

- Assume a same checking order is used for every node of a 2D grid in a DBFS algorithm. It suffices to analyze the average sensitivity of the DBFS algorithm with three orders.
- The DBFS algorithm with one of the three orders has average sensitivity at most 2 for $2 \leq m \leq n$. Since a 2D grid can be rotated, there is a DBFS algorithm with average sensitivity at most 2 for $2 \leq m, n$. The DBFS algorithms with the other two orders have average sensitivity at most $1 + O(\frac{m}{n} + \frac{n}{m})$ for $2 \leq m, n$, and at most 3 for $m = n$.
- The RBFS algorithm has average sensitivity at most $1 + \frac{m}{2n} + \frac{n-1}{2(m-1)}$ for $2 \leq m \leq n$, $1 + \frac{m-1}{2(n-1)} + \frac{n}{2m}$ for $m \geq n$, and at most 2 for $m = n$.

Our results on $n_1 \times n_2 \times n_3$ 3D grids:

- There is a DBFS algorithm with average sensitivity at most $\frac{2}{3} + \frac{4(n_1-1)}{3n_2(n_3-1)} + \frac{2(n_2-1)}{3(n_3-1)} + \frac{2}{3(n_3-1)}$ for $2 \leq n_3 \leq n_1, n_2$, and at most $\frac{4}{3} + \frac{2}{n}$ for $n_1 = n_2 = n_3 = n$.
- The RBFS algorithm has average sensitivity at most $\frac{2}{3} + \frac{2(n_1-1)}{3n_2(n_3-1)} + \frac{2(n_2-1)}{3n_1(n_3-1)} + \frac{2n_3}{3n_1 n_2} + \frac{2}{3(n_3-1)} + \frac{2}{3n_2} + \frac{2}{3n_1}$ for $2 \leq n_3 \leq n_1, n_2$, and at most $\frac{2}{3} + \frac{4}{n}$ for $n_1 = n_2 = n_3 = n$.

In the rest of the paper, Sect. 2 gives the preliminaries. Our results on 2D and 3D grids are presented in Sects. 3 and 4, respectively. The final section concludes the paper.

2 Preliminaries

We consider undirected simple connected graphs without edge weights. A graph G has node set $V(G) = \{v_1, .., v_{|V(G)|}\}$ and edge set $E(G) = \{e_1, .., e_{|E(G)|}\}$.

Assume that a solution for a problem in G is represented by a subset of $E(G)$. A subgraph of G obtained by removing a subset E' of edges of $E(G)$ is denoted by $G \setminus E'$. For a solution B in G and a solution B' in $G \setminus E'$, the Hamming distance between B and B' is defined as $d_{\text{Ham}}(B, B') = |B \Delta B'|$, $B \Delta B' = (B \setminus B') \cup (B' \setminus B)$ is the symmetric difference between B and B'. For an algorithm A, let $\mathcal{A}(G)$ and $\mathcal{A}(G \setminus E')$ be the sets of solutions of A on G and $G \setminus E'$, respectively. For a randomized algorithm A, let p_f and q_g be the probabilities that solutions $B_f \in \mathcal{A}(G)$ and $B'_g \in \mathcal{A}(G \setminus E')$ are outputted by A on G and $G \setminus E'$, respectively. The earth mover's distance [19] between $\mathcal{A}(G)$ and $\mathcal{A}(G \setminus E')$ is

$$d_{\text{EM}}(\mathcal{A}(G), \mathcal{A}(G \setminus E')) = \min \sum_{B_f \in \mathcal{A}(G)} \sum_{B'_g \in \mathcal{A}(G \setminus E')} z_{fg} d_{\text{Ham}}(B_f, B'_j),$$

$$\text{subject to} \sum_{B_f \in \mathcal{A}(G)} z_{fg} = q_g \text{ and } \sum_{B'_g \in \mathcal{A}(G \setminus E')} z_{fg} = p_f.$$

The average sensitivity of an algorithm A on a graph G is

$$\text{AS}(A) = \mathbb{E}_{e \in E(G)}[d_{\text{EM}}(\mathcal{A}(G), \mathcal{A}(G \setminus \{e\}))].$$

For a deterministic algorithm A which computes a unique solution B on G and a unique solution B_e on $G \setminus \{e\}$, $d_{\text{EM}}(\mathcal{A}(G), \mathcal{A}(G \setminus \{e\})) = d_{\text{Ham}}(B, B_e)$ and the average sensitivity of A is

$$\text{AS}(A) = \mathbb{E}_{e \in E(G)} d_{\text{Ham}}(B, B_e) = \frac{1}{|E(G)|} \sum_{e \in E(G)} d_{\text{Ham}}(B, B_e).$$

A 2-dimensional (2D) $m \times n$ grid is a graph G with node set

$$V(G) = \{(i, j) | 0 \le i \le m-1, 0 \le j \le n-1\}$$

and edge set

$$E(G) = \{\{(i, j), (i', j')\} | (|i - i'| = 1, j = j') \text{ or } (i = i', |j - j'| = 1)\}.$$

For a node $u = (i, j)$ of G, we call node $(i-1, j)$ the T (top) neighbor, $(i+1, j)$ the B (bottom) neighbor, $(i, j-1)$ the L (left) neighbor and $(i, j+1)$ the R (right) neighbor of u (see Fig. 1).

A 3-dimensional (3D) $n_1 \times n_2 \times n_3$ grid is a graph G with node set

$$V(G) = \{(i, j, k) | 0 \le i \le n_1 - 1, 0 \le j \le n_2 - 1, 0 \le k \le n_3 - 1\}$$

and edge set

$$\begin{aligned} E(G) = \{\{(i, j, k), (i', j', k')\} | \ & (|i - i'| = 1, j = j', k = k') \\ & \text{or } (i = i', |j - j'| = 1, k = k') \\ & \text{or } (i = i', j = j', |k - k'| = 1)\}. \end{aligned}$$

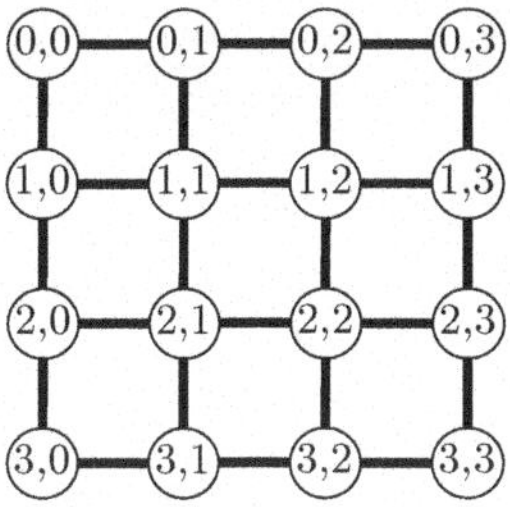

Fig. 1. A 4×4 2D grid. For node $u = (1, 1)$, node $(0, 1)$ is the T neighbor, $(2, 1)$ is the B neighbor, $(1, 0)$ is the L neighbor and $(1, 2)$ is the R neighbor of u.

An outline of the DBFS and RBFS algorithms is as follows: Given a graph G and a source s, for $0 \leq r \leq |V(G)| - 1$, let L_r be the set of nodes which is reachable from s by a path of r edges but not reachable by any path of $r - 1$ edges. Notice that $L_0 = \{s\}$ and L_1 is the set of neighbors of s. Let $E_{r-1,r} = \{\{u, v\}|u \in L_{r-1}, v \in L_r\}$ be the set of edges between a node of L_{r-1} and a node of L_r, where $0 \leq r \leq |V(G)| - 1$.

Algorithm DBFS(G, s) /* Deterministic BFS
$L_r = \emptyset$ for $1 \leq r \leq |V(G)| - 1$; $L_0 = \{s\}$; $r = 1$; BFS tree $T(G) = \emptyset$;
while $V(T(G)) \neq V(G)$ **do**
  for every node $u \in L_{r-1}$ in the first-in-first-out order **do**
    for each edge $\{u, v\} \in E_{r-1,r}$ in a fixed order **do**
      if $v \notin L_r$ **then** $L_r = L_r \cup \{v\}$ and $T(G) = T(G) \cup \{\{u, v\}\}$;
    end-for
  end-for
  $r = r + 1$;
end-while
Output $T(G)$ and terminate.

Algorithm RBFS(G, s) /* Randomized BFS
$L_r = \emptyset$ for $1 \leq r \leq |V(G)| - 1$; $L_0 = \{s\}$; $r = 1$; BFS tree $T(G) = \emptyset$;
while $V(T(G)) \neq V(G)$ **do**
  while $E_{r-1,r} \neq \emptyset$ **do**
    select an edge $\{u, v\}$ from $E_{r-1,r}$ uniformly at random ($u \in L_{r-1}$);
    if $v \notin L_r$ **then** $L_r = L_r \cup \{v\}$ and $T(G) = T(G) \cup \{\{u, v\}\}$;
    remove $\{u, v\}$ from $E_{r-1,r}$;
  end-while
  $r = r + 1$;
end-while
Output $T(G)$ and terminate.

3 Average Sensitivity of BFS Algorithms on 2D Grids

Let G be an $m \times n$ 2D grid and $s = (i, j)$ be a source node in G. The earth mover's distance $d_{\mathrm{EM}}(\mathcal{A}(G), \mathcal{A}(G \setminus \{e\}))$ depends on the location of the removed edge e. We partition the edges of G based on $s = (i, j)$ into the following sets and compute $d_{\text{EM}}(\mathcal{A}(G), \mathcal{A}(G \setminus \{e\}))$ for the edges in each set:

- $E_T = \{\{(i', j), (i' + 1, j)\} | 0 \leq i' < i\}$,
- $E_B = \{\{(i', j), (i' + 1, j)\} | i \leq i' < m - 1\}$,
- $E_L = \{\{(i, j'), (i, j' + 1)\} | 0 \leq j' < j\}$,
- $E_R = \{\{(i, j'), (i, j' + 1)\} | j \leq j' < n - 1\}$, and
- $\tilde{E} = E(G) \setminus (E_T \cup E_L \cup E_B \cup E_R)$.

3.1 Checking Orders of DBFS Algorithm

Assume that every node u of G has four edges $\{u, v\}$ incident to it (for nodes with fewer than four incident edges, we add dummy edges). For every node u of G, assume a same order is used to check the four edges incident to u in DBFS algorithms. Then there are $4! = 24$ different checking orders. Because a 2D grid can be rotated and flipped, the checking orders can be partitioned into dihedral groups for a polygon of four sides. Let S be the set of all checking orders and D_4 be the dihedral group of four rotational symmetries and four reflection symmetries on the 2D grids. Then there are $|S|/|D_4| = 24/8 = 3$ different groups for the checking orders. Every order in a group can be obtained from each of the other orders in the same group by rotating and/or flipping the grid. Thus, a same analysis can be applied to every order in a same group. We consider the following three orders in DBFS algorithms for the average sensitivity analysis:

- **Dimension Order** Check the four edges incident to $u = (i, j)$ in the order of their dimensions, that is, in the order of **T-B-L-R** ($(i - 1, j)$-$(i + 1, j)$-$(i, j - 1)$-$(i, j + 1)$) neighbors. This order represents the following orders in a same group,
  T-B-L-R, R-L-T-B, B-T-R-L, L-R-B-T (rotation of grid from T-B-L-R)
  T-B-R-L, R-L-B-T, B-T-L-R, L-R-T-B (flip of grid from T-B-L-R)
- **Clockwise Order** Check the four edges incident to u in the order of **T-R-B-L**. This order represents the following orders in a same group,
  T-R-B-L, R-B-L-T, B-L-T-R, L-T-R-B (rotation of grid from T-R-B-L)
  T-L-B-R, L-B-R-T, B-R-T-L, R-T-L-B (flip of grid from T-R-B-L)
- **Reading Order** Check the four edges incident to u in the order of **T-L-R-B**. This order represents the following orders in a same group,
  T-L-R-B, R-T-B-L, B-R-L-T, L-B-T-R (rotation of grid from T-L-R-B)
  T-R-L-B, L-T-B-R, B-L-R-T, R-B-T-L (flip of grid from T-L-R-B)

The DBFS algorithm with dimension order, clockwise order and reading order are called DBFS-D, DBFS-C and DBFS-R, respectively. Figure 2 gives the BFS trees computed by DBFS-D, DBFS-C and DBFS-R.

3.2 Average Sensitivity of DBFS Algorithms

We first analyze the average sensitivity of DBFS algorithm with dimension order (DBFS-D). Given source $s = (i, j)$ in G and $G \setminus \{e\}$, algorithm DBFS-D outputs a

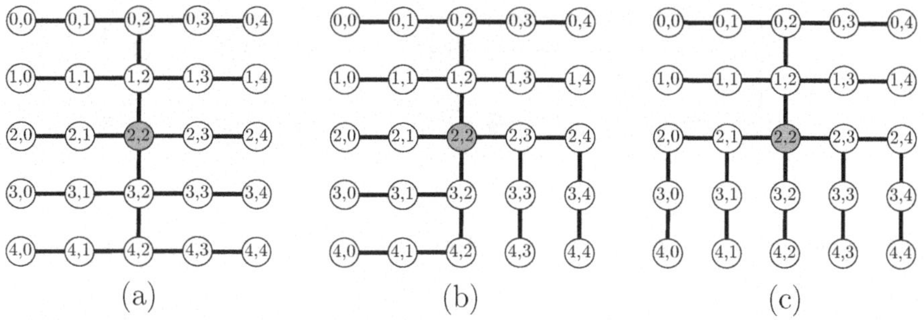

Fig. 2. Source $s = (2, 2)$. (a) BFS tree by DBFS-D, (b) BFS tree by DBFS-C, (c) BFS tree by DBFS-R.

unique BFS tree $T(G)$ and a unique BFS tree $T(G\setminus\{e\})$. The average sensitivity of DBFS-D is

$$\mathbb{E}_{e\in E(G)} d_{\text{Ham}}(T(G), T(G\setminus\{e\})) = \frac{1}{|E(G)|}\sum_{e\in E(G)} d_{\text{Ham}}(T(G), T(G\setminus\{e\}).$$

The edge set $E(G)$ can be partitioned into three subsets $E_T \cup E_B$, $E_L \cup E_R$ and $\tilde{E}$. We analyze the sum of $d_{\text{Ham}}(T(G), T(G\setminus\{e\})$ on edges in the three subsets in the following three lemmas, respectively.

Lemma 1. *For DBFS-D on* $s = (i, j)$,
$\sum_{e\in E_T\cup E_B} d_{Ham}(T(G), T(G\setminus\{e\})) \le 2i(i+1) + 2(m-i-1)(m-i)$.

Proof: We give an outline of the proof. Assume edge $e = \{(i', j), (i'+1, j)\} \in E_T$ is removed for $0 \le i' < i$ and $0 \le j \le n-1$. Intuitively, edges in $T(G)\Delta T(G\setminus\{e\})$ are incident to nodes in columns $j-1, j, j+1$ and rows x with $0 \le x \le i'+1$ only, implying $d_{\text{Ham}}(T(G), T(G\setminus\{e\})) \le c(i'+1)$ for some constant c.

More specifically, for $0 < j < n-1$ and each node (x, j), $0 \le x \le i'$, edges $\{(x, j), (x+1, j)\}$ and $\{(x, j), (x, j+1)\}$ of $T(G)$ are not in $T(G)\setminus\{e\}$, edges $\{(x, j-1), (x+1, j-1)\}$ and $\{(x, j+1), (x+1, j+1)\}$ not in $T(G)$ are in $T(G\setminus\{e\})$; all other edges of $T(G)$ are in $T(G\setminus\{e\})$. Therefore, $d_{\text{Ham}}(T(G), T(G\setminus\{e\})) = 4(i'+1)$ (see Fig. 3). Similarly, $d_{\text{Ham}}(T(G), T(G\setminus\{e\})) = 2(i'+1)$ for $j = 0$ or $j = n-1$. Hence,

$$\sum_{e\in E_T} d_{\text{Ham}}(T(G), T(G\setminus\{e\})) \le \sum_{i'=0}^{i-1} 4(i'+1) = 2i(i+1).$$

By a similar analysis, we have

$$\sum_{e\in E_B} d_{\text{Ham}}(T(G), T(G\setminus\{e\})) \le 2(m-i)(m-i-1).$$

□

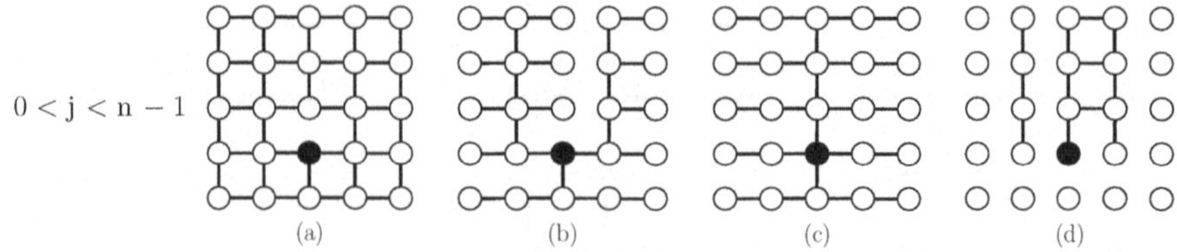

Fig. 3. Remove edge $e \in E_T$, source s is black. (a) $G \setminus \{e\}$, (b) $T(G \setminus \{e\})$, (c) $T(G)$, (d) $T(G) \Delta T(G \setminus \{e\})$.

By a similar analysis for Lemma 1, we have the next two lemmas.

Lemma 2. *For DBFS-D on* $s = (i, j)$,
$\sum_{e \in E_L \cup E_R} d_{Ham}(T(G), T(G \setminus \{e\}) = 2(n-1)$.

Lemma 3. *For DBFS-D on* $s = (i, j)$,
$\sum_{e \in \tilde{E}} d_{Ham}(T(G), T(G \setminus \{e\}) = |E(G)| - m - n + 2$.

Theorem 1. *The DBFS-D has average sensitivity at most 2 for any source node* $s = (i, j)$ *in an* $m \times n$ *grid with* $2 \leq m \leq n$.

Proof: From Lemmas 1–3,

$$\begin{aligned} \sum_{e \in E(G)} d_{\text{Ham}}(T(G), T(G \setminus \{e\})) &= \sum_{e \in E_T \cup E_B} d_{\text{Ham}}(T(G), T(G \setminus \{e\})) \\ + \sum_{e \in E_L \cup E_R} & d_{\text{Ham}}(T(G), T(G \setminus \{e\})) + \sum_{e \in \tilde{E}} d_{\text{Ham}}(T(G), T(G \setminus \{e\})) \\ &\leq 2i(i+1) + 2(m-i-1)(m-i) + 2(n-1) + |E(G)| - m - n + 2 \\ &= |E(G)| + 2(i(i+1) + (m-i-1)(m-i)) + n - m. \end{aligned} \tag{1}$$

Function $f(i) = i(i+1) + (m-i)(m-i-1)$ is parabolic and has maximum value $f(0) = f(m-1) = m(m-1)$ for $0 \leq i \leq m-1$. From this, Eq. (1) and $|E(G)| = m(n-1) + n(m-1) = 2mn - m - n$, we have

$$\begin{aligned} \text{AS(DBFS-D)} &= \frac{1}{|E(G)|} \sum_{e \in E(G)} d_{\text{Ham}}(T(G), T(G \setminus \{e\}) \\ &\leq 1 + \frac{2\,m(m-1) + n - m}{|E(G)|} = 1 + \frac{2\,m^2 - 3\,m + n}{2mn - m - n}. \end{aligned}$$

Since $2 \leq m \leq n$, $n - m \leq m(n-m) = mn - m^2$, implying $m^2 - m + n \leq mn$ and $2m^2 - 2m + 2n \leq 2mn$. From this, $2m^2 - 3m + n \leq 2mn - m - n$. Therefore, the average sensitivity of DBFS-D is at most 2 for any source node. □

Corollary 1. *There is a DBFS algorithm with average sensitivity at most 2 for any source node in an* $m \times n$ *grid with* $2 \leq m, n$.

Proof: Given a grid, we can rotate the grid to get grid G with $m \leq n$ and then apply DBFS-D. The corollary follows from Theorem 1. □

Theorem 1 holds for any source $s = (i, j)$. It can be shown that DBFS-D has average sensitivity at most $\frac{3}{2} + O(\frac{1}{n})$ for $s = (i, j)$ with $i = \lceil \frac{m-1}{2} \rceil$.

For the DBFS algorithms with clockwise order (DBFS-C) and reading order (DBFS-R), we obtain the following results.

Theorem 2. *The DBFS-C has average sensitivity at most* $1 + \frac{m-1}{n-1} + \frac{n-1}{m-1}$ *for any source node* $s = (i, j)$ *in an* $m \times n$ *grid with* $2 \leq m, n$*, at most* $O(1)$ *for* $m = \Theta(n)$ *and at most 3 for* $m = n$.

Theorem 3. *The DBFS-R has average sensitivity at most* $1 + \frac{m-1}{n-1} + \frac{n-1}{m-1}$ *for any source node* $s = (i, j)$ *in an* $m \times n$ *grid with* $2 \leq m, n$*, at most* $O(1)$ *for* $m = \Theta(n)$ *and at most 3 for* $m = n$.

3.3 Average Sensitivity of RBFS on 2D Grids

Recall that $\mathcal{A}(G)$ is the set of BFS trees of G and $\mathcal{A}(G \setminus \{e\})$ is the set of BFS trees of $G \setminus \{e\}$ by the RBFS algorithm. The average sensitivity of the RBFS algorithm is

$$\mathbb{E}_{e \in E(G)}[d_{\mathrm{EM}}(\mathcal{A}(G), \mathcal{A}(G \setminus \{e\})] = \frac{1}{|E(G)|} \sum_{e \in E(G)} d_{\mathrm{EM}}(\mathcal{A}(G), \mathcal{A}(G \setminus \{e\})).$$

The sum of earth mover's distances $d_{\mathrm{EM}}(\mathcal{A}(G), \mathcal{A}(G \setminus \{e\}))$ for $e \in E_T \cup E_B \cup E_L \cup E_R$ is given in Lemma 4 and the sum for $e \in \tilde{E}$ is given in Lemma 5.

We first introduce some notations. For a node $v \in L_r$, $1 \leq r$, in G, each edge $\{u, v\}$ with $u \in L_{r-1}$ is included in a $T(G) \in \mathcal{A}(G)$ by RBFS with the same probability. We call such an edge a *candidate edge* for v and let $C(v, G)$ be the set of candidate edges for v. For a subset $U \subseteq V(G)$, let $C(U, G) = \cup_{v \in U} C(v, G)$. Similarly, for $v \in L_r$ in $G \setminus \{e\}$, let $C(v, G \setminus \{e\})$ be the set of candidate edges for v, and for a subset $U \subseteq V(G \setminus \{e\})$, $C(U, G \setminus \{e\}) = \cup_{v \in U} C(v, G \setminus \{e\})$.

Lemma 4. *For a source node* $s = (i, j)$, $\sum_{e \in E_T \cup E_B \cup E_L \cup E_R} d_{EM}(\mathcal{A}(G), \mathcal{A}(G \setminus \{e\})) \leq i(i+1) + (m-i-1)(m-i) + \frac{m-1}{2} + j(j+1) + (n-j-1)(n-j) + \frac{n-1}{2}$.

Proof: We give an outline of the proof. For a source node $s = (i, j)$, assume an edge $e = \{(i', j), (i'+1, j)\}$, $0 \leq i' < i$, in E_T is removed. Let $U = \{(x, j-1), (x, j), (x, j+1) | 0 \leq x \leq i'\}$. Then for $v \in U$, $C(v, G) \neq C(v, G \setminus \{e\})$, and for $v \notin U$, $C(v, G) = C(v, G \setminus \{e\})$. The earth mover's distance $d_{\mathrm{EM}}(\mathcal{A}(G), \mathcal{A}(G \setminus \{e\}))$ can be upper bounded by the earth mover's distances between the set of subgraphs of $T(G) \in \mathcal{A}(G)$ induced by $C(U, G)$ and the set of subgraphs of $T(G \setminus \{e\}) \in \mathcal{A}(G \setminus \{e\})$ induced by $C(U, G \setminus \{e\})$.

More specifically, for each x of $0 \leq x \leq i'$, $C((x, j-1), G)$ has two edges $\{(x, j-1), (x+1, j-1)\}$ and $\{(x, j-1), (x, j)\}$, $C((x, j), G)$ has one

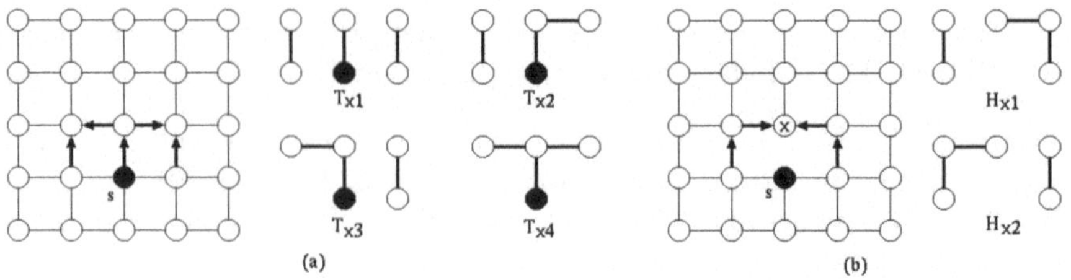

Fig. 4. Edges in $C(U,G)$ and $C(U,G\setminus\{e\})$ are oriented from L_{r-1} to L_r. (a) $C(U,G)$ for $x=i'$, and $T_{x1},T_{x2},T_{x3},T_{x4}$. (b) $C(U,G\setminus\{e\})$ for $x=i'$, and H_{x1},H_{x2}.

edge $\{(x,j),(x+1,j)\}$, and $C((x,j+1),G)$ has two edges $\{(x,j),(x,j+1)\}$ and $\{(x,j+1),(x+1,j+1)\}$. There are four subgraphs $T_{x1},T_{x2},T_{x3},T_{x4}$ in $T(G)\cap C(U,G)$ (see Fig. 4 (a)), each of them is generated with probability $\frac{1}{4}$.

For $x=i'$, $C((x,j-1),G\setminus\{e\})$ has one edge $\{(x,j-1),(x+1,j-1)\}$, $C((x,j),G\setminus\{e\})$ has two edges $\{(x,j-1),(x,j)\}$ and $\{(x,j),(x,j+1)\}$, and $C((x,j+1),G\setminus\{e\})$ has one edge $\{(x,j+1),(x+1,j+1)\}$. There are two subgraphs H_{x1} and H_{x2} in $T(G\setminus\{e\})\cap C(U,G\setminus\{e\})$ (see Fig. 4 (b)), each of them is generated with probability $\frac{1}{2}$. It can be shown that $d_{\mathrm{EM}}(\{T_{x1},T_{x2},T_{x3},T_{x4}\},\{H_{x1},H_{x2}\})\le 2.5$. For every x with $0<x<i'$, there are three subgraphs in $T(G\setminus\{e\})\cap C(U,G\setminus\{e\})$, and the earth mover's distance between $\{T_{x_1},T_{x_2},T_{x_3},T_{x_4}\}$ and the three subgraphs of $T(G\setminus\{e\})\cap C(U,G\setminus\{e\})$ is at most 2. Therefore, $d_{\mathrm{EM}}(\mathcal{A}(G),\mathcal{A}(G\setminus\{e\})\le 2(i'+1)+0.5$ and

$$\sum_{e\in E_T} d_{\mathrm{EM}}(\mathcal{A}(G),\mathcal{A}(G\setminus\{e\}))\le i(i+1)+\frac{i}{2}.$$

By a similar analysis for $e\in E_B$, $e\in T_L$ and $e\in E_R$, we get the lemma. □

Lemma 5. *For a source node $s=(i,j)$,*
$\sum_{e\in\tilde{E}} d_{EM}(\mathcal{A}(G),\mathcal{A}(G\setminus\{e\}))\le|E(G)|-m-n+2.$

Theorem 4. *The RBFS algorithm on an arbitrary source node in a $m\times n$ grid has average sensitivity at most $1+\frac{m}{2n}+\frac{n-1}{2(m-1)}$ for $2\le m\le n$, $1+\frac{m-1}{2(n-1)}+\frac{n}{2m}$ for $m\ge n$, and 2 for $m=n$.*

Proof: We give an outline of the proof. From Lemmas 4–5,

$$\sum_{e\in E(G)} d_{\mathrm{EM}}(\mathcal{A}(G),\mathcal{A}(G\setminus\{e\}))\le|E(G)|+i(i+1)+(m-i-1)(m-i)$$
$$+j(j+1)+(n-j-1)(n-j)-\frac{m-1}{2}-\frac{n-1}{2}. \tag{2}$$

From Eq. (2), we can get the average sensitivity of the RBFS algorithm

$$\mathrm{AS(RBFS)}\le 1+\frac{m(m-1)}{2n(m-1)}+\frac{n(n-1)}{2n(m-1)}=1+\frac{m}{2n}+\frac{n-1}{2(m-1)}.$$

for $m \leq n$ and

$$\mathrm{AS(RBFS)} \leq 1 + \frac{m-1}{2(n-1)} + \frac{n}{2m}.$$

for $m \geq n$.

For $m = \Theta(n)$, AS(RBFS) $= O(1)$, and for $m = n$, AS(RBFS) ≤ 2. □

Theorem 4 holds for any source node $s = (i, j)$. It is easy to show that the RBFS algorithm has average sensitivity at most $1 + \frac{m}{4n} + \frac{n-1}{4(m-1)}$ for $s = (i, j)$ with $i = \lceil \frac{m-1}{2} \rceil$ and $j = \lceil \frac{n-1}{4(m-1)} \rceil$.

4 Average Sensitivity of BFS Algorithms on 3D Grids

Let G be an $n_1 \times n_2 \times n_3$ 3D grid and $s = (i, j, k)$ be a source node in G. Let $T(G)$ and $T(G \setminus \{e\})$ be the BFS trees rooted at s in G and $G \setminus \{e\}$, respectively.

4.1 Average Sensitivity of DBFS on 3D Grids

Motivated by the DBFS algorithm with dimension order (DBFS-D) which has the smallest average sensitivity on 2D grids, we analyze the average sensitivity of DBFS-D on 3D grids. For a source $s = (i, j, k)$, we partition the edges of G into the following sets:

- $E_T = \{\{(i', j, k), (i'+1, j, k)\} | 0 \leq i' < i\}$,
- $E_B = \{\{(i', j, k), (i'+1, j, k)\} | i \leq i' < n_1 - 1\}$,
- $E_{L^*} = \{\{(i', j', k), (i', j'+1, k)\} | 0 \leq i' \leq n_1 - 1, 0 \leq j' < j\}$,
- $E_{R^*} = \{\{(i', j', k), (i', j'+1, k)\} | 0 \leq i' \leq n_1 - 1, j \leq j' < n_2 - 1\}$,
- $\tilde{E} = E(G) \setminus (E_T \cup E_B \cup E_{L^*} \cup E_{R^*})$.

For each node $u = (i, j, k)$ in G, DBFS-D checks neighbors of u in the order of $(i-1, j, k), (i+1, j, k), (i, j-1, k), (i, j+1, k), (i, j, k-1), (i, j, k+1)$. The DBFS-D computes a $T(G)$ which contains the BFS tree rooted at s by the DBFS-D on the $n_1 \times n_2$ 2D grid (subgraph of G induced by node set $\{(i', j', k) | 0 \leq i' \leq n_1 - 1, 0 \leq j' \leq n_2 - 1\}$) and all vertical edges $\{(i', j', k'), (i', j', k' \pm 1)\}$, $0 \leq i' \leq n_1 - 1, 0 \leq j' \leq n_2 - 1, 0 < k' < n_3 - 1$.

The average sensitivity of DBFS-D is

$$\mathbb{E}_{e \in E(G)} d_{\mathrm{Ham}}(T(G), T(G \setminus \{e\})) = \frac{1}{|E(G)|} \sum_{e \in E(G)} d_{\mathrm{Ham}}(T(G), T(G \setminus \{e\}).$$

The sum of $d_{\mathrm{Ham}}(T(G), T(G \setminus \{e\}))$ for $e \in E_T \cup E_B$, the sum for $e \in E_{L^*} \cup E_{R^*}$ and the sum for $e \in \tilde{E}$ are given in Lemmas 6–8, respectively.

Lemma 6. *For DBFS-D on a source* $s = (i, j, k)$,
$\sum_{e \in E_T \cup E_B} d_{Ham}(T(G), T(G \setminus \{e\}) \leq 4i(i+1) + 4(n_1 - i - 1)(n_1 - i)$.

Lemma 7. *For DBFS-D on a source* $s = (i, j, k)$,
$\sum_{e \in E_{L^*} \cup E_{R^*}} d_{Ham}(T(G), T(G \setminus \{e\}) \leq 2n_1(j(j+1) + (n_2 - j - 1)(n_2 - j) + (n_2 - 1))$.

Lemma 8. *For DBFS-D on source* $s = (i, j, k)$, $\sum_{e \in \tilde{E}} d_{Ham}(T(G), T(G \setminus \{e\})) = 2n_1n_2(n_3 - 1)$.

Theorem 5. *The DBFS-D on an arbitrary source in an* $n_1 \times n_2 \times n_3$ *grid* G *with* $2 \leq n_3 \leq n_1, n_2$ *has average sensitivity at most* $\frac{2}{3} + \frac{4(n_1-1)}{3n_2(n_3-1)} + \frac{2(n_2-1)}{3(n_3-1)} + \frac{2}{3(n_3-1)}$ *for* $2 \leq n_3 \leq n_1, n_2$, *and at most* $\frac{4}{3} + \frac{2}{n}$ *for* $n_1 = n_2 = n_3 = n$.

Proof: We give an outline of the proof. From Lemmas 6–8, we can get

$$\sum_{e \in E(G)} d_{\text{Ham}}(T(G), T(G \setminus \{e\})) = \sum_{e \in E_T \cup E_B} d_{\text{Ham}}(T(G), T(G \setminus \{e\}))$$
$$\leq 4n_1(n_1 - 1) + 2n_1n_2(n_2 - 1) + 2n_1(n_2 - 1) + 2n_1n_2(n_3 - 1). \tag{3}$$

From Eq. (3), we have

$$\begin{aligned} \text{AS(DBFS-D)} &= \frac{1}{|E(G)|} \sum_{e \in E(G)} d_{\text{Ham}}(T(G), T(G \setminus \{e\})) \\ &\leq \frac{2}{3} + \frac{4(n_1 - 1)}{3n_2(n_3 - 1)} + \frac{2(n_2 - 1)}{3(n_3 - 1)} + \frac{2}{3(n_3 - 1)}. \end{aligned}$$

For $n_1 = n_2 = n_3 = n$,

$$\text{AS(DBFS-D)} \leq \frac{4}{3n} + \frac{2}{3} + \frac{2}{3n} + \frac{2}{3} = \frac{4}{3} + \frac{2}{n}.$$

□

4.2 Average Sensitivity of RBFS on 3D Grids

For a source $s = (i, j, k)$ in G, we partition $E(G)$ into the following sets:

- $E_i = \{\{(i', j, k), (i' + 1, j, k)\} | 0 \leq i' < n_1 - 1\}$,
- $E_j = \{\{(i, j', k), (i, j' + 1, k)\} | 0 \leq j' < n_2 - 1\}$,
- $E_k = \{\{(i, j, k'), (i, j, k' + 1)\} | 0 \leq k' < n_3 - 1\}$, and
- $\tilde{E} = E(G) \setminus (E_i \cup E_j \cup E_k)$.

The average sensitivity of the RBFS algorithm is

$$\mathbb{E}_{e \in E(G)}[d_{\text{EM}}(\mathcal{A}(G), \mathcal{A}(G \setminus \{e\})] = \frac{1}{|E(G)|} \sum_{e \in E(G)} d_{\text{EM}}(\mathcal{A}(G), \mathcal{A}(G \setminus \{e\})).$$

The sum of earth mover's distances $d_{\text{EM}}(\mathcal{A}(G), \mathcal{A}(G \setminus \{e\}))$ for $e \in E_i \cup E_j \cup E_k$ is given in Lemma 9 and the sum for $e \in \tilde{E}$ is given in Lemma 10.

Lemma 9. *For a source* $s = (i, j, k)$, $\sum_{e \in E_i \cup E_j \cup E_k} d_{EM}(A(G), A(G \setminus \{e\})) \leq 2(i(i + 1) + (n_1 - i - 1)(n_1 - i) + j(j + 1) + (n_2 - j - 1)(n_2 - j) + k(k + 1) + (n_3 - k - 1)(n_3 - k))$.

Lemma 10. *For a source* $s = (i, j, k)$, $\sum_{e \in \tilde{E}} d_{EM}(A(G), A(G \setminus \{e\})) \leq 2((n_1 - 1)(n_2 - 1) + (n_1 - 1)(n_3 - 1) + (n_2 - 1)(n_3 - 1) + (n_1 - 1)(n_2 - 1)(n_3 - 1))$.

Theorem 6. *The RBFS algorithm on* $n_1 \times n_2 \times n_3$ *3D grids has average sensitivity at most* $\frac{2}{3} + \frac{2(n_1-1)}{3n_2(n_3-1)} + \frac{2(n_2-1)}{3n_1(n_3-1)} + \frac{2n_3}{3n_1n_2} + \frac{2}{3(n_3-1)} + \frac{2}{3n_2} + \frac{2}{3n_1}$ *for* $2 \leq n_3 \leq n_1, n_2$, *and at most* $\frac{2}{3} + \frac{4}{n}$ *for* $n_1 = n_2 = n_3 = n$.

5 Conclusion

Our analysis shows that BFS algorithms have average sensitivities of small constants on 2D and 3D grid graphs, significantly better than $\Theta(|V(G)|)$ on arbitrary graphs. For $m \times n$ 2D grids, DBFS-D has average sensitivity at most 2 for $2 \leq m \leq n$, and is considered more stable than the RBFS algorithm when $m \ll n$, while the RBFS algorithm has a smaller average sensitivity than DBFS-D for more than half of the source nodes when $m = n$. For 3D grids, the RBFS algorithm has a smaller average sensitivity than the DBFS algorithm. For a 3D grid, our results suggest to rotate the grid to have the dimension of the smallest value as n_3 to obtain a small average sensitivity of BFS algorithms. Our results also indicate that randomization of the BFS algorithm can reduce the average sensitivity of BFS algorithms on 2D and 3D grids in many cases. It is worth exploring the k-sensitivity of BFS algorithms for $k > 1$ on grids and $k \geq 1$ on other practical networks.

Acknowledgment. The authors thank anonymous reviewers for their constructive comments. The research was partially supported by NSERC discovery grant.

References

1. Alwan, H., Agarwal, A.: A survey on fault tolerant routing techniques in wireless sensor networks. In: Proceedings of the 2009 International Conference on Sensor Technologies and Applications, pp. 366–371 (2009)
2. Bousquet, O., Klochkov, Y., Zhivotovskiy, N.: Sharper bounds for uniformly stable algorithms. In: Proceedings of Conference on Learning Theory, pp. 610–626. PMLR (2020)
3. Deng, K., Zhang, Q., Zhang, H., Xiao, P., Chen, J.: Optimal emergency evacuation route planning model based on fire prediction data. Mathematics **10**(17), 3146 (2022)
4. Diakonikolas, I., Raghavendra, P., Servedio, R.A., Tan, L.-Y.: Average sensitivity and noise sensitivity of polynomial functions. SIAM J. Comput. **43**(1), 231–253 (2014)
5. Fornberg, B., Piret, C.: A stable algorithm for flat radial basis functions on a sphere. SIAM J. Sci. Comput. **30**(1), 60–80 (2008)
6. Friedgut, E.: Boolean functions with low average sensitivity depend on few coordinates. Combinatorica **18**(1), 27–35 (1998)

7. Hara, S., Yoshida, Y.: Average sensitivity of decision tree learning. In: Proceedings of the 11th International Conference on Learning Representations (ICLR 2023) (2023)
8. Kartoirono, S., Riadi, I., Furizal, F., Azhari, A.: Improved breadth first search for public transit line search optimization. Mob. Forensics **5**(1), 12–22 (2022)
9. Kumabe, S., Yoshida, Y.: Average sensitivity of dynamic programming. In: Proceedings of the 33rd Symposium on Discrete Algorithms (SODA 2022), pp. 1925–1961 (2022)
10. Kumabe, S., Yoshida, Y.: Average sensitivity of the knapsack problem. In: Proceedings of the 30th Annual European Symposium on Algorithms (ESA 2022), pp. 75:1–75:14 (2022)
11. McLaughlin, A., Bader, D.A.: Fast execution of simultaneous breadth-first search on sparse graphs. In: Proceedings of the 2015 IEEE International Conference on Parallel and Distributed Systems, pp. 14–17 (2015)
12. Murai, S., Yoshida, Y.: Sensitivity analysis of centralities on unweighted networks. In: Proceedings of the 2019 World Wide Web Conference (WWW 2019), pp. 1332–1342 (2019)
13. Navya, P., Ranjith, R.: Analysis of path planning algorithms for service robots in hospital environment. In: Proceedings of the 2021 International Conference on Computing Communication and Networking Technologies, pp. 1–6 (2021)
14. Ng, A.Y., Zheng, A.X., Jordan, M.I.: Stable algorithms for link analysis. In: Proceedings of the 24th Annual International ACM SIGIR Conference on Research and Development in Information Retrieval, pp. 258–266 (2001)
15. Panigrahi, P.K., Bisoy, S.K., Tripathy, H.: Intelligent path planning with improved BFS (I-BFS) strategy for mobile robot in RFID equipped unknown environment. In: Proceedings of International Conference on Computational Intelligence and Data Engineering, pp. 237–249 (2022)
16. Peng, P., Yoshida, Y.: Average sensitivity of spectral clustering. In: Proceedings of the 2020 ACM SIGKDD International Conference on Knowledge Discovery & Data Mining (KDD 2020), pp. 1925–1961 (2020)
17. Plonka, G., Tasche, M.: Fast and numerically stable algorithms for discrete cosine transforms. Linear Algebra Appl. **394**, 309–345 (2005)
18. Rossman, B.: Average sensitivity of bounded depth formulas. Comput. Complex. **27**, 209–223 (2018)
19. Rubner, Y., Tomasi, C., Guibas, L.J.: A metric for distributions with applications to image database. In: Proceedings of the 1998 IEEE International Conference on Computer Vision, pp. 59–66 (1998)
20. Sturtevant, N.: An argument for large-scale breadth-first search for game design and content generation via a case study of fling. In: Proceedings of the 2013 Artificial Intelligence in the Game Design Process Workshop, pp. 28–33 (2013)
21. Varma, N., Yoshida, Y.: Average sensitivity of graph algorithms. In: Proceedings of the 32nd ACM-SIAM Symposium on Discrete Algorithms (SODA 2021), pp. 684–703 (2021)
22. Vásconez, J.P., et al.: Comparison of path planning methods for robot navigation in simulated agricultural environment. Procedia Comput. Sci. **220**, 898–903 (2023)
23. Vavasis, S.A.: Stable numerical algorithms for equilibrium systems. SIAM J. Matrix Anal. Appl. **15**(4), 1108–1131 (1994)
24. Yoshida, Y., Ito, S.: Average sensitivity of Euclidean k-clustering. Adv. Neural. Inf. Process. Syst. **35**, 32487–32489 (2022)

25. Yoshida, Y., Zhou, S.: Sensitivity analysis of the maximum matching problem. In: Proceedings of the 12th Innovations in Theoretical Computer Science Conference (ITCS 2021), pp. 58:1-58:20 (2021)
26. Yu, P.-D., Tan, C.W., Fu, H.-L.: Epidemic source detection in contact tracking networks: epidemic centrality in graphs and message-passing algorithms. IEEE J. Sel. Top. Signal Process. **16**(2), 234–249 (2022)

Permanent of Bipartite Graphs in Terms of Determinants

Surabhi Chakrabartty[1(✉)] and Ranveer Singh[2]

[1] Indian Institute of Science Education and Research Berhampur, Berhampur, India
surabhic20@iiserbpr.ac.in

[2] Indian Institute of Technology Indore, Indore, India
ranveer@iiti.ac.in

Abstract. Computing the permanent of a $(0, 1)$-matrix is a well-known $\#P$-complete problem. In this paper, we present an expression for the permanent of a bipartite graph in terms of the determinant of the graph and its subgraphs, obtained by successively removing rows and columns corresponding to vertices involved in vertex-disjoint $4k$-cycles. Our formula establishes a general relationship between the permanent and the determinant for any bipartite graph. Since computing the permanent of a biadjacency matrix is equivalent to counting the number of its perfect matchings, this approach also provides a more efficient method for counting perfect matchings in certain types of bipartite graphs.

Keywords: Permanent · Bipartite graphs · Perfect matching · $4k$-cycles

1 Introduction

A graph $G = (V, E)$ consists of a vertex set $V = \{v_1, \ldots, v_n\}$ and an edge set E. It is said to be simple if it does not contain loops or multiple edges between a pair of vertices, and it is said to be undirected if the edges have no direction associated with them. In this paper, we will be concerned with simple and undirected graphs only. Adjacency matrix $A(G)$ of a graph G with n vertices is an $n \times n$ matrix defined by $A(G) = (a_{ij})$ where

$$a_{ij} = \begin{cases} 1, & \text{if there is an edge between } v_i \text{ and } v_j, \\ 0, & \text{otherwise.} \end{cases}$$

A matching in a graph is a subset of edges such that no two edges share a common vertex. A perfect matching is a matching that covers every vertex of the graph. It is evident that a graph with an odd number of vertices cannot have a perfect matching. While determining the existence of a perfect matching in a general graph can be done efficiently [4], counting the number of perfect matchings is computationally very hard.

H. Fernau and B. Zhu (Eds.): IWOCA 2025, LNCS 15885, pp. 188–200, 2025.
https://doi.org/10.1007/978-3-031-98740-3_14

A graph is said to be bipartite if its vertex set can be divided into two disjoint sets, such that no two vertices within the same set are adjacent. Given a bipartite graph $G = (U, V, E)$, its biadjacency matrix is a $|U| \times |V|$ matrix defined by $B(G) = (b_{ij})$ where

$$b_{ij} = \begin{cases} 1, & \text{if there is an edge between } u_i \in U \text{ and } v_j \in V, \\ 0, & \text{otherwise.} \end{cases}$$

The determinant and permanent of an $n \times n$ matrix $M = (a_{ij})$ are defined as

$$\det(M) = \sum_{\sigma \in S_n} \operatorname{sgn}(\sigma) \prod_{i=1}^{n} a_{i,\sigma(i)},$$

and

$$\operatorname{per}(M) = \sum_{\sigma \in S_n} \prod_{i=1}^{n} a_{i,\sigma(i)},$$

respectively, where S_n is the set of all permutations of $\{1, 2, \ldots, n\}$, the sign function $\operatorname{sgn}(\sigma)$ is 1, if σ is even and -1 if σ is odd, respectively. While the determinant of any matrix can be computed in $\mathcal{O}(n^{2.37})$, computing the permanent is known to be $\#P$-complete even for a $(0, 1)$-matrix [14]. The determinant and permanent of a graph are defined as the determinant and permanent associated with its adjacency matrix, respectively. We denote them by $\det(G)$ and $\operatorname{per}(G)$.

The Pólya Permanent Problem [9] asks under what conditions the signs of certain nonzero entries in a $(0, 1)$-matrix A can be adjusted to obtain a matrix B such that $\operatorname{per}(A) = \det(B)$. This problem is known to be equivalent to twenty-three other graph-theoretic problems [7,10], including the problem of counting the number of perfect matchings in bipartite graphs. For a detailed survey and a complete characterization of minor-closed graph classes where perfect matchings can be computed efficiently, see [13]. While computing the permanent of a general matrix M is intractable in general, when M is the biadjacency matrix of a bipartite planar graph, the permanent can be computed in polynomial time by reducing it to the Pfaffian of a suitably constructed skew-symmetric matrix, as established by Kasteleyn [6].

Beyond its role in graph theory, the permanent also has important applications in quantum computing [1]. In Boson sampling, it helps determine the probability amplitudes of photon distributions in linear optical networks. It also shows up in statistical physics, particularly in models like the dimer model and monomer-dimer model, which are used to study phase transitions and other physical phenomena.

Writing the permanent as a function of the determinant has been a long-standing problem, more generally referred to as the Permanent vs. Determinant problem. In this paper, we aim to establish a relationship between the permanent and determinant of bipartite graphs using graph-theoretic concepts. As a corollary of this relation, we also observe the fact that the absence of $4k$-cycles in a bipartite graph guarantees that the permanent and determinant have the same absolute value. This formulation will not only provide insights into their general relationship but will also facilitate more efficient computation of the permanent for certain types of bipartite graphs.

A cycle of length l is a sequence of distinct vertices $u_1, u_2, \dots, u_l$ such that $u_i \sim u_{i+1}$ for $i = 1, \dots, l-1$ and $u_l \sim u_1$. A $4k$-cycle is a cycle with a length that is a multiple of four. Likewise, a $(4k+2)$-cycle is a cycle whose length exceeds a multiple of four by two. Since bipartite graphs do not contain odd-length cycles, the only cycles they can contain are $4k$-cycles and $(4k+2)$-cycles. We denote the set of all $4k$-cycles and $(4k+2)$-cycles of a graph G by $C_{4k}(G)$ and $C_{4k+2}(G)$, respectively.

We use the notation T_z to represent any ordered tuple of z mutually vertex-disjoint $4k$-cycles. Note that T_1 simply denotes any $4k$-cycle in the graph, as a single cycle is always vertex-disjoint by definition.

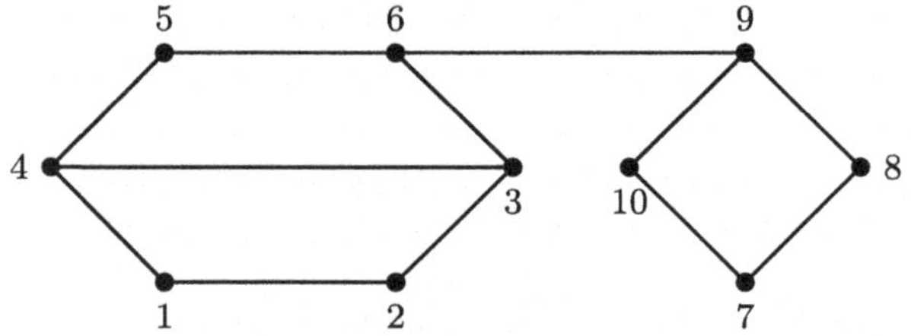

Fig. 1. Example of a bipartite graph

Example 1. Consider the graph G shown in Fig. 1. The cycles in G are:

$C_1 = (1, 2, 3, 4, 1), C_2 = (3, 4, 5, 6, 3), C_3 = (7, 8, 9, 10, 7), C_4 = (1, 2, 3, 6, 5, 4, 1),$

Then the set of $4k$-cycles in G is given by

$$C_{4k}(G) = \{C_1, C_2, C_3\},$$

and the set of $(4k+2)$-cycles is

$$C_{4k+2}(G) = \{C_4\}.$$

Now, the single-cycle tuples are:

$$T_1^1 = C_1, \quad T_1^2 = C_2, \quad T_1^3 = C_3.$$

The two-cycle vertex-disjoint tuples are:

$$T_2^1 = (C_1, C_3), \quad T_2^2 = (C_2, C_3), \quad T_2^3 = (C_3, C_1), \quad T_2^4 = (C_3, C_2).$$

The rest of the paper is organized as follows: In Sect. 2, we give the graph-theoretic interpretation of the permanent by introducing Sachs subgraphs. In the main result (Sect. 3), we demonstrate how the determinant of the submatrices obtained after removing rows and columns corresponding to $4k$-cycles can be used to compute the permanent of a bipartite graph. Finally, in the discussion (Sect. 4), we provide a few conditions such that the permanent of such a type of bipartite graph can be computed efficiently.

2 Permanent Using Sachs Subgraphs

A Sachs subgraph is a subgraph in which every component is either a cycle or an individual edge. We observe that, for a bipartite graph, there is no Sachs subgraph on an odd number of vertices, as bipartite graphs do not contain odd-length cycles. We will use the notation U_i to denote any Sachs subgraph of G on i vertices.

In the graph shown in Fig. 1, some possible Sachs subgraphs on six vertices are

$$\begin{aligned}
&U_6^1 = \{C_4\}, \quad U_6^2 = \{C_1 \cup (5,6)\}, \quad U_6^3 = \{C_1 \cup (7,8)\},\\
&U_6^4 = \{C_2 \cup (1,2)\}, \quad U_6^5 = \{C_3 \cup (3,4)\},\\
&U_6^6 = \{(1,4) \cup (6,9) \cup (7,8)\}, \quad U_6^7 = \{(1,2) \cup (3,4) \cup (5,6)\}.
\end{aligned}$$

In total, there are 94 Sachs subgraphs of G on six vertices.

Some possible Sachs subgraphs of G on four vertices are

$$\begin{aligned}
&U_4^1 = \{C_1\}, \quad U_4^2 = \{C_2\}, \quad U_4^3 = \{C_3\}.\\
&U_4^4 = \{(2,3) \cup (9,10)\}, \quad U_4^5 = \{(3,6) \cup (1,2)\}.
\end{aligned}$$

In total, there are 51 Sachs subgraphs of G on four vertices. Furthermore, each Sachs subgraph of G on two vertices contains exactly one edge of the graph G.

Hence, from this example, we observe that in a bipartite graph, the existence of a Sachs subgraph on i vertices, implies the existence of Sachs subgraph on $2, 4, \ldots, i-2$ vertices. This is because if the Sachs subgraph has even one cycle, then we can decompose it to disjoint edges or copies of K_2 by removing alternate edges, and if the Sachs subgraph has no cycles and comprises edges only, then we can remove the edges one by one to obtain smaller Sachs subgraphs. We will use this fact for the construction of different Sachs subgraphs from the known ones.

2.1 Permanent Using the Determinant of Subgraphs

We recall the formula for computing the determinant and permanent using Sachs subgraphs in the following way [11,12]:

$$\det(G) = \sum_{U_n} (-1)^{n-p(U_n)} 2^{c(U_n)} \text{ and } \operatorname{per}(G) = \sum_{U_n} 2^{c(U_n)}, \tag{1}$$

where the summation is taken over all the Sachs subgraphs U_n of G on n (where $n = |V|$) vertices, $p(U_n)$ denotes the number of components in U_n, and $c(U_n)$ denotes the number of cycles in U_n.

As observed earlier, in a bipartite graph, there is no Sachs subgraph on an odd number of vertices; thus, $\det(G) = \mathrm{per}(G) = 0$ when n is odd [2,8]. Thus, from now on, we will consider n to be even.

Our goal is to express the permanent of any bipartite graph G in terms of determinants, thereby establishing a general relationship between the two. Equation (1) can be written as

$$\det(G) = \sum_{U_n} (-1)^{s+t+r} 2^{s+t} \text{ and } \mathrm{per}(G) = \sum_{U_n} 2^{s+t}, \tag{2}$$

respectively, where U_n varies over the set of Sachs subgraphs of G on n vertices, s denotes the number of $4k$-cycles in U_n, t denotes the number of $(4k+2)$-cycles in U_n and r denotes the number of edges in U_n. That is, U_n can be expressed in the following way:

$$U_n = \{C_{4k_1} \cup \cdots \cup C_{4k_s}\} \cup \{C_{4l_1+2} \cup \cdots \cup C_{4l_t+2}\} \cup \underbrace{\{K_2 \cup \cdots \cup K_2\}}_{r \text{ times}},$$

for some positive integers $k_1, \ldots, k_s, l_1, \ldots, l_t$. We use the notation K_2 to represent any edge; however, it is important to note that in our case, each K_2 is distinct and does not share any vertices with another K_2 in the same Sachs subgraph. As U_n represents a Sachs subgraph on n vertices, we must have the following equality:

$$n = 4(k_1 + \cdots + k_s) + 4(l_1 + \cdots + l_t) + 2(t + r).$$

As we have taken n to be even, we have

$$\frac{n}{2} \equiv t + r \pmod 2.$$

Therefore, using Eq. (2), determinant reduces to

$$\det(G) = \sum_{U_n} (-1)^{s+\frac{n}{2}} 2^{s+t}. \tag{3}$$

Determinant of Subgraphs Using 4*k*-Cycles. We will use the notation $\det(G \backslash R)$ to denote the determinant of the submatrix obtained by removing the rows and columns corresponding to the vertices in the $4k$-cycle R of G. Now, we use Formula (3) to obtain:

$$\sum_{R_1 \in C_{4k}(G)} \det(G \backslash R_1) = \sum_{R_1 \in C_{4k}(G)} \sum_{W_{n-|V(R_1)|}} (-1)^{s' + \frac{n-|V(R_1)|}{2}} 2^{s'+t'},$$

where the last summation is over all the Sachs subgraph $W_{n-|V(R_1)|}$ of $G \backslash R_1$, s' denotes the number of $4k$-cycles in $W_{n-|V(R_1)|}$ and t' denotes the number of $(4k+2)$-cycles in $W_{n-|V(R_1)|}$.

Using the fact that $(-1)^{\frac{n-|V(R_1)|}{2}} = (-1)^{\frac{n}{2}}$ (since $|V(R_1)|$ is a multiple of 4), and the one-to-one correspondence between the Sachs subgraph U_n in G containing R_1 and the Sachs subgraph $W_{n-|V(R_1)|}$ in $G\backslash R_1$, and noting that $s = s' + 1$ (since s represents the number of $4k$-cycles in U_n and always includes one more—namely, R_1—than the number s' in $W_{n-|V(R_1)|}$), we obtain

$$\sum_{R_1 \in C_{4k}(G)} \det(G\backslash R_1) = \sum_{R_1 \in C_{4k}(G)} \sum_{\substack{U_n \text{ containing} \\ R_1}} (-1)^{s-1+\frac{n}{2}} 2^{s-1+t}. \tag{4}$$

We will iteratively apply Eq. (4) by using two disjoint $4k$-cycles, then three, then four, and so on, while computing the determinant for each resulting reduced subgraph.

3 The Main Result

Theorem 1. *Let G be a bipartite graph with n vertices. If G contains at most m vertex-disjoint $4k$-cycles then*

$$\operatorname{per}(G) = \begin{cases} (-1)^{n/2} \displaystyle\sum_{z=0}^{m} \frac{4^z}{z!} \sum_{T_z} \det(G\backslash T_z), & \text{if } n \text{ is even,} \\ 0, & \text{if } n \text{ is odd,} \end{cases} \tag{5}$$

where the sum varies over all possible ordered tuples T_z of z mutually vertex-disjoint $4k$-cycles in G.

Proof. For a bipartite graph with an odd number of vertices, the permanent is zero, as follows trivially from our earlier observations about the non-existence of a Sachs subgraph on an odd number of vertices in a bipartite graph. Therefore, it remains to prove the case when the number of vertices is even.

Recall from Eq. (4) that

$$\begin{aligned} \sum_{T_1} \det(G\backslash T_1) &= \sum_{R_1 \in C_{4k}(G)} \det(G\backslash R_1) \\ &= \sum_{R_1 \in C_{4k}(G)} \sum_{\substack{U_n \text{ containing} \\ R_1}} (-1)^{s-1+\frac{n}{2}} 2^{s+t-1}, \end{aligned}$$

where the first summation on the right hand side, is taken over all $4k$-cycles R_1 of G, the second summation is over all Sachs subgraphs U_n of G on n vertices that contain R_1; s denotes the number of $4k$-cycles in U_n, and t denotes the number of $(4k+2)$-cycles in U_n. Notice that on the right-hand side of the equation, each Sachs subgraph is counted exactly as many times as the number of $4k$-cycles it contains. Hence,

$$\sum_{T_1} \det(G\backslash T_1) = (-1)^{\frac{n}{2}} \sum_{j=1,2,3,\ldots,m} (-1)^{j-1} j 2^{j-1} \sum_{\substack{U_n \text{ containing} \\ \text{exactly } j \text{ } 4k\text{-cycles}}} 2^t.$$

Similarly, using Eq. (4) again, we observe that for each $R_1 \in C_{4k}(G)$, we have

$$\sum_{R_2 \in C_{4k}(G \backslash R_1)} \det(G \backslash R_1 \backslash R_2) = (-1)^{\frac{n-|V(R_1)|}{2}} \sum_{R_2 \in C_{4k}(G \backslash R_1)} \sum_{\substack{W_{n-|V(R_1)|} \\ \text{containing } R_2}} (-1)^{s'-1} 2^{s'+t'-1}, \quad (6)$$

where the last summation is taken over all the Sachs subgraph $W_{n-|V(R_1)|}$ of $G \backslash R_1$ that contain R_2, s' denotes the number of $4k$-cycles in $W_{n-|V(R_1)|}$, and t' denotes the number of $(4k+2)$-cycles in $W_{n-|V(R_1)|}$. Following the same trick as before, we note that in the above equation's right-hand side, each Sachs subgraph is summed exactly as many times as the number of $4k$-cycles it contains. Hence,

$$\begin{aligned}
\sum_{T_2} \det(G \backslash T_2) &= \sum_{R_1 \in C_{4k}(G)} \sum_{R_2 \in C_{4k}(G \backslash R_1)} \det(G \backslash R_1 \backslash R_2) \\
&= (-1)^{\frac{n}{2}} \sum_{R_1 \in C_{4k}(G)} \sum_{j=1,2,\ldots,m} (-1)^{j-1} j 2^{j-1} \sum_{\substack{W_{n-|V(R_1)|} \text{containing} \\ \text{exactly } j \text{ } 4k\text{-cycles}}} 2^{t'} \\
&= (-1)^{\frac{n}{2}} \sum_{R_1 \in C_{4k}(G)} \sum_{W_{n-|V(R_1)|}} (-1)^{s'-1} s' 2^{s'+t'-1}.
\end{aligned}$$

Using the fact that there is a one-to-one correspondence between the Sachs subgraph U_n in G containing R_1 and the Sachs subgraph $W_{n-|V(R_1)|}$ in $G \backslash R_1$, and noting that $s = s' + 1$ (since s represents the number of $4k$-cycles in U_n containing R_1 and hence starts from 1, always containing one $4k$-cycle more—namely, R_1—than $W_{n-|V(R_1)|}$), we obtain

$$\begin{aligned}
\sum_{T_2} \det(G \backslash T_2) &= (-1)^{\frac{n}{2}} \sum_{R_1 \in C_{4k}(G)} \sum_{\substack{U_n \text{ containing} \\ R_1}} (-1)^{s-2}(s-1)2^{s+t-2} \\
&= (-1)^{\frac{n}{2}} \sum_{j=2,3,4\ldots,m} (-1)^{j-2}(j-1)j2^{j-2} \sum_{\substack{U_n \text{ containing} \\ \text{exactly } j \text{ } 4k\text{-cycles}}} 2^t \\
&= (-1)^{\frac{n}{2}} \sum_{j=2,3,4\ldots,m} (-1)^{j-2} 2! \binom{j}{j-2} 2^{j-2} \sum_{\substack{U_n \text{ containing} \\ \text{exactly } j \text{ } 4k\text{-cycles}}} 2^t.
\end{aligned}$$

Similarly, it can be verified by mathematical induction on z that for any $z \in \{0, 1, 2, \ldots, m\}$,

$$\sum_{T_z} \det(G \backslash T_z) = (-1)^{\frac{n}{2}} \sum_{j=z}^{m} (-1)^{j-z} z! \binom{j}{j-z} 2^{j-z} \sum_{\substack{U_n \text{containing} \\ \text{exactly } j \text{ } 4k\text{-cycles}}} 2^t. \quad (7)$$

Next, for each value of $z \in \{0, 1, 2, \ldots, m\}$, we multiply Eq. (7) by the coefficient $\frac{4^z}{z!}$ and sum over all values of z to obtain

$$\begin{aligned}\sum_{z=0}^{m} \frac{4^z}{z!} \sum_{T_z} \det(G \backslash T_z) &= (-1)^{\frac{n}{2}} \sum_{z=0}^{m} \frac{4^z}{z!} \sum_{j=z}^{m} (-1)^{j-z} z! \binom{j}{j-z} 2^{j-z} \sum_{\substack{U_n \text{ containing} \\ \text{exactly } j \text{ } 4k\text{-cycles}}} 2^t \\ &= (-1)^{\frac{n}{2}} \sum_{z=0}^{m} \sum_{j=z}^{m} (-1)^{j-z} \binom{j}{j-z} 2^{j+z} \sum_{\substack{U_n \text{ containing} \\ \text{exactly } j \text{ } 4k\text{-cycles}}} 2^t \\ &= (-1)^{\frac{n}{2}} \sum_{j=0}^{m} 2^j \sum_{z=0}^{j} (-1)^{j-z} \binom{j}{j-z} 2^z \sum_{\substack{U_n \text{ containing} \\ \text{exactly } j \text{ } 4k\text{-cycles}}} 2^t.\end{aligned}$$

By using the binomial theorem, we have:

$$\begin{aligned}\sum_{z=0}^{j} (-1)^{j-z} \binom{j}{j-z} 2^z &= (-1+2)^j \\ &= 1.\end{aligned}$$

Hence,

$$\begin{aligned}\sum_{z=0}^{m} \frac{4^z}{z!} \sum_{T_z} \det(G \backslash T_z) &= (-1)^{\frac{n}{2}} \sum_{j=0}^{m} 2^j \sum_{\substack{U_n \text{ containing} \\ \text{exactly } j \text{ } 4k\text{-cycles}}} 2^t \\ &= (-1)^{\frac{n}{2}} \sum_{U_n} 2^{s+t} \\ &= (-1)^{\frac{n}{2}} \operatorname{per}(G).\end{aligned}$$

This completes the proof. □

Corollary 1. *If a bipartite graph G is free of $4k$-cycles then*

$$\operatorname{per}(G) = |\det(G)|. \tag{8}$$

In simpler words, the absence of $4k$-cycles in a bipartite graph guarantee that the permanent and determinant have the same absolute value.

Theorem 2. *Let G be a bipartite graph and let G_i denote an induced subgraph of G on i vertices. Then G contains at most m vertex-disjoint $4k$-cycles if and only if*

$$\operatorname{per}(G_i) = (-1)^{i/2} \sum_{z=0}^{m} \frac{4^z}{z!} \sum_{T_z} \det(G_i \backslash T_z), \tag{9}$$

holds for all such G_i, where $i \in \{0, 2, 4, \ldots\}$, and the inner sum varies over all possible ordered tuples T_z of z mutually vertex-disjoint $4k$-cycles in G_i.

Proof. Since for each even i, G_i is itself a bipartite graph on i vertices, the forward implication follows directly from Theorem 1.

Let us prove the backward implication. Given that Eq. (9) holds, we must show that G does not contain more than m vertex-disjoint $4k$-cycles. It is equivalent to showing that none of the Sachs subgraphs of G contains more than m $4k$-cycles.

Let us assume that a Sachs subgraph contains at least $(m+1)$ $4k$-cycles. Now, by keeping any $m+1$ $4k$-cycles of this Sachs subgraph and removing the alternate edges from the other $4k$-cycles, we can derive a Sachs subgraph containing exactly $m+1$ $4k$-cycles.

For all $i \in \{0, 2, 4, \dots\}$, we have:

$$\operatorname{per}(G_i) = (-1)^{i/2} \sum_{z=0}^{m} \frac{4^z}{z!} \sum_{T_z} \det(G_i \backslash T_z). \tag{10}$$

Rewriting in terms of Sachs' subgraphs, we have

$$\begin{aligned}\sum_{j=0}^{n} 2^j \sum_{\substack{U_i \text{ containing} \\ \text{exactly } j \text{ } 4k\text{-cycles}}} 2^t &= \sum_{z=0}^{m} \frac{4^z}{z!} \sum_{j=z}^{n} (-1)^{j-z} z! \binom{j}{j-z} 2^{j-z} \sum_{\substack{U_i \text{ containing} \\ \text{exactly } j \text{ } 4k\text{-cycles}}} 2^t \\ &= \sum_{j=0}^{n} 2^j \sum_{z=0}^{\min(j,m)} (-1)^{j-z} \binom{j}{j-z} 2^z \sum_{\substack{U_i \text{ containing} \\ \text{exactly } j \text{ } 4k\text{-cycles}}} 2^t.\end{aligned} \tag{11}$$

For $j \leq m$, we use the binomial theorem to obtain

$$\begin{aligned}\sum_{z=0}^{j} (-1)^{j-z} \binom{j}{j-z} 2^z &= (-1+2)^j \\ &= 1.\end{aligned} \tag{12}$$

For $j = m+1$, the binomial theorem results in

$$\begin{aligned}\sum_{z=0}^{m} (-1)^{m+1-k} \binom{m+1}{m+1-k} 2^z &= (-1+2)^{m+1} - 2^{m+1} \\ &= 1 - 2^{m+1}.\end{aligned} \tag{13}$$

We substitute Eq. (12) into Eq. (11) to obtain:

$$\begin{aligned}\sum_{j=0}^{n} 2^j \sum_{\substack{U_i \text{ containing} \\ \text{exactly } j \text{ } 4k\text{-cycles}}} 2^t = &\sum_{j=0}^{m} 2^j \sum_{\substack{U_i \text{ containing} \\ \text{exactly } j \text{ } 4k\text{-cycles}}} 2^t \\ &+ \sum_{j=m+1}^{n} 2^j \sum_{z=0}^{m} (-1)^{j-z} \binom{j}{j-z} 2^z \sum_{\substack{U_i \text{ containing} \\ \text{exactly } j \text{ } 4k\text{-cycles}}} 2^t.\end{aligned}$$

On simplification, we have

$$\sum_{j=m+1}^{n} 2^j (1 - \sum_{z=0}^{m} (-1)^{j-z} \binom{j}{j-z} 2^z) \sum_{\substack{U_i \text{ containing} \\ \text{exactly } j \text{ } 4k\text{-cycles}}} 2^t = 0.$$

Substituting the value from Eq. (13), we obtain

$$\sum_{j=m+2}^{n} 2^j (1 - \sum_{z=0}^{m} (-1)^{j-z} \binom{j}{j-z} 2^z) \sum_{\substack{U_i \text{ containing} \\ \text{exactly } j \text{ } 4k\text{-cycles}}} 2^t + 2^{2m+2} \sum_{\substack{U_i \text{ containing} \\ \text{exactly } m+1 \text{ } 4k\text{-cycles}}} 2^t = 0. \tag{14}$$

From Eq. (14), it follows that there must exist a Sachs subgraph of G with more than $m+1$ $4k$-cycles. If no such subgraph exists, the first sum in Eq. (14) would be zero. Consequently, the last term must also be zero, as the right-hand side is zero. This contradicts the assumption that a Sachs subgraph exists with exactly $m+1$ $4k$-cycles. Hence, we conclude that a Sachs subgraph of G with at least $m+2$ $4k$-cycles exists. By a similar construction as in our earlier arguments, this further implies the existence of a Sachs subgraph with exactly $m+2$ $4k$-cycles.

Among all Sachs subgraphs of G containing exactly $m+2$ $4k$-cycles, we choose the Sachs subgraph containing the least number of vertices and call it U'_q. Also, observe that U'_q must not contain any $(4k+2)$-cycles or edges due to minimality of q. Hence,

$$U'_q = C_1 \cup C_2 \cup \cdots \cup C_{m+2},$$

for some vertex-disjoint $C_1, C_2, \ldots, C_{m+2} \in C_{4k}(G)$.

Let $p = \mid V(C_1) \mid + \mid V(C_2) \mid + \cdots + \mid V(C_{m+1}) \mid < q$. We know that Eq. (14) is satisfied for all $i \in \{0, 2, 4, \ldots\}$, so it should be satisfied for $i = p$ as well. As $p < q$, there is no Sachs subgraph of G on p vertices which contains more than $m+1$ $4k$-cycles. Hence, by setting $i = p$ in Eq. (14), we obtain

$$2^{2m+2} \sum_{\substack{U_p \text{containing} \\ \text{exactly } m+1 \text{ } 4k\text{-cycles}}} 2^t = 0.$$

This is a contradiction to our assumption since there exists a Sachs subgraph $U'_p = U'_q \backslash C_{m+2} = C_1 \cup C_2 \cup \cdots \cup C_{m+1}$ and hence the sum is non-zero. Thus, no Sachs subgraph of G contains more than m $4k$-cycles. Therefore, the backward implication of Theorem 2 holds. This completes the proof. □

Example 2. Consider the bipartite graph G in Fig. 1. Since C_1 and C_2 share a common vertex, only one of them can be included at a time in the tuple of vertex-disjoint $4k$-cycles along with C_3; hence, the maximum number of vertex-disjoint $4k$-cycles in G is 2. Therefore, the permanent of G can be computed using Eq. (5) with $m = 2$.

$$\begin{aligned}\text{per}(G) &= (-1)^{n/2} \sum_{z=0}^{2} \frac{4^z}{z!} \sum_{T_z} \det(G \backslash T_z) \\ &= (-1)^{n/2} \left[\det(G) + 4 \sum_{T_1} \det(G \backslash T_1) + \frac{4^2}{2!} \sum_{T_2} \det(G \backslash T_2) \right]. \end{aligned} \tag{15}$$

as n is even. Now,

$$\det(G) = 0.$$

Recall that $C_{4k}(G) = \{C_1, C_2, C_3\}$. Hence,

$$\begin{aligned}\sum_{T_1} \det(G \backslash T_1) &= \sum_{R_1 \in C_{4k}(G)} \det(G \backslash R_1) \\ &= \det(G \backslash C_1) + \det(G \backslash C_2) + \det(G \backslash C_3) \\ &= 0 + 0 + (-1) \\ &= -1.\end{aligned}$$

and,

$$\begin{aligned}\sum_{T_2} \det(G \backslash T_2) &= \sum_{R_1 \in C_{4k}(G)} \sum_{R_2 \in C_{4k}(G \backslash R_1)} \det(G \backslash R_1 \backslash R_2) \\ &= \det(G \backslash C_1 \backslash C_3) + \det(G \backslash C_2 \backslash C_3) \\ &\quad + \det(G \backslash C_3 \backslash C_1) + \det(G \backslash C_3 \backslash C_2) \\ &= (-1) + (-1) + (-1) + (-1) \\ &= -4.\end{aligned}$$

Hence, using (15), we get:

$$\begin{aligned}\text{per}(G) &= (-1)^{\frac{10}{2}} \{0 + 4(-1) + \frac{4^2}{2!}(-4)\} \\ &= 36.\end{aligned}$$

4 Discussion

Theorem 1 can be utilized to compute the permanent for any bipartite graph. Since permanent is closely related to the count of perfect matching, this formula could also be used to compute the number of perfect matching of bipartite graphs, where the biadjacency matrix itself serves as the adjacency matrix of another bipartite graph. The formula consists of two summations. The first summation runs from 1 to m, where m is the maximum number of vertex-disjoint $4k$-cycles. Since m can be at most $n/4$, this summation contributes at most

$n/4$ terms. However, the second summation is taken over all z-tuples of mutually vertex-disjoint $4k$-cycles, which can grow exponentially with the size of the graph, making it computationally challenging. Thus, the main challenges in this process include efficiently identifying all $4k$-cycles within the graph and ensuring that their number does not grow too large, as it directly impacts computational efficiency.

These challenges can be addressed by identifying classes of bipartite graphs that contain only a limited number of $4k$-cycles and allow efficient enumeration of these cycles. Trees, acyclic structures, and cycle graphs are particularly easy to handle using this formula due to their simpler structures. In general, enumerating all cycles in a graph is an NP-hard problem, but recent advancements [3, 5] have shown that all short cycles up to length $2g - 2$ in a bipartite graph can be enumerated efficiently. Therefore, it is worth investigating bipartite graphs that are characterized by the presence of only short cycles.

The bipartite cactus graph, a connected graph that is known for sharing at most one vertex between any two cycles, along with the condition:

$$g > \frac{n + \binom{c}{2} + c}{c + 2},$$

where g, n, c are the number of girth, the number of vertices, and the number of cycles of length g, respectively, is one such candidate where the theorem can be used efficiently. It would be valuable to identify more classes of bipartite graphs where this formula can be applied efficiently.

Acknowledgment. We thank Hitesh Wankhede, Noga Alon, and RB Bapat for their valuable suggestions. We also thank the anonymous reviewers for their insightful comments and acknowledge financial support from the DST INSPIRE grant.

References

1. Aaronson, S., Arkhipov, A.: The computational complexity of linear optics. Theory Comput. **9**, 143–252 (2013). https://doi.org/10.4086/TOC.2013.V009A004
2. Borowiecki, M.: On spectrum and per-spectrum of graphs. Publications de l'Institut Mathématique. Nouvelle Série **38** (1985)
3. Dehghan, A., Banihashemi, A.H.: Counting short cycles in bipartite graphs: a fast technique/algorithm and a hardness result. IEEE Trans. Commun. **68**(3), 1378–1390 (2020). https://doi.org/10.1109/TCOMM.2019.2962397
4. Hopcroft, J.E., Karp, R.M.: An $n^{5/2}$ algorithm for maximum matchings in bipartite graphs. SIAM J. Comput. **2**(4), 225–231 (1973). https://doi.org/10.1137/0202019
5. Karimi, M., Banihashemi, A.H.: Message-passing algorithms for counting short cycles in a graph. IEEE Trans. Commun. **61**(2), 485–495 (2013). https://doi.org/10.1109/TCOMM.2012.100912.120503
6. Kasteleyn, P.W.: Graph Theory and Crystal Physics. Graph Theory and Theoretical Physics, pp. 43–110. Academic Press (1967)
7. McCuaig, W.: Pólya's permanent problem. Electron. J. Comb. **11**(1) (2004). https://doi.org/10.37236/1832

8. Merris, R., Rebman, K.R., Watkins, W.: Permanental polynomials of graphs. Linear Algebra Appl. **38**, 273–288 (1981). https://doi.org/10.1016/0024-3795(81)90026-4
9. Pólya, G.: Aufgabe 424. Archiv der Mathematik und Physik **20**, 271 (1913)
10. Robertson, N., Seymour, P.D., Thomas, R.: Permanents, pfaffian orientations, and even directed circuits. Ann. Math. **150**(1), 929–975 (1999). https://doi.org/10.2307/121059
11. Shi, Y., Dehmer, M., Li, X., Gutman, I.: Graph Polynomials. 1st edn. Chapman and Hall/CRC, Boca Raton (2017). https://doi.org/10.1201/9781315367996
12. Tang, Z., Li, Q., Deng, H.: On the permanental polynomial and permanental sum of signed graphs. Discrete Math. Lett. **10**, 14–20 (2022). https://doi.org/10.47443/dml.2022.005
13. Thilikos, D.M., Wiederrecht, S.: Killing a vortex. J. ACM **71**(4), 56 (2024). https://doi.org/10.1145/3664648. Article 27
14. Valiant, L.G.: The complexity of computing the permanent. Theor. Comput. Sci. **8**, 189–201 (1979). https://doi.org/10.1016/0304-3975(79)90044-6

Improved Approximation for Unpopularity in (3, 3)-Hypergraph Matching with One-Sided Preferences

Yashdeep Singh(✉) and Sushanta Karmakar

Department of Computer Science and Engineering, Indian Institute of Technology Guwahati, Guwahati 781039, India
{yashdeep18,sushantak}@iitg.ac.in

Abstract. Given two sets of distinct items and a group of people with combined preferences for one item from each set, how can we optimally assign pairs of items to individuals? To solve this problem, we examine the popular matching problem in a 3-uniform 3-partite hypergraph. Here, the first set represents agents, whereas the second and third sets correspond to two different types of items. Agents express preferences over hyperedges, whereas the items in the other sets do not have any preferences. A matching M is considered popular if there is no alternative matching M' such that more agents prefer M' to matching M. Finding a popular matching, if it exists, is computationally hard. In such a case, an approximation is required. A common measure for such an approximation is called the unpopularity factor for a matching M, denoted as $u(M)$. It is defined as the largest ratio $|A_2|/|A_1|$, where A_1 represents the set of agents who prefer matching M over any other M', and A_2 represents the set of agents who prefer M' over M. In this paper, we provide an approximation algorithm that produces a 3-uniform 3-partite hypergraph matching M such that $u(M) \leq d^2$ where d is the maximum degree of an agent. This work improves on a previously known result in the same hypergraph model, where the unpopularity factor was bounded by $1.5 \times 3^{\frac{d}{2}}$. Furthermore, we also show that $u(M) < n$. The approximation algorithm takes $O(d(n^2 + m))$ time to produce the matching M where n is the total number of agents and m is the total number of hyperedges.

1 Introduction

A 3-uniform hypergraph $H = (V, E)$ consists of vertices V and hyperedges E, with each hyperedge consisting of exactly three vertices. A 3-uniform 3-partite hypergraph goes beyond this by dividing the vertices into three distinct sets, ensuring that each hyperedge contains exactly one vertex from each set. Consider a 3-uniform 3-partite hypergraph $H = (A \cup B \cup C, E)$, where A, B, and C are disjoint subsets of the vertices, and every hyperedge $e \in E$ contains one vertex from each of these subsets. Each subset (A, B, or C) contains n vertices. The elements in A, known as agents, have preference lists for hyperedges, whereas

H. Fernau and B. Zhu (Eds.): IWOCA 2025, LNCS 15885, pp. 201–213, 2025.
https://doi.org/10.1007/978-3-031-98740-3_15

items in B and C do not have preference lists. A matching is an ensemble of tuples (a, b, c), in which $a \in A$, $b \in B$, and $c \in C$, ensuring that every component appears in at most one tuple. When it comes to choosing between matchings, agents participate in a vote to select the more preferred one. An agent prefers a matching M over M' if it is part of M but not M', or if it is included in both, it prefers the hyperedge in M to that in M'. The notation $\phi(M, M')$ specifies the number of agents who prefer matching M over matching M'.

Definition 1. *(Kavitha et al. [12]) A matching M is called a popular matching if $\phi(M, M') \geq \phi(M', M)$ for every matching M'.*

In this paper, we refer to a 3-uniform 3-partite hypergraph as a $(3, 3)$-hypergraph. In a $(3, 3)$-hypergraph having one-sided preferences, a popular matching may or may not exist. Let us have a $(3, 3)$-hypergraph $H = (A \cup B \cup C, E)$ where $A = \{a_1, a_2, a_3\}$, $B = \{b_1, b_2, b_3\}$, $C = \{c_1, c_2, c_3\}$ and hyperedges $e_{3i} = (a_i, b_1, c_1)$, $e_{3i-1} = (a_i, b_2, c_2)$ and $e_{3i-2} = (a_i, b_3, c_3)$ for $1 \leq i \leq 3$. Assume that each a_i prefers e_{3i} over e_{3i-1} and e_{3i-1} over e_{3i-2}. The possible maximal matchings, considering symmetry, are $M_1 = \{e_3, e_5, e_7\}$, $M_2 = \{e_1, e_6, e_8\}$ and $M_3 = \{e_2, e_4, e_9\}$. In this situation, M_1 is more popular than M_3, M_2 is more popular than M_1, and M_3 is more popular than M_1. This scenario exemplifies the absence of a popular matching in a $(3, 3)$-hypergraph.

The popular matching problem determines whether an instance admits a popular matching. We have shown that this problem is NP-hard in a $(3, 3)$-hypergraph [15]. In such cases, seeking a suitable alternative to popular matching would be desirable: the unpopularity factor is one such measure. The unpopularity factor of a matching M is defined as the maximum ratio $|A_2|/|A_1|$, where A_1 and A_2 are the sets of agents that prefer M over any other matching M' and M' over M, respectively. The objective is to find a matching with a low unpopularity factor. We have given a polynomial-time algorithm that provides a matching M with an unpopularity factor bounded by $1.5 \times 3^{d/2}$ where d is the maximum degree of an agent [16]. In this paper, we improve this bound.

1.1 Our Contribution

We utilized the popular matching problem defined for the $(3, 3)$-hypergraph model in [15]. In this model, the input contains a set of hyperedges along with the preference lists of all agents. Our objective is to find a matching with bounded unpopularity. In a bipartite graph with one-sided preferences, Huang et al. [9] proposed an algorithm that gives a matching M with $u(M) \leq d$ where d is the maximum degree of a node. For a given $(3, 3)$-hypergraph and the preference lists of all the agents, we propose an algorithm that generates a matching M with $u(M) \leq d^2$ where d is the maximum degree of an agent, defined as the number of hyperedges the agent participates in. Our algorithm takes $O(d(n^2 + m))$ time to generate a matching M in a $(3, 3)$-hypergraph where n and m are the total number of agents and hyperedges, respectively. Furthermore, we show that the unpopularity factor of the matching M is always bounded by $n-1$, which implies $u(M) \leq d^2 \leq n - 1$.

1.2 Related Work

Gärdenfors [4] first introduced popular matching in bipartite graphs. Two variants of the popular matching problem have since been examined in bipartite graphs: one-sided and two-sided preference models [2]. Abraham et al. [1] introduced the first polynomial-time algorithm for the one-sided preference model, while McCutchen [13] conceptualized the unpopularity factor. Huang et al. [9] proposed an $O(km\sqrt{n})$-time algorithm to generate a matching M with an unpopularity factor $u(M) \leq d-1$, where d is the maximum degree of an agent. Huang and Kavitha [7] presented a polynomial-time algorithm to identify maximum-size popular matchings in a two-sided preference model. Kavitha [10] proposed an algorithm that results in a matching M of size k with $u(M) \leq 2$, and another matching with $u(M) \leq 2k$ for scenarios where one side permits ties up to length k [11]. In the roommate problem for non-bipartite graphs, the popular matching problem is NP-hard [5]. Huang and Kavitha [8] proposed an algorithm that gives a matching M with $u(M) \leq O(\log |V|)$. Cseh and Peters [3] examined three-dimensional popular matchings with cyclic preferences, establishing its NP-hardness. We demonstrated NP-hardness of popular matching problem in $(3,3)$-hypergraph where agents have preferences over the hyperedges [15]. Furthermore, we designed an approximation algorithm for 3, 3-hypergraphs producing a matching M with $u(M) \leq 1.5 \times 3^{\frac{d}{2}}$, where d is the maximum degree of an agent [16].

2 Preliminaries

2.1 Definition and Notations for the (3, 3)-Hypergraph

In a given $(3,3)$-hypergraph $H = (A \cup B \cup C, E)$, each agent $a \in A$ maintains a preference list for all the hyperedges incident upon it. The members of set B and set C do not have any preferences over the incident hyperedges. For a matching M, let $M(a)$ denote a hyperedge e if the hyperedge contains the agent a and $e \in M$. If the preference list $p(a) = \langle e_1, e_2, ..., e_d \rangle$, we interpret that agent a prefers e_i over e_j where $1 \leq i < j \leq d$. This preference is denoted by $r(e_i) < r(e_j)$, where $r(e_i)$ denotes the rank of hyperedge e_i for the agent a. The rank of a hyperedge represents its position in the agent's preference list, with a lower rank indicating a higher preference. Let M_1 and M_2 be two matchings in a $(3,3)$-hypergraph. To measure the relative popularity between M_1 and M_2, an election is held. Each agent casts a vote from the set $\{1, 0, -1\}$ in favor of a matching. For an agent a, if $r(M_1(a)) < r(M_2(a))$ or is matched in M_1 and unmatched in M_2, then the agent votes 1 for M_1. Similarly, if $r(M_1(a)) > r(M_2(a))$ or is matched in M_2 and not matched in M_1, then the agent a votes -1 for M_1. If $r(M_1(a)) = r(M_2(a))$, agent a votes 0. From M_1 to M_2, an agent experiences a *promotion* when it is reassigned to a preference of higher rank and a *demotion* when it is reassigned to a lower rank or left unmatched. We denote the vote of agent a_i for M_1 over

M_2 as $v_{a_i}(M_1, M_2)$. Thus, we can write;

$$v_{a_i}(M_1, M_2) = \begin{cases} 1 & r(M_1(a_i)) < r(M_2(a_i)) \text{ if matched} \\ & \text{in both matchings or matched in } M_1 \\ & \text{and unmatched in } M_2 \\ -1 & r(M_1(a_i)) > r(M_2(a_i)) \text{ if matched} \\ & \text{in both matchings or matched in } M_2 \\ & \text{and unmatched in } M_1 \\ 0 & \text{otherwise} \end{cases}$$

Unpopularity Factor. There exist some instances of (3, 3)-hypergraphs that do not have any popular matching (as shown in Sect. 1), and we have also shown that the popular matching problem is NP-hard in a (3, 3)-hypergraph so the next best thing is to find a matching that is close to popular matching. McCutchen [14] defined a criterion to measure the unpopularity of a matching as follows:

Given two matchings M and M', let us define $\phi(M, M')$ and the comparison function $\delta(M, M')$:

$$\phi(M, M') = |\{a_i \in A : v_{a_i}(M, M') = 1\}|$$

$$\delta(M, M') = \begin{cases} \frac{\phi(M', M)}{\phi(M, M')} & \text{if } \phi(M, M') > 0 \\ 1 & \text{if} \phi(M', M) = \phi(M, M') = 0 \\ \infty & \text{otherwise} \end{cases}$$

For the unpopularity factor of a matching M, denoted as $u(M)$, we take the maximum of $\delta(M, M')$ where the maximum is taken over all other matchings M' (where $M' \neq M$).

$$u(M) = \max_{M'} \delta(M, M')$$

For a popular matching M, $u(M) = 1$. The objective is to find a matching M where the unpopularity factor is as small as possible.

3 Approximation Scheme

Previously, we have given an algorithm that gives a bounded unpopularity matching M with $u(M) \leq 1.5 \times 3^{d/2}$ where d is the maximum degree of an agent [16]. In this paper, we use that algorithm as a foundation and refer to the matching it produces as M', to avoid confusion with the final output. This matching M' serves as an intermediate solution, which we use as input to a new algorithm proposed in the paper that computes a matching M with $u(M) \leq d^2$.

Algorithm 1. Input: $H = (A \cup B \cup C, E)$, Preference lists; Output: M' [16]

Create $P = \{P_1, P_2, \ldots, P_d\}$, where P_i contains all hyperedges of rank i.
Initialize $i = 1$, $M' = \emptyset$, $S = \{(a_1, y_1, z_1), (a_2, y_2, z_2), ..., (a_n, y_n, z_n)\}$; ▷ S is the set of last resort hyperedges.
while $i \leq d$ **and** $|M'| < n$ **do**
  Find a maximum (3,3)-hypergraph matching M'' in P_i; ▷ using a maximum matching algorithm for bipartite graph using set B and C as distinct bipartite sets.
  $M' = M' \cup M''$;
  for each $e \in M''$ **do**
    Remove all hyperedges e' from $P_{i+1}...P_d$ where $e \cap e' \neq \emptyset$;
  end for
  $i = i + 1$;
end while
$M' = M' - S$
Return M';

3.1 The First Algorithm (Algorithm 1)

We employ the greedy hypergraph matching algorithm introduced in our previous work [16]. This algorithm prioritizes hyperedges of higher rank and ensures that each agent has at least one available option; this is done by incorporating last resort hyperedges (equivalent to being unmatched). It iteratively constructs a maximum matching over rank-based partitions of hyperedges, leveraging bipartite matching techniques [6] with an overall time complexity of $O(d(n^{1.5} + m))$. The resulting matching serves as the input for our proposed algorithm.

3.2 Dependency Graph for the Matching M'

To understand the properties and the structure of the matching M' produced by Algorithm 1, we construct a dependency graph G'. For each hyperedge $e_i(a_x, b_y, c_z)$ in M' we add a corresponding vertex $v_i(a_x, b_y, c_z)$ in G', where $e_i.a = a_x = v_i.a$, $e_i.b = b_y = v_i.b$ and $e_i.c = c_z = v_i.c$. Let us assume that $B' \subset B$ and $C' \subset C$ are the sets of items that do not belong to M'. For each node $b_i \in B'$, we add a node b_i in G'. Similarly, for each node $c_i \in C'$, we add a node c_i in G'. The graph G' contains three types of directed edges: *bc*-edge, *b*-edge, and *c*-edge. For a vertex $v_{i_1} \in G'$, given that $r(e_{i_1}) = k$, we add the edges as follows: for the agent $v_{i_1}.a$, (i) we add a *bc*-edge from v_{i_1} to v_{i_2} if $(v_{i_1}.a, v_{i_2}.b, v_{i_2}.c) \in P_j$ where $j < k$ (example shown in Fig. 1), (ii) we add a *b*-edge from v_{i_1} to v_{i_2} if $(v_{i_1}.a, v_{i_2}.b, c') \in P_j$ where $j < k$ and $v_{i_2}.c \neq c'$. If $c' \neq v_{i_1}.c$, then we also add a *c*-edge to a vertex v_{i_3} if $c' = v_{i_3}.c$ or to a vertex c_i if $c' = c_i$ where $c_i \in C'$. (iii) Similarly, we add a *c*-edge from v_{i_1} to another vertex in G'. Some examples shown in Fig. 4 represent different types of *b*-edges and *c*-edges. We say v_{i_1} depends on a vertex v' if there is a directed edge from v_{i_1} to v'.

The graph G' represents a matching where higher-ranked hyperedges are prioritized. We say, an agent a is at a level l for $1 \leq l \leq d$ if $r(M'(a)) = l$. We say, a level l_1 is higher than another level l_2 if $l_1 < l_2$. Let us call the items from set B and C as resources. An agent a_i at level l can only be promoted if the necessary resources, held by agents at higher levels, become available. This must be the case; otherwise, the higher-ranked hyperedge of a_i would have been selected for the matching. The edges in G' can be directed from lower to higher levels or vice versa, but there will be no edge between nodes in the same level. When an edge starts from a lower level and ends at a higher level, we say it is an *upward dependency*. Similarly, when an edge starts from a higher level and ends at a lower level, we say it is a *downward dependency*. If dependency for both resources is upward, then it is a complete upward dependency. Similarly, we can have a complete downward dependency, although this type of dependency is not possible for the dependency graph G'. If out of two resources, one resource is unmatched (belongs to either B' or C') and the other resource is matched and there is an upward dependency for it, then this is a case of complete upward dependency. In G', two types of dependencies exist: (i) complete upward dependency (different cases are shown in Fig. 1, Fig. 3, Fig. 4) and (ii) mixed dependency involving one upward and one downward dependency (shown in Fig. 2). Consider a scenario where a hyperedge (a_i, b, c) at level l or higher is not selected in M'. This indicates that one or both resources are unavailable. If both resources are unavailable at level l then the dependency for both is upward. Conversely, if one of the resources is available, it may be utilized at level l or a lower level by some agent for matching, potentially leading to a mixed dependency.

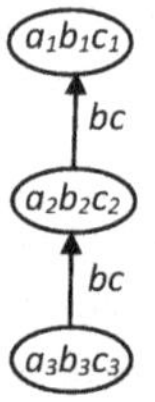

Fig. 1. Given that $\{(a_1, b_1, c_1), (a_2, b_2, c_2)\} \in M'$, $(a_2, b_1, c_1) \notin M'$ and $r(a_2, b_1, c_1) < r(a_2, b_2, c_2)$. Agent a_2 can promote itself by occupying the resources b_1 and c_1 from a_1. Similarly a_3 can promote itself if $r(a_3, b_2, c_2) < r(a_3, b_3, c_3)$ where $(a_3, b_3, c_3) \in M'$ and $(a_3, b_2, c_2) \notin M'$.

3.3 Cases of Possible Promotions Within G'

Agents may require one or two resources to promote themselves. The following cases are possible: (1) If an agent needs one resource for promotion, it can be obtained from one hyperedge (agent a_2 in Fig. 4b). (2) If an agent needs two

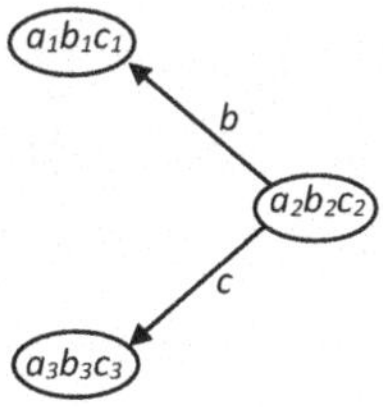

Fig. 2. A case of mixed dependency: Consider three hyperedges in M' such that $r(a_1, b_1, c_1) < r(a_2, b_2, c_2) < r(a_3, b_3, c_3)$. Let $r(a_2, b_1, c_3) < r(a_2, b_2, c_2)$, where $(a_2, b_1, c_3) \notin M'$, a_2 may be promoted if a_1 and a_3 release their resources.

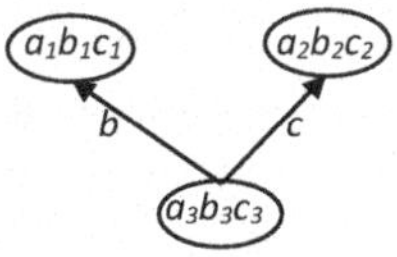

Fig. 3. Let a hyperedge $(a_3, b_1, c_2) \notin M'$ have a rank higher than $(a_3, b_3, c_3) \in M'$. In this scenario, a_3 gets promoted using b_1 and c_2.

resources for promotion, then those resources can be obtained in any one of the following ways: (i) both resources from one hyperedge (shown in Fig 1), or (ii) each from two different hyperedges (shown in Fig. 2, Fig. 3), or (iii) one resource from one hyperedge and another from an unmatched resource (shown in Fig. 4a).

In the case of a promotion requiring one resource, there is exactly one demotion for a promotion; whereas in the case of a promotion requiring two resources, there are at most two demotions for a promotion. For example, in Fig. 1, to promote a_2 and a_3, there is one demotion of agent a_1. At most k agents can have dependencies in the same manner, resulting in $(d-1)$ promotions at the cost of one demotion. In Fig. 3, the promotion of agent a_3 costs the demotions of agents a_1 and a_2. If this pattern follows for more agents, then it would create an upside-down binary tree where d promotions require $(d+1)$ demotions. Figure 4 represents three cases where the demotion of one agent can trigger the promotion of two agents. In Fig. 4a, the relationship of three agents is shown where $\{e_1(a_1, b_1, c_1), e_2(a_2, b_2, c_2), e_3(a_3, b_3, c_3)\} \in M'$ and $\{e_4(a_2, b_1, c_4), e_5(a_3, b_4, c_1)\} \notin M'$, and $r(e_1) < r(e_2)$, $r(e_1) < r(e_3)$, $r(e_4) < r(e_2)$ and $r(e_5) < r(e_3)$. Agents a_2 and a_3 can promote themselves using c_4, b_4 respectively and the resources of agent a_1. Figure 4bshows a case where a_2 needs only one resource from a_1 as it relinquishes only b_2 during promotion from (a_2, b_2, c_2) to (a_2, b_1, c_2). Figure 4c represents a case where both a_2 and a_3 are promoted using one resource each. The case shown in Fig. 4 is the worst-case in terms of unpopularity factor, as it increases the promotions exponentially. There can be $(2^d - 2)$ promotions possible at the cost of one demotion of agent a_1.

Lemma 1. *There is no downward or same-level dependency in the dependency graph G' for the matching M'.*

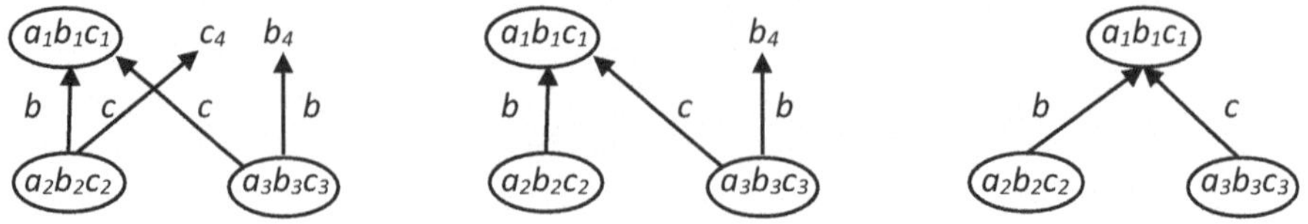

(a) Case 1 (b) Case 2 (c) Case 3

Fig. 4. Bad Cases

Proof. Let there be an agent a at level l that has a downward or same-level dependency. This implies that some resources are present at level l or a lower level that can promote a, implying that those resources must not be utilized at a higher level. Since the algorithm prioritizes higher ranked hyperedges for the matching, the algorithm would have picked the higher ranked hyperedge of the agent a, contradicting the assumption. □

3.4 The Second Algorithm (Algorithm 2)

Algorithm 1 generates a matching M', which serves as an input for Algorithm 2). The primary objective of Algorithm 2) is to eliminate bad cases from the matching M' while preserving the upward dependency for each agent in the dependency graph. In each iteration, the algorithm identifies two distinct cases: (1) for an agent a_α at level l, the algorithm searches two agents that can be promoted using the resources of a_α, and (2) during this process, certain resources are relinquished by the promoted agents, allowing the algorithm to identify additional agents who can be promoted using these newly available resources at each level.

Lemma 2. *The time complexity of Algorithm 2 is $O(dn^2)$, and the total runtime of both algorithms is $O(d(n^2 + m))$.*

Proof. At each level, we iterate through all lower-level agents for promotions. In Step 3 (Algorithm 2), we search for an agent that can be promoted using the available resources, where $O(n)$ time is taken to look for an agent and $O(n)$ time is needed to check whether it can be promoted from the set of available resources. Overall it takes $O(n^2)$ time. In Step 4, we look for any agent having a non-upward dependency on the agent promoted in Step 3; this takes a maximum $O(n)$ time. In Steps 5 to 8, for each agent at a level, we search for two agents that can be promoted, which takes $O(n^2)$ time. Step 10 is similar to Step 4 and takes $O(n)$ time. Hence, the time complexity of Algorithm Algorithm 2 is $O(dn^2)$. The total run time of both algorithms is $O(d(n^{1.5} + m) + dn^2) = O(d(n^2 + m))$. □

Lemma 3. *There is no downward/same-level dependency in the dependency graph for matching M.*

Algorithm 2. Input: G', M', Preference lists P; Output: M

1: Initialize $l = 1$.
2: Do the following steps for each level l starting from level 1 to $d + 1$.
3: Search for agents at level $l, l+1, l+2, \cdots, d+1$ that can be promoted using available (unmatched) resources. Promote those agents.
4: If some promoted agent a' uses only one available resource, search for another agent a of the highest rank with a complete dependency on a' that is not upward. Promote a using the resources of a'.
5: For each agent a_α at level l, search for an agent at level $l+1, l+2, \cdots, d+1$ that can be promoted by one of the resources of a_α.
6: Let an agent a_β be found at level l_1. (where $l < l_1 \leq d + 1$)
7: Search for an agent on level $l_1, l_1 + 1, l_1 + 2, \cdots, d + 1$ that can be promoted by using the other resource of a_α.
8: Let an agent a_γ be found at level l_2. (where $l_1 \leq l_2 \leq d + 1$)
9: Promote a_β and a_γ.
10: If a_β and/or a_γ do not use any available resource, then search for an agent a' with the highest rank with a complete dependency on a_β or a_γ that is not upward. Promote a' using the resources of a_β or a_γ.

Proof. Let there be an agent a. A downward or same-level dependency from the agent a represents that the agent can be promoted by using the resources of other agents, which are at the level of a or at a lower level. We have shown in Lemma 1 that the matching M' does not have such dependencies. To prove that such dependencies do not exist in M, we use proof by contradiction. We assume that in Algorithm 2, we make changes that introduce some dependency that can be downward or at the same level. The Algorithm 2 does two types of promotions.

Case 1: (In Step 9). In this case two agents a_β and a_γ get promoted by using the resources of a_α. Let us assume that after this step, a downward or same-level dependency arises. Let us take a_β for instance. Let the agent a_β be at level l_2, and it got promoted to level l_1 where $l_1 < l_2$. Suppose agent a is at level $l_3 \leq l_1$ such that it has a dependency on a_β. The agent a_β could have been promoted in two ways:

1. Let a_β pick one resource from a_α and another resource from the available resources. The agent a is accessed before a_β by the algorithm for the same resource since it is at a higher level l_3, whereas a_β was at level l_2 where $l_3 < l_2$. Then, the algorithm would have picked the agent a for the promotion rather than a_β. Hence, the assumption is false.
2. Let a_β pick one resource from a_α and keep its other resource. In this case, a downward or same-level dependency may exist. For example let $(a_\beta, b_2, c_2) \in M'$ and later it got promoted to (a_β, b_1, c_2). Suppose b_1, c_2 are needed for a to promote itself. Hence, there will be a dependency from a to a_β. In this situation, a has a complete dependency on a_β. There can not be a downward/same-level dependency for a single resource from such a to a_β. If

there is a single resource dependency, then it would imply that the algorithm did not pick a for promotion in Step 9, even though it is at a higher level than a_β (before promotion), which is not possible. Step 10 of Algorithm 2 identifies the highest level agent that has a downward or same-level dependency on a_β. Once a gets promoted, that dependency is gone. If some other agent having a rank higher than a gets promoted, then the dependency of a changes to the newly promoted higher-ranked agent and hence becomes an upward dependency. Hence, the assumption is false.

Case 2: (In Step 3). When the resources are available, the algorithm prioritizes the higher-ranked agents for promotion, ensuring that no downward or same-level dependency is introduced. Suppose an agent a' at level l_2 is promoted to level l_1 using Step 3, and there is an agent a on level l_3 where $l_3 \leq l_1$ such that a has a dependency on a'. There are two possible cases:

1. Suppose a' requires two resources for promotion, say b' and c'. These resources become available when an agent relinquishes them upon promotion. This implies that at least one of b' or c' should be associated with an agent x that has a rank higher than l_2. The agent a can not be at a level higher than x in M'. When x leaves its resources, the algorithm looks for agents from subsequent lower levels for promotion, and given that a is at a level higher than a', the algorithm will pick a. This is a contradiction.
2. Let a' take one resource from the available resources. There can be an agent with a downward or same-level dependency on a', similar to sub-case 2 of Case 1, which will be resolved by the algorithm in Step 4. □

A matching is considered Pareto efficient if no other matching exists that matches one or more vertices to a more preferred choice without reducing the preference level of any other vertex. In the following lemma, we show that the matching produced by our algorithm is Pareto efficient.

Lemma 4. *The matching M is Pareto efficient.*

Proof. Let us assume that the matching M is not Pareto efficient. Under this assumption, there exists a set of agents $X = \{x_1, x_2, ..., x_s\}$, where $1 \leq s \leq n$, such that promotion of agents in X does not require demotion of any agent. Two cases are possible: (Case 1) these agents may exchange some of their resources to get promoted, (Case 2) they find some unmatched resources and use them to promote themselves. Let $E_X = \{e_{x_1}, e_{x_2}, ..., e_{x_s}\}$, where $E_X \subseteq M$, be a set of hyperedges that contains agents $X = \{x_1, x_2, ..., x_s\}$. Let e_{x_i} have the highest rank in the set E_X. In Case 1, it is not possible to promote agent x_i since at least one resource of higher-ranked hyperedge of x_i is not present in E_X, as it must belong to a hyperedge $e \in M$ such that $r(e) < r(e_{x_i})$. This is due to the fact that there are no downward or same-level dependencies in the dependency graph. It implies that at least one of those resources must be part of the matching M with an agent that does not belong to X. Case 2 is also not possible since the algorithm looks for such agents at each level. If an agent x_i exists that can be promoted using unmatched resources, the algorithm would have matched x_i with those resources, resulting in a higher-ranked hyperedge of x_i in M. This contradicts the initial assumption; consequently, the set X can not exist in the matching M. □

Corollary 1. *The matching M has an unpopularity factor $u(M) \leq n-1$.*

Proof. Let there be a matching N such that n_2 number of agents prefer N over M and n_1 number of agents prefer M over N. Given that M is a Pareto efficient matching, at least one agent is demoted to promote one or more agents. This implies that $n_1 \geq 1$ and $n_2 \leq n-1$. Consequently, for any alternative matching N, the popularity comparison function $\delta(N, M) = \frac{n_2}{n_1} \leq \frac{n-1}{1}$. This implies that the unpopularity factor of the matching M is bounded by $u(M) \leq n-1$. □

Lemma 5. *There does not exist an agent a_α in M such that a_β and a_γ can promote themselves using the resources of a_α.*

Proof. Algorithm 2 iteratively finds each agent a_α where two other agents can be promoted using the resources of a_α. When the algorithm iterates on a level l, it finds all such agents and promotes them. When the algorithm iterates at a level $l' > l$, it does not reassign the matches of the agents that belong to the higher levels. Therefore, such agents can not exist in the final matching M. □

3.5 Analysis for the Unpopularity Factor

In the given (3, 3)-hypergraph, for each agent a_i, we add a last resort hyperedge (a_i, y_i, z_i). In this setup, all maximal matchings are perfect in terms of cardinality. We create a line graph $L(H)$ for the given hypergraph H. The vertices of $L(H)$ are the hyperedges of H, and there is an edge between a pair of vertices in the line graph if the corresponding pair in the hyperedges has a nonempty intersection. A matching of H is an independent set of the line graph $L(H)$.

The algorithm has given the matching M. Consider another matching N such that A_1 is a set of agents that prefer M over N, and A_2 is a set of agents that prefer N over M. In matching N, we reassign all agents from set A_1 to their last resort. This reassignment does not change A_1 or A_2. Let I_M and I_N be the independent sets of $L(H)$ corresponding to matching M and N, respectively. We create a graph G as follows: the nodes of the graph G are the vertices of $L(H)$ that contain the symmetric difference of I_M and I_N ($I_M \Delta I_N$). A pair of nodes in G is connected by an edge if the corresponding pair of nodes is connected by an edge in $L(H)$. The graph G can be a connected graph if all vertices of $L(H)$ belong exclusively to M or N. This implies that each agent is promoted or demoted from M to N. So each agent will prefer M over N or vice-versa. Let G have connected components $G_1, G_2, \cdots, G_k$. The Lemma 4 states that the matching M is a Pareto-efficient matching; therefore, each G_i must demote at least one agent from matching M so that some other agents in G_i can be promoted. This puts a limit on the number of components of G and it can be shown that $k \leq |A_1|$. Let us assume that a component G^* has only one agent from A_1.

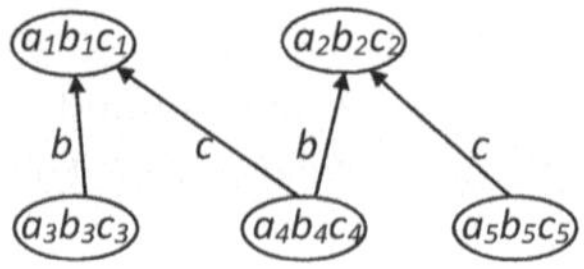

Fig. 5. Let $r(a_3, b_1, c_2) < r(a_3, b_3, c_3)$, $r(a_4, b_2, c_1) < r(a_4, b_4, c_4)$ and $r(a_5, b_5, c_2) < r(a_5, b_5, c_5)$. In this scenario, agent a_3 needs one resource from a_1 for promotion, a_4 needs one each from a_1 and a_2, and a_5 needs one from a_2 for its promotion.

Lemma 6. *The unpopularity factor of M in G^* is $u(M) \leq d$.*

Proof. Let a_α be an agent in G^* that prefers M over N, which is demoted to its last resort in N. Let e_1 be a node in G^* that contains agent a_α matched to its last resort (unmatched). The node e_1 can have only one neighbor, which should contain the hyperedge containing the same agent a_α; let us call that node e_2. The node e_2 belongs to M and the node e_1 belongs to N. The node e_2 may have two more neighbors, as each node contains three items, one from each set A, B and C. Let us assume that e_3 and e_4 are neighbors of e_2 and belong to N. Since each agent is being promoted except a_α, this implies that e_3 and e_4 should have one more neighbor each, e_5 and e_6 respectively, such that $e_5.a = e_3.a = a_\beta$ and $e_6.a = e_4.a = a_\gamma$. This relation implies that agents a_β and a_γ are being promoted using the resources of a_α. The Lemma 5 confirms that this case is not possible, implying that e_2 can not have two neighbors besides e_1. This implies that e_2 has only one neighbor e_3 in addition to e_1. The agent a_β is promoted from e_5 to e_3. Here e_3 and e_5 belong to N and M respectively. Similar to e_2, e_5 can also have one more neighbor; let us call that node e_7. Similarly, e_7 can have one more neighbor e_8 in addition to e_5, where the agent is promoted from e_8 to e_7, where e_8 and e_7 belong to M and N, respectively. Given that the dependencies are only upward, the rank relation in M should be: $r(e_2) < r(e_5) < r(e_8) < \ldots$ and so on. This relation implies that this sequence can go up to the highest rank d. This implies that one demotion can promote at most d number of agents, hence the unpopularity factor M in G^* is $u(M) \leq d$. □

Theorem 1. *The unpopularity factor for matching M is $u(M) \leq d^2$.*

Proof. Let us assume that a component G_i has two agents a_1 and a_2 from A_1. We have established that no pair of agents can be promoted using the resources of one agent, but that does not imply that three agents can not be promoted using the resources of two agents. For example, consider five agents $a_1, a_2, .., a_5$ where a_1 and a_2 lose their resources and a_3, a_4 and a_5 promote themselves as shown in Fig. 5. Similarly, $k+1$ agents can be promoted using the resources of k agents ($1 < k < \frac{n}{2}$). Let a connected component have k agents from A_1. To maximize promotions, let us assume that these k agents are at level 1 in M. A total of $k+1$ agents can be promoted by using the resources in the subsequent lower level. Similarly, when these $k+1$ agents are promoted, they relinquish their resources and, in turn, help $k+2$ agents from the next lower level to be promoted. Hence, the sequence becomes: $(k+1, k+2, \cdots, k+d)$. The total number of promotions is $\sum_{i=1}^{d}(k+i) = kd + d(d+1)/2$ and the unpopularity factor becomes $(dk + d(d+1)/2)/k$. To maximize this value, k can be as low as 2. Hence the total unpopularity factor becomes $(d^2 + 5d)/4$, which is less than d^2 for $d \geq 2$ (necessary condition). Since the unpopularity factor is bounded by the maximum unpopularity factor of M in any connected component and given that a connected component with one agent demotion has an unpopularity factor of at most d for the matching M (from Lemma 6), it follows that the overall unpopularity factor for matching M is $u(M) \leq d^2$. □

References

1. Abraham, D.J., Irving, R.W., Kavitha, T., Mehlhorn, K.: Popular matchings. SIAM J. Comput. **37**(4), 1030–1045 (2007)
2. Cseh, Á.: Popular matchings. Trends Comput. Soc. Choice, **105**(3) (2017)
3. Cseh, Á., Peters, J.: Three-dimensional popular matching with cyclic preferences. AAMAS, 309–317 (2022)
4. Gärdenfors, P.: Match making: assignments based on bilateral preferences. Syst. Res. **20**, 166–173 (1975)
5. Gupta, S., Misra, P., Saurabh, S., Zehavi, M.: Popular matching in roommates setting is NP-hard. ACM Trans. Comput. Theory **13**(2) (2021). Article 9
6. Hopcroft, J.E., Karp, R.M.: An $n^{2.5}$ algorithm for maximum matchings in bipartite graphs. SIAM J. Comput. **2**(4), 225–231 (1973)
7. Huang, C.C., Kavitha, T.: Popular matchings in the stable marriage problem. Inf. Comput. **222**, 180–94 (2013)
8. Huang, C.C., Kavitha, T.: Near-popular matchings in the roommates problem. SIAM J. Discret. Math. **27**(1), 43–62 (2013)
9. Huang, C.C., Kavitha, T., Michail, D., Nasre, M.: Bounded unpopularity matchings. Algorithmica **61**(3), 738–757 (2011)
10. Kavitha, T.: Popular matchings of desired size. In: International Workshop on Graph-Theoretic Concepts in Computer Science, pp. 306–317 (2018)
11. Kavitha, T.: Stable matchings, one-sided ties, and approximate popularity. Algorithmica, 1–33 (2024)
12. Kavitha, T., Mestre, J., Nasre, M.: Popular mixed matchings. In: International Colloquium on Automata, Languages, and Programming, pp. 574–584 (2009)
13. McCutchen, R.M.: The least-unpopularity-factor and least-unpopularity-margin criteria for matching problems with one-sided preferences. In: Latin American Symposium on Theoretical Informatics, pp. 593–604. Springer, Heidelberg (2008)
14. McDermid, E., Irving, R.W.: Popular matchings: structure and algorithms. J. Comb. Optim. **22**(3), 339–58 (2008)
15. Singh, Y., Karmakar, S.: The popular matching problem in a (3,3)-hypergraph with one-sided preferences. In: 2023 Eleventh International Symposium on Computing and Networking Workshops (CANDARW), pp. 121–127. IEEE (2023)
16. Singh, Y., Karmakar, S.: Bounded unpopularity matching in a (3, 3)-hypergraph with one-sided preferences. In: 2024 Twelfth International Symposium on Computing and Networking Workshops (CANDARW), pp. 204–210. IEEE (2024)

String Algorithms

Inverting Parameterized Burrows-Wheeler Transform

Shogen Kawanami, Kento Iseri, and I. Tomohiro(✉)

Kyushu Institute of Technology, 680-4 Kawazu, Iizuka, Fukuoka 820-8502, Japan
{kawanami.syogen518,iseri.kento210}@mail.kyutech.jp,
tomohiro@ai.kyutech.ac.jp

Abstract. The Burrows-Wheeler Transform (BWT) of a string is an invertible permutation of the string, which can be used for data compression and compact indexes for string pattern matching. Ganguly et al. [SODA, 2017] introduced the parameterized BWT (pBWT) to design compact indexes for parameterized matching (p-matching), a variant of string pattern matching with parameter symbols introduced by Baker [STOC, 1993]. Although the pBWT was inspired by the BWT, it is not obvious whether the pBWT itself is invertible or not. In this paper we show that we can retrieve the original string (up to renaming of parameter symbols) from the pBWT of length n in $O(n^2)$ time and $O(n)$ space.

Keywords: Parameterized pattern matching · Parameterized Burrows-Wheeler Transform · Inversion algorithm

1 Introduction

A *parameterized string* (*p-string*) consists of two different types of symbols, static symbols (s-symbols) and parameter symbols (p-symbols). Let Σ_s and Σ_p denote the sets of s-symbols and p-symbols, respectively. Two p-strings x and y are said to *parameterized match* (*p-match*) if and only if x can be transformed into y by applying a bijection on Σ_p to every occurrence of p-symbols in x. For example with $\Sigma_s = \{\mathtt{A}, \mathtt{B}\}$ and $\Sigma_p = \{\mathtt{x}, \mathtt{y}, \mathtt{z}\}$, two p-strings `xAyxzzyB` and `zAxzyyxB` p-match because `xAyxzzyB` can be transformed into `zAxzyyxB` by replacing `x`, `y` and `z` with `z`, `x` and `y`, respectively. P-matching was introduced by Baker to address a variant of string pattern matching with renamable symbols in software maintenance and plagiarism detection [3–5]. Later it finds wider applications such as detecting similar images and isomorphic graphs, and algorithms for p-matching have been extensively studied (see a recent survey [21] and references therein).

Almost all efficient p-matching algorithms are based on the *prev-encoding* [3] that replaces every occurrence of a p-symbol in a p-string with the distance to the previous occurrence of the same p-symbol, or with "0" if there is no previous occurrence. By definition, two p-strings p-match if and only if their prev-encoded strings are identical. For example, the p-matching p-strings `xAyxzzyB`

H. Fernau and B. Zhu (Eds.): IWOCA 2025, LNCS 15885, pp. 217–227, 2025.
https://doi.org/10.1007/978-3-031-98740-3_16

and zAxzyyxB are both prev-encoded to 0A03014B. Also, the prefixes of the prev-encoding do not change when symbols are appended. These properties enable us to extend suffix trees and suffix arrays for indexes of p-matching: For a p-string T of length n, the parameterized suffix tree (pST) [3] is the compacted trie of the prev-encoded suffixes of T, and the parameterized suffix array (pSA) [9], denoted by $\tau[1..n]$ in this paper, stores the starting positions of prev-encoded suffixes in their lexicographic order. On the other hand, the suffixes of the prev-encoding may change when symbols are prepended, which makes some algorithms for p-matching more complicated than those for exact string pattern matching.

The Burrows-Wheeler Transform (BWT) [7] of a string is an invertible permutation of the string. The BWT was originally introduced for data compression but has become the basis of compact or compressed indexes for string pattern matching such as the FM-index [10] and the r-index [12]. A notable property of the BWT is that it enables us to implement the *LF-mapping* that maps the lexicographic rank of a suffix of the original string to the rank of the suffix longer by one.

Inspired by the FM-index, Ganguly et al. [13] introduced the parameterized BWT (pBWT) to design a compact index for p-matching. For a p-string T of length n over $(\Sigma_s \cup \Sigma_p)$, the pBWT L of T is a string of length n over an alphabet of size $\sigma = |\Sigma_s \cup \Sigma_p|$, and thus, within compact space of $O(n \log \sigma)$ bits. For any position i with $\tau[i] \neq 1$, $\mathsf{L}[i]$ holds the necessary and sufficient information to turn the prev-encoding of $\mathsf{T}[\tau[i]..n]$ into that of $\mathsf{T}[\tau[i] - 1..n]$ (see Sect. 2 for its formal definition). So it is not difficult to see that the prev-encoding of T is retrieved from L if the LF-mapping is provided. However it is not obvious whether we can retrieve the information of the LF-mapping from L alone because existing pBWT-based indexes [13,17,19] use other data structures together with L to implement the LF-mapping.

In this paper, we show that we can retrieve the information of the LF-mapping and T (up to renaming of p-symbols) from L in $O(n^2)$ time and $O(n)$ space. To the best of our knowledge, the invertibility of pBWTs has not been explored before and our result paves the way to using pBWTs as a compression method for p-strings.

1.1 Related Work

The problem we tackled in this work can also be seen as a reverse engineering problem for string data structures, which have been actively studied to gain a deeper understanding of data structures (e.g., [6,8,16,18,20,23]). For those related to p-matching, there are studies on parameterized border arrays [15] and parameterized suffix and LCP arrays [1,2]. Since pBWTs sometimes act differently from BWTs, a deeper understanding of its properties would be a key to further development in pBWT-based data structures like [14].

2 Preliminaries

2.1 Notations and Tools

An integer interval $\{i, i+1, \ldots, j\}$ is denoted by $[i..j]$, where $[i..j]$ represents the empty interval if $i > j$. Also $[i..j)$ denotes $[i..j-1]$.

Let Σ be an ordered finite *alphabet*. An element of Σ^* is called a *string* over Σ. The length of a string w is denoted by $|w|$. The empty string ε is the string of length 0, that is, $|\varepsilon| = 0$. Let $\Sigma^+ = \Sigma^* - \{\varepsilon\}$ and $\Sigma^k = \{w \in \Sigma^* \mid |w| = k\}$ for any non-negative integer k. The concatenation of two strings x and y is denoted by $x \cdot y$ or simply xy. When a string w is represented by the concatenation of strings x, y and z (i.e., $w = xyz$), then x, y and z are called a *prefix*, *substring*, and *suffix* of w, respectively. A substring x of w is called *proper* if $x \neq w$.

The i-th symbol of a string w is denoted by $w[i]$ for $1 \leq i \leq |w|$, and the substring of a string w that begins at position i and ends at position j is denoted by $w[i..j]$ for $1 \leq i \leq j \leq |w|$, i.e., $w[i..j] = w[i]w[i+1] \cdots w[j]$. For convenience, let $w[i..j] = \varepsilon$ if $j < i$. For two strings x and y, let $lcp(x, y)$ denote the length of the longest common prefix between x and y. We consider the lexicographic order over Σ^* by extending the strict total order $<$ defined on Σ: x is lexicographically smaller than y (denoted as $x < y$) if and only if either x is a proper prefix of y or $x[lcp(x, y) + 1] < y[lcp(x, y) + 1]$ holds.

We will use the following result for dynamic strings as a tool:

Lemma 1 **([22]).** *For a positive integer U, a string over an alphabet $[1..U]$ can be dynamically maintained while supporting insertion/deletion of a symbol to/from any position of the string as well as random access in $(m + o(m)) \lg U$ bits of space and $O(\frac{\lg m}{\lg \lg m})$ query and update times, where m is the length of the current string.*

2.2 Parameterized Strings

Let Σ_s and Σ_p denote two disjoint sets of symbols. We call a symbol in Σ_s a *static symbol* (*s-symbol*) and a symbol in Σ_p a *parameter symbol* (*p-symbol*). For a technical reason, Σ_s is considered to be disjoint with the set of integers and we assume that an s-symbol is lexicographically smaller than any integer. Let $\sigma_s = |\Sigma_s|$, $\sigma_p = |\Sigma_p|$ and $\sigma = \sigma_s + \sigma_p$. A *parameterized string* (*p-string*) is a string over $(\Sigma_s \cup \Sigma_p)$.

For any p-string w of length n, the *prev-encoding* $\langle w \rangle$ of w is the string in $(\Sigma_s \cup [0..n))^n$ such that, for any i $(1 \leq i \leq n)$,

$$\langle w \rangle[i] = \begin{cases} w[i] & \text{if } w[i] \in \Sigma_s, \\ 0 & \text{if } w[i] \in \Sigma_p \text{ and } w[i] \text{ does not appear in } w[1..i-1], \\ i - j & \text{otherwise,} \end{cases}$$

where j is the largest position in $[1..i-1]$ with $w[i] = w[j]$. By prev-encoding, all the occurrences of a p-symbol are connected by a single chain on the array storing the distance to the previous occurrence of the same p-symbol, which ends with the symbol 0 at the leftmost occurrence of the p-symbol. Let $|\langle w\rangle|_0$ denote the the number of 0 in $\langle w\rangle$, which is equal to the number of distinct p-symbols in w. For example, $\langle w\rangle = \texttt{0A03014B}$ and $|\langle w\rangle|_0 = 3$ for $w = \texttt{xAyxzzyB}$ with $\Sigma_s = \{\texttt{A}, \texttt{B}\}$ and $\Sigma_p = \{\texttt{x}, \texttt{y}, \texttt{z}\}$.

The following properties hold for prev-encoding $\langle w\rangle$ of a p-string w of length n:

- A p-string w' p-matches w if and only if $\langle w'\rangle = \langle w\rangle$.
- For any $1 \le i \le n$, $\langle w\rangle[1..i] = \langle w[1..i]\rangle$.
- If $c \in \Sigma_s$, then $\langle cw\rangle = c\langle w\rangle$.
- If $c \in \Sigma_p$ does not appear in w, then $\langle cw\rangle = 0\langle w\rangle$.
- If $c \in \Sigma_p$ and d is the leftmost occurrence of c in w, then $\langle cw\rangle = 0 \cdot \langle w\rangle[1..d-1] \cdot d \cdot \langle w\rangle[d+1..n]$.

It is easy to see that the prev-encoding has enough information to retrieve the p-strings that p-match the original p-string and we can compute the lexicographically smallest one in linear time:

Lemma 2. *Given prev-encoding W of a p-string w of length n, we can compute the lexicographically smallest p-string w' that p-matches w in $O(n)$ time.*

Proof. For increasing positions i, we determine $w'[i]$ as follows:

- If $W[i] \in \Sigma_s$, set $w'[i] = W[i]$.
- If $W[i] = 0$, set $w'[i]$ to be the lexicographically smallest p-symbol that has not been used.
- If $W[i] = d \in [1..n)$, set $w'[i] = w'[i-d]$.

Since the p-strings that p-match w differ up to renaming of p-symbols and we choose lexicographically smaller p-symbols in the order of their leftmost occurrences, w' becomes the lexicographically smallest. It is clear that it runs in linear time. □

Throughout this paper, we consider a p-string T of length n over $(\Sigma_s \cup \Sigma_p)$ with the condition that $\mathsf{T}[n] = \$$, where $\$$ is the smallest s-symbol that does not appear anywhere else in T. For any integer i, let T_i denote the cyclic rotation of T such that $\mathsf{T}_i = \mathsf{T}[i'..n]\mathsf{T}[1..i'-1]$ for $i' \in [1..n]$ with $i' \equiv i \mod n$. Let τ be the permutation of $[1..n]$ such that $\langle \mathsf{T}_{\tau[i]}\rangle$ is the lexicographically i-th string in $\{\langle \mathsf{T}_{i'}\rangle\}_{i'=1}^{n}$. Due to the existence of end-marker $\$$, τ is equivalent to the *parameterized suffix array (pSA)* of T.

The *parameterized Burrows Wheeler Transform (pBWT)* L of T is a string of length n over the alphabet $(\Sigma_s \cup [1..\sigma_p])$ such that, for any i $(1 \le i \le n)$,

$$\mathsf{L}[i] = \begin{cases} \mathsf{T}_{\tau[i]}[n] & \text{if } \mathsf{T}_{\tau[i]}[n] \in \Sigma_s, \\ |\langle \mathsf{T}_{\tau[i]}[1..j]\rangle|_0 & \text{otherwise,} \end{cases}$$

where j is the smallest position in $\mathsf{T}_{\tau[i]}$ with $\mathsf{T}_{\tau[i]}[n] = \mathsf{T}_{\tau[i]}[j]$.

Note that $\mathsf{L}[i]$ holds the information to turn $\langle \mathsf{T}_{\tau[i]}\rangle[1..\ell]$ into $\langle \mathsf{T}_{\tau[i]-1}\rangle[1..\ell+1]$:

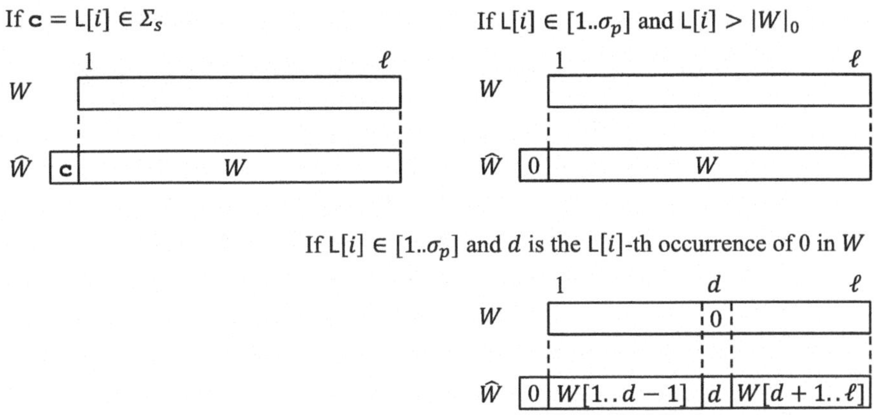

Fig. 1. Illustrations for the cases of Proposition 1.

Table 1. An example of $\tau[i]$, $LF[i]$ and $\mathsf{L}[i]$ for a p-string T = xyxzzzxxyx\$ with $\Sigma_s = \{\$\}$ and $\Sigma_p = \{\mathtt{x}, \mathtt{y}, \mathtt{z}\}$.

i	T_i	$\langle \mathsf{T}_i \rangle$	$\tau[i]$	$LF[i]$	$\mathsf{L}[i]$	$\langle \mathsf{T}_{\tau[i]} \rangle$	$\mathsf{T}_{\tau[i]}$
1	xyxzzxxyx$	002013162$	10	2	1	$002013162	$xyxzzxxyx
2	yxzzxxyx$x	00013162$2	9	3	2	0$20201316	x$xyxzzxxy
3	xzzxxyx$xy	0013102$24	8	7	2	00$2420131	yx$xyxzzxx
4	zzxxyx$xyx	010102$242	2	8	2	00013162$2	yxzzxxyx$x
5	zxxyx$xyxz	00102$2429	5	9	1	00102$2429	zxxyx$xyxz
6	xxyx$xyxzz	0102$24201	3	4	3	0013102$24	xzzxxyx$xy
7	xyx$xyxzzx	002$242013	7	10	1	002$242013	xyx$xyxzzx
8	yx$xyxzzxx	00$2420131	1	1	$	002013162$	xyxzzxxyx$
9	x$xyxzzxxy	0$20201316	4	6	2	010102$242	zzxxyx$xyx
10	$xyxzzxxyx	$002013162	6	5	3	0102$24201	xxyx$xyxzz

Proposition 1. *For any position i and $\ell \in [0..n)$, let $W = \langle \mathsf{T}_{\tau[i]} \rangle[1..\ell]$ and $\hat{W} = \langle \mathsf{T}_{\tau[i]-1} \rangle[1..\ell{+}1]$. The following statements hold (see Fig. 1 for illustration):*

- *If $\mathsf{L}[i] \in \Sigma_s$, then $\hat{W} = \mathsf{L}[i]W$.*
- *If $\mathsf{L}[i] \in [1..\sigma_p]$ and $\mathsf{L}[i] > |W|_0$, then $\hat{W} = 0W$.*
- *If $\mathsf{L}[i] \in [1..\sigma_p]$ and d is the $\mathsf{L}[i]$-th occurrence of 0 in W, then $\hat{W} = 0 \cdot W[1..d-1] \cdot d \cdot W[d+1..\ell]$.*

The *LF-mapping* LF for a p-string T is the permutation of $[1..n]$ such that $\tau[LF[i]] \equiv \tau[i] - 1 \mod n$.

Table 1 shows an example of $\tau[i]$, $LF[i]$ and $\mathsf{L}[i]$.

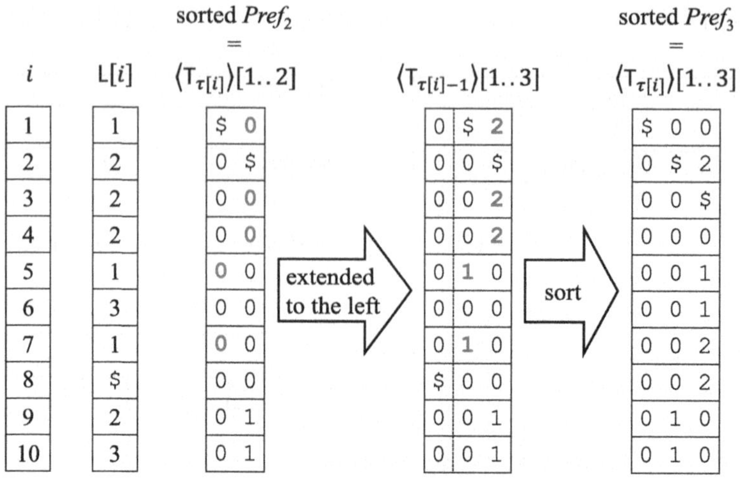

Fig. 2. Illustration for the computing process from $\langle \mathsf{T}_{\tau[i]} \rangle[1..2]$ to $\langle \mathsf{T}_{\tau[i]} \rangle[1..3]$ for our running example $\mathsf{T} = \mathtt{xyxzzxxyx\$}$ of Table 1. When $\langle \mathsf{T}_{\tau[i]} \rangle[1..2]$ is extended to the left according to Proposition 1, the 0s in red are turned into the distance to the beginning position in $\langle \mathsf{T}_{\tau[i]-1} \rangle[1..3]$.

3 Algorithms

In this section, we present our algorithms for the inversion problem of pBWTs.

We first present a simple algorithm to see that pBWTs are invertible.

Theorem 1. *Given the pBWT* L *of a p-string* T *of length* n*, we can compute in* $O(n^3)$ *time and* $O(n^2)$ *space a p-string that p-matches* T*.*

Proof. For any $\ell \geq 0$ let $Pref_\ell$ denote the multiset of prev-encoded strings of the form $\langle \mathsf{T}_i \rangle[1..\ell]$, i.e., $Pref_\ell = \{\!\{ \langle \mathsf{T}_i \rangle[1..\ell] \}\!\}_{i=1}^{n}$. We show how to compute $Pref_{\ell+1}$ from $Pref_\ell$ in $O(n\ell) = O(n^2)$ time. First we sort $Pref_\ell$ in $O(n\ell) = O(n^2)$ time by radix sort with radix of size $\Theta(n)$. Then the i-th element of the sorted $Pref_\ell$ is equivalent to $\langle \mathsf{T}_{\tau[i]} \rangle[1..\ell]$. Since Proposition 1 enables us to compute $\langle \mathsf{T}_{\tau[i]-1} \rangle[1..\ell+1]$ from $\langle \mathsf{T}_{\tau[i]} \rangle[1..\ell]$ in $O(\ell)$ time, we get $Pref_{\ell+1}$ in $O(n\ell) = O(n^2)$ time. Figure 2 shows an example of the steps of the process.

Repeating the above process from $Pref_0 = \{\!\{ \varepsilon \}\!\}_{i=1}^{n}$ to $Pref_n = \{\!\{ \langle \mathsf{T}_i \rangle \}\!\}_{i=1}^{n}$, we get $\langle \mathsf{T} \rangle$ in $O(n^3)$ time. During the process we only maintain the latest $Pref_\ell$ (discarding old ones), which takes $O(n^2)$ space to store. Given $\langle \mathsf{T} \rangle$, we can compute a p-string that p-matches T using Lemma 2 in $O(n)$ time. □

Next we show how to improve the algorithm to run in $O(n^2)$ time and $O(n)$ space. From now on $Pref_\ell$ is conceptual as we do not store them explicitly. Instead we consider an integer array G_ℓ of length n to record in $G_\ell[i]$ the tentative lexicographic rank of $\langle \mathsf{T}_{\tau[i]-1} \rangle$ at the moment we have processed $Pref_\ell$ so that the lexicographic rank of $\langle \mathsf{T}_{\tau[i]-1} \rangle[1..\ell+1]$ can essentially be determined by radix sort with keys of the form $(G_\ell[i], \langle \mathsf{T}_{\tau[i]-1} \rangle[\ell+1])$. Our algorithm gradually refines

G_ℓ toward the LF-mapping while increasing ℓ. When G_ℓ becomes a permutation, a p-string that p-matches T is obtained using Lemma 3:

Lemma 3. *Given constant-time access to the pBWT* L *and LF-mapping* LF *for a p-string* T *of length* n, *we can compute in* $O(n\frac{\lg \sigma_p}{\lg\lg \sigma_p})$ *time a p-string* T' *that p-matches* T, *where* $\sigma_p \le n$ *is the number of p-symbols used in* T.

Proof. We determine $\mathsf{T}'[k]$ for $k = n, n-1, \ldots, 1$ while keeping track of the position i with $\tau[i] = k$ using the formula $\tau[LF[i]] = k - 1 \mod n$. Recall that our starting point for $k = n$ is $i = 1$ and $\mathsf{T}'[n] = \$$.

Suppose that we have determined $\mathsf{T}'[k..n]$, which contains h_k distinct p-symbols from $\{x_h\}_{h=1}^{h_k}$. At the moment we also have maintained a dynamic string for the permutation Q of $\{x_h\}_{h=1}^{h_k}$ sorted in the increasing order of their leftmost occurrences in $\mathsf{T}'[k..n]$. Then we determine $\mathsf{T}'[k-1]$ and update Q as follows:

- If $\mathsf{L}[i] \in \Sigma_s$, we set $\mathsf{T}'[k-1] = \mathsf{L}[i]$. We do nothing for Q.
- If $\mathsf{L}[i] \in [1..\sigma_p]$ and $\mathsf{L}[i] > h_k$, we use a new p-symbol x_{h_k+1} for $\mathsf{T}'[k-1]$. We insert x_{h_k+1} into the beginning of Q.
- If $\mathsf{L}[i] \in [1..\sigma_p]$ and $\mathsf{L}[i] \le h_k$, we set $\mathsf{T}'[k-1] = Q[\mathsf{L}[i]]$. We delete $Q[\mathsf{L}[i]]$ in Q and reinsert it into the beginning of Q.

Since $\mathsf{T}'[k]$ is chosen according to Proposition 1 so that $\langle \mathsf{T}'[k..n] \rangle = \langle \mathsf{T}[k..n] \rangle$, the resulting T' p-matches T. Initially Q is set to be ε and there are $O(n)$ insert/delete queries for Q of length at most σ_p, which takes a total of $O(n\frac{\lg \sigma_p}{\lg\lg \sigma_p})$ time using a dynamic string of Lemma 1. Meanwhile other tasks can be done in $O(n)$ time. □

Theorem 2. *Given the pBWT* L *of a p-string* T *of length* n, *we can compute in* $O(n^2)$ *time and* $O(n)$ *space a p-string that p-matches* $\langle \mathsf{T} \rangle$.

Proof. For any $W \in \mathit{Pref}_\ell$ let $I(W)$ be the maximal interval of p-encoded strings that are prefixed by W in the sorted Pref_ℓ, i.e., $I(W) = [|\{\!\{X \in \mathit{Pref}_\ell \mid X < W\}\!\}| + 1..|\{\!\{X \in \mathit{Pref}_\ell \mid X \le W\}\!\}|]$. Since $LF[i] \in I(\langle \mathsf{T}_{\tau[i]-1} \rangle[1..\ell])$, we can determine $LF[i]$ if the interval is a singleton.

For some $\ell \in [1..n)$, suppose that $G_\ell[i]$ is set to the smallest integer of $I(\langle \mathsf{T}_{\tau[i]-1} \rangle[1..\ell])$. Suppose also that we have two arrays Z_ℓ and E_ℓ of length n each such that $Z_\ell[i] = |\langle \mathsf{T}_{\tau[i]} \rangle[1..\ell-1]|_0$ and $E_\ell[i] = \langle \mathsf{T}_{\tau[i]} \rangle[\ell]$. See Fig. 3 for an example. Below we show how to compute $G_{\ell+1}$, $Z_{\ell+1}$ and $E_{\ell+1}$ using G_ℓ, Z_ℓ and E_ℓ in $O(n)$ time.

By Proposition 1 we can compute $\langle \mathsf{T}_{\tau[i]-1} \rangle[\ell+1]$ in $O(1)$ time as

$$\langle \mathsf{T}_{\tau[i]-1} \rangle[\ell+1] = \begin{cases} \ell & \text{if } Z_\ell[i] = \mathsf{L}[i] - 1 \text{ and } E_\ell[i] = 0, \\ E_\ell[i] & \text{otherwise.} \end{cases}$$

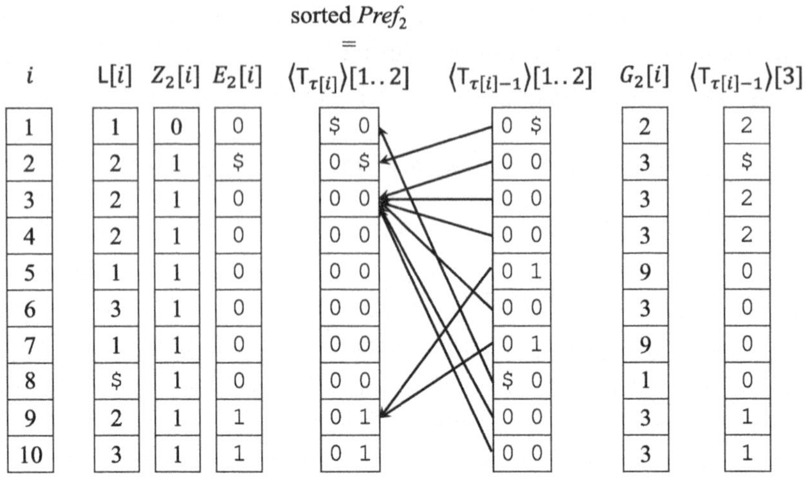

Fig. 3. Illustration of $G_2[i]$, $Z_2[i]$, $E_2[i]$ and $\langle \mathsf{T}_{\tau[i]-1}\rangle[3]$ for our running example $\mathsf{T} =$ xyxzzxxyx\$ of Table 1. $G_2[i]$ points to the smallest position at which $\langle \mathsf{T}_{\tau[i]-1}\rangle[1..2]$ appears in the sorted $Pref_2$. Remark that we do not have $\langle \mathsf{T}_{\tau[i]}\rangle[1..2]$ and $\langle \mathsf{T}_{\tau[i]-1}\rangle[1..2]$ explicitly. Still we can compute $\langle \mathsf{T}_{\tau[i]-1}\rangle[3]$ from $\mathsf{L}[i]$, $Z_2[i]$ and $E_2[i]$. By sorting the pairs of the form $(G_2[i], \langle \mathsf{T}_{\tau[i]-1}\rangle[3])$, we can obtain the refined intervals with prefixes of length 3.

We then sort the set $\{i\}_{i=1}^{n}$ of indexes with keys of the form $(G_\ell[i], \langle \mathsf{T}_{\tau[i]-1}\rangle[\ell+1])$ by radix sort in $O(n)$ time and space. While incrementing position k on the sorted list S of indexes, we keep track of the smallest index k' such that $(G_\ell[S[k']], \langle \mathsf{T}_{\tau[S[k']]-1}\rangle[\ell+1]) = (G_\ell[S[k]], \langle \mathsf{T}_{\tau[S[k]]-1}\rangle[\ell+1])$, and set $G_{\ell+1}[S[k]] = k'$. Also, $E_{\ell+1}[k] = \langle \mathsf{T}_{\tau[S[k]]-1}\rangle[\ell+1]$ and $Z_{\ell+1}[k] = Z_\ell[k] + b$, where $b = 1$ if $\langle \mathsf{T}_{\tau[S[k]]-1}\rangle[\ell+1] = 0$ and otherwise $b = 0$. Since all entries of $G_{\ell+1}$, $Z_{\ell+1}$ and $E_{\ell+1}$ are filled in a single scan on S, it takes $O(n)$ time.

The algorithm starts the above iterations from $\ell = 1$. First, we compute the sorted $Pref_1$ in $O(n)$ time using the same procedure with the algorithm of Theorem 1, which allows us to compute $E_1[i]$ and $G_1[i]$ for every i. Obviously Z_1 is the zero initialized array. We then increment ℓ until G_ℓ becomes a permutation, which takes $O(n^2)$ time in total. Since the permutation is equivalent to the LF-mapping, we can get a p-string that p-matches T in $O(n^2)$ time using Lemma 3. During the process we only maintain the latest $G_{\ell+1}$, $Z_{\ell+1}$ and $E_{\ell+1}$ (discarding old ones), which takes $O(n)$ space. □

4 Discussion

In this paper, we showed that pBWTs are invertible in $O(n^2)$ time and $O(n)$ space. The algorithm can easily be modified to work on a variant of pBWTs defined with prev$_\infty$-encoding of [19] in which the leftmost occurrence of a p-symbol is replaced with the largest symbol ∞ instead of 0. It is known that the

prev_∞-encoding has a better property than the original prev-encoding to work on pBWTs: On pBWTs with prev_∞-encoding, it holds that $LF[i] < LF[j]$ for any positions $i < j$ with $\mathsf{L}[i] = \mathsf{L}[j]$. The modified version of our algorithm runs in $O(n^2)$ time and $O(n)$ space as it does not take advantage of the property of prev_∞-encoding.

An obvious open question is whether $o(n^2)$-time inversion is possible. A possible direction for improvement is to reduce the amount of steps needed to infer the LF-mapping. For example, there are some cases where we are able to pinpoint $LF[i]$ even if $I(\langle \mathsf{T}_{\tau[i]-1}\rangle[1..\ell])$ contains two or more integers, using the following lemma.[1]

Lemma 4. *For any* $\ell \in [1..n]$ *and* $i \in [1..n)$, *let* $W = \langle \mathsf{T}_{\tau[i]}\rangle[1..\ell-1]$ *and* $\hat{W} = \langle \mathsf{T}_{\tau[i]-1}\rangle[1..\ell]$. *If* $\mathsf{L}[i] \in \Sigma_s$ *or* $\mathsf{L}[i] \le |W|_0$, *then it holds that*

$$LF[i] = |\{X \in \mathit{Pref}_{\ell+1} \mid X < \hat{W}\}| + |\{j \in [1..i] \mid \langle \mathsf{T}_{\tau[j]-1}\rangle[1..\ell] = \hat{W}\}|.$$

Proof. It is enough to show that $\langle \mathsf{T}_{\tau[j]-1}\rangle < \langle \mathsf{T}_{\tau[j']-1}\rangle$ for any positions $j < j'$ with $\langle \mathsf{T}_{\tau[j]-1}\rangle[1..\ell] = \langle \mathsf{T}_{\tau[j']-1}\rangle[1..\ell] = \hat{W}$. Let $k = \mathit{lcp}(\mathsf{T}_{\tau[j]}, \mathsf{T}_{\tau[j']})$. Note that k is at least $\ell - 1$ since $\langle \mathsf{T}_{\tau[j]}\rangle[1..\ell-1] = \langle \mathsf{T}_{\tau[j']}\rangle[1..\ell-1] = W$ is necessary to have $\langle \mathsf{T}_{\tau[j]-1}\rangle[1..\ell] = \langle \mathsf{T}_{\tau[j']-1}\rangle[1..\ell] = \hat{W}$. Also k is at most $n-1$ due to $\mathsf{T}_{\tau[j]} \ne \mathsf{T}_{\tau[j']}$.

If $\mathsf{L}[i] \in \Sigma_s$, it follows from Proposition 1 that $\langle \mathsf{T}_{\tau[j]-1}\rangle = \mathsf{L}[j]\langle \mathsf{T}_{\tau[j]}\rangle[1..n-1]$ and $\langle \mathsf{T}_{\tau[j]-1}\rangle = \mathsf{L}[j']\langle \mathsf{T}_{\tau[j']}\rangle[1..n-1]$. If $\mathsf{L}[i] \le |W|_0$, it follows from Proposition 1 that $\langle \mathsf{T}_{\tau[j]-1}\rangle = \hat{W} \cdot \langle \mathsf{T}_{\tau[j]}\rangle[\ell..n-1]$ and $\langle \mathsf{T}_{\tau[j']-1}\rangle = \hat{W} \cdot \langle \mathsf{T}_{\tau[j']}\rangle[\ell..n-1]$ with $\hat{W}$ being obtained by modifying the $\mathsf{L}[i]$-th 0 in W and prepend 0. In either case, we have $\mathit{lcp}(\langle \mathsf{T}_{\tau[j]-1}\rangle, \langle \mathsf{T}_{\tau[j']-1}\rangle) = k+1$ and $\langle \mathsf{T}_{\tau[j]-1}\rangle[k+1] = \langle \mathsf{T}_{\tau[j]}\rangle[k] < \langle \mathsf{T}_{\tau[j']}\rangle[k] = \langle \mathsf{T}_{\tau[j']-1}\rangle[k]$, which implies that $\langle \mathsf{T}_{\tau[j]-1}\rangle < \langle \mathsf{T}_{\tau[j']-1}\rangle$. $\square$

Using properties of this kind, one might hope to improve to $O(n\sigma_p)$ time, which is often seen in time complexities for p-matching algorithms like direct construction of pSAs [11].

Acknowledgements. Tomohiro I was supported by JSPS KAKENHI (Grant Numbers 22K11907 and 24K02899) and JST AIP Acceleration Research JPMJCR24U4, Japan.

References

1. Amir, A., Kondratovsky, E., Landau, G.M., Marcus, S., Sokol, D.: Reconstructing parameterized strings from parameterized suffix and LCP arrays. Theor. Comput. Sci. **981**, 114230 (2024)

[1] The property has been used in pBWT-based data structures (e.g., see cases (b) and (c1) of Lemma 1 of [13]). We rephrase it to fit our context and give a proof for completeness.

2. Amir, A., Kondratovsky, E., Marcus, S., Sokol, D.: Linear time reconstruction of parameterized strings from parameterized suffix and LCP arrays for constant-sized alphabets. In: Proceedings of the 31st International Symposium on String Processing and Information Retrieval (SPIRE) 2024. Lecture Notes in Computer Science, vol. 14899, pp. 1–15. Springer (2024)
3. Baker, B.S.: A theory of parameterized pattern matching: algorithms and applications. In: Kosaraju, S.R., Johnson, D.S., Aggarwal, A. (eds.) Proceedings of the 25th Annual ACM Symposium on Theory of Computing (STOC), pp. 71–80. ACM (1993). https://doi.org/10.1145/167088.167115
4. Baker, B.S.: Parameterized pattern matching: algorithms and applications. J. Comput. Syst. Sci. **52**(1), 28–42 (1996). https://doi.org/10.1006/jcss.1996.0003
5. Baker, B.S.: Parameterized duplication in strings: algorithms and an application to software maintenance. SIAM J. Comput. **26**(5), 1343–1362 (1997). https://doi.org/10.1137/S0097539793246707
6. Bannai, H., Inenaga, S., Shinohara, A., Takeda, M.: Inferring strings from graphs and arrays. In: Rovan, B., Vojtáš, P. (eds.) MFCS 2003. LNCS, vol. 2747, pp. 208–217. Springer, Heidelberg (2003). https://doi.org/10.1007/978-3-540-45138-9_15
7. Burrows, M., Wheeler, D.J.: A block-sorting lossless data compression algorithm. Technical report, HP Labs (1994)
8. Cazaux, B., Rivals, E.: Reverse engineering of compact suffix trees and links: a novel algorithm. J. Discrete Algorithms **28**, 9–22 (2014)
9. Deguchi, S., Higashijima, F., Bannai, H., Inenaga, S., Takeda, M.: Parameterized suffix arrays for binary strings. In: Proceedings of the Prague Stringology Conference (PSC) 2008, pp. 84–94 (2008). http://www.stringology.org/event/2008/p08.html
10. Ferragina, P., Manzini, G.: Opportunistic data structures with applications. In: Proceedings of the 41st Annual Symposium on Foundations of Computer Science (FOCS) 2000, pp. 390–398 (2000). https://doi.org/10.1109/SFCS.2000.892127
11. Fujisato, N., Nakashima, Y., Inenaga, S., Bannai, H., Takeda, M.: Direct linear time construction of parameterized suffix and LCP arrays for constant alphabets. In: Brisaboa, N.R., Puglisi, S.J. (eds.) SPIRE 2019. LNCS, vol. 11811, pp. 382–391. Springer, Cham (2019). https://doi.org/10.1007/978-3-030-32686-9_27
12. Gagie, T., Navarro, G., Prezza, N.: Optimal-time text indexing in BWT-runs bounded space. In: In Proceedings of the 29th Annual ACM-SIAM Symposium on Discrete Algorithms (SODA) 2018, pp. 1459–1477 (2018)
13. Ganguly, A., Shah, R., Thankachan, S.V.: pBWT: achieving succinct data structures for parameterized pattern matching and related problems. In: Proceedings of the 28th Annual ACM-SIAM Symposium on Discrete Algorithms (SODA) 2017, pp. 397–407 (2017). https://doi.org/10.1137/1.9781611974782.25
14. Ganguly, A., Shah, R., Thankachan, S.V.: Fully functional parameterized suffix trees in compact space. In: Bojanczyk, M., Merelli, E., Woodruff, D.P. (eds.) Proc. 49th International Colloquium on Automata, Languages, and Programming, (ICALP) 2022. LIPIcs, vol. 229, pp. 65:1–65:18. Schloss Dagstuhl - Leibniz-Zentrum für Informatik (2022). https://doi.org/10.4230/LIPIcs.ICALP.2022.65
15. Tomohiro, I., Inenaga, S., Bannai, H., Takeda, M.: Verifying and enumerating parameterized border arrays. Theor. Comput. Sci. **412**(50), 6959–6981 (2011). https://doi.org/10.1016/j.tcs.2011.09.008, http://www.sciencedirect.com/science/article/pii/S0304397511007742
16. Tomohiro, I., Inenaga, S., Bannai, H., Takeda, M.: Inferring strings from suffix trees and links on a binary alphabet. Discret. Appl. Math. **163**, 316–325 (2014)

17. Iseri, K., Tomohiro, I., Hendrian, D., Köppl, D., Yoshinaka, R., Shinohara, A.: Breaking a barrier in constructing compact indexes for parameterized pattern matching. In: Proceedings of the 51st International Colloquium on Automata, Languages, and Programming (ICALP) 2024. LIPIcs, vol. 297, pp. 89:1–89:19. Schloss Dagstuhl - Leibniz-Zentrum für Informatik (2024)
18. Kärkkäinen, J., Piatkowski, M., Puglisi, S.J.: String inference from longest-common-prefix array. Theor. Comput. Sci. **942**, 180–199 (2023)
19. Kim, S., Cho, H.: Simpler FM-index for parameterized string matching. Inf. Process. Lett. **165**, 106026 (2021). https://doi.org/10.1016/j.ipl.2020.106026
20. Kucherov, G., Tóthmérész, L., Vialette, S.: On the combinatorics of suffix arrays. Inf. Process. Lett. **113**(22–24), 915–920 (2013)
21. Mendivelso, J., Thankachan, S.V., Pinzón, Y.J.: A brief history of parameterized matching problems. Discret. Appl. Math. **274**, 103–115 (2020). https://doi.org/10.1016/j.dam.2018.07.017
22. Munro, J.I., Nekrich, Y.: Compressed data structures for dynamic sequences. In: Bansal, N., Finocchi, I. (eds.) ESA 2015. LNCS, vol. 9294, pp. 891–902. Springer, Heidelberg (2015). https://doi.org/10.1007/978-3-662-48350-3_74
23. Starikovskaya, T.A., Vildhøj, H.W.: A suffix tree or not a suffix tree? J. Discrete Algorithms **32**, 14–23 (2015)

A Space-Efficient Algorithm for Longest Common Almost Increasing Subsequence of Two Sequences

Md. Tanzeem Rahat[1], Md. Manzurul Hasan[1], and Debajyoti Mondal[2(✉)]

[1] Department of Computer Science, American International University-Bangladesh, Dhaka, Bangladesh
{tanzeem,manzurul}@aiub.edu

[2] Department of Computer Science, University of Saskatchewan, Saskatoon, Canada
dmondal@cs.usask.ca

Abstract. Let A and B be two number sequences of length n and m, respectively, where $m \leq n$. Given a positive number δ, a common almost increasing sequence $s_1 \ldots s_k$ is a common subsequence for both A and B such that for all $2 \leq i \leq k$, $s_i + \delta > \max_{1 \leq j < i} s_j$. The LCaIS problem seeks to find the longest common almost increasing subsequence (LCaIS) of A and B. An LCaIS can be computed in $O(nm\ell)$ time and $O(nm)$ space [Ta, Shieh, Lu (TCS 2021)], where ℓ is the length of the LCaIS of A and B. In this paper we first give an $O(nm\ell)$-time and $O(n + m\ell)$-space algorithm to find LCaIS, which improves the space complexity. We then design an $O((n+m)\log n + \mathcal{M}\log\mathcal{M} + \mathcal{C}\ell)$-time and $O(\mathcal{M}(\ell + \log\mathcal{M}))$-space algorithm, which is faster when the number of matching pairs $\mathcal{M}$ and the number of compatible matching pairs $\mathcal{C}$ are in $o(nm/\log m)$.

Keywords: Longest Common Subsequence · Almost Increasing Subsequence · String Algorithms

1 Introduction

The *longest increasing subsequence* (LIS) problem is a classical problem in computer science that has been extensively studied over the past few decades. The problem seeks for the longest subsequence of a given sequence in which the elements are arranged in strictly increasing order. Schensted [11] employed a dynamic programming approach that efficiently computes an LIS in $O(n \log k)$ time, where n is the length of the input sequence and k is the length of the longest increasing subsequence. A natural extension of the LIS problem is the *longest common increasing subsequence* (LCIS) problem, where given two number sequences, one needs to find a longest increasing subsequence that must also be common to both the given sequences. The problem of finding a longest common subsequence (LCS) of two sequences and its variants appear in many practical applications such as matching DNA sequences, analyzing stock market, and so on. Although there exists a quadratic-time algorithm for LCIS [13], an $O(n^{2-\epsilon})$-time algorithm, where

H. Fernau and B. Zhu (Eds.): IWOCA 2025, LNCS 15885, pp. 228–241, 2025.
https://doi.org/10.1007/978-3-031-98740-3_17

$\epsilon > 0$, cannot exist unless the Strong Exponential Time Hypothesis (SETH) fails [1,4]. Consequently, research has focused on developing algorithms with output-dependent running times. Kutz et al. [9] proposed an algorithm to compute LCIS in $O((n + m \cdot l) \log \log \sigma + \text{Sort})$ time, where n and m are the lengths of the input sequences, l is the length of the LCIS, σ is the size of the alphabet, and Sort is the time required to sort the sequences. More recently, subquadratic-time algorithms have been developed with running times $O((n \log \log n)^2 / \log^{1/6} n)$ [6] and $O(n^2 \log \log n / \sqrt{\log n})$ [2], respectively.

The *longest almost increasing sequence* (LaIS) problem, introduced by Elmasry [7], is a relaxed version of the LIS problem, where the goal is to find a longest subsequence of a given sequence that is "almost" increasing. Specifically, for a sequence $A = (s_1, s_2, \ldots, s_n)$ and a constant $\delta > 0$, the LaIS problem seeks to construct a longest possible subsequence $(s_1, s_2, \ldots, s_l)$ of A such that for all $2 \le i \le k$, $s_i + \delta > \max_{1 \le j < i} s_j$. If we allow δ to be 0, then the LaIS problem becomes equivalent to the LIS problem. The LaIS problem helps to deviate from strict monotonicity, making it applicable in scenarios where noise or fluctuations are present in the data. An interesting example is stock prices that might exhibit an almost increasing (or decreasing) trend without being strictly increasing. Elmasry [7] provided an algorithm that solves the LaIS problem in $O(n \log l)$ time, which is asymptotically optimal for the comparison-tree model.

The *longest common almost increasing subsequence* (LCaIS) problem [10,12], combines the challenges of the LaIS and LCIS problems. The LCaIS problem takes two number sequences A and B, along with a constant $\delta > 0$, and seeks to find the longest common subsequence of A and B that is almost increasing. Formally, the subsequence $(s_1, s_2, \ldots, s_l)$ must satisfy the condition $s_i + \delta > \max_{1 \le j < i} s_j$ for all $2 \le i \le k$. Moosa, Rahman and Zohora [10] claimed an $O(n^2)$-space and $O(n(n + \delta^2))$-time algorithm for computing LCaIS for two permutations, each of length n, which they also claimed to be applicable for inputs that do not contain repeated elements. Ta, Shieh, and Lu [12] identified a flaw in this algorithm and presented a dynamic programming algorithm that can compute an LCaIS in $O(nm\ell)$ time and $O(nm)$ space, where ℓ is the length of the LCaIS. Bhuiyan, Alam, and Rahman [3] gave several approaches to compute LCaIS, but all take $O(nm)$ space if an LCaIS needs to be computed. They also gave an $O(nm + \mathcal{M}(p + \ell))$-time and $O((\mathcal{M} + m)\delta)$-space algorithm, where $\mathcal{M}$ is the number of matching pairs and p is the longest common subsequence of the input sequences. This yields faster running time compared to [12] in some specific cases, e.g., when input sequences do not contain any repeated element.

Contribution. In this paper we show how to compute an LCaIS in $O(nm\ell)$ time and $O(n + m\ell)$ space, where ℓ is the length of the LCaIS. Note that all previously known algorithms take $O(nm)$ space [3,12]. We adapt a seminal work of Hirschberg [8] to obtain this improved space complexity. We then design an $O((n + m) \log n + \mathcal{M} \log \mathcal{M} + \mathcal{C}\ell)$-time and $O(\mathcal{M}(\ell + \log \mathcal{M}))$-space algorithm, which is faster when $\mathcal{M}$ and $\mathcal{C}$ are in $o(nm/\log m)$. Here $\mathcal{M}$ is the number of matching pairs between the two input sequences and $\mathcal{C}$ is the number of compatible matching pairs, where two matching pairs are called *compatible* if they form a common subsequence in the given input sequences. To design this algorithm, we transform the LCaIS problem into the problem of finding a special path in a directed graph. The vertices of the graph correspond to the matching pairs, and the edges are formed between compatible matching pairs that also satisfy the almost increasing condition. We then show an efficient way to find a path that corresponds to an LCaIS. This algorithm appears to be interesting when we observe that all previously known algorithms take $\Omega(nm)$ time [3,12] and there are cases when $\mathcal{C}$ can be asymptotically smaller than $\mathcal{M}$.

The rest of the paper is structured as follows. Section 2 gives an overview of Hirschberg's Algorithm [8] for computing LCS in $O(n+m)$ space. Section 3 presents our proposed algorithm for computing LCaIS. Section 4 discusses the case when we consider compatible matching pairs. Finally, Sect. 5 concludes the paper.

2 Overview of Hirschberg's Algorithm [8] for LCS

Let $A[1\ldots n]$ and $B[1\ldots m]$ be two number sequences of length n and m, respectively. By $\mathcal{L}(i,j)$ we denote the LCS length of $A[1\ldots i]$ and $B[1\ldots j]$, where $0 \le i \le n$ and $0 \le j \le m$. By $\mathcal{L}'(i,j)$ we denote the LCS length of $A[i+1\ldots n]$ and $B[j+1\ldots m]$, where $1 \le i+1 \le n$ and $1 \le j+1 \le m$. Given two number sequences S_1 and S_2, we denote their concatenation as $S_1 \circ S_2$.

Hirschberg [8] gave an $O(nm)$-time and $O(n+m)$-space algorithm for the LCS problem. The idea is to first design an $O(nm)$-time linear-space algorithm (Algorithm 1) that computes $L[j]$, i.e., the LCS length for every $A[1\ldots n]$ and $B[1\ldots j]$, where $1 \le j \le m$. An LCS is then computed using a divide-and-conquer approach (Algorithm 2) within the same time and space, where Algorithm 2 is used to decompose the problem into smaller subproblems.

Algorithm 1 iterates over the sequence A and at the start of an iteration i, it assumes that the LCS length $\mathcal{L}(i-1,j)$ is already computed for $0 \le j \le m$ and stored in an array $prev[1\ldots m]$. In the ith iteration, it computes $L[1\ldots m]$ by iterating over j from 1 to m. If $A[i] \ne B[j]$, then $L[j]$ is assigned $\max\{\mathcal{L}(i,j-1), \mathcal{L}(i-1,j)\}$. Here $\mathcal{L}(i,j-1)$ is already computed at the $(j-1)$th iteration and $\mathcal{L}(i-1,j)$ can be accessed from $prev[j]$. If $A[i]=B[j]$, then $L[j]$ is the same as $\mathcal{L}(i-1,j-1)+1$, i.e., $prev[j-1]+1$. Note here that $\mathcal{L}(i-1,j)$ and $\mathcal{L}(i,j-1)$ cannot be larger than $\mathcal{L}(i,j)$.

Algorithm 1: ComputeL($A[1\ldots n], B[1\ldots m], L[1\ldots m]$)

$L[0\ldots m] \leftarrow [0,\ldots,0]$;
for $i \leftarrow 1$ **to** n **do**
    $prev[0\ldots m] \leftarrow L[0,\ldots,m]$;
    for $j \leftarrow 1$ **to** m **do**
        if $A[i]=B[j]$ **then** $L[j] \leftarrow prev[j-1]+1$;
        else $L[j] \leftarrow \max\{L[j-1], prev[j]\}$;
    end
end

The divide-and-conquer approach for constructing the LCS is based on the observation that for every $0 \le i \le n$ there exists some j, where $0 \le j \le m$, such that the LCS length is equal to $M(j) = \mathcal{L}(i,j) + \mathcal{L}'(i,j)$. Therefore, it suffices to find the index j that maximizes $M(j)$ and divide the problem of constructing LCS into two subproblems. It may initially appear that the above equation should use $\mathcal{L}'(i+1, j+1)$ instead of $\mathcal{L}'(i,j)$, but note that $\mathcal{L}'(i,j)$ denotes the LCS length of $A[i+1 \dots n]$ and $B[j+1 \dots m]$.

Algorithm 2: LCS($A[1 \dots n], B[1 \dots m], S$)

if $m = 0$ **then** $S \leftarrow$ empty sequence, return;
if $n = 1$ **then**
    if $\exists j$ *where* $A[1] = B[j]$ **then** $S \leftarrow A[1]$;
    else $S \leftarrow$ empty sequence;
    return;
$i \leftarrow \lfloor n/2 \rfloor$;
ComputeL($A[1 \dots i], B[1 \dots m], L[1 \dots m]$);
ComputeL($A[i+1 \dots n], B[1 \dots m], L'[1 \dots m]$);
Determine an index j that maximizes $L[j] + L'[j]$; // Problem decomposition
LCS($A[1 \dots i], B[1 \dots j], S_1$);
LCS($A[i+1 \dots n], B[j+1 \dots m], S_2$);
$S \leftarrow S_1 \circ S_2$

3 Finding LCaIS in $O(nm\ell)$ Time and $O(n + m\ell)$ Space

We now show that the divide-and-conquer framework of Hirschberg [8] can be adapted to compute LCaIS in $O(nm\ell)$ time and $O(n+m\ell)$ space, where ℓ is the length of the LCaIS for input sequences $A[1 \dots n]$ and $B[1 \dots m]$.

The idea is to design two subroutines (Sect. 3.1)—one that computes the LCaIS for $A[1 \dots n]$ and $B[1 \dots j]$ ending at $B[j]$ and the other that computes the LCaIS for $A[1 \dots n]$ and $B[j \dots m]$ starting at $B[j]$, where $1 \le j \le m$. We also ensure that these LCaIS possess additional properties. The former should minimize the largest element used in the LCaIS, while the latter should maximize the smallest element. We use the former subroutine to find LCaIS for the sequences $A[1 \dots \lfloor n/2 \rfloor]$ and $B[1 \dots m]$, and the latter subroutine to compute LCaIS for the sequences $A[\lfloor n/2 \rfloor + 1 \dots n]$ and $B[1 \dots m]$. We use the solutions of these subroutines to divide the problem of computing an LCaIS into solving two subproblems (Sect. 3.2). Specifically, we find a suitable index j that ensures that a common increasing subsequence ending at $B[j]$ can be concatenated with a common increasing subsequence starting at $B[j]$ to form the desired LCaIS.

3.1 Deciding LCaIS Length When Minimizing Largest Element

We first design an $O(nm\ell)$-time and $O(m\ell)$-space subroutine similar to Algorithm 1 that computes $P'[j,r]$, where $1 \le j \le m$ and $1 \le r \le \ell$, and $P'[j,r]$ is True if and only if there exists an LCaIS for $A[1\dots n]$ and $B[1\dots j]$ of length r ending with $B[j]$. We will also compute $Q'[j,r]$, i.e., the smallest number such that there exists such an LCaIS for $A[1\dots n]$ and $B[1\dots j]$ where the largest element is $Q'[j,r]$. If $P'[j,r]$ is False, then $Q'[j,r]$ is set to ∞.

To compute $P'[j,r]$ and $Q'[j,r]$, we will use the following lemma.

Lemma 1. *Let $\mathcal{P}'(i,j,r)$ be True if there exists an LCaIS for $A[1\dots i]$ and $B[1\dots m]$ of length r ending with $B[j]$. For $r > 0$, if $\mathcal{P}'(i,j,r)$ is True, then let $\mathcal{Q}'(i,j,r)$ denote the smallest number such that there exists an LCaIS ending with $B[j]$ of length r such that no element of the LCaIS is larger than $\mathcal{Q}'(i,j,r)$. Then the following equalities hold.*

$$\mathcal{P}'(i,j,r) = \begin{cases} \displaystyle\bigvee_{\substack{1\le k<j \\ A[i]+\delta>\mathcal{Q}'(i-1,k,r-1)}} \mathcal{P}'(i-1,k,r-1), & \text{if } A[i]=B[j] \\ \mathcal{P}'(i-1,j,r), & \text{otherwise.} \end{cases}$$

$$\mathcal{Q}'(i,j,r) = \begin{cases} \displaystyle\min_{\substack{1\le k<j \\ A[i]+\delta>\mathcal{Q}'(i-1,k,r-1) \\ \mathcal{P}'(i-1,k,r-1)=True}} \max\left(A[i], \mathcal{Q}'(i-1,k,r-1)\right), & \text{if } A[i]=B[j] \\ \mathcal{Q}'(i-1,j,r), & \text{otherwise.} \end{cases}$$

Proof. Consider first the case of computing $\mathcal{P}'(i,j,r)$.

If $A[i] \ne B[j]$, then LCaIS cannot match $A[i]$ and $B[j]$. Hence there exists an LCaIS of length r ending at $B[j]$, i.e., $\mathcal{P}'(i,j,r)$ is True, if and only if there exists an LCaIS of length r (ending at $B[j]$) for the sequences $A[1\dots i-1]$ and $B[1\dots j]$, i.e., if $\mathcal{P}'(i-1,j,r)$ is True.

If $A[i] = B[j]$, then we assume that $A[i]$ has been matched to $B[j]$ in the LCaIS. If there exists an LCaIS of length r, then there must be an LCaIS σ_k of length $r-1$ for $A[1\dots i-1]$ and $B[1\dots k]$ for some k where $1 \le k < j$ such that $A[i]$ can be appended to σ_k to form an LCaIS of length r. However, to append $A[i]$ to σ_k, σ_k must be compatible with $A[i]$, i.e., the largest element of σ_k must be smaller than $A[i]+\delta$. Hence we can restrict our attention to those indices where $A[i]+\delta > \mathcal{Q}'(i-1,k,r-1)$. Since $\mathcal{Q}'(i-1,k,r-1)$ minimizes the largest element over all σ_k of length $r-1$, if there exists some compatible σ_k, then we must have $A[i]+\delta > \mathcal{Q}'(i-1,k,r-1)$ where $\mathcal{P}'(i-1,k,r-1)$ is True.

Consider now the case of computing $\mathcal{Q}'(i,j,r)$.

If $A[i] \ne B[j]$, then LCaIS cannot match $A[i]$ and $B[j]$. Therefore, if there exists an LCaIS of length r ending with $B[j]$, then that would also be an LCaIS of length r (ending with $B[j]$) for the sequences $A[1\dots i-1]$ and $B[1\dots j]$. Since $\mathcal{Q}'(i-1,j,r)$ minimizes the largest element over all LCaIS of length r that ends with $B[j]$, $\mathcal{Q}'(i,j,r)$ can be obtained from $\mathcal{Q}'(i-1,j,r)$.

If $A[i] = B[j]$, then similar to the computation of $\mathcal{P}'(i, j, r)$, we can restrict our attention to the candidate indices k where $1 \leq k < j$ and $A[i] + \delta > \mathcal{Q}'(i-1, k, r-1)$. In addition, we can only consider the cases when $\mathcal{P}'(i-1, k, r-1)$ is True because we want the final LCaIS to be of length r. Consider one such LCaIS $\sigma_k \circ A[i]$. Since $B[j](= A[i])$ is the last element of the LCaIS, we find the largest element by taking the maximum of $A[i]$ and $\mathcal{Q}'(i-1, k, r-1)$. Finally, we take the minimum over all candidate indices k. □

We now design a subroutine MinimizeLargestElem based on Lemma 1 that computes $P'[j, \ell']$ and $Q'[j, \ell']$ for every j from 1 to m and for every ℓ' from 1 to ℓ.

Implementation Details of MinimizeLargestElem. We iterate over the sequence A and at the start of an iteration i, we assume that $P'[j, 0 \dots \ell']$ is already computed for $A[1 \dots i-1]$ and $B[1 \dots m]$ where $0 \leq j \leq m$, and ℓ' is the length of the LCaIS found so far. In the implementation of P', for each index j, we maintain a list of tuples $\mathcal{L}_j$. Let (r, a_r, b_r) be the rth tuple of $\mathcal{L}_j$. Then $a_r = \mathcal{P}'(i-1, j, r)$ indicates whether there exists an LCaIS of length r (ending at $B[j]$), and if so, then b_r is equal to $\mathcal{Q}'(i-1, j, r)$. In the ith iteration, we compute $P'[j, 0 \dots \ell'+1]$ for $A[1 \dots i]$ and $B[1 \dots m]$ by iterating j from 1 to m.

To compute $P'[j, 0 \dots \ell'+1]$ for $A[1 \dots i]$ and $B[1 \dots j]$, we iterate over $\mathcal{L}_1 \dots \mathcal{L}_{j-1}$ and maintain a list of tuples $\mathcal{T}$ where the rth tuple holds the information about whether an LCaIS of length r has been seen at some $\mathcal{L}_k$, where $k < j$ and $A[i] + \delta > \mathcal{Q}'(i-1, k, r-1)$. If so, and if $A[i] = B[j]$ then we update the tuples $\mathcal{L}_j$ based on the recurrence relation of Lemma 1. If we find an LCaIS of length $\ell'+1$, then we append a new tuple to $\mathcal{L}_j$ and update ℓ', i.e., the length of the LCaIS found so far. Algorithm 3 gives a pseudocode for MinimizeLargestElem. The variable $L[j]$ implements $\mathcal{L}_j$, and the arrays *current* and *curMinElem* implement the role of $\mathcal{T}$. Table 1 gives an example computation for two specific sequences.

We next compute $X'[j, r]$, $1 \leq j \leq m$, that indicates whether there exists an LCaIS of length r for $A[1 \dots n]$ and $B[1 \dots j]$ (without specifying the ending element). Similarly, we compute $Y'[j, r]$, $1 \leq j \leq m$, which minimizes the maximum element of the LCaIS. It is straightforward to compute $X'[j, r]$ and $Y'[j, r]$, where $1 \leq j \leq m$, in $O(m\ell)$ time by iterating over the tuples of each list $\mathcal{L}_j$.

Let $\mathcal{P}''(i, j, r)$ be True if there exists an LCaIS for $A[i+1 \dots n]$ and $B[1 \dots m]$ of length r that starts at $B[j]$. For $r > 0$, if $\mathcal{P}''(i, j, r)$ is True, then let $\mathcal{Q}''(i, j, r)$ be the largest number such that there exists an LCaIS (with first element $B[j]$) of length r such that no element of the LCaIS is smaller than $\mathcal{Q}''(i, j, r)$. One can design a subroutine MaximizeSmallestElem symmetric to MinimizeLargestElem for computing $X''[j, r]$, $1 \leq j \leq m$, that indicates whether there exists an LCaIS of length r for $A[1 \dots n]$ and $B[j+1 \dots m]$ (without specifying the starting element) and $Y'[j, r]$ that maximizes the smallest element.

Algorithm 3: MinimizeLargestElem($A[1 \ldots n], B[1 \ldots m], \delta, X', Y'$)

```
Initialize L[0...m], where L[i] with 0 ≤ i ≤ m is a doubly linked list of tuples.;
Set L[i], where 0 ≤ i ≤ m, to (0, 0, ∞) ;       // Maintain count(L[j]) in a
    variable to retrieve the number of tuples in O(1) time
currentMax ← 0 ;                                // LCaIS length found so far
for i ← 1 to n do
    attemptNewMax ← currentMax + 1;
    for z ← 0 to attemptNewMax do
        current[z] ← 0 ;   // Initialize to track current to process A[i]
        curMinElem[z] ← ∞
    for j ← 1 to m do
        for z ← 0 to attemptNewMax;                 // Iterate over L[j]
          simultaneously to retrieve L[j][z] or L[j][z + 1] in O(1) time
         do
            if z ≥ count(L[j]) then
                L[j].append(z, 0, ∞) ;
                r, p, q ← z, 0, ∞
            else r, p, q ← L[j][z] ;
        if A[i] = B[j] then
            if r = 0 then
                if count(L[j]) ≤ 1 then
                    L[j].append(1, 1, A[i]);   // No past choices to consider
                    currentMax ← max(currentMax, 1)
                else
                    L[j][r + 1] ← (1, 1, A[i]) ;   // LCaIS length 1; take A[i]
            else
                if current[r] = 1;      // Length-r LCaIS exists; build r + 1
                 then
                    if count(L[j]) ≤ r + 1 then
                        L[j].append(r + 1, 1, max(A[i], curMinElem[r]));
                        currentMax ← max(currentMax, r + 1)
                    else
                        a, b, d ← L[j][r + 1];
                        L[j][r+1] ← (r+1, 1, min(d, max(A[i], curMinElem[r])))
            /* Update curMinElem if LCaIS of length r exists               */
            current[r] ← (current[r] ∨ p);
            if p = 1 then
                curMinElem[r] ← min(curMinElem[r], q)
        else
            if (A[i] + δ > q) then
                current[r] ← (current[r] ∨ p);
                if p = 1 then
                    curMinElem[r] ← min(curMinElem[r], q)
Compute X' and Y' from L
```

Table 1. Computation of L with $A = [3, 1, 4, 5, 2, 4, 5, 1]$, $B = [4, 2, 3, 1, 2, 5, 3, 1]$ and $\delta = 3$.

	$r = 0$	$r = 1$	$r = 2$	$r = 3$	$r = 4$
$L[1]$	(0, 0, ∞)	**(1, 1, 4)**	(2, 0, ∞)	(3, 0, ∞)	(4, 0, ∞)
$L[2]$	(0, 0, ∞)	(1, 1, 2)	**(2, 1, 4)**	(3, 0, ∞)	(4, 0, ∞)
$L[3]$	(0, 0, ∞)	**(1, 1, 3)**	(2, 0, ∞)	(3, 0, ∞)	(4, 0, ∞)
$L[4]$	(0, 0, ∞)	(1, 1, 1)	**(2, 1, 2)**	(3, 0, ∞)	(4, 0, ∞)
$L[5]$	(0, 0, ∞)	(1, 1, 2)	(2, 1, 2)	**(3, 1, 3)**	(4, 0, ∞)
$L[6]$	(0, 0, ∞)	(1, 1, 5)	(2, 1, 5)	(3, 1, 5)	**(4, 1, 5)**
$L[7]$	(0, 0, ∞)	**(1, 1, 3)**	(2, 0, ∞)	(3, 0, ∞)	(4, 0, ∞)
$L[8]$	(0, 0, ∞)	(1, 1, 1)	(2, 1, 1)	(3, 1, 2)	**(4, 1, 3)**

3.2 Divide and Conquer

We now design a divide-and-conquer method to construct the LCaIS, where we will use MINIMIZELARGESTELEM and MAXIMIZESMALLESTELEM to divide the problem into smaller subproblems.

We will use the following lemma.

Lemma 2. *Let i be an index where $0 \leq i \leq n$. Assume that X', Y' have been computed for the sequences $A[1 \ldots i]$ and $B[1 \ldots m]$, and X'', Y'' have been computed for $A[i+1 \ldots n]$ and $B[1 \ldots m]$. Then the LCaIS length of $A[1 \ldots n]$ and $B[1 \ldots m]$ is equal to* $\max_{\substack{0 \leq j \leq m \\ X'[j,r']=X''[j,r'']=True \\ Y''[j,r'']+\delta > Y'[j,r']}} (r' + r'')$.

Proof. Let S be an LCaIS of $A[1 \ldots n]$ and $B[1 \ldots m]$. By $|S|$, we denote the length of S. We can write S as $S' \circ S''$, where $0 \leq j \leq m$, S' is the LCaIS of $A[1 \ldots i]$ and $B[1 \ldots j]$, and S'' is the LCaIS of $A[i+1 \ldots n]$ and $B[j+1 \ldots m]$. Hence $X'[j, |S'|]$ and $X''[j, |S''|]$ must be True. Let u and w be the largest element in S' and smallest element in S'', respectively. Then $u \geq Y'[j, |S'|]$ and $Y''[j, |S''|] \geq w$. Since $w + \delta > u$, we must have $Y''[j, |S''|] + \delta > Y'[j, |S'|]$. Therefore, we must have an index j with $r' = |S'|$ and $r'' = |S''|$ such that $|S| = r' + r''$. However, we still need to prove that the maximum value of $(r' + r'')$ over all options is equal to $|S|$.

Assume for a contradiction that there exists some j such that $X'[j, r'] = X''[j, r''] = True$, $Y''[j, r''] + \delta > Y'[j, r']$, and $(r' + r'') > |S|$. Note that $Y'[j, r']$ minimizes the largest element, whereas $Y''[j, r'']$ maximizes the smallest element. Therefore, for an element u corresponding to the sequence $X'[j, r']$ and for an element w corresponding to the sequence $X''[j, r'']$, we have $w + \delta \geq Y''[j, r''] + \delta > Y'[j, r'] \geq u$. In this case, the LCaIS of $A[1 \ldots i]$ and $B[1 \ldots j]$ and the LCaIS of $A[i+1 \ldots n]$ and $B[j+1 \ldots m]$ can be concatenated to obtain an LCaIS of $A[1 \ldots n]$ and $B[1 \ldots m]$ with a larger length than $|S|$, which contradicts the optimality of S. □

We now can determine an index j based on Lemma 2 to decompose the problem into two smaller subproblems. The pseudocode for the divide-and-conquer approach is given in Algorithm 4, where S corresponds to the LCaIS.

Algorithm 4: LCaIS($A[1 \dots n], B[1 \dots m], \delta, S$)

1 **if** $m = 0$ **then** $S \leftarrow$ empty sequence, return;
2 **if** $n = 1$ **then**
3   **if** $\exists j$ *where* $A[1] = B[j]$ **then** $S \leftarrow A[1]$;
4   **else** $S \leftarrow$ empty sequence;
5   return;
6 $i \leftarrow \lfloor n/2 \rfloor$;
7 MinimizeLargestElem($A[1 \dots i], B[1 \dots m], \delta, X', Y'$);
8 MaximizeSmallestElem($A[i+1 \dots n], B[1 \dots m], \delta, X'', Y''$);
9 Determine an index j that maximizes $(r' + r'')$ satisfying the conditions $0 \le j \le m$, $X'[j, r'] = X''[j, r''] = True$ and $Y''[j, r''] + \delta > Y'[j, r']$;
10 Remove the elements that are larger than $Y'[j]$ and smaller than $Y''[j]$ from $B[1 \dots j]$ and $B[j+1 \dots m]$, respectively;
11 LCaIS($A[1 \dots i], B[1 \dots j], \delta, S_1$);
12 LCaIS($A[i+1 \dots n], B[j+1 \dots m], \delta, S_2$);
13 $S \leftarrow S_1 \circ S_2$

3.3 Running Time and Space Complexity

Let $T(n, m)$ be the running time and let ℓ be the LCaIS length for $A[1 \dots n]$ and $B[1 \dots n]$. Lines 7 and 8 of Algorithm 4 take $O(nm\ell)$ time. Line 9 can be computed in $O(mr'r'')$ time, where r' and r'' are the lengths of the longest subsequences returned by the two subroutines, and hence in $O(nm\ell)$ time. Line 10 takes $O(m)$ time. The subproblems on Lines 11 and 12 are of size $T(n/2, j)$ and $T(n/2, m - j)$, respectively. Assume that there exists a constant d such that $T(n, m) \le d \cdot nm\ell$ for sufficiently large n and m. Therefore,

$$
\begin{aligned}
T(n, m) &= T(n/2, j) + T(n/2, m - j) + O(nm\ell) \\
&\le dnj\ell/2 + dn(m - j)\ell/2 + O(nm\ell) \\
&\le dnm\ell/2 + d'nm\ell, && \text{where } d' \text{ is a constant} \\
&\le dnm\ell/2 + (d/2)nm\ell, && \text{by setting } d \text{ to be } 2d' \\
&\le dnm\ell, && \text{where } d = 2d' \text{ and } m, n \ge 1.
\end{aligned}
$$

The space complexity of Algorithm 3 is dominated by the lists that are used to implement P' and Q'. Since there are $O(m)$ lists, each of size $O(\ell)$, the space complexity becomes $O(m\ell)$. However, we also need to consider the space complexity for Algorithm 4. Lines 7–10 use a temporary space of size $O(m\ell)$. Therefore, it suffices to consider the space required for the recursive calls. The number of recursive calls, i.e., $R(n, m)$ can be written as $R(n/2, j) + R(n/2, m - j) + 1$. Assume that there exists a constant d such that $R(n, m) \leq dn - 1$ for sufficiently large n. Then

$$\begin{aligned} R(n, m) &= R(n/2, j) + R(n/2, m - j) + 1 \\ &\leq dn/2 + dn/2 - 1 \\ &\leq dn - 1, \qquad\qquad \text{where } d = 2 \text{ and } n \geq 1. \end{aligned}$$

We thus obtain the following theorem.

Theorem 1. *Given two number sequences $A[1 \ldots n]$ and $B[1 \ldots m]$ where $m \leq n$, and a positive constant δ, an LCaIS for these sequences can be computed in $O(nm\ell)$ time and $O(n + m\ell)$ space, where ℓ is the length of the LCaIS.*

4 LCaIS with Few Compatible Matching Pairs

In this section we show that the LCaIS can be computed faster for the case when the number of matching pairs $\mathcal{M}$ and the number of compatible matching pairs $\mathcal{C}$ are in $o(nm/\log m)$. For each $B[k]$ we precompute the set of all indices $match(k)$ such that $A[i] = B[k]$, where $i \in match(k)$. We can do this in $O((n + m) \log n + \mathcal{M})$ time by first sorting the numbers in A and then searching the elements of B using binary search. Figure 1(a) shows an input and their matching pairs.

The idea now is to form a directed acyclic graph G and formulate the LCaIS problem as finding a path in the graph. The graph G consists of $\mathcal{M}$ nodes $v_1, \ldots, v_{\mathcal{M}}$ corresponding to the matching pairs where vertices are ordered so that they first correspond to all the matching pairs for $B[1]$, and then all the matching pairs for $B[2]$, and so on.

Let $\mathcal{M}_i$ be the matching pair corresponding to v_i and let e_i be the number corresponding to v_i. The graph G has a directed edge $\overrightarrow{(v_i, v_j)}$ where $1 \leq i < j \leq \mathcal{M}$ if and only if $e_j + \delta > e_i$, and $\mathcal{M}_i, \mathcal{M}_j$ are compatible, i.e., form a common subsequence for A and B. Figure 1(c) illustrates an example for G where there is no edge from v_1 to v_2. This is because despite $e_2 + \delta > e_1$, the matching pairs $\mathcal{M}_1, \mathcal{M}_2$ do not form a common subsequence for A and B. We maintain the points (k, i), where $1 \leq k \leq m$ and $i \in match(k)$, in a 2D range tree data structure so that given v_i, its outgoing edges can be found in $O(\log \mathcal{M} + c_i)$ time, where c_i is the number of pairs that are compatible with $\mathcal{M}_i$. This is done by searching (using the range tree data structure) for the points with both coordinates higher than the point represented by v_i and then removing the candidates that do not meet the almost increasing condition. In Fig. 1(b), there are three pairs compatible with vertex v_1 but only two of them satisfy the almost increasing condition. Hence v_1 has two outgoing edges. A range tree data structure can be computed in $O(\mathcal{M} \log \mathcal{M})$ time and space [5].

We maintain a list of pairs at each v_i, where $1 \leq i \leq \mathcal{M}$. A tuple (r, w, v) at v_i indicates that there exists an LCaIS of length r that ends with the matching pair $\mathcal{M}_i$, where the largest element used in the LCaIS is w and the matching

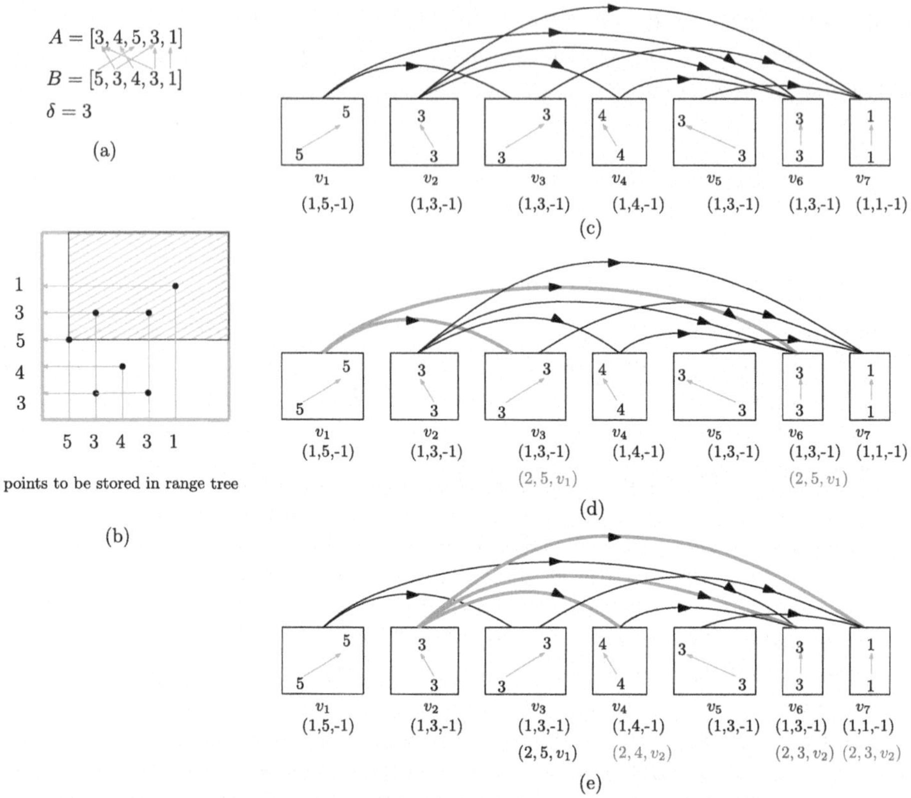

Fig. 1. Illustration for the computation of LCaIS length.

pair just before v_i is v. During our algorithm, we will minimize w over all LCaIS of length r that ends with matching pair $\mathcal{M}_i$.

Initially, each v_i contains the tuple $(1, e_i, -1)$ indicating that there exists an LCaIS of length 1 that ends with matching pair $\mathcal{M}_i$, where the largest element used is e_i and the matching pair before $\mathcal{M}_i$ is null. We now iterate an index i from 1 to m and for each v_i, we consider the directed edges $\overrightarrow{(v_i, v_j)}$ with $i < j \leq m$. When processing an edge $\overrightarrow{(v_i, v_j)}$ we check whether the tuples at v_j can be updated. Let (r_i, w_i, v) be a tuple at v_i. We now consider the following cases.

- If $e_j + \delta > w_i$, then we have a potential for updating the tuples at v_j.
  - If v_j does not contain any tuple that starts with $1 + r_i$, then we can insert the tuple $(1 + r_i, \max(w_i, e_j), v_i)$. This is because we now can form an LCaIS of length $1 + r_i$ that ends with $\mathcal{M}_j$. Figure 1(d) shows two such insertions in red.
  - If v_j already contains a tuple $(1 + r_i, w_j, v')$, then we check whether the condition $\max(w_i, e_j) < w_j$ holds. If so, then we update the tuple with

$(1+r_i, \max(w_i, e_j), v_i)$. This is because we found a better LCaIS of length $1 + r_i$, i.e., the maximum element of the LCaIS computed by appending $\mathcal{M}_j$ has a lower value. Figure 1(e) shows such an update in blue.
- If $e_j + \delta \leq w_i$, then we do not update any tuple at v_j. This is because $\mathcal{M}_j$ cannot be added to form a longer LCaIS.

Theorem 2. *Given two number sequences $A[1 \ldots n]$ and $B[1 \ldots m]$ where $m \leq n$, and a positive constant δ, an LCaIS for these sequences can be computed in $O((n+m)\log n + \mathcal{M}\log\mathcal{M} + \mathcal{C}\ell)$ time and $O(\mathcal{M}(\ell + \log\mathcal{M}))$ space, where ℓ is the length of the LCaIS, $\mathcal{M}$ is the number of matching pairs, and $\mathcal{C}$ is the number of compatible matching pairs.*

Proof. We now show that at the end of processing all the edges, there must be a tuple corresponding to an LCaIS. Furthermore, we can construct the LCaIS by repeatedly following the parent. To this end, it suffices to prove the following.

Claim: By the time we start processing v_i, where $1 \leq i \leq \mathcal{M}$, it already stores a tuple (r, w, v) at v_i if and only if there exists an LCaIS of length r with largest element w that ends with $\mathcal{M}_i$. Here w is the minimum possible, i.e., there does not exist another LCaIS of length r that ends with $\mathcal{M}_i$ with largest element smaller than w. Furthermore, we can construct such an LCaIS by taking e_i and repeatedly following the parent starting at v.

The case when $i = 1$ is straightforward because v_1 only contains the tuple $(1, e_1, -1)$ and no LCaIS of length larger than 1 that ends at $\mathcal{M}_i$ is possible. Assume now that the claim holds for any $1 \leq k < i$, and we have reached v_i. This means all the edges $\overrightarrow{(v_k, v_i)}$ have been processed already.
If no such edge $\overrightarrow{(v_k, v_i)}$ exists then v_i contains the tuple $(1, e_i, -1)$ and no LCaIS of length larger than 1 that ends at $\mathcal{M}_i$ is possible. Otherwise, assume an edge $\overrightarrow{(v_k, v_i)}$, where v_k has a tuple (r_k, w_k, v).
Since $k < i$, by induction hypothesis there exists an LCaIS of length r_k that ends with $\mathcal{M}_k$ and w_k is the smallest integer such that there exists an LCaIS of length r_k that ends with $\mathcal{M}_k$. Furthermore, we can construct such an LCaIS by taking $\mathcal{M}_k$ and repeatedly following the parent starting at v. If $e_i + \delta > w_k$, then we either insert or update an existing tuple at v_i. Therefore, after processing all the edges $\overrightarrow{(v_k, v_i)}$, v_i contains a tuple (r_i, w_i, v_x) if and only if there exists an LCaIS of length r_i with last element $\mathcal{M}_i$ where w_i minimizes the largest element over all such LCaIS. By induction hypothesis, v_x contains a tuple $(r_i - 1, w_x, v')$. Since $e_i + \delta > w_x$, we can find an LCaIS of length r_i by taking $\mathcal{M}_i$ and repeatedly following the parent links.

We now analyze the time and space complexity. We do not need to create the graph explicitly. For each v_i we first find its compatible matching pairs in $O(\log\mathcal{M} + c_i)$ time and then determine the outgoing edges of v_i within the same running time. Hence we need $O(\sum_{1 \leq i \leq \mathcal{M}}(\log\mathcal{M} + c_i)) = O(\mathcal{M}\log\mathcal{M} + \mathcal{C})$ time to determine the edges of the graph. Each edge $\overrightarrow{(v_i, v_j)}$ is processed exactly once and updating the tuples at v_j takes $O(\ell)$ time, where ℓ is the length of the LCaIS of A and B. Therefore, the total time to process all edges is $O(\mathcal{C}\ell)$. It is straightforward to construct the LCaIS in $O(\ell)$ time by following the parent links. Therefore, together with the time for determining matching pairs, computing the range tree data structure, and determining the graph edges, the overall running time becomes $O((n+m)\log n + \mathcal{M}\log\mathcal{M} + \mathcal{C}\ell)$.

The space needed corresponds to the range tree data structure plus the total number of tuples. The space for the former is $O(\mathcal{M}\log\mathcal{M})$, while the latter requires $O(\mathcal{M}\ell)$ space. Hence the space complexity is $O(\mathcal{M}(\ell + \log\mathcal{M}))$. □

5 Conclusion

We gave an algorithm that given two sequences of lengths n and m with $m \leq n$, finds an LCaIS in $O(nm\ell)$-time and $O(n + m\ell)$-space, where ℓ is the length of the LCaIS. A natural avenue of research would be to improve the running time. Can we find some logarithmic factor improvement employing the techniques used in designing subquadratic-time algortihms for LCIS [2,6]? It would also be interesting to examine the case when $\ell = o(\log n)$ to obtain faster algorithms by potentially leveraging bit-vector operations. We gave another $O((n+m)\log n + \mathcal{M}\log\mathcal{M} + \mathcal{C}\ell)$-time and $O(\mathcal{M}(\ell + \log\mathcal{M}))$-space algorithm which is faster when the number of matching pairs $\mathcal{M}$ and the number of compatible matching pairs $\mathcal{C}$ are in $o(nm/\log m)$. One may attempt to further improve the time and space complexities under these conditions.

Acknowledgment. The research is supported in part by the Natural Sciences and Engineering Research Council of Canada (NSERC). We thank M. Sohel Rahman for our stimulating initial discussions and the anonymous reviewers for their feedback, which improved the presentation of the paper.

References

1. Abboud, A., Backurs, A., Williams, V.V.: Tight hardness results for lcs and other sequence similarity measures. In: 2015 IEEE 56th Annual Symposium on Foundations of Computer Science, pp. 59–78. IEEE (2015)
2. Agrawal, A., Gawrychowski, P.: A faster subquadratic algorithm for the longest common increasing subsequence problem. In: 31st International Symposium on Algorithms and Computation (ISAAC 2020), pp. 4–1. Schloss Dagstuhl–Leibniz-Zentrum für Informatik (2020)
3. Bhuiyan, M., Alam, M.R., Rahman, M.S.: Computing the longest common almost-increasing subsequence. Theor. Comput. Sci. **930**, 157–178 (2022)
4. Bringman, K., Künnemann, M.: Multivariate fine-grained complexity of longest common subsequence. In: Proceedings of the Twenty-Ninth Annual ACM-SIAM Symposium on Discrete Algorithms, pp. 1216–1235. SIAM (2018)
5. De Berg, M.: Computational Geometry: Algorithms and Applications. Springer, Heidelberg (2000)

6. Duraj, L.: A sub-quadratic algorithm for the longest common increasing subsequence problem. In: 37th International Symposium on Theoretical Aspects of Computer Science (STACS 2020), pp. 41–1. Schloss Dagstuhl–Leibniz-Zentrum für Informatik (2020)
7. Elmasry, A.: The longest almost-increasing subsequence. In: Thai, M.T., Sahni, S. (eds.) COCOON 2010. LNCS, vol. 6196, pp. 338–347. Springer, Heidelberg (2010). https://doi.org/10.1007/978-3-642-14031-0_37
8. Hirschberg, D.S.: A linear space algorithm for computing maximal common subsequences. Commun. ACM **18**(6), 341–343 (1975)
9. Kutz, M., Brodal, G.S., Kaligosi, K., Katriel, I.: Faster algorithms for computing longest common increasing subsequences. J. Disc. Algor. **9**(4), 314–325 (2011)
10. Moosa, J.M., Rahman, M.S., Zohora, F.T.: Computing a longest common subsequence that is almost increasing on sequences having no repeated elements. J. Disc. Algor. **20**, 12–20 (2013)
11. Schensted, C.: Longest increasing and decreasing subsequences. Can. J. Math. **13**, 179–191 (1961)
12. Ta, T.T., Shieh, Y.K., Lu, C.L.: Computing a longest common almost-increasing subsequence of two sequences. Theor. Comput. Sci. **854**, 44–51 (2021)
13. Yang, I.H., Huang, C.P., Chao, K.M.: A fast algorithm for computing a longest common increasing subsequence. Inf. Process. Lett. **93**(5), 249–253 (2005)

Fast Pattern Matching with Epsilon Transitions

Nicola Cotumaccio(✉)

Department of Computer Science, University of Helsinki, Helsinki, Finland
nicola.cotumaccio@helsinki.fi

Abstract. In the String Matching in Labeled Graphs (SMLG) problem, we need to determine whether a pattern string appears on a given labeled graph or a given automaton. Under the Orthogonal Vectors hypothesis, the SMLG problem cannot be solved in subquadratic time [ICALP 2019]. In typical bioinformatics applications, pattern matching algorithms should be both fast and space-efficient, so we need to determine useful classes of graphs on which the SLMG problem can be solved efficiently.

In this paper, we improve on a recent result [STACS 2024] that shows how to solve the SMLG problem in linear time on the compressed representation of Wheeler generalized automata, a class of string-labeled automata that extend de Bruijn graphs. More precisely, we show how to remove the assumption that the automata contain no ϵ-transitions (namely, edges labeled with the empty string), while retaining the same time and space bounds. This is a significant improvement because ϵ-transitions add considerable expressive power (making it possible to jump to multiple states for free) and capture the complexity of regular expressions (through Thompson's construction for converting a regular expression into an equivalent automaton). We prove that, to enable ϵ-transitions, we only need to store two additional bitvectors that can be constructed in linear time.

Keywords: Pattern matching · Wheeler automata · Burrows-Wheeler transform · FM-index

1 Introduction

Pattern matching is one of the fundamental problems in computer science. In the 1970s, two elegant linear-time solutions were proposed. One solution is based on the Knuth-Morris-Pratt (KMP) algorithm [28]; the other solution is based on suffix trees [37]. Twenty years later, several authors extensively studied the *String Matching in Labeled Graphs (SMLG)* problem, which is the natural generalization of pattern matching to a graph setting [1,5,30,31,33,35]. Loosely speaking, the SMLG problem can be defined as follows: given a directed graph whose nodes (or edges) are labeled with strings and given a pattern string, decide

H. Fernau and B. Zhu (Eds.): IWOCA 2025, LNCS 15885, pp. 242–255, 2025.
https://doi.org/10.1007/978-3-031-98740-3_18

whether the pattern can be read by following a path on the graph and concatenating the labels encountered along the way. Since every text can be seen as a graph consisting of a single path, the SMLG problem is a natural extension of the classical pattern matching problem. Unfortunately, the elegant and general solutions based on the KMP algorithm and suffix trees do not admit a simple generalization to the graph setting. Amir et al. [5] showed that the (exact) SMLG problem on arbitrary graphs can be solved in $O(\mathfrak{e} + me)$ time, where e is the number of edges in the graph, m is the length of the pattern, and $\mathfrak{e}$ is the total length of all labels in the graph. In the (approximate) variant where one allows errors in the graph, the problem becomes NP-hard [5], so generally errors are only allowed in the pattern. Recently, some conditional lower bounds showed that the SMLG problem is inherently more complex than the classical pattern matching problem. Equi et al. [20,21] showed that, given an arbitrary graph, the SMLG cannot be solved in $O(me^{1-\epsilon})$ or $O(m^{1-\epsilon}e)$ time (for every $\epsilon > 0$), unless the Orthogonal Vectors hypothesis is not true.

Due to the sheer increase in the amount of data (and in particular of genomic data), in applications (and especially in bioinformatics) we need algorithms that are both fast and space-efficient. A polynomial algorithm for the SMLG problem running in $O(me)$ time cannot be satisfactory, so the SMLG problem has been restricted to classes of graphs on which it can be solved more efficiently. For example, the SMLG problem can be solved in linear time on *Elastic Founder graphs* under some non-restrictive assumptions [22,36]. Elastic Founder graphs are used to represent multiple sequence alignments (MSA), a central model of biological evolution.

We mentioned that we also strive for space-efficient algorithms. The pattern matching problem *on texts* has been revolutionized by the invention of the Burrows-Wheeler Transform [10] and the FM-index [24], which allow solving pattern matching queries efficiently on *compressed* text, thus establishing a new paradigm in bioinformatics. Over the last decade, these ideas were extended to labeled graphs and automata. On the popular class of *Wheeler automata*, the SMLG problem can be solved in linear time while only storing a compressed representation of the Wheeler automaton [3,25]. Note that a labeled graph and an automaton are essentially the same object: an automaton is simply a labeled graph with an initial state and a set of final states. However, we will generally speak of automata rather than graphs because the results in this area often have relevant applications in formal language theory, and because some (algorithmic) properties of labeled graphs are tied to determinism. For example, the powerset construction applied to a Wheeler non-deterministic automaton leads to a linear blow-up in the number of states of the equivalent deterministic automaton, and the equivalent deterministic automaton is Wheeler [3]; on arbitrary non-deterministic automata, the blow-up can be exponential. A special case of Wheeler automata are de Bruijn graphs [9], which are used to perform Eulerian sequence assembly [7,27,34].

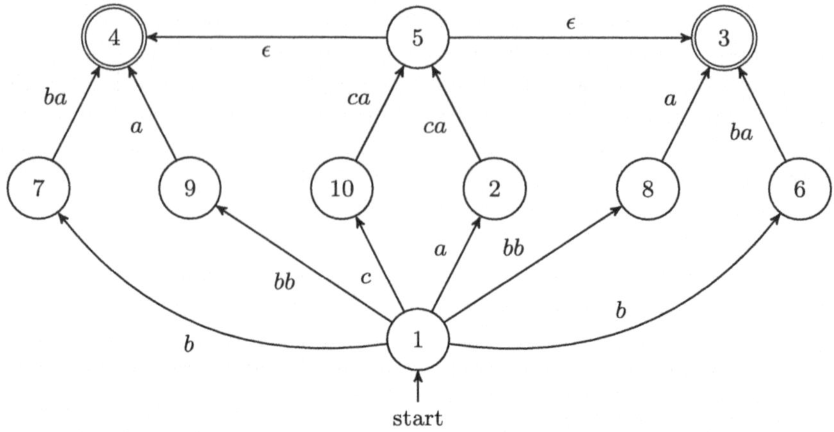

Fig. 1. A Wheeler 2-GNFA. The states are numbered following the Wheeler order.

Our Results. Recently, the notion of Wheelerness was extended to *generalized automata* (see Fig. 1), namely, automata where edges can be labeled not only with characters, but also strings [14]. This is a natural step, because in data compression and bioinformatics many common data structures are based on string-labeled graphs, such as Patricia trees, suffix trees and pangenomes (as well as Elastic Founder graphs) [6,29,32]. We say that a generalized non-deterministic finite automaton (GNFA) is an *r-GNFA* if all edge labels have length at most r (so a GNFA is a conventional NFA if and only if it is an 1-GNFA). Let m, e and $\mathfrak{e}$ as above, and let $\sigma = |\Sigma|$. Consider the following result.

Theorem 1. *Let $\mathcal{A}$ be a Wheeler r-GNFA, and assume that $r = O(1)$. Then, we can encode $\mathcal{A}$ by using $\mathfrak{e} \log \sigma(1+o(1))+O(e)$ bits so that later on, given a pattern $\alpha \in \Sigma^*$ of length m, we can solve the SMLG problem on $\mathcal{A}$ in $O(m \log\log \sigma)$ time. Within the same time bound, we can also decide whether α is accepted by $\mathcal{A}$.*

Theorem 1 states that on Wheeler GNFAs we can solve pattern matching queries in linear time (for constant alphabets), thus overcoming the lower bound of Equi et al. while only storing a compressed representation of $\mathcal{A}$. The case $r = 1$ corresponds to conventional Wheeler automata [25]. This theorem was proved in [14] under the crucial assumption that $\mathcal{A}$ contains no ϵ-transitions (that is, edges labeled with the empty string ϵ): a typical pattern matching algorithm runs in time proportional to the pattern length m, which is not suitable in the context of ϵ-transitions because the empty string has length equal to zero.

In this paper, we show that, surprisingly, we can achieve the time and space bounds in Theorem 1 *even in the presence of ϵ-transitions*, without imposing any restrictions on the automaton $\mathcal{A}$. More precisely, we will show that we only need to store two additional bitvectors. Intuitively, these bitvectors capture the range of states that we can reach by following ϵ-transitions. We also show that these bitvectors can be built in linear time (Theorem 2). In addition to proving the result in [14] in full generality, supporting ϵ-transitions provides some specific advantages.

- ϵ-transitions add considerable expressive power to the set of all strings captured by $\mathcal{A}$ because they enable jumping from one state to another for free (one can potentially follow multiple ϵ-transitions at once for free).
- ϵ-transitions already have a crucial role in the context of conventional automata. Wheeler automata and related automata capture the complexity of regular expression under standard operators such as union and complementation [15], and Thompson's algorithm, the standard construction for converting a regular expression into an automaton, heavily relies on ϵ-transitions [26].

The paper is organized as follows. Section 2 introduces our notation and our first definitions. Section 3 outlines the proof in [14] for GNFAs without ϵ-transitions. In Sect. 4, we prove Theorem 1. In Sect. 5, we prove Theorem 2 (linear-time construction of the two bitvectors). Section 6 presents our conclusions.

Due to space constraints, some proofs can be found in the arXiv version of the paper.

2 Preliminaries

First Definitions. Let V be a set. We say that a (binary) relation $\leq$ on V is a *partial* order if $\leq$ is reflexive, antisymmetric and transitive. For every $u, v \in V$, we write $u < v$ if $(u \leq v) \wedge (u \neq v)$. A partial order $\leq$ is a *total* order if for every $u, v \in V$ we have $(u \leq v) \vee (v \leq u)$. We say that $U \subseteq V$ is $\leq$*-convex* if, for every $u, v, z \in V$, from $u < v < z$ and $u, z \in U$ we obtain $v \in U$.

Let Σ be a finite alphabet. We denote by Σ^* the set of all finite strings on Σ, where $\epsilon \in \Sigma^*$ is the empty string. For $i \geq 0$, let $\Sigma^i \subseteq \Sigma^*$ be the set of all strings of length i on the alphabet Σ. If $\alpha, \beta \in \Sigma^*$, we write $\alpha \dashv \beta$ if and only if α is a suffix of β (equivalently, if and only if there exists $\beta' \in \Sigma^*$ such that $\beta = \beta'\alpha$). If $\alpha \in \Sigma^*$ and $0 \leq k \leq |\alpha|$, let $p(\alpha, k)$ and $s(\alpha, k)$ be the prefix and the suffix of α of length k, respectively. We assume that there exists a fixed total order $\preceq$ on the alphabet Σ (in our examples, we always assume $a \prec b \prec c \prec \dots$), and $\preceq$ is extended *co-lexicographically* to Σ^* (that is, for every $\alpha, \beta \in \Sigma^*$ we have $\alpha \prec \beta$ if and only if the reverse string α^R is lexicographically smaller than the reverse string β^R).

A *generalized non-deterministic finite automaton (GNFA)* is a 4-tuple $\mathcal{A} = (Q, E, s, F)$, where Q is a finite set of states, $E \subseteq Q \times Q \times \Sigma^*$ is a finite set of string-labeled edges, $s \in Q$ is the initial state and $F \subseteq Q$ is a set of final states (see Fig. 1). Moreover, we assume that each $u \in Q$ is reachable from the initial state and is co-reachable, that is, it is either final or allows reaching a final state. This is a standard assumption because all states that do not satisfy this requirement can be removed without changing the language recognized by the automaton. We say that a GNFA $\mathcal{A} = (Q, E, s, F)$ is a *GNFA without ϵ-transitions* if for every $(u, v, \rho) \in E$ we have $\rho \neq \epsilon$. A *generalized deterministic finite automaton (GDFA)* is a GNFA without ϵ-transitions $\mathcal{A} = (Q, E, s, F)$ such that for every pair of *distinct* edges $(u, v, \rho), (u', v', \rho') \in E$, if $u = u'$, then ρ is not a prefix of ρ' and ρ' is a not a prefix of ρ (in particular, $\rho \neq \rho'$).

Let $\mathcal{A} = (Q, E, s, F)$ be a GNFA. For $r \geq 0$, we say that $\mathcal{A}$ is an *r-GNFA* if each edge label has length at most r (namely, $|\rho| \leq r$ for every $(u, v, \rho) \in E$). For $U \subseteq Q$ and $\rho \in \Sigma^*$, let $\mathtt{out}(U, \rho)$ be the number of edges labeled with ρ that leave states in U, and let $\mathtt{in}(U, \rho)$ be the number of edges labeled with ρ that enter states in U. For every $u \in Q$ let $\lambda(u)$ be the set of all strings in Σ^* labeling an edge reaching u. Moreover, for every $u \in Q$, let I_u be the set of all strings that can be read from the initial state to u by concatenating edge labels. In other words, for every $\alpha \in \Sigma^*$ we have $\alpha \in I_u$ if and only, for some $t \geq 1$, there exist $u_1, \ldots, u_t \in Q$ and $\alpha_1, \ldots, \alpha_{t-1} \in \Sigma^*$ such that (i) $u_1 = s$ and $u_t = u$, (ii) $(u_i, u_{i+1}, \alpha_i) \in E$ for every $1 \leq i \leq t-1$, and (iii) $\alpha = \alpha_1 \alpha_2 \ldots \alpha_{t-1}$ (in particular, $\epsilon \in I_s$). As observed in [14], if $\mathcal{A} = (Q, E, s, F)$ is a GDFA and $u, v \in Q$, with $u \neq v$, then $I_u \cap I_v = \emptyset$.

Let $\mathcal{A}$ be a GNFA. The *String Matching in Labeled Graphs (SMLG)* problem for GNFAs is defined as follows: build a data structure that encodes $\mathcal{A}$ such that, given a string α, we can efficiently compute the set of all states reached by a path suffixed by α.

Our main result (Theorem 1) holds for polynomial alphabets (namely, $\sigma \leq e^{O(1)}$) in the word RAM model with words of size $w \in \Theta(\log N)$ bits, where N is the input size.

Wheeler GNFAs. Let $\mathcal{A} = (Q, E, s, F)$ be a GNFA. Let $\preceq_{\mathcal{A}}$ be the relation on Q such that, for every $u, v \in Q$, we have $u \preceq_{\mathcal{A}} v$ if and only if $(\forall \alpha \in I_u)(\forall \beta \in I_v)((\{\alpha, \beta\} \not\subseteq I_u \cap I_v) \to (\alpha \prec \beta))$. As observed in [14], in general $\preceq_{\mathcal{A}}$ is only a *preorder*, that is, it is a reflexive and transitive relation, but it need not be antisymmetric. If $\mathcal{A}$ is a GDFA, we know that $u \neq v$ implies $I_u \cap I_v = \emptyset$, so $\preceq_{\mathcal{A}}$ is a partial order, and it is the reflexive relation such that, for every distinct $u, v \in Q$ we have $u \prec_{\mathcal{A}} v$ if and only if $(\forall \alpha \in I_u)(\forall \beta \in I_v)(\alpha \prec \beta)$.

Let $\mathcal{A} = (Q, E, s, F)$ be a GNFA. We say that $\mathcal{A}$ is *Wheeler* [14] if there exists a total order $\leq$ on Q such that:

- *(Axiom 1)* For every $u, v \in Q$, if $u \leq v$, then $u \preceq_{\mathcal{A}} v$.
- *(Axiom 2)* s comes first in the total order $\leq$.
- *(Axiom 3)* For every $(u', u, \rho), (v', v, \rho') \in E$, if $u < v$ and ρ' is not a strict suffix of ρ, then $\rho \preceq \rho'$.
- *(Axiom 4)* For every $(u', u, \rho), (v', v, \rho) \in E$, if $u < v$, then $u' \leq v'$.

We say that $\leq$ is a *Wheeler order* on $\mathcal{A}$. (see Fig. 1 for an example).

Notice that, if $\mathcal{A}$ is a GNFA without ϵ-transitions, then Axiom 2 follows from Axiom 1 because $\epsilon \in I_s$ and for every $u \in Q \setminus \{s\}$ we have $\epsilon \notin I_u$. Moreover, if $\mathcal{A}$ is a GDFA, then Axioms 2, 3, 4 follow from Axiom 1 (see [14]). Lastly, if $\mathcal{A}$ is a conventional NFA, then Axiom 1 follows from Axiom 2, 3, 4 (see [14]) and we retrieve the usual *local* definition of Wheeler NFA [3, Corollary 3.1]: a (conventional) NFA is Wheeler if there exists a total order $\leq$ on Q such that (i) s comes first in the total order $\leq$, (ii) for every $(u', u, a), (v', v, b) \in E$, if $u < v$, then $a \preceq b$ and (iii) for every $(u', u, a), (v', v, a) \in E$, if $u < v$, then $u' \leq v'$.

If $\mathcal{A} = (Q, E, s, F)$ is a GNFA and $\leq$ is a Wheeler order on $\mathcal{A}$, we write $Q = \{Q[1], Q[2], \dots, Q[|Q|]\}$, where $Q[1] < Q[2] < \dots < Q[|Q|]$. If $1 \leq i \leq j \leq |Q|$, let $Q[i, j] = \{Q[i], Q[i+1], \dots, Q[j-1], Q[j]\}$, and if $i > j$, let $Q[i, j] = \emptyset$.

3 How to Solve Pattern Matching Queries

In [14], Theorem 1 was proved under the assumption that $\mathcal{A}$ is a Wheeler GNFA without ϵ-transitions. This section briefly summarizes the proof in [14] to highlight the technical contributions required to remove the assumption that $\mathcal{A}$ does not have ϵ-transitions. We start with a definition.

Definition 1. *Let $\mathcal{A} = (Q, E, s, F)$ be a Wheeler GNFA, and let $\alpha \in \Sigma^*$. Define:*

- $G^{\prec}(\alpha) = \{u \in Q \mid (\forall \beta \in I_u)(\beta \prec \alpha)\}$;
- $G_{\dashv}(\alpha) = \{u \in Q \mid (\exists \beta \in I_u)(\alpha \dashv \beta)\}$;
- $G^{\prec}_{\dashv}(\alpha) = G^{\prec}(\alpha) \cup G_{\dashv}(\alpha) = \{u \in Q \mid (\forall \beta \in I_u)(\beta \prec \alpha) \vee (\exists \beta \in I_u)(\alpha \dashv \beta)\}$.

The set $G_{\dashv}(\alpha)$ is the set of states that the SMLG problem must return on input α. The following lemma shows, among other properties, that $G_{\dashv}(\alpha) = Q[|G^{\prec}(\alpha)|+1, |G^{\prec}_{\dashv}(\alpha)|]$. This result was proved in [14] for Wheeler GNFAs without ϵ-transitions, but it can be extended readily to arbitrary Wheeler GNFAs.

Lemma 1. *Let $\mathcal{A} = (Q, E, s, F)$ be a Wheeler GNFA, let $\leq$ be a Wheeler order on $\mathcal{A}$, and let $\alpha \in \Sigma^*$. Then:*

1. $G^{\prec}(\alpha) \cap G_{\dashv}(\alpha) = \emptyset$.
2. $G_{\dashv}(\alpha)$ *is $\leq$-convex.*
3. *If $u, v \in Q$ are such that $u < v$ and $v \in G^{\prec}(\alpha)$, then $u \in G^{\prec}(\alpha)$. In other words, $G^{\prec}(\alpha) = Q[1, |G^{\prec}(\alpha)|]$.*
4. *If $u, v \in Q$ are such that $u < v$ and $v \in G^{\prec}_{\dashv}(\alpha)$, then $u \in G^{\prec}_{\dashv}(\alpha)$. In other words, $G^{\prec}_{\dashv}(\alpha) = Q[1, |G^{\prec}_{\dashv}(\alpha)|]$.*
5. $G_{\dashv}(\alpha) = Q[|G^{\prec}(\alpha)| + 1, |G^{\prec}_{\dashv}(\alpha)|]$.

Since $G_{\dashv}(\alpha) = Q[|G^{\prec}(\alpha)| + 1, |G^{\prec}_{\dashv}(\alpha)|]$, to compute $G_{\dashv}(\alpha)$ we only need to compute $|G^{\prec}(\alpha)|$ and $|G^{\prec}_{\dashv}(\alpha)|$. To this end, we recursively compute $|G^{\prec}(p(\alpha, k))|$ and $|G^{\prec}_{\dashv}(p(\alpha, k))|$ for every $0 \leq k \leq |\alpha|$, and the conclusion follows for $k = |\alpha|$. Note that, for $k = 0$ we have $p(\alpha, k) = \epsilon$, and we have $|G^{\prec}(\epsilon)| = 0$ and $|G^{\prec}_{\dashv}(\epsilon)| = |Q|$ because ϵ is the smallest string in Σ^* and ϵ is a suffix of every string in Σ^*.

We are only left with showing how to compute $|G^{\prec}(\alpha)|$ and $|G^{\prec}_{\dashv}(\alpha)|$, assuming that we have already computed $|G^{\prec}(p(\alpha, k))|$ and $|G^{\prec}_{\dashv}(p(\alpha, k))|$ for every $0 \leq k < |\alpha|$.

Let us describe the approach in [14] for Wheeler GNFAs without ϵ-transitions. First, let us show how to compute $|G^{\prec}(\alpha)|$. Fix any $Q[j] \in Q$. We know that $Q[j] \in G^{\prec}(\alpha)$ if and only if for every $\beta \in I_{Q[j]}$ we have $\beta \prec \alpha$. As a consequence, if $Q[j] \in G^{\prec}(\alpha)$, then for every $\rho \in \lambda(Q[j])$ we must have $\rho \prec \alpha$. Moreover, if ρ is a suffix of α—say $\rho = s(\alpha, k)$—then for every $u' \in Q$ such that $(u', Q[j], \rho) \in E$ we must have $u' \in G^{\prec}(p(\alpha, |\alpha| - k))$. Crucially, we have $k \geq 1$ (and not $k \geq 0$) because we are considering a GNFA without ϵ-transitions, so we have already computed $|G^{\prec}(p(\alpha, |\alpha| - k))|$. Axiom 4 ensures that all edges labeled $s(\alpha, k)$ "respect" the Wheeler order $\leq$ so, if $f_k = \mathtt{out}(Q[1, |G^{\prec}(p(\alpha, |\alpha| - k))|], s(\alpha, k))$, we have $u' \in G^{\prec}(p(\alpha, |\alpha| - k))$ for every $u' \in Q$ such that $(u', Q[j], s(\alpha, k)) \in E$ if and only if $\mathtt{in}(Q[1, j], s(\alpha, k)) \leq f_k$. Intuitively, we obtain the following lemma (see [14]).

Lemma 2. *Let $\mathcal{A} = (Q, E, s, F)$ be a Wheeler r-GNFA without ϵ-transitions, let $\leq$ be a Wheeler order on $\mathcal{A}$, and let $\alpha \in \Sigma^*$, with $\alpha \neq \epsilon$. For $0 < k < \min\{r+1, |\alpha|\}$, let $f_k = \mathtt{out}(Q[1, |G^{\prec}(p(\alpha, |\alpha| - k))|], s(\alpha, k))$. Then, $|G^{\prec}(\alpha)|$ is the largest integer $0 \leq j \leq |Q|$ such that:*

- $\mathtt{in}(Q[1, j], s(\alpha, k)) \leq f_k$, *for every* $0 < k < \min\{r + 1, |\alpha|\}$;
- $\rho \prec \alpha$ *for every* $\rho \in \Sigma^k \cap \lambda(Q[h])$, *for every* $1 \leq h \leq j$ *and for every* $0 < k < r + 1$.

Now, let us show how to compute $|G^{\preceq}_{\dashv}(\alpha)|$. Let $G^*(\alpha)$ be the set of all states reached by an edge labeled with a string suffixed by α. Formally:

$$G^*(\alpha) = \{u \in Q \mid (\exists \rho \in \lambda(u))(\alpha \dashv \rho)\}.$$

Notice that $G^*(\alpha) \subseteq G_{\dashv}(\alpha)$. Indeed, pick $u \in G^*(\alpha)$. Then, there exists $\rho \in \lambda(u)$ such that $\alpha \dashv \rho$. In particular, there exists $u' \in Q$ such that $(u', u, \rho) \in E$. Pick any $\beta \in I_{u'}$. Then, $\beta\rho \in I_u$. From $\alpha \dashv \rho$ we conclude $\alpha \dashv \beta\rho$, so $u \in G_{\dashv}(\alpha)$.

Consider any $Q[j] \in Q$. Then, we have $Q[j] \in G^{\preceq}_{\dashv}(\alpha)$ if and only if $Q[j] \in G^{\prec}(\alpha)$ or $Q[j] \in G_{\dashv}(\alpha)$. Now, we have $Q[j] \in G^{\prec}(\alpha)$ if and only if $j \leq |G^{\prec}(\alpha)|$, and we have already computed $|G^{\prec}(\alpha)|$. We consider a GNFA without ϵ-transitions, so we have $Q[j] \in G_{\dashv}(\alpha)$ if and only if either $Q[j] \in G^*(\alpha)$, or there exists $1 \leq k \leq r$ and $u' \in G_{\dashv}(p(\alpha, |\alpha| - k))$ such that $(u', Q[j], s(\alpha, k)) \in E$. Axiom 4 ensures that we can compute $|G^{\preceq}_{\dashv}(\alpha)|$ as follows (see [14]).

Lemma 3. *Let $\mathcal{A} = (Q, E, s, F)$ be a Wheeler r-GNFA without ϵ-transitions, let $\leq$ be a Wheeler order on $\mathcal{A}$, and let $\alpha \in \Sigma^*$, with $\alpha \neq \epsilon$. For $0 < k < \min\{r + 1, |\alpha|\}$, let $f_k = \mathtt{out}(Q[1, |G^{\prec}(p(\alpha, |\alpha| - k))|], s(\alpha, k))$ and let $g_k = \mathtt{out}(Q[1, |G^{\preceq}_{\dashv}(p(\alpha, |\alpha| - k))|], s(\alpha, k))$. Then, $g_k \geq f_k$ for every $0 < k < \min\{r + 1, |\alpha|\}$. Moreover, $|G^{\preceq}_{\dashv}(\alpha)|$ is the maximum among:*

- $|G^{\prec}(\alpha)|$;
- *the largest integer* $0 \leq i \leq |Q|$ *such that, if* $i \geq 1$, *then* $Q[i] \in G^*(\alpha)$;
- *the smallest integer* $0 \leq j \leq |Q|$ *such that, for every* $0 < k < \min\{r+1, |\alpha|\}$ *for which* $g_k > f_k$, *we have* $\mathtt{in}(Q[1, j], s(\alpha, k)) \geq g_k$.

For $r = O(1)$ there exists a compressed data structure of $\mathfrak{e} \log \sigma(1 + o(1)) + O(e)$ bits such that we can determine each condition in Lemma 2 and Lemma 3 in $O(\log \log \sigma)$ time (for example, for a given $0 < k < \min\{r + 1, |\alpha|\}$, we can compute the largest integer $0 \leq j \leq |Q|$ such that $\mathtt{in}(Q[1, j], s(\alpha, k)) \leq f_k$ in $O(\log \log \sigma)$ time) [14]. We conclude that, assuming that we have already computed $|G^{\prec}(p(\alpha, k))|$ and $|G^{\prec}_{\dashv}(p(\alpha, k))|$ for every $0 \leq k < |\alpha|$, we can compute $|G^{\prec}(\alpha)|$ and $|G^{\prec}_{\dashv}(\alpha)|$ in $O(\log \log \sigma)$ time. This yields the time bound $O(m \log \log \sigma)$ in Theorem 1 (where $m = |\alpha|$) for Wheeler GNFAs without ϵ-transitions.

4 Pattern Matching with ϵ-Transitions

Let us show how to extend the results of Sect. 3 to arbitrary Wheeler GNFAs (possibly with ϵ-transitions). First, let us show that Lemma 2 is not true if ϵ-transitions are allowed. In Fig. 1, for $\alpha = cba$, the largest integer $0 \leq j \leq |Q|$ such that the two conditions of Lemma 2 are true is 4 (note that $|G^{\prec}(cb)| = 9$, $f_1 = 3$, $|G^{\prec}(c)| = 9$, $f_2 = 2$), but $|G^{\prec}(\alpha)| = 2 \neq 4$. This is due to the ϵ-transitions $(5, 3, \epsilon)$ and $(5, 4, \epsilon)$ and intuitively, to obtain a correct variant of Lemma 2, we need to remember the range of states that we can reach by following ϵ-transitions in a backward fashion.

Let $\mathcal{A} = (Q, E, s, F)$ be a GNFA. Given $u, v \in Q$, we say that there exists an *ϵ-walk* from u to v if, for some $t \geq 1$ there exist $z_1, z_2, \ldots, z_t \in Q$ such that (i) $z_1 = u$ and $z_t = v$ and (ii) $(z_i, z_{i+1}, \epsilon) \in E$ for every $1 \leq i \leq t - 1$ (in particular, for every $u \in Q$ there exists an ϵ-walk from u to u).

We can now define $A_{\max}$ and $A_{\min}$, the two crucial arrays to obtain our result.

Definition 2. *Let $\mathcal{A} = (Q, E, s, F)$ be a Wheeler GNFA, and let $\leq$ be a Wheeler order on $\mathcal{A}$. Let $A_{\max}[1, |Q|]$ ($A_{\min}[1, |Q|]$, respectively) be the array such that, for every $1 \leq i \leq |Q|$, we have that $A_{max}[i]$ ($A_{\min}[i]$, respectively) is the largest (smallest, respectively) integer $1 \leq j \leq |Q|$ for which there exists an ϵ-walk from $Q[j]$ to $Q[i]$.*

Note that for every $1 \leq i \leq |Q|$ we have $A_{\min}[i] \leq i \leq A_{\max}[i]$ because there exists an ϵ-walk from $Q[i]$ to $Q[i]$. In Fig. 1 we have $A_{\max}[3] = A_{\max}[4] = 5$ and $A_{\max}[i] = i$ for $i \notin \{3, 4\}$.

We will use $A_{\max}$ and $A_{\min}$ to prove Theorem 1, but we will not explicitly store $A_{\max}$ and $A_{\min}$ (which would require $|Q| \log |Q|$ bits per array), because otherwise we could not achieve the space bound of $\mathfrak{e} \log \sigma(1 + o(1)) + O(e)$ bits that holds for Wheeler GNFAs without ϵ-transitions.

Consider again Fig. 1. We have observed that, for $\alpha = cba$, the largest integer $0 \leq j \leq |Q|$ such that the two conditions of Lemma 2 are true is 4. Intuitively, since $A_{\max}[4] > 4$, then we must have $|G^{\prec}(\alpha)| \leq 3$. We also know that $A_{\max}[3] > 3$, so we must have $|G^{\prec}(\alpha)| \leq 2$. Since $A_{\max}[2] = 2$, then we can conclude $|G^{\prec}(\alpha)| = 2$. We can now formally prove an extension of Lemma 2 to arbitrary Wheeler GNFAs.

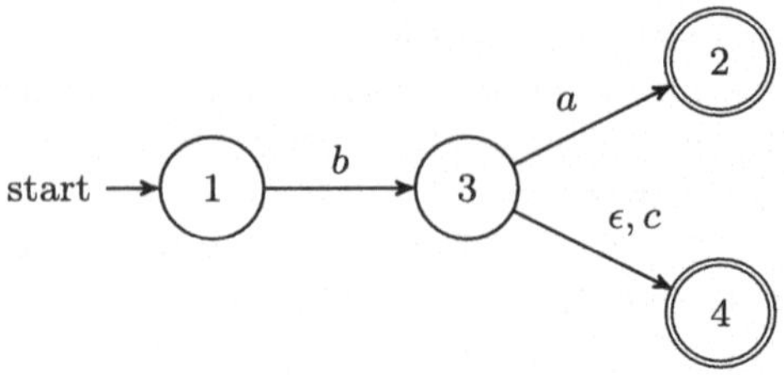

Fig. 2. A Wheeler 1-GNFA showing that in Lemma 5 we need to distinguish between the case $h^* = |G^{\prec}(\alpha)|$ and the case $h^* > |G^{\prec}(\alpha)|$. The states are numbered following the Wheeler order.

Lemma 4. *Let $\mathcal{A} = (Q, E, s, F)$ be a Wheeler r-GNFA, let $\leq$ be a Wheeler order on $\mathcal{A}$, and let $\alpha \in \Sigma^*$, with $\alpha \neq \epsilon$. For $0 < k < \min\{r+1, |\alpha|\}$, let $f_k = \mathtt{out}(Q[1, |G^{\prec}(p(\alpha, |\alpha| - k))|], s(\alpha, k))$. Let j^* be the largest integer $0 \leq j \leq |Q|$ such that:*

- *$\mathtt{in}(Q[1, j], s(\alpha, k)) \leq f_k$, for every $0 < k < \min\{r+1, |\alpha|\}$;*
- *$\rho \prec \alpha$ for every $\rho \in \Sigma^k \cap \lambda(Q[h])$, for every $1 \leq h \leq j$ and for every $0 < k < r+1$.*

Then, $|G^{\prec}(\alpha)|$ is the largest integer $0 \leq t \leq j^*$ such that, if $t \geq 1$, then $A_{\max}[t] = t$.

Similarly, we will now prove an extension of Lemma 3 to arbitrary Wheeler GNFAs. We will now use the array $A_{\min}$ (and not $A_{\max}$). Intuitively, Lemma 3 could return a value of $|G^{\preceq}_{\dashv}(\alpha)|$ *smaller* (and not larger) than the correct value, so we will use $A_{\min}$ to determine $|G^{\preceq}_{\dashv}(\alpha)|$. However, we must be careful. In Lemma 5, we will introduce the integer h^* and we will distinguish between the case $h^* = |G^{\prec}(\alpha)|$ and the case $h^* > |G^{\prec}(\alpha)|$. If $h^* = |G^{\prec}(\alpha)|$, we can prove that $G_{\dashv}(\alpha) = \emptyset$ and so we must have $|G^{\preceq}_{\dashv}(\alpha)| = h^* = |G^{\prec}(\alpha)|$. If $h^* > |G^{\prec}(\alpha)|$, then we can proceed similarly to Lemma 4. Notice that we cannot collapse the two cases into a single case by always returning the smallest integer $h^* \leq t \leq |Q|$ such that, if $t < |Q|$, then $A_{\min}[t+1] = t+1$. In Fig. 2, for $\alpha = bb$, we have $h^* = |G^{\prec}(\alpha)| = 3 = |G^{\preceq}_{\dashv}(\alpha)|$ and the smallest integer $h^* \leq t \leq |Q|$ such that, if $t < |Q|$, then $A_{\min}[t+1] = t+1$ is equal to 4, not 3.

Lemma 5. *Let $\mathcal{A} = (Q, E, s, F)$ be a Wheeler r-GNFA, let $\leq$ be a Wheeler order on $\mathcal{A}$, and let $\alpha \in \Sigma^*$, with $\alpha \neq \epsilon$. For $0 < k < \min\{r+1, |\alpha|\}$, let $f_k = \mathtt{out}(Q[1, |G^{\prec}(p(\alpha, |\alpha| - k))|], s(\alpha, k))$ and let $g_k = \mathtt{out}(Q[1, |G^{\preceq}_{\dashv}(p(\alpha, |\alpha| - k))|], s(\alpha, k))$. Then, $g_k \geq f_k$ for every $0 < k < \min\{r+1, |\alpha|\}$. Let h^* be the maximum among:*

- *$|G^{\prec}(\alpha)|$;*
- *the largest integer $0 \leq i \leq |Q|$ such that, if $i \geq 1$, then $Q[i] \in G^*(\alpha)$;*
- *the smallest integer $0 \leq j \leq |Q|$ such that, for every $0 < k < \min\{r+1, |\alpha|\}$ for which $g_k > f_k$, we have $\mathtt{in}(Q[1, j], s(\alpha, k)) \geq g_k$.*

Then, $h^* \geq |G^{\prec}(\alpha)|$. *Moreover:*

- *If* $h^* = |G^{\prec}(\alpha)|$, *then* $|G^{\preceq}_{\dashv}(\alpha)| = h^* = |G^{\prec}(\alpha)|$.
- *If* $h^* > |G^{\prec}(\alpha)|$, *then* $|G^{\preceq}_{\dashv}(\alpha)| \geq h^* > |G^{\prec}(\alpha)|$, *and* $|G^{\preceq}_{\dashv}(\alpha)|$ *is the smallest integer* $h^* \leq t \leq |Q|$ *such that, if* $t < |Q|$, *then* $A_{\min}[t+1] = t+1$.

Given a bit array $B[1,n]$, we can define the following queries: (i) $\mathsf{access}(B,i)$: given $1 \leq i \leq n$, return $B[i]$; (ii) $\mathsf{rank}_1(B,i)$: given $1 \leq i \leq n$, return $|\{1 \leq h \leq i \mid B[h] = 1\}|$; (iii) $\mathsf{select}_1(B,i)$: given $1 \leq i \leq \mathsf{rank}_1(B,n)$, return the unique $1 \leq j \leq n$ such that $B[j] = 1$ and $|\{1 \leq h \leq j \mid B[h] = 1\}| = i$. A bit array B of length n can be stored using a data structure (called *bitvector*) of $n + o(n)$ bits that supports access, rank and select queries in $O(1)$ time and can be built from B in $O(n)$ time [32].

We are now ready to prove Theorem 1 for arbitrary Wheeler GNFAs. We store the same data structure of $\mathfrak{e} \log \sigma(1 + o(1)) + O(e)$ bits used in [14] for Wheeler GNFAs without ϵ-transitions. Moreover, let $B_{\max}[1,|Q|]$ ($B_{\min}[1,|Q|]$, respectively) the bit array such, for every $1 \leq i \leq |Q|$, we have $B_{\max}[i] = 1$ ($B_{\min}[i] = 1$, respectively) if and only if $A_{\max}[i] = i$ ($A_{\min}[i] = i$, respectively). We can store $B_{\max}$ and $B_{\min}$ as bitvectors using $2|Q| + o(|Q|)$ bits, which is absorbed by the $O(e)$ term (we have $e \geq |Q| - 1$ because every state is reachable from the initial state). To compute $G_{\dashv}(\alpha)$, we proceed as explained in Sect. 3, but we use Lemma 4 and Lemma 5 instead of Lemma 2 and Lemma 3.

- To compute $|G^{\prec}(\alpha)|$, we first compute the value j^* of Lemma 4 by proceeding as in Lemma 2. By Lemma 4, if $\mathsf{rank}_1(B_{\max}, j^*) = 0$, we conclude $|G^{\prec}(\alpha)| = 0$; otherwise, we conclude $|G^{\prec}(\alpha)| = \mathsf{select}_1(B_{\max}, \mathsf{rank}_1(B_{\max}, j^*))$.
- To compute $|G^{\preceq}_{\dashv}(\alpha)|$, we first compute the value h^* of Lemma 5 by proceeding as in Lemma 3. By Lemma 5, if $h^* = |G^{\prec}(\alpha)|$, then we conclude $|G^{\preceq}_{\dashv}(\alpha)| = h^*$. Now assume that $h^* > |G^{\prec}(\alpha)|$. If $\mathsf{rank}_1(B_{\min}, |Q|) = \mathsf{rank}_1(B_{\min}, h^*)$, then $|G^{\preceq}_{\dashv}(\alpha)| = |Q|$; otherwise, we conclude $|G^{\preceq}_{\dashv}(\alpha)| = \mathsf{select}_1(B_{\min}, \mathsf{rank}_1(B_{\min}, h^*) + 1) - 1$.

5 Construction

In Sect. 4, we showed that by storing a bitvector representation of $B_{\max}$ and $B_{\min}$ we can prove Theorem 1 for an arbitrary Wheeler GNFA $\mathcal{A} = (Q, E, s, F)$. Let us show that we can build $B_{\max}$ and $B_{\min}$ in $O(e)$ time, where $e = |E|$ (recall that $e \geq |Q| - 1$). We focus on $B_{\max}$ because the proof for $B_{\min}$ is analogous. For every $1 \leq i \leq |Q|$, we have $B_{\max}[i] = 1$ if and only if $A_{\max}[i] = i$, so we only have to show how to build $A_{\max}$ in $O(e)$ time.

Let $\mathcal{A} = (Q, E, s, F)$ be a Wheeler GNFA, with Wheeler order $\leq$. We have the following properties.

- *(Property 1)* For every $t \geq 2$, there exist no pairwise distinct $u_1, u_2, \ldots, u_t \in Q$ such that $(u_i, u_{i+1}, \epsilon) \in E$ for every $1 \leq i \leq t-1$ and (u_t, u_1, ϵ). Indeed, suppose for the sake of a contradiction that such $u_1, u_2, \ldots, u_t \in Q$ exist. In particular, $u_1 \neq u_2$. Suppose that $u_1 < u_2$ (the case $u_2 < u_1$ leads to a contradiction analogously). Since $(u_t, u_1, \epsilon), (u_1, u_2, \epsilon) \in E$, from Axiom 4 we obtain $u_t \leq u_1$, and we have $u_t < u_1$ because the u_i's are pairwise distinct. Since $(u_{t-1}, u_t, \epsilon), (u_t, u_1, \epsilon) \in E$, we analogously obtain $u_{t-1} < u_t$. By iterating the same argument, we obtain $u_{t-2} < u_{t-1}$, ..., $u_2 < u_3$. We conclude $u_2 < u_3 < \cdots < u_{t-2} < u_{t-1} < u_t < u_1 < u_2$ and so $u_2 < u_2$, a contradiction.

- *(Property 2)* Let us prove that for every $1 \leq i \leq |Q|$, we have:

$$A_{\max}[i] = \max\{i, \max\{A_{\max}[j] \mid 1 \leq j \leq |Q|, j \neq i, (Q[j], Q[i], \epsilon) \in E\}\}.$$

($\geq$) First, we have $A_{\max}[i] \geq i$ because there exists an ϵ-walk from $Q[i]$ to $Q[i]$. Next, for every $1 \leq j \leq |Q|$ such that $j \neq i$ and $(Q[j], Q[i], \epsilon) \in E$, we have $A_{\max}[i] \geq A_{\max}[j]$ because there exists an ϵ-walk from $Q[A_{\max}[j]]$ to $Q[j]$, so there exists an ϵ-walk from $Q[A_{\max}[j]]$ to $Q[i]$.
($\leq$) If $A_{\max}[i] = i$, we are done. Now assume that $A_{\max}[i] \neq i$. We know that there exists an ϵ-walk from $Q[A_{\max}[i]]$ to $Q[i]$, so there exists $1 \leq j \leq Q$ such that $j \neq i$, $(Q[j], Q[i], \epsilon) \in E$ and there exists an ϵ-walk from $Q[A_{\max}[i]]$ to $Q[j]$, which implies $A_{\max}[i] \leq A_{\max}[j]$.

We are now ready to prove the main result of this section.

Theorem 2. *Let $\mathcal{A} = (Q, E, s, F)$ be a Wheeler GNFA. Then, in $O(e)$ time we can build $A_{\max}$ and $A_{\min}$.*

Proof. We present only the proof for $A_{\max}$ because the proof for $A_{\min}$ is analogous. Consider an array $A[1, |Q|]$. At the beginning of the algorithm, we have $A[i] = i$ for every $1 \leq i \leq |Q|$, and at the end of the algorithm we will have $A[i] = A_{\max}[i]$ for every $1 \leq i \leq |Q|$. We perform a depth-first search on the transpose of $\mathcal{A}$ (that is, we follow the edges in a backward fashion). In other words, when we process a node u, we consider all nodes v such that for some $\rho \in \Sigma^*$ we have $(v, u, \rho) \in E$. At any time, every node is white, gray or black. A white node can only become gray and a gray node can only become black. Every node is white at the beginning of the algorithm, and every node is black at the end of the algorithm.

The algorithm considers all nodes $Q[i]$. If $Q[i]$ is white, we paint $Q[i]$ gray and we consider all j's such that $j \neq i$ and $(Q[j], Q[i], \epsilon) \in E$. For every such node $Q[j]$, if $Q[j]$ is black and $A[j] > A[i]$, we set $A[i] \leftarrow A[j]$, and if $Q[j]$ is white, we recursively process $Q[j]$ and when the recursion is complete, if $A[j] > A[i]$, we set $A[i] \leftarrow A[j]$. After processing every j, we paint $Q[i]$ black and we process the next $Q[i]$.

The algorithm runs in $O(e)$ time because every edge is processed once. Note that after a node $Q[i]$ becomes black, the value $A[i]$ is no longer updated. Moreover, when we consider all j's such that $j \neq i$ and $(Q[j], Q[i], \epsilon) \in E$, no such j can be gray by Property 1. As a consequence, by Property 2 we conclude that at the end of the algorithm we have $A[i] = A_{\max}[i]$ for every $1 \leq i \leq |Q|$, which can be seen by considering all i's in the order in which they become black. □

6 Conclusions and Future Work

In this paper, we have shown how to achieve linear-time pattern matching on arbitrary Wheeler GNFAs within compressed space (Theorem 1). To this end, we have introduced two arrays $A_{\max}$ and $A_{\min}$ that can be built in linear time (Theorem 2). Our result subsumes all previous approaches based on the Burrows-Wheeler transform and the FM-index, including labeled trees [23], de Bruijn graphs [9] and conventional Wheeler graphs [3,25], which have led to a fruitful line of research [2,4,8,11,13,16,18,19].

There are several potential directions for future research. The ideas behind Wheeler NFAs have been extended to arbitrary NFAs [12,17], so the next natural step is to extend the same paradigm to arbitrary GNFAs (however, on arbitrary NFAs and GNFAs we cannot achieve linear-time pattern matching because of Equi et al.'s lower bound mentioned in the introduction). It is still an open problem to characterize the class of all regular languages recognized by some Wheeler GNFA, even though it is known that this class is strictly larger than the class of all regular languages recognized by some Wheeler NFA [14]. Moreover, in this paper we focused on ϵ-transitions, so it should be possible to extend the results in [15] on the complexity of regular expression by leveraging Thompson's construction for converting a regular expression into an equivalent automaton [26], which heavily relies on ϵ-transitions.

Acknowledgements. Funded by the Helsinki Institute for Information Technology (HIIT).

References

1. Akutsu, T.: A linear time pattern matching algorithm between a string and a tree. In: Apostolico, A., Crochemore, M., Galil, Z., Manber, U. (eds.) CPM 1993. LNCS, vol. 684, pp. 1–10. Springer, Heidelberg (1993). https://doi.org/10.1007/BFb0029792
2. Alanko, J., Cotumaccio, N., Prezza, N.: Linear-time minimization of Wheeler DFAs. In: 2022 Data Compression Conference (DCC), pp. 53–62. IEEE (2022)
3. Alanko, J., D'Agostino, G., Policriti, A., Prezza, N.: Regular languages meet prefix sorting. In: Proceedings of the Fourteenth Annual ACM-SIAM Symposium on Discrete Algorithms, pp. 911–930. SIAM (2020)
4. Alanko, J.N., Cenzato, D., Cotumaccio, N., Kim, S.H., Manzini, G., Prezza, N.: Computing the LCP array of a labeled graph. In: 35th Annual Symposium on Combinatorial Pattern Matching (2024)
5. Amir, A., Lewenstein, M., Lewenstein, N.: Pattern matching in hypertext. J. Algorithms **35**(1), 82–99 (2000). https://doi.org/10.1006/jagm.1999.1063
6. Baaijens, J.A., et al.: Computational graph pangenomics: a tutorial on data structures and their applications. Nat. Comput. **21**(1), 81–108 (2022). https://doi.org/10.1007/s11047-022-09882-6
7. Bankevich, A., et al.: SPAdes: a new genome assembly algorithm and its applications to single-cell sequencing. J. Comput. Biol. **19**(5), 455–477 (2012). https://doi.org/10.1089/cmb.2012.0021. pMID: 22506599

8. Becker, R., Cotumaccio, N., Kim, S.-H., Prezza, N., Tosoni, C.: Encoding co-lex orders of finite-state automata in linear space. In: 36th Annual Symposium on Combinatorial Pattern Matching, p. 17 (2025)
9. Bowe, A., Onodera, T., Sadakane, K., Shibuya, T.: Succinct de Bruijn graphs. In: Raphael, B., Tang, J. (eds.) WABI 2012. LNCS, pp. 225–235. Springer, Heidelberg (2012). https://doi.org/10.1007/978-3-642-33122-0_18
10. Burrows, M., Wheeler, D.J.: A block-sorting lossless data compression algorithm. Technical report (1994)
11. Conte, A., Cotumaccio, N., Gagie, T., Manzini, G., Prezza, N., Sciortino, M.: Computing matching statistics on Wheeler DFAs. In: 2023 Data Compression Conference (DCC), pp. 150–159. IEEE (2023)
12. Cotumaccio, N.: Graphs can be succinctly indexed for pattern matching in $O(|E|^2 + |V|^{5/2})$ time. In: 2022 Data Compression Conference (DCC), pp. 272–281 (2022). https://doi.org/10.1109/DCC52660.2022.00035
13. Cotumaccio, N.: Prefix sorting DFAs: a recursive algorithm. In: 34th International Symposium on Algorithms and Computation (ISAAC 2023), pp. 22–1. Schloss Dagstuhl–Leibniz-Zentrum für Informatik (2023)
14. Cotumaccio, N.: A Myhill-Nerode theorem for generalized automata, with applications to pattern matching and compression. In: 41st International Symposium on Theoretical Aspects of Computer Science (STACS 2024). Schloss Dagstuhl–Leibniz-Zentrum für Informatik (2024)
15. Cotumaccio, N., D'Agostino, G., Policriti, A., Prezza, N.: Co-lexicographically ordering automata and regular languages - part I. J. ACM **70**(4) (2023). https://doi.org/10.1145/3607471
16. Cotumaccio, N., Gagie, T., Köppl, D., Prezza, N.: Space-time trade-offs for the LCP array of Wheeler DFAs. In: International Symposium on String Processing and Information Retrieval, pp. 143–156. Springer (2023)
17. Cotumaccio, N., Prezza, N.: On indexing and compressing finite automata. In: Proceedings of the Thirty-Second Annual ACM-SIAM Symposium on Discrete Algorithms, SODA 2021, pp. 2585–2599. Society for Industrial and Applied Mathematics, USA (2021)
18. Cotumaccio, N., Trubiani, C.: Convex petri nets. In: Köhler-Bussmeier, M., Moldt, D., Rölke, H. (eds.) Proceedings of the International Workshop on Petri Nets and Software Engineering 2024 Co-located with the 45th International Conference on Application and Theory of Petri Nets and Concurrency (PETRI NETS 2024), 24–25 June 2024, Geneva, Switzerland. CEUR Workshop Proceedings, vol. 3730. CEUR-WS.org (2024)
19. Cotumaccio, N.: Improved circular dictionary matching. In: 36th Annual Symposium on Combinatorial Pattern Matching (CPM 2025), pp. 1–18. Schloss Dagstuhl–Leibniz-Zentrum für Informatik (2025)
20. Equi, M., Grossi, R., Mäkinen, V., Tomescu, A.I.: On the complexity of string matching for graphs. In: Baier, C., Chatzigiannakis, I., Flocchini, P., Leonardi, S. (eds.) 46th International Colloquium on Automata, Languages, and Programming, ICALP 2019, 9–12 July 2019, Patras, Greece. LIPIcs, vol. 132, pp. 55:1–55:15. Schloss Dagstuhl - Leibniz-Zentrum für Informatik (2019). https://doi.org/10.4230/LIPIcs.ICALP.2019.55
21. Equi, M., Mäkinen, V., Tomescu, A.I.: Graphs cannot be indexed in polynomial time for sub-quadratic time string matching, unless SETH fails. In: Bureš, T., et al. (eds.) SOFSEM 2021. LNCS, vol. 12607, pp. 608–622. Springer, Cham (2021). https://doi.org/10.1007/978-3-030-67731-2_44

22. Equi, M., Norri, T., Alanko, J., Cazaux, B., Tomescu, A.I., Mäkinen, V.: Algorithms and complexity on indexing founder graphs. Algorithmica **85**(6), 1586–1623 (2023). https://doi.org/10.1007/s00453-022-01007-w
23. Ferragina, P., Luccio, F., Manzini, G., Muthukrishnan, S.: Compressing and indexing labeled trees, with applications. J. ACM **57**(1) (2009). https://doi.org/10.1145/1613676.1613680
24. Ferragina, P., Manzini, G.: Indexing compressed text. J. ACM **52**(4), 552–581 (2005). https://doi.org/10.1145/1082036.1082039
25. Gagie, T., Manzini, G., Sirén, J.: Wheeler graphs: A framework for BWT-based data structures. Theor. Comput. Sci. **698**, 67–78 (2017). https://doi.org/10.1016/j.tcs.2017.06.016. Algorithms, Strings and Theoretical Approaches in the Big Data Era (In Honor of the 60th Birthday of Professor Raffaele Giancarlo)
26. Hopcroft, J.E., Motwani, R., Ullman, J.D.: Introduction to Automata Theory, Languages, and Computation, 3rd edn., Pearson International Edition. Addison-Wesley (2007)
27. Idury, R.M., Waterman, M.S.: A new algorithm for DNA sequence assembly. J. comput. biol.: J. Comput. Mol. Cell Biol. **2**(2), 291–306 (1995)
28. Knuth, D.E., Morris, J.H., Jr., Pratt, V.R.: Fast pattern matching in strings. SIAM J. Comput. **6**(2), 323–350 (1977). https://doi.org/10.1137/0206024
29. Mäkinen, V., Belazzougui, D., Cunial, F., Tomescu, A.I.: Genome-Scale Algorithm Design: Bioinformatics in the Era of High-Throughput Sequencing. Cambridge University Press (2023)
30. Manber, U., Wu, S.: Approximate string matching with arbitrary costs for text and hypertext. In: Advances In Structural And Syntactic Pattern Recognition, pp. 22–33. World Scientific (1992)
31. Navarro, G.: Improved approximate pattern matching on hypertext. Theor. Comput. Sci. **237**(1–2), 455–463 (2000). https://doi.org/10.1016/S0304-3975(99)00333-3
32. Navarro, G.: Compact Data Structures: A Practical Approach. Cambridge University Press (2016)
33. Park, K., Kim, D.K.: String matching in hypertext. In: Galil, Z., Ukkonen, E. (eds.) CPM 1995. LNCS, vol. 937, pp. 318–329. Springer, Heidelberg (1995). https://doi.org/10.1007/3-540-60044-2_51
34. Pevzner, P.A., Tang, H., Waterman, M.S.: An Eulerian path approach to DNA fragment assembly. Proc. Natl. Acad. Sci. **98**(17), 9748–9753 (2001). https://doi.org/10.1073/pnas.171285098
35. Rautiainen, M., Marschall, T.: Aligning sequences to general graphs in $O(V + mE)$ time. bioRxiv (2017). https://doi.org/10.1101/216127
36. Rizzo, N., Mäkinen, V.: Linear time construction of indexable Elastic Founder graphs. In: Bazgan, C., Fernau, H. (eds.) Combinatorial Algorithms, pp. 480–493. Springer, Cham (2022)
37. Weiner, P.: Linear pattern matching algorithms. In: 14th Annual Symposium on Switching and Automata Theory (SWAT 1973), pp. 1–11. IEEE (1973)

Reconstructing Sets of Strings from Their k-Way Projections: Algorithms & Complexity

(Extended Abstract)

Elise Tate[1] and Joshua A. Grochow[1,2](✉)

[1] Department of Computer Science, University of Colorado Boulder, Boulder, USA
{jordan.tate,jgrochow}@colorado.edu
[2] Department of Mathematics, University of Colorado Boulder, Boulder, USA

Abstract. Graphs are a powerful tool for analyzing large data sets, but many real-world phenomena involve interactions that go beyond the simple pairwise relationships captured by a graph. In this paper we introduce and study a simple combinatorial model to capture higher order dependencies from an algorithms and computational complexity perspective. Specifically, we introduce the String Set Reconstruction problem, which asks when a set of strings can be reconstructed from seeing only the k-way projections of strings in the set. This problem is distinguished from genetic reconstruction problems in that we allow projections from any k indices and we maintain knowledge of those indices, but not which k-mer came from which string. We give several results on the complexity of this problem, including hardness results, inapproximability, and parametrized complexity.

Our main result is the introduction of a new algorithm for this problem using a modified version of overlap graphs from genetic reconstruction algorithms. A key difference we must overcome is that in our setting the k-mers need not be contiguous, unlike the setting of genetic reconstruction. We exhibit our algorithm's efficiency in a variety of experiments, and give high-level explanations for how its complexity is observed to scale with various parameters. We back up these explanation with analytic approximations and further detailed experiments. We also consider the related problems of: whether a single string can be reconstructed from the k-way projections of a given set of strings, and finding the largest k at which we get no information about the original data set from its k-way projections (i.e., the largest k for which it is "k-wise independent").

Keywords: Higher-order interactions · Algorithms · Hypergraphs

1 Introduction

Your friend has a complex data set, but you can only view a subset at a time. Can you figure out what their data set is? There are many different precise instantiations of this general type of question. For example, the ability to extrapolate

H. Fernau and B. Zhu (Eds.): IWOCA 2025, LNCS 15885, pp. 256–269, 2025.
https://doi.org/10.1007/978-3-031-98740-3_19

from an incomplete data set gets at the core of many tasks in machine learning and (artificial) intelligence. As another example, this question can also be used to understand when—and which—higher-order interactions are most relevant to a given phenomenon about a complex, multi-part system.

In this paper we consider a novel problem that is a combinatorial interpretation of this question, which we refer to as the STRING SET RECONSTRUCTION problem. Consider an $m \times n$ matrix. For each fixed set of k out of the n columns, we get a set of length-k substrings, called a k-way projection. Can all the k-way projections be used to reassemble the original data set? How does this change as we change the value of k? In this framework, we know which columns we are viewing, but we only see the *set* of those length-k strings. In particular, we do not know which length-k string came from which original length-n string.

This problem is natural from the perspective of studying systems by their higher-order interactions (beyond pairwise) (see, e.g., the surveys [1,2,9,23]). In particular, the STRING SET RECONSTRUCTION problem can be viewed as a particular instance of asking when k-th order interactions suffice to completely determine a given data set. This may be especially beneficial in the context of data sets that one may collect in real-world experiments, where collecting a data set dense enough for traditional machine learning algorithms is difficult or impossible. One example that was inspirational in our approach to this problem is a data set of conflicts among a group of macaques [5,16]. This data set was collected through over one hundred hours of field work recording every fight that occurred among a group of 48 adult macaques, resulting in 1087 recorded fights. Each fight is recorded as a binary string of length 48 where each individual macaque corresponds to an index value which is a 1 if that macaque was in the fight, and a 0 if it was not. Even though this data set took considerable effort to collect, it is clear that 1087 strings out of the 2^{48} possible strings is a very small portion, and to collect enough data to change that would be impossible—indeed, even if they could be observed for an arbitrary long period of time, it's not clear that anywhere close to all 2^{48} possible fights would arise. A second example we have considered include disease spread on a floor of a dorm, where each student is either a 0/1 depending on if they are healthy or ill on a given week, and each string corresponds to one week in the semester. Yet another example is a psychological survey on several hundred participants and their perception of discrimination, where each participant corresponds to one string and each digit of the string is their answer for a particular question. All of these examples are data sets with relatively few data points (compared to the maximum possible given the dimension of the data set) but where higher-order interactions come into play.

In this paper, our main results are two-fold. In terms of complexity, we give a nearly-complete complexity classification of the problems we study, from the viewpoints of worst-case complexity, (in)approximability, and parametrized complexity. In terms of algorithms, we introduce a new algorithmic technique for these problems, generalizing the notion of "overlap graphs" from genetic reconstruction [19]. We demonstrate the efficiency of our algorithm in practice through a series of experiments, as well as heuristic mathematical analysis of how the runtime is observed to scale with various parameters.

A string x is contained in our reconstruction from the k-way projections (k-reconstruction for short) iff all of its size-k substrings (not necessarily contiguous) are contained in the corresponding k-way projection of our data set. That is, given a set of strings S and another string x, x is in the k-reconstruction of S iff for each $I \subseteq [n]$, there is a $y \in S$ such that $y|_I = x|_I$, where $y|_I$ denotes the characters in y at the positions indexed by the set I. As such, for any value of k we will recover all of the strings in the original data set. It is possible, however, that for smaller values of k we may recover additional strings as well: there may be strings that do not contradict any of our substrings in their k-way projections, and as such they will be included in our reconstruction. We can either think of k as a parameter that we increase from 1 upward to eliminate more and more potential strings, or one that we decrease from n downward until we start being unable to eliminate strings outside of the input set.

This same concept shows up under the name "k-limit" or "local limit" in the proof of certain circuit complexity lower bounds [8,12], but we are not aware of prior studies of the algorithmic problems we study.

Within this analytical framework, perfect reconstruction is just one extreme. In the next several subsections we discuss several related problems that are interesting both for their potential applications in data analysis, and for the algorithmic questions they raise. There are also several variations to these problems that are significant both in terms of the computational complexity of the problem, as well as the types of algorithms that can be used to tackle the problem.

1.1 The Problems we Study

In this paper we assume that a set S of strings of length n is given as part of the input.[1] For simplicity, throughout this paper we focus on the alphabet $\{0, 1\}$, though the exact characters are irrelevant. Almost everything in the paper immediately generalizes to alphabets of arbitrary fixed size; when this is not the case we will note it.

Point of Perfect Reconstruction. The most apparent question, and the one that we will primarily focus on throughout this paper, is determining the value of k that allows us to perfectly reconstruct our original data set. We give this value a name:

Definition 1. *Given* $S \subseteq \{0, 1\}^n$, $x \in \{0, 1\}^n$ *is* k-reconstructible *from* S *if for every* $I \subseteq [n]$ *with* $|I| = k$, $\exists s \in S$ *such that* $s|_I = x|_I$. *The* k-reconstruction *of* S *is the set of all strings* k*-reconstructible from* S*, denoted* $Recon_k(S)$. *A set* S *is*

[1] Although it would also be interesting to study the variant where only the k-way projections are given, and not the set S, the complexity can be quite different, and for consistency in this paper we focus on the case where S is given, leaving the other setting for the full version of the paper and future work.

called k-reconstructible if $S = Recon_k(S)$. The point of perfect reconstruction is the least integer k such that $S = Recon_k(S)$.

We will refer to the computational problem—given S, compute the point of perfect reconstruction—as the PERFECT RECONSTRUCTION problem. This value is an interesting structural parameter of the set of strings S, offering some insight into higher-order interactions in S. See Fig. 1 for an example.

1	2	3
0	0	1
0	1	1
1	0	0

→

1	2
0	0
0	1
1	0

2	3
0	1
1	1
0	0

1	3
0	1
1	0

Fig. 1. An example of a set S of three length-3 strings whose point of perfection reconstruction is < 3. To the left is the starting data set $S = \{001, 011, 100\}$, where the strings correspond to rows after the first, and the indices 1,2,3 are listed in the first row. To the right are the 2-way projections for index sets $\{1, 2\}$, $\{2, 3\}$, and $\{1, 3\}$, respectively. From the $\{1, 3\}$ projection, we see that indices 1 and 3 completely define one another: $x_1 = 1 - x_3$. This, coupled with the $\{1, 2\}$ projection, allows us to fill in the 2nd column and perfectly reconstruct the original data set.

There are sets whose point of perfect reconstruction is n, the largest possible value. For example, $S = \{10^{n-1}, 010^{n-2}, 0010^{n-3}, \ldots, 0^{n-1}1\}$ (the "standard basis vectors" of an n-dimensional vector space). Since 0^{n-1} appears in all $(n-1)$-way projections, we have that $0^n \in Recon_{n-1}(S) \backslash S$.

Computing the point of perfect reconstruction can be done by a brute-force, exponential-time algorithm. Namely, the algorithm takes each possible string of length n and generates all $\binom{n}{k}$ of its substrings, and for each check those substrings against the list of substrings from the original data set (in the same k positions) to see if it is in the input set or not. The runtime of this procedure is $O(2^n m \binom{n}{k})$, where m is the number of starting strings.

Point of no Information (k-Wise Independence). To better understand the dependence on k, we can look at the "opposite extreme" of the point of perfect reconstruction, namely, the largest k for which the k-way projections give essentially *no information* about the original set of strings. To clarify what we mean by "no information", we begin with a straightforward observation:

Observation 1. *Let $S \subseteq A^n$, and for $i \in [n]$ let $A_i = S|_{\{i\}}$ be the 1-projections. Then S is 1-reconstructible iff $S = A_1 \times A_2 \times \cdots \times A_n$, iff $|S| = \prod_{i=1}^{n} |A_i|$.*

In particular, every character occurs in every position if and only if the 1-reconstruction of S is equal to all of A^n. This leads us to the following definition:

Definition 2. *Given a set of strings $S \subseteq A^n$ over an alphabet A, the* point of no information *is the largest k such that the k-reconstruction of S is A^n.*

A probability distribution on $\{0,1\}^n$ that is k-wise independent is supported on a set S of strings whose point of no information is $\geq k$. The converse need not hold, however. A tighter connection with k-wise independence can be obtained if one requires not only that the k-way projection sets contain all 2^k possibilities, but that each such possibility comes from the same number of strings in the original set; equivalently, looking at the k-way projection *multi*-sets, rather than just sets. We leave exploring this connection further for future work.

There will always be no information at $k = 1$ unless there is some column that doesn't use all the characters; the point of no information may also be much larger, as in the case of S be the set of length-n strings of even parity, where the point of no information is $k = n - 1$.

As with the point of perfect reconstruction, there is a simple brute-force algorithm for this problem. The point of no information is at least k iff every length-k string occurs in every k-way projection. Thus we can simply check each of the $\binom{n}{k}$ lists of substrings to make sure each set contains all possible 2^k substrings. The runtime is $O\left(\binom{n}{k}|S|\right)$.

String Containment. When we are trying to determine the difference between perfect reconstruction and nearly perfect reconstruction, we come across a critical question: given some string x, how can we determine if x is in $Recon_k(S)$? We will refer to this as STRING CONTAINMENT problem. More precisely, given some string x and a data set S, can we determine some k-length substring that is present in x but not in any string in S? Such a substring exists iff $x \notin Recon_k(S)$. The STRING NON-CONTAINMENT problem turns out to be isomorphic to the classic HITTING SET problem (Theorem 2), in a way that preserves approximability and multi-parameter complexity.

1.2 Motivating Our Algorithmic Technique with a Larger Example

We introduce our new algorithm with a larger example that exhibits several of the phenomena we've seen so far. Figure 2 shows a data set S where $k = 2$ contains no information, $k = 3$ has some information, and $k = 4$ perfectly reconstructs S: $\{0,1\}^5 = Recon_1(S) = Recon_2(S) \supsetneq Recon_3(S) \supsetneq Recon_4(S) = S$.

$k = 2$: One can verify by "brute force inspection" that all $\binom{5}{2}$ 2-way projections contain all 4 possible substrings of length 2.

$k = 3$: As there are only 7 strings, $k = 3$ must contain some constraints or information, as there aren't enough strings to exhibit all 8 possible substrings of size 3. We illustrate the connection with HITTING SET from Sect. 2 here, by showing that $x = 11110 \in Recon_3(S) \setminus S$, so $k = 3$ is not the point of perfect reconstruction. x differs from $01110 \in S$ only in position 1, so any index set I such that $x|_I \notin S|_I$ must have $1 \in I$. Similarly, x differs from $11111 \in S$ in index 5, so we have $\{1,5\}$ in any distinguishing index set I. The three possibilities are

$I = 125$, 135, or 145, and the substring of x for each of these is 110. For $I = 125$ or 135, $11100 \in S$ has restriction 110, agreeing with x; and similarly for $I = 145$, $10010 \in S$ agrees with x. Thus $x \in Recon_3(S)$.

$k = 4$: To show $Recon_4(S) = S$, we give an example of our main algorithmic technique: the overlap graph. See Fig. 2. Using the ordering $1, 2, 3, 4, 5$ on the columns, the overlap graph will have levels, one for each cyclically contiguous set of k indices in our ordering: $[1, 2, 3, 4], [2, 3, 4, 5], [3, 4, 5, 1], \ldots$. The nodes in each level correspond to the projections of the strings in S onto each size-k index set. There is a directed edge between nodes at one level and the next if they agree on the overlapping $k - 1$ indices. From this setup, the length-n directed cycles in this graph correspond bijectively to the set of strings that would be reconstructed if only looking at these particular k-way projections (the overlap graph only sees n out of the $\binom{n}{k}$ possible k-way projections). The overlap graph may have extra cycles that can be eliminated by other k-subsets of indices not used in the graph—which can be done using calls to HITTING SET—but the overlap graph does not eliminate any k-reconstructible strings. In Fig. 2, since the number of directed cycles (yellow highlights) is the same as the number of strings we started with, at $k = 4$ all strings outside S are excluded by using the k-way projections in this overlap graph, so $Recon_4(S) = S$. We give more algorithmic details of the overlap graph in Sect. 3.

1	2	3	4	5
0	0	1	0	0
0	1	0	0	1
1	0	0	1	0
1	1	1	1	1
0	0	0	1	1
1	1	1	0	0
0	1	1	1	0

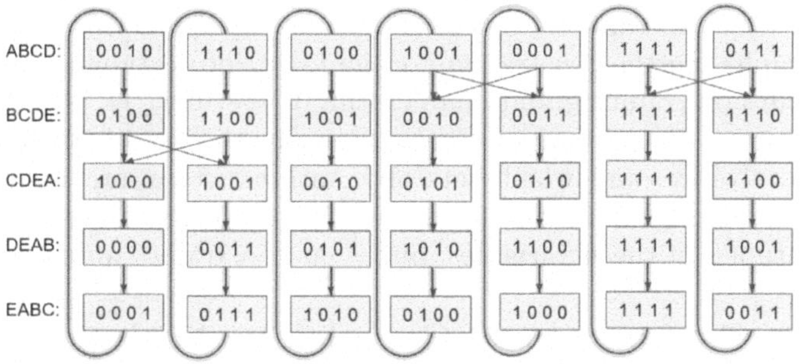

Fig. 2. An example containing 7 strings of length 5 with the indices enumerated across the top of the table, along with one of its overlap graphs for $k = 4$. The cycles of length 5 are highlighted yellow; the number of 5-cycles is the same as $|S|$, so $k = 4$ is the point of perfect reconstruction. The arrangement of vertices into columns is not inherent, but merely for convenience. (Color figure online)

1.3 Related Work

Overlap graphs are commonly used in DNA sequence assembly methods (see, e.g., [19]). The core concept is to make a graph where each node represents one sequence read, and edges are drawn between nodes with matching prefix/suffix pairs. When using this to reconstruct a DNA sequence, you are generally constructing a single string from pieces where you do not have the location information for those pieces. In contrast, we have the location information but do not know how many sequences we are producing.

We must address these critical differences in our adaptation of the overlap graph. First, we are reassembling many strings (of the same length) rather than only one of unknown length; second, we maintain knowledge of our indices; third, we have access to all substrings instead of only contiguous ones. These first two features allow us to develop a more rigid structure in our overlap graphs when compared to their bioinformatics counterparts, while the third feature will come into play in our preprocessing steps (see the full version [22]).

The benefit of knowing the indices of our substrings is observed in the layered structure of our overlap graph. In overlap graphs seen in bioinformatics, there is no layered structure and edges are drawn between two nodes with overlaps of all sizes. By knowing our indices we can instead draw edges only between adjacent layers of the overlap graph without losing information. We also benefit from complete saturation of our overlap graph; we can observe every window of indices rather than risking there being a gap, as there might be in DNA sequence reassembly. This means we can be confident that our overlap graph contains all of our information.

We can also consider the problem of computing $Recon_k(S)$ from S. The set $Recon_k(S)$ consists precisely of the solutions to a corresponding k-CSP instance, in which the underlying constraint hypergraph is the complete k-uniform hypergraph, and the constraints are the k-way projections. This connects to the literature on counting and enumerating CSPs (e.g., [4,7,18,24]). However, we note that in the context of CSPs, typically the set of allowed constraints is fixed in advance, and which sets of k variables they are applied to varies with the instance. In contrast, in our setting, the constraints are always applied to all k-tuples of the n variables, but what constraints we use depend on the input.

2 Computational Complexity Analysis

Preliminaries for Complexity Results. We assume the reader is familiar with standard complexity classes $\mathsf{NP}, \mathsf{coNP}$ and the polynomial hierarchy (e.g., [4,20], as well as the multi-parameter complexity classes FPT and $\mathsf{W}[2]$ (e.g., [6]). The complexity class DP [17] consists of those languages L such that there are two languages $L_1, L_2 \in \mathsf{NP}$ with $L = L_1 \setminus L_2 = \{x : x \in L_1 \text{ and } x \notin L_2\}$. For the Strong Exponential Time Hypothesis (SETH) see [3,13]. For the Unique Games Conjecture see [14]. Two sets $A, B \subseteq \Sigma^*$ are *p-isomorphic* if there is a bijection $f\colon \Sigma^* \to \Sigma^*$ such that both f and f^{-1} are computable in polynomial time, and $f(A) = B$. In this case we write $A \cong_p B$.

String (Non-)Containment is Hitting Set. We will show that String Non-Containment is essentially the same problem as Hitting Set, from the viewpoints of worst-case, approximation, fixed-parameter, and fine-grained complexity. The Hitting Set decision problem is: given a collection S of subsets of a finite universe U, and an integer k, does there exist a set $H \subseteq U$ of size k such that every set in S contains at least one element of H?

To state some of our hardness results, we use an additional parameter, beyond k. Namely, given a set $S \subseteq \{0,1\}^n$ of strings, let

$$d(S) := \max_{x \in \{0,1\}^n} \min_{s \in S} d_{Hamming}(x, s) = \max_x d_{Hamming}(x, S),$$

where $d_{Hamming}(x, s)$ denotes the Hamming distance between x and s.

Theorem 2. String Non-Containment *is essentially the same problem as* Hitting Set. *In particular:*

1. *The two are p-isomorphic, and thus* String Containment *is* coNP*-complete.*
2. *The minimum k such that $x \notin Recon_k(S)$ can be $d(S)$-approximated in polynomial time. If UGC holds, the factor of $d(S)$ is optimal.*
3. *Finding the exact minimum k for which $x \notin Recon_k(S)$ is* DP*-complete.*
4. String Containment*, when parametrized by k and $d(S)$, is in* FPT.
5. *When parametrized only by k,* String Containment *is* coW[2]*-complete.*
6. *SETH is equivalent to: For all $\varepsilon > 0$, there is a d such that* String Containment *with $d(S) \leq d$ is not solvable in time $O(2^{(1-\varepsilon)n})$.*

The proof occurs in the full version [22].

No Information. The point of no information is the point at which every k-window has all 2^k unique substrings. This problem is in coNP because if a given k is not the point of no information, you can provide the window and a missing substring as witness, which could be checked by checking each of the m substrings.

Perfect Reconstruction. Perfect Reconstruction appears to lie somewhere between the first and second levels of the polynomial hierarchy.

There is no known quick guaranteed verification that a given k is the point of perfect reconstruction, because this would require confirming that every possible string is either in the data set or has some window of k indices that is not observed in our data set. The negative is similarly difficult to prove; to verify that the point of perfect reconstruction is greater than k, we would need to provide at least one string that does not have a window of size k missing from our set, which would require proving there is no hitting set of that size.

There is a higher-level version of Hitting Set that matches our problem, built using the String Containment reduction of Theorem 2. We essentially want to know the maximum, over all strings $x \in \{0,1\}^n$ where $x \notin S$ of the smallest size k of a hitting set showing that $x \notin Recon_k(S)$. However, note that the different Hitting Set instances for different x are closely related: we can move between the instance for x and the instance for y by "toggling" the inclusion or exclusion of each index i such that $x_i \neq y_i$ from each set to be hit.

This is very similar to the structure of min-max problems proven to be $\Pi_2\mathsf{P}$-complete [15]. However there is one critical difference that leads us to believe this problem might not actually be hard for the second level of PH. The PERFECT RECONSTRUCTION problem can be fully defined by a single instance of HITTING SET, and all other subproblems are modifications of that original instance as above, whereas in other problems that are hard for the second level of the polynomial hierarchy that we are aware of, the subproblems can be unrelated.

However, the complexity class $\Pi_2\mathsf{P}[\star k, 2]$, introduced by de Haan and Szeider in the context of multi-parameter complexity [10,11], seems like a closer fit for PERFECT RECONSTRUCTION. Rather than define this class intrinsically, we define it by one of its standard complete problems: An instance is a quantified Boolean formula of the form $\forall x \exists y \varphi$ where φ has weft 2, and an integer k (in the multi-parameter setting, k is standardly used as the parameter as well), and the question is to decide whether for every assignment to x, there is an assignment to y *of Hamming weight* k, such that $\varphi(x, y) = 1$.

Proposition 1. *Testing whether a given set of strings is perfectly reconstructible from its k-way projections is in* $\Pi_2\mathsf{P}[\star k, 2]$.

We leave open the question completeness:

Open Question 1. *Is testing whether* $S = Recon_k(S)$ *complete for* $\Pi_2\mathsf{P}[\star k, 2]$ *under fpt reductions?*

A positive answer would imply that there is no fpt-reduction from PERFECT RECONSTRUCTION to UNSAT unless there is a subexponential reduction from $\exists\forall$DNF-SAT to SAT [10, Cor. 40].

Proof. (Proof sketch of Proposition 1*).* We can naturally translate a given instance of our problem by considering X to be a particular string we are checking, and Y the chosen k indices to confirm whether that string is in our reconstructed set or not. If $S = Recon_k(S)$, then for all strings X there must be some choice of indices where X disagrees with every string in S (note that quantifying over all strings in the original data set can be done in polynomial time in the input size, so it does not "count" as an additional unbounded quantifier), OR the string X must be in S. We can use this to build a boolean formula in a natural way, and observe that construction produces a boolean formula of weft 2. This completes the proof.

Observation 3. *Testing whether* 1 *is the point of perfect reconstruction of a set of* m *strings of length* n *can be done in linear time* $O(nm)$ *on a RAM.*

See the full version [22] for the proof, which involves a careful accounting of the arithmetic operations needed to verify the condition from Observation 1.

Theorem 4. *Over* $\{0, 1\}$, *testing whether* 2 *is the point of perfect reconstruction can be done in time* $O(n^2(m + \log n))$ *for sets of* m *strings of length* n.

See the full version [22] for the full proof. The proof here is more involved: it first involves reducing to 2-SAT, and then using Feder's [7] algorithm for enumerating 2-SAT solutions with polynomial delay as a subroutine. We leave as an open question:

Open Question 2 *What is the complexity of deciding whether a given set of strings over an alphabet of fixed size ≥ 3 is perfectly reconstructible from its 2-way projections? In particular, it is in* P*? Is it* coNP*-complete?*

As a polynomial-delay enumeration algorithm for binary CSPs over an alphabet of size $a \geq 3$ would be able to solve the a-coloring problem on graphs, no such algorithm exists unless P = NP. Thus, while it is conceivable that Theorem 4 could be extended to larger alphabets, doing so requires a different approach than in our current proof.

3 Overlap Graph Algorithm

In Sect. 1.2 we introduced the concept of an overlap graph as adapted to fit our problem. Here we explain our efficient algorithms for constructing and using the overlap graph.

In order to construct our overlap graph, we must first choose an ordering of our indices. Since we can use all substrings of length k, including those with non-consecutive indices, this allows us to reorder our indices with no loss of information. We discuss our current methods for choosing an ordering in the full version [22]; this is an area where further improvements may be possible.

After choosing our ordering, we will then create rows in our overlap graph using consecutive substrings that overlap by $k-1$. Our first row will contain all substrings from our data set observed at indices 1 through k; the second row will contain all substrings at indices 2 through $k+1$ and so on, cycling around, viz. the last row corresponds to indices $\{n, 1, 2, \ldots, k-1\}$. After populating all rows of our overlap graph we will draw edges from the first row to the second row where indices 2 through k are identical; we will similarly do the same for all adjacent rows in the overlap graph for their $k-1$ length overlaps until we eventually draw edges from the final row of the graph back to the first row.

To find the edges between row i and $i+1$: for each k-mer in row i, we take its length-$(k-1)$ suffix s, and then check whether $s0$ (resp., $s1$) are present in row $i+1$. If the maximum number of k-mers in a row is t, this takes at most $O(t \log t)$ time if the strings in a given row are stored in sorted order, so that they can be found by binary search. With m strings of length n, the overall time to construct this graph is thus $O(nmk + t \log t) \leq O(nmk + k2^k)$.

The strings that are consistent with these n windows of size k are precisely the length-n cycles in the overlap graph, so our next step is to compute the set of all such cycles. Because of the layered structure of our overlap graph, each such cycle must contain exactly one node from each row, which we take advantage of in order to find all such cycles quickly. Finding these cycles can be done using modifications of breadth-first or depth-first search; one simple way to do so is, for each vertex v in the first layer, use BFS to depth n to find all directed paths of length exactly n back to v.

After we have reconstructed these strings, we separate out the strings that were not in our initial data set and use a HITTING SET subroutine to determine which of them are in the k-reconstruction and, therefore, whether or not the Point of Perfect Reconstruction is at most k. For our experimental evaluation, even with a brute force HITTING SET algorithm our overlap graph method demonstrates significant speedup over a simply greedy method once n is large enough (roughly $n \geq 8$).

4 Experimental Evaluation

We ran a series of experiments to calculate the k-reconstruction of a set, comparing our overlap graph algorithm (including all needed calls to HITTING SET, which we solve using a simple brute-force algorithm) with a simple greedy algorithm (see Figs. 3 and 4). The greedy algorithm we consider builds up the strings starting from a size-k window, ensuring compatibility with the data set each time it extends the strings by a single index. Although simple, this is still significantly faster than brute-forcing over all 2^n strings; see the full version [22] for details of the greedy algorithm.

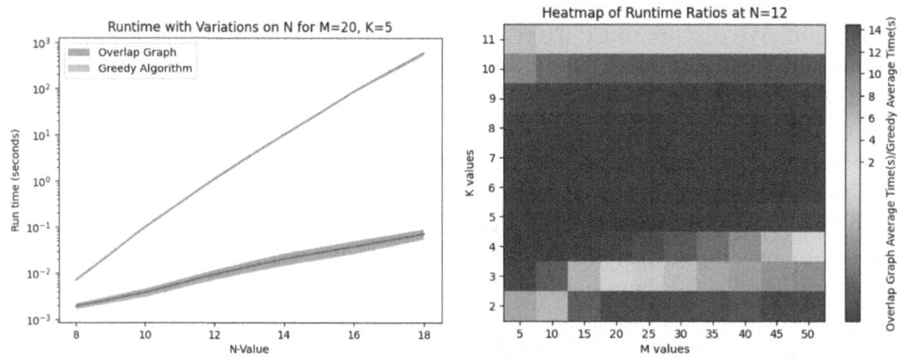

(a) At $m = 20$ strings and window size $k = 5$, our overlap graph algorithm is exponentially faster for $n \geq 8$.

(b) Our overlap graph algorithm is faster than the greedy algorithm for most values of m, k.

Fig. 3. Comparing our overlap algorithm with a simple greedy algorithm.

All experiments were conducted using C++ on a machine with 8.0 GB of RAM and an 11th Gen Intel(R) Core(TM) i7-11800H processor, compiled using `g++-11` and with the C++20 standard (`-std=c++20`) and no additional optimization options. Each experiment began by randomly generating a data set and then testing the time to completion for our overlap graph-based algorithm,

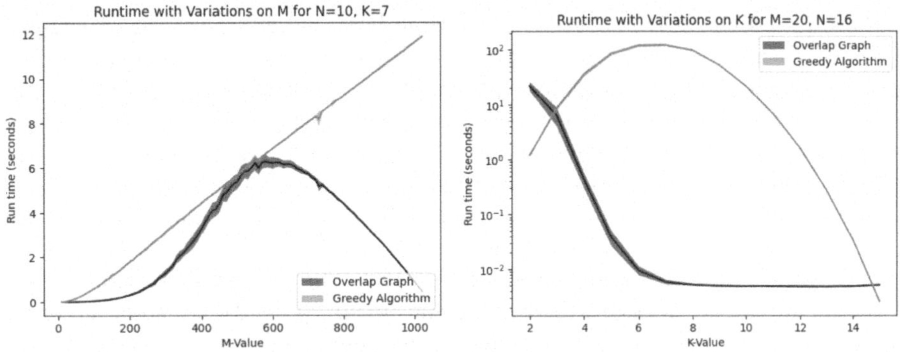

(a) At length $n = 10$ and window size $k = 7$, our overlap graph algorithm is faster for essentially all values of $m \in [1, 1024]$.

(b) At $m = 20$ strings of length $n = 16$, our overlap graph algorithm is faster for almost all values of k.

Fig. 4. Comparing our overlap algorithm with a simple greedy algorithm.

and then the time to completion for the greedy algorithm. Each combination of parameters m, n, and k shown were evaluated over 30 trials. The code we used, as well as the randomly generated data sets, can be found at [21].

Although exhaustive exploration of all combinations of (n, m, k) was infeasible, we first fixed $m = 20$ and $k = 5$ and saw how the performance varied with n. In Fig. 3(a) see a clear trend that for fixed (m, k), as n gets large our overlap graph algorithm becomes faster. We then fixed $n = 12$ and explored for which pairs (m, k) our algorithm out-performed the greedy algorithm and by how much. Figure 3(b) demonstrates that for most values of (m, k), our overlap graph algorithm significantly outperforms the greedy algorithm.

The exceptions are when $k = n - 1$, and separately when $m \gtrsim k2^k$. At $k = n - 1$, the overlap graph is in fact using all $\binom{n}{k} = \binom{n}{n-1} = n$ windows, so it never needs to call Hitting Set, and the greedy algorithm only needs to check n windows as well. The comparison here is simply because of the overhead in the basic operations of the two algorithms. In the case of small k, when $m \gtrsim k2^k$, we are essentially at a point of no information ($k2^k$ comes from the coupon collector problem; also in the full version [22] we validate this experimentally), and in this case the overlap graph algorithm has to call Hitting Set $2^n - m$ times, whereas the greedy algorithm only reconstructs strings "from the bottom up" that are guaranteed to appear in the k-reconstruction. We back up these intuitive explanations for the observed behavior with some analytic estimates in the full version [22].

5 Conclusion and Future Directions

In this paper we introduced a number of computational questions in the study of higher-order interactions, proposed a new algorithm that is efficient in practice for these problems, and analyzed its runtime both analytically and experimentally. In addition to open questions posed within our work (e.g. Open Questions 1 and 2), these tools open up many further questions for exploration.

We have come across several variations of these questions in our research, not yet explored. For example, what if we allow incomplete strings in our data set? Perhaps the data is meant to represent a survey response where a participant did not answer every question, or a system only part of which could be observed. In the case of strings with missing characters, what even is a good definition of "perfect reconstruction"? Can our algorithms be adapted to this setting?

Larger alphabet sizes are likely to come up in applications, such as survey response data with numerous options, or modeling disease transmission with multiple states (infected, recovered, asymptomatic). The basic idea of the overlap graph and many of the results in this paper still hold for arbitrary alphabet sizes, but the complexity increases with the alphabet size. Can we do better?

Another direction we find particularly promising is leveraging sparsity in the data, since most real data sets will be quite sparse. In the full version [22] we show how sparsity can have an immediate impact on some of the questions we study here, and it would be interesting to take further advantage of sparsity.

Acknowledgments. The authors would like to thank Jessica Flack for inspiring conversations that eventually led to this project. JAG would like to thank Gülce Kardeş for teaching him about k-limits in [8,12]. Both authors' work was funded in part by NSF CAREER award CCF-2047756, and E. T. was additionally funded by funds from the CU Boulder Interdisciplinary Quantitative Biology program.

Disclosure of Interests. The authors have no competing interests to declare that are relevant to the content of this article.

References

1. Battiston, F., et al.: Networks beyond pairwise interactions: structure and dynamics. Phys. Rep. **874**, 1–92 (2020). https://doi.org/10.1016/j.physrep.2020.05.004
2. Bick, C., Gross, E., Harrington, H.A., Schaub, M.T.: What are higher-order networks? SIAM Rev. **65**(3), 686–731 (2023). https://doi.org/10.1137/21M1414024
3. Calabro, C., Impagliazzo, R., Paturi, R.: The complexity of satisfiability of small depth circuits. In: Chen, J., Fomin, F.V. (eds.) Parameterized and Exact Computation, pp. 75–85. Springer, Heidelberg (2009)
4. Creignou, N., Khanna, S., Sudan, M.: Complexity classifications of Boolean constraint satisfaction problems. SIAM Monographs on Discrete Mathematics and Applications, vol. 7. SIAM, Philadelphia (2001). https://doi.org/10.1137/1.9780898718546
5. Daniels, B.C., Krakauer, D.C., Flack, J.C.: Sparse code of conflict in a primate society. PNAS **109**(35), 14259–14264 (2012). https://doi.org/10.1073/pnas.1203021109
6. Downey, R.G., Fellows, M.R.: Fundamentals of Parameterized Complexity. Springer, London (2013). https://doi.org/10.1007/978-1-4471-5559-1
7. Feder, T.: Network flow and 2-satisfiability. Algorithmica **11**(3), 291–319 (1994). https://doi.org/10.1007/BF01240738

8. Göös, M., Riazanov, A., Sofronova, A., Sokolov, D.: Top-down lower bounds for depth-four circuits. In: 64th IEEE Annual Symposium on Foundations of Computer Science, FOCS 2023, pp. 1048–1055. IEEE (2023). https://doi.org/10.1109/FOCS57990.2023.00063
9. Grochow, J.A.: Beyond pairwise: higher-order interactions in complex systems. John Templeton Foundation Research Review (2022). www.templeton.org/discoveries/complexity. Produced by the Santa Fe Institute
10. de Haan, R., Szeider, S.: Parameterized complexity classes beyond para-NP. J. Comput. Syst. Sci. **87**, 16–57 (2017). https://doi.org/10.1016/J.JCSS.2017.02.002
11. de Haan, R., Szeider, S.: A compendium of parameterized problems at higher levels of the polynomial hierarchy. Algorithms **12**(9) (2019). https://doi.org/10.3390/a12090188
12. Håstad, J., Jukna, S., Pudlák, P.: Top-down lower bounds for depth-three circuits. Comput. Complex. **5**(2), 99–112 (1995). https://doi.org/10.1007/BF01268140
13. Impagliazzo, R., Paturi, R.: On the complexity of k-SAT. J. Comput. Syst. Sci. **62**(2), 367–375 (2001). https://doi.org/10.1006/JCSS.2000.1727
14. Khot, S.: On the power of unique 2-prover 1-round games. In: Proceedings of 34th ACM Symposium on the Theory of Computing (STOC), pp. 767–775. ACM (2002). https://doi.org/10.1145/509907.510017
15. Ko, K.I., Lin, C.L.: On the complexity of min-max optimization problems and their approximation. In: Du, D.Z., Pardalos, P.M. (eds.) Minimax and Applications, pp. 219–239. Springer, Boston (1995). https://doi.org/10.1007/978-1-4613-3557-3_15
16. Lee, E.D., Daniels, B.C., Flack, J.C., Krakauer, D.C.: Capturing collective conflict dynamics with sparse social circuits. arXiv:1406.7720 [cs.SI] (2014)
17. Papadimitriou, C., Yannakakis, M.: The complexity of facets (and some facets of complexity). J. Comput. Syst. Sci. **28**(2), 244–259 (1984). https://doi.org/10.1016/0022-0000(84)90068-0
18. Petke, J.: Bridging constraint satisfaction and Boolean satisfiability. Artificial Intelligence: Foundations, Theory, and Algorithms. Springer, Cham (2015). https://doi.org/10.1007/978-3-319-21810-6
19. Rizzi, R., Beretta, S., Patterson, M., Pirola, Y., Previtali, M., Della Vedova, G., Bonizzoni, P.: Overlap graphs and de Bruijn graphs: data structures for *de novo* genome assembly in the big data era. Quant. Biol. **7**(4), 278–292 (2019). https://doi.org/10.1007/s40484-019-0181-x
20. Sipser, M.: Introduction to the Theory of Computation, 3rd edn. Cengage Learning (2012)
21. Tate, E., Grochow, J.A.: Code and data for "reconstructing sets of strings from their k-way projections: algorithms & complexity" (this paper). https://git.cutheory.dev/jetate/String_Reassembly/src/branch/main/IWOCA_Code
22. Tate, E., Grochow, J.A.: Reconstructing sets of strings from their k-way projections: algorithmis & complexity. To appear on the arXiv (2025)
23. Torres, L., Blevins, A.S., Bassett, D., Eliassi-Rad, T.: The why, how, and when of representations for complex systems. SIAM Rev. **63**(3), 435–485 (2021). https://doi.org/10.1137/20M1355896
24. Živný, S.: The complexity of valued constraint satisfaction problems. Cognitive Technologies. Springer, Berlin (2012). https://doi.org/10.1007/978-3-642-33974-5

Games: Complexity and Combinatorics

The Closed Geodetic Game: Algorithms and Strategies

Antoine Dailly[1,2(✉)], Harmender Gahlawat[1,3], and Zin Mar Myint[4]

[1] Université Clermont Auvergne, CNRS, Mines de Saint-Étienne, Clermont-Auvergne-INP, LIMOS, 63000 Clermont-Ferrand, France

[2] Université Clermont Auvergne, INRAE, UR TSCF, 63000 Clermont-Ferrand, France

antoine.dailly@inrae.fr

[3] Laboratoire G-SCOP (CNRS, Univ. Grenoble Alpes), Grenoble, France

[4] Indian Institute of Technology Dharwad, Dharwad, India

Abstract. The *geodetic closure* of a set S of vertices of a graph G is the set of all vertices in shortest paths between pairs of vertices of S. A set S of vertices in a graph is *geodetic* if its geodetic closure contains all the vertices of the graph.

Several authors have studied variants of games around constructing geodetic sets. The most studied of those, Geodetic Game, was introduced by Harary in 1984 and developed by Buckley and Harary in 1985. It is an *achievement* game: both players construct together a geodetic set by alternately adding vertices to the set, the winner being the one who plays last. However, this version of the game allows the players to select vertices that already are in the geodetic closure of the current set.

We study the more natural variant, called Closed Geodetic Game, where the players alternate adding to S vertices that are not in the geodetic closure of S. This variant was also introduced in another, less-noticed, paper by Buckley and Harary in 1985, and only studied since then in the context of trees by Araujo *et al.* in 2024. We provide a full characterization of the Sprague-Grundy values (equivalence values for games) of graph classes such as paths and cycles, of the outcomes of the Cartesian product of several graphs in function of their individual outcomes, and give polynomial-time algorithms to determine the Sprague-Grundy values in cactus and block graphs.

Keywords: Geodetic Set · Geodetic Game · Polynomial-time algorithms · Cartesian Product · Block Graphs · Cactus Graphs · Combinatorial Games

1 Introduction

1.1 A History of Geodetic Games

Let $G(V, E)$ be a simple, undirected graph with vertex set V and edge set E. Given two vertices $u, v \in V$, let $\mathcal{I}(u, v)$ denote the set of vertices that lie on some

H. Fernau and B. Zhu (Eds.): IWOCA 2025, LNCS 15885, pp. 273–286, 2025.
https://doi.org/10.1007/978-3-031-98740-3_20

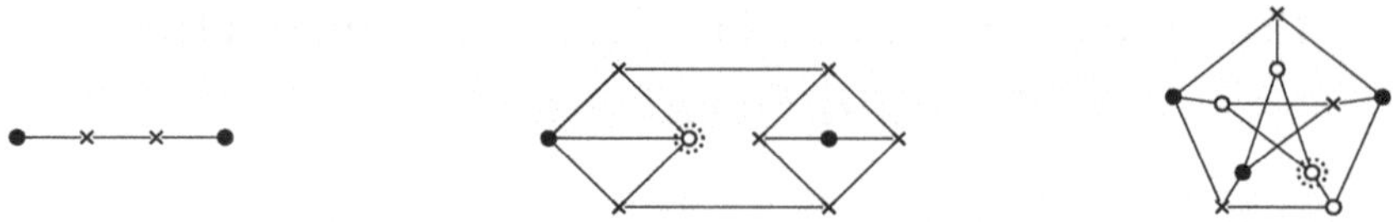

Fig. 1. Black vertices belong to the set S, crossed vertices belong to the geodetic closure (S), white vertices are uncovered vertices yet. However, if we add the dotted vertices to S, then $V = (S)$, *i.e.*, S becomes a geodetic set.

shortest path with u and v as endpoints. Given a subset $S \subseteq V$ of vertices, its *geodetic closure*, denoted by (S), is defined as the set of vertices in all shortest paths between every pair of vertices of S. More formally, $(S) = \bigcup_{u,v \in S} \mathcal{I}(u, v)$.

A set S of vertices of $G(V, E)$ is called a *geodetic set of G* if $(S) = V$ (see Fig. 1 for examples). The problem of finding a geodetic set of minimum size was defined in the 1980s [6,9,18,19] and has been well-studied and is known to be NP-hard for general graphs [3,18]. The problem has been also well-studied on graph classes and displays quite an interesting behaviour. In particular, the problem stays NP-hard even for subcubic partial grids [11], for interval graphs [10], and admits a polynomial time algorithm for solid grids [11], outerplanar graphs [21], and proper interval graphs [14].

Interestingly, the game version of the geodetic number was introduced in parallel of its combinatorial counterpart. The first mention of a game based around geodetic sets was in 1984, by Harary [17]: two players alternate adding new vertices of a graph $G(V, E)$ to a set S, the game ends when $(S) = V$. Two winning conditions are proposed: in the *achievement* setting, the player who plays the last move wins; while they lose in the *avoidance* setting. Achievement and avoidance correspond to the *normal* and *misère* settings of combinatorial games, respectively. We will focus exclusively on achievement games.

With respect to combinatorial game theory, the first question to ask is, **given an input graph, decide which player wins**. This is called computing the *outcome* of the game. Note that, since the game is *impartial* (both players are indistinguishable), there are only two possible outcomes: either the first player wins (denoted by $\mathcal{N}$) or the second player wins (denoted by $\mathcal{P}$). Furthermore, when playing on disjoint connected components, each component is independent (selecting a vertex in a component has no influence whatsoever on the other components). If two components are $\mathcal{P}$, then, playing on their disjoint union is also $\mathcal{P}$ (the second player always applies their winning strategy on the component on which the first player played a move); if one component is $\mathcal{P}$ and the other is $\mathcal{N}$, then the first player wins by applying their winning strategy on the $\mathcal{N}$ component, becoming the second player on the disjoint union of two $\mathcal{P}$ components. However, outcomes are not enough to solve the disjoint union of two $\mathcal{N}$ components: we need to use *Sprague-Grundy values*, which are equivalence classes for games (and can be mapped to nonnegative integers). More details will be given in Sect. 1.3. Hence, a stronger question is, **for a given input graph, compute its Sprague-Grundy value**.

1.2 Geodetic Game, Closed Geodetic Game and Other Related Variants

In his seminal paper [17], Harary discussed several possible rulesets for games based around the notion of geodesics. Two of those rulesets were then expanded in two 1985 papers by Buckley and Harary, and constitute the main variants. All games are played on a graph $G(V, E)$.

In the first paper [8], the authors define Geodetic Game. In this game, two players alternately add vertices to a set S, until $(S) = V$. They give the outcome of the game for complete graphs, cycles, (generalized) wheels, complete bipartite graphs and n-cubes. This line of research was expanded later in [22] which improved on the wheels' result, and in [20], which studied complete multipartite graphs, hypercubes, and graphs with a unique minimum-size geodetic set (such as trees, split graphs or coronas).

In the second paper [7], they introduce Closed Geodetic Game. In this game, the players alternatively add to S vertices that are not in (S) as it is at the moment they are selecting it. They study the same graph classes as in [8]. Surprisingly, this more natural variant was left unstudied for decades, probably due to the difficulty of access of [7] compared to [8]. Recently, Araujo *et al.* [2] studied several geodetic game variants, including Closed Geodetic Game (which they called *geodesic closed interval game*), for which they provided a linear-time algorithm computing the Sprague-Grundy value of a tree.

Note that Geodetic Game and Closed Geodetic Game can have different outcomes on even very simple graphs. One such example is illustrated in Fig. 2. Since the players are allowed to pick covered vertices in Geodetic Game, the parity of the graph (and of covered vertices) plays a bigger role in determining the outcome compared to Closed Geodetic Game. Indeed, such is the case for Proposition 6 in [20], where graphs with a unique minimum-size geodetic set are solved due to parity, since the simili-passing moves (adding to S vertices that are in the current geodetic closure) are available.

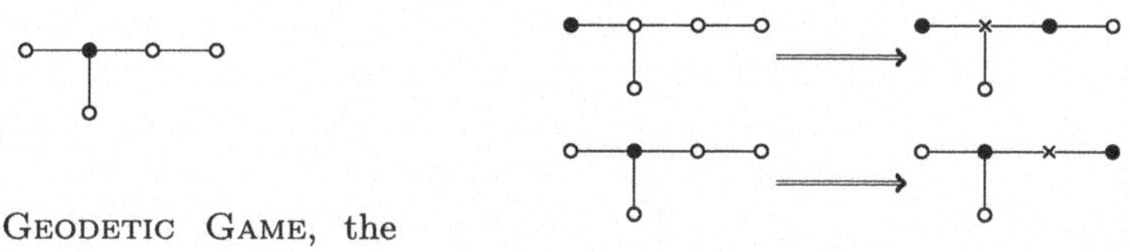

(a) For Geodetic Game, the first player wins by picking the black vertex, ensuring that every vertex will be in S (if the second player picks a leaf, they pick the degree 2 vertex, and conversely).

(b) For Closed Geodetic Game, the second player always has an answer to the first player's first move, ensuring that exactly two moves will remain (other possible moves for the first player are treated by symmetry or similarity).

Fig. 2. An example of a graph with different outcomes for Geodetic Game and Closed Geodetic Game.

Other variants were later introduced, such as the *hull game*, where the *convex hull* (instead of the geodetic closure) of S is considered [4,15]. The outcomes for trees, cycles and complete graphs with rays were determined in [15], while [4] focused on Sprague-Grundy values for (among others) block graphs, lattice graphs, (generalized) wheels and complete multipartite graphs, as well as algebraic properties of the game.

Note that those games can be generalized as hypergraph games [23], where two players alternate selecting vertices of a hypergraph (potentially modifying it or under conditions) until those satisfy some condition related to the hyperedges. In particular, in GEODETIC GAME, the players add hyperedges corresponding to the shortest paths between the selected vertices, until every vertex is in at least one hyperedge.

1.3 Game Definitions

GEODETIC GAME and CLOSED GEODETIC GAME are impartial games, a subset of combinatorial games which have been extensively studied. Since both players have the same possible moves, impartial games can have two possible *outcomes*: either the first player has a winning strategy ($\mathcal{N}$) or the second player does ($\mathcal{P}$). A position P' that can be reached by a move from a position P is called an *option* of P, the set of options of P is denoted by $\text{opt}(P)$. If every option of P is $\mathcal{N}$, then P is $\mathcal{P}$; conversely, if any option is $\mathcal{P}$, then P is $\mathcal{N}$ (the winning strategy for the first player is to play to a $\mathcal{P}$ option). A refinement of outcomes are the *Sprague-Grundy values* [16,25] (also called *nim-values* or *nimbers*), denoted by $\mathcal{G}(P)$ for a given position P and which can be mapped to nonnegative integers. The value of a position can be computed inductively, with $\mathcal{G}(P) = \text{mex}(\{\mathcal{G}(P') : P' \in \text{opt}(P)\})$, where $\text{mex}(X)$ (for a set X of nonnegative integers) is the smallest nonnegative integer not in X. A position P is $\mathcal{P}$ if and only if $\mathcal{G}(P) = 0$. Two positions P and P' are said to be *equivalent*, denoted by $P \equiv P'$, if $\mathcal{G}(P) = \mathcal{G}(P')$.

The *disjoint sum* of two positions P and Q, denoted by $P + Q$, is a position where each player chooses to play on one of the two positions at their turn. When one of them is finished, the game continues on the other one; the last player to play wins. For CLOSED GEODETIC GAME, a disjoint sum represents the disjoint union of two graphs. Computing the Sprague-Grundy value of a disjoint sum $P + Q$ can be done by applying the so-called *nim-sum*, which is the bitwise XOR: $\mathcal{G}(P + Q) = \mathcal{G}(P) \oplus \mathcal{G}(Q)$. An example is given below for $7 \oplus 10$.

$$\begin{array}{rcr l}
7 & \rightarrow & 111 & \\
10 & \rightarrow & \underline{1010} & \\
 & & 1101 & \rightarrow 13
\end{array}$$

For more information and history about combinatorial games, we refer the reader to [1,5,12,24].

Due to the definition of the sum of positions, we can limit our study to connected graphs: studying a graph with several connected components is equivalent to studying each of the components independently and applying the nim-sum. Furthermore, we will always consider that both players play optimally. Optimal play is understood as applying a winning strategy using as few rounds as possible if possible, and trying to make the game last as long as possible if no winning strategy exists. We will also make use of the following notation:

Notation 1. *Let G be a graph. We also denote the game position by G. Furthermore, let S be a subset of vertices of G, then the position denoted by G, S is the graph G where the vertices in S have already been selected.*

1.4 Structure of the Paper

We will begin in Sect. 2 by proving general results on CLOSED GEODETIC GAME as well as studying how to compute the outcome for the Cartesian product of several graphs in function of the outcomes of the individual graphs. Then, in Sect. 3, we will study some graph classes where we can predict exactly which vertices will be picked. Afterwards, Sect. 4 will be focused on highly symmetric graphs such as paths, cycles and grids, and we compute the Sprague-Grundy values for the former two. Finally, in Sect. 5, we present polynomial-time algorithms based on dynamic programing computing the Sprague-Grundy values of block graphs and cacti. Due to space restrictions, proofs of results marked with (∗) are omitted; they can be found in the full version of the paper [13].

2 General Results and Cartesian Product

In this section, we present a few general results that help the study of CLOSED GEODETIC GAME, be it by stating that some vertices have to be selected, by helping decompose a graph, or by showing how a well-known graph product affects the outcome of graphs.

First, we note that, in CLOSED GEODETIC GAME, due to the fact that vertices can be in the geodetic closure $\langle S \rangle$ without being in the geodetic set S only if they are on some shortest path between two vertices, some vertices always have to be selected. For example, a leaf (*i.e.*, a degree 1 vertex), can never be in a shortest path between two other vertices. More generally, consider *simplicial* vertices, that is, vertices whose neighbourhood induces a complete subgraph.

Lemma 1. *In* CLOSED GEODETIC GAME, *all simplicial vertices of the graph will have to be selected.*

Proof. A simplicial vertex can never be on a shortest path between two other vertices, since all its neighbours are fully connected to each other. Hence, it can never be in a geodetic closure $\langle S \rangle$ if it is not in S itself. □

Articulation points, that is, vertices whose removal disconnects the graph (also called *cut vertices* or *separating vertices*), are important for the study of CLOSED GEODETIC GAME. The following lemma will often be used to allow for decompositions:

Lemma 2. *Let G be a graph with an articulation point u, such that components $G_1, \ldots, G_k$ are attached at u. In* Closed Geodetic Game, $\mathcal{G}(G, \{u\}) = \mathcal{G}(G_1, \{u\}) \oplus \ldots \oplus \mathcal{G}(G_k, \{u\})$.

Proof. When u is selected, any selected vertex x in G_i will never help cover any vertex in G_j for $j \neq i$. Indeed, all the shortest paths from x to any vertex in G_j will go through u, so, from the point of view of G_j selecting x after u has no impact. Hence, each $G_i, \{u\}$ is an independent position, and we are playing on their disjoint union. This is depicted on Fig. 3. □

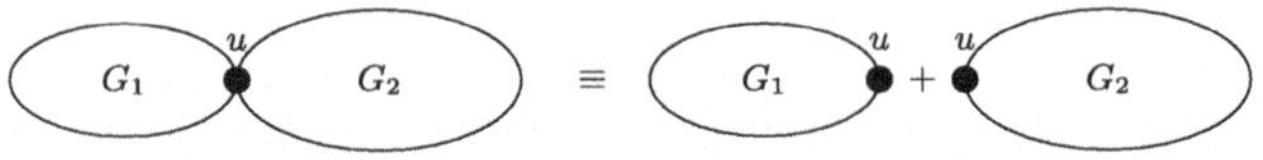

Fig. 3. Selecting an articulation point is equivalent to splitting the graph and playing independently on the components.

We also focus on a classical question whenever studying games: if we know the outcomes of two graphs, can we obtain the outcome of some composition of those graphs? This question stems from the disjoint sum, which is managed through the nim-sum of Sprague-Grundy values. We study how the Cartesian product of graphs interacts with the outcomes for Closed Geodetic Game.

Given two graphs G and H, their *Cartesian product* $G \square H$ has vertices of the form (u, v) where $u \in V(G)$ and $v \in V(H)$ such that two vertices (u_1, v_1) and (u_2, v_2) are adjacent in $G \square H$ if

- $u_1 = u_2$ and v_1 is adjacent to v_2 in H, or
- $v_1 = v_2$ and u_1 is adjacent to u_2 in G.

Theorem 1. *Let $G = G_1 \square G_2 \square \cdots \square G_n$ be the Cartesian product of n graphs. The graph G is $\mathcal{N}$ for* Closed Geodetic Game *if and only if G_i is $\mathcal{N}$ for* Closed Geodetic Game *for all $i \in \{1, 2, \cdots, n\}$.*

Proof. First, we will show that it is enough to prove the statement for $n = 2$. To this end, suppose that we have proved $H = H_1 \square H_2$ is $\mathcal{N}$ for Closed Geodetic Game if and only if both H_1 and H_2 are $\mathcal{N}$. On the one hand, let $G = G_1 \square G_2 \square \cdots \square G_n$ where G_i is $\mathcal{N}$ for all $i \in \{1, 2, \cdots, n\}$. In this case, we can use our assumption of $H = H_1 \square H_2$ is $\mathcal{N}$ if and only if both H_1 and H_2 are $\mathcal{N}$, recursively, to conclude that G is $\mathcal{N}$. On the other hand, let $G = G_1 \square G_2 \square \cdots \square G_n$ where G_i is $\mathcal{P}$ for some $i \in \{1, 2, \cdots, n\}$. Let $G' = G_1 \square \cdots G_{i-1} \square G_{i+1} \square \cdots G_n$. Observe that $G' \square G_i = G$. Hence, we can use our assumption of $H = H_1 \square H_2$ is $\mathcal{N}$ if and only if both H_1 and H_2 are $\mathcal{N}$ to conclude that G is $\mathcal{P}$. Therefore, it is enough to prove the statement of our theorem for $n = 2$, and henceforth we will focus on this simpler statement. We will need the next two useful claims.

Claim 1 (*). Let $G = G_1 \square G_2$, $x_1, x_2 \in V(G_1)$, $y_1, y_2 \in V(G_2)$, a and b respectively be the distance between x_1 and x_2 in G_1 and the distance between y_1 and y_2 in G_2. Then the distance between (x_1, y_1) and (x_2, y_2) is $a + b$ in G.

Claim 2. Let $G = G_1 \square G_2$, $S_1 \subseteq V(G_1)$, $S_2 \subseteq V(G_2)$, and $S = S_1 \times S_2 \subseteq V(G)$. We have $(u, v) \in (S)$ if and only if $(u, v) \in (S_1) \times (S_2)$.

Proof of Claim 2. In one direction, assume that $(u, v) \in (S)$. Thus, there exists $(x_1, y_1), (x_2, y_2) \in S$ such that (u, v) lies on some shortest path P between (x_1, y_1) and (x_2, y_2). Let P' be the subpath of P that connects (x_1, y_1) to (u, v). Similarly, let P'' be the subpath of P that connects (u, v) to (x_2, y_2). Note that both P' and P'' are shortest paths.

Denote the length of a path X (that is, its number of edges) by $\mathrm{lt}(X)$. By Claim 1, we know that there are shortest paths P_1' in G_1 and P_2' in G_2 connecting x_1 to u and y_1 to v, respectively, satisfying $\mathrm{lt}(P_1') + \mathrm{lt}(P_2') = \mathrm{lt}(P')$. Similarly, we know that there are shortest paths P_1'' and P_2'' in G_2 connecting u to x_2 and v to y_2, respectively, satisfying $\mathrm{lt}(P_1'') + \mathrm{lt}(P_2'') = \mathrm{lt}(P'')$. Let P_1 (resp., P_2) be the path obtained through concatenation of P_1' and P_1'' (resp., P_2' and P_2''). Thus, we must have $\mathrm{lt}(P_1) + \mathrm{lt}(P_2) = \mathrm{lt}(P)$. This implies that P_1 and P_2 are shortest paths due to Claim 1. Thus, $(u, v) \in (S)$ implies $(u, v) \in (S_1) \times (S_2)$.

In the other direction, assume that $(u, v) \in (S_1) \times (S_2)$. Thus, there are $x_1, x_2 \in S_1$ such that u lies on the path P_1 connecting x_1 and x_2. Moreover, without loss of generality, we may assume that the path is the concatenation of the subpaths P_1' connecting x_1 and u and P_1'' connecting u and x_2. Similarly, there are $y_1, y_2 \in S_2$ such that v lies on the path P_2 connecting x_1 and x_2 where P_2 is the concatenation of the subpaths P_2' connecting y_1 and v and P_2'' connecting v and y_2. By Claim 1, we know that there are two paths P' connecting (x_1, y_1) and (u, v) and P'' connecting (u, v) and (x_2, y_2) in G, and satisfying $\mathrm{lt}(P_1') + \mathrm{lt}(P_2') = \mathrm{lt}(P')$ and $\mathrm{lt}(P_1'') + \mathrm{lt}(P_2'') = \mathrm{lt}(P'')$. In particular, this implies that P' and P'' are shortest paths.

Let P be the path obtained by concatenation of P' and P''. Observe that P connects (x_1, y_1) and (x_2, y_2) and passes through the vertex (u, v). Note that

$$\mathrm{lt}(P) = \mathrm{lt}(P') + \mathrm{lt}(P'') = \mathrm{lt}(P_1') + \mathrm{lt}(P_2') + \mathrm{lt}(P_1'') + \mathrm{lt}(P_2'') = \mathrm{lt}(P_1) + \mathrm{lt}(P_2).$$

Thus, by Claim 1, P must be a shortest path. ○

In one direction, assume that G_1 and G_2 are $\mathcal{N}$. We show that $G = G_1 \square G_2$ is also $\mathcal{N}$.

We will play the game simultaneously on G, G_1 and G_2. The vertex sets that will be constructed during the games are S, S_1 and S_2, respectively. We will maintain $S = S_1 \times S_2$. The strategy of the first player on G will be derived from the strategies of the first player on G_1 and G_2. The second player will play on G, and the corresponding moves will be replicated on G_1 and G_2.

Let us explain first how a move is replicated or translated. At any point in the game, if a player selects the vertex (u, v) in G, we know that $(u, v) \notin (S)$ at the time it is selected. However, by Claim 2, it is possible to have either $u \in (S_1)$ or $v \in (S_2)$, but not both. If $u \in (S_1)$, then the selection of (u, v) corresponds to a "dummy move" in G_1. If $u \notin (S_1)$, then the selection of (u, v) corresponds to selecting u in G_1. Similarly, one can translate the moves on G to G_2.

The selection of u and v in G_1 and G_2, respectively, corresponds to the selection of (u, v) in G. Note that either $u \notin (S_1)$ or $v \notin (S_2)$ at the time of their respective selection, implying $(u, v) \notin (S) = (S_1) \times (S_2)$ due to Claim 2, and thus selecting (u, v) is a valid move.

The first player will begin the game by selecting vertices u_1, v_1 in G_1, G_2, respectively, according to the corresponding winning strategies. In G, the vertex (u_1, v_1) is selected by the first player. After that the players take turns to make their moves. Let us assume that the second player has made a move by selecting the vertex (u, v) in G. This moves translates to moves in G_1, G_2. However due to Claim 2, we know that at most one of these translated moves can be a dummy move. Without loss of generality assume that the second player's move translates to a dummy move in G_1, and a valid move in G_2. In that case, the first player makes a dummy move in G_1 and provides a response by selecting v' in G_2 according to winning strategy. These moves are translated to G as selection of the vertex (u, v'), which is a valid move due to Claim 2. If the second player's move in G translates to valid moves in both G_1 and G_2, then the first player responds in both the graphs according to corresponding winning strategies by selecting u', v' in G_1, G_2, respectively. This translates to the valid move of selecting (u', v') in G by the first player due to Claim 2.

Hence, the first player always has a valid response to the second player's move. Thus, G is $\mathcal{N}$.

In the other direction, assume that G_2 is $\mathcal{P}$. We show that $G = G_1 \square G_2$ is also $\mathcal{P}$. In this case, the second player will simulate the game on both graphs, selecting (u, v) in $G \square H$ where u is either the winning move (if G_1 is $\mathcal{P}$ or the first player played a non-optimal move) or the same vertex that the first player selected, and v is their winning move in H for replying to the first player. The game will end in $G \square H$ when both G and H will be over, and thus the second player will always have an answer to the first player's moves. This is consistent due to Claim 2. □

3 She Loves Me-She Loves Me Not Situations

She Loves Me-She Loves Me Not is one of the simplest combinatorial games: the two players alternate removing blossoms from a flower until it is empty. It has become a shorthand for easy, binary combinatorial games where every possible move from the initial position will be played. Note that, in those cases, the Sprague-Grundy values alternate between 0 (for an even number of moves) and 1 (for an odd number of moves), since no real strategy is involved. Some complex combinatorial games can be equivalent to She Loves Me-She Loves Me Not depending on the initial position. It is in particular the case for Closed Geodetic Game in some cases:

Proposition 1 (∗). *A game of* Closed Geodetic Game *on the complete graph K_n or a star $K_{1,n}$ will see all its vertices selected.*

(a) On this graph, the second player wins: if the first player selects a leaf, they select the adjacent internal vertex, and conversely. There is then a pairing strategy for the second player.

(b) On this graph, the first player wins by picking the black vertex (this can be shown with case analysis).

Fig. 4. Two graphs of same order (8) with unique minimum-size geodetic sets of same size (5) but different outcomes for CLOSED GEODETIC GAME.

Note that those two graphs have a unique minimum-size geodetic set, so one could think that such parity results could be obtained on any graph family with a unique, well-identified minimum-size geodetic set on an underlying structure. However, such a general structural results would be difficult to achieve, as illustrated by Fig. 4: those two graphs are based on the same underlying structure (a triangle with leaves), have the same order and unique minimum-size geodetic sets of the same size, but different outcomes for CLOSED GEODETIC GAME.

For other graphs, the game is not fully reducible to a SHE LOVES ME-SHE LOVES ME NOT, but is still close since, under optimal play, we can determine exactly which vertices will be selected.

Proposition 2 (∗). *Let $n \geq 2$. A game of* CLOSED GEODETIC GAME *on the complete bipartite graph $K_{2,n}$ will see all the vertices of its larger part selected.*

We can also compute the Sprague-Grundy values of complete bipartite graphs, for which one part will always be fully selected:

Proposition 3 (∗). *Let m and n be two integers such that $m, n \geq 2$. If m and n have the same parity, then, $\mathcal{G}(K_{m,n}) = 0$. Otherwise, $\mathcal{G}(K_{m,n}) = 2$.*

4 Symmetry Strategies

The outcome of cycles can be fully derived from a symmetry strategy. Furthermore, we can also characterize the Sprague-Grundy value of a cycle:

Theorem 2 (∗). *For any positive integer n, $\mathcal{G}(C_n) = n \bmod 2$.*

We can also characterize the Sprague-Grundy value of a cycle with one or two selected vertices:

Proposition 4 (∗). *For any positive integer k:*

- $\mathcal{G}(C_{2k}, \{u\}) = k$ *and* $\mathcal{G}(C_{2k+1}, \{u\}) = 0$ *for any vertex u;*
- $\mathcal{G}(C_{2k+1}, \{u_0, u_i\}) = k + 1 - i$ *for* $i \in \{1, \ldots, k\}$ *(denoting the vertices by $u_0, \ldots, u_{2k}$ in the cyclic order).*

Theorem 1 combined with Theorem 4 already give us the outcome for multidimensional grids, but we can also prove it with a symmetry strategy:

Theorem 3 **(∗).** Closed Geodetic Game *has outcome* $\mathcal{N}$ *for a multidimensional grid if and only if all its dimensions are odd.*

Paths are another graph class where the outcome can be derived from a symmetry strategy. Going further, we hereby give a full characterization of the Sprague-Grundy values of paths (which complements the possible computation using the algorithm for trees from Araujo *et al.* [2]). Note that, while the value is expected, the proof itself is more involved.

Theorem 4 **(∗).** *For any positive integer* n, $\mathcal{G}(P_n) = n \bmod 2$.

Proof (Sketch). If n is even, the second player has an easy symmetry strategy. If n is odd, the first player wins by picking the middle vertex. We then prove that no option of P_n has Sprague-Grundy value 1, by showing that all of those options have themselves an option with Sprague-Grundy value 1. In order to do this, we prove that $\mathcal{G}(P_{4m+2}, \{u_{2m+1}\}) = 1$, and exhibit such options for each option of P_n. □

5 Sprague-Grundy Values of Block Graphs and Cacti

In their paper [2, Algorithm 2], Araujo *et al.* presented a linear-time algorithm for computing the Sprague-Grundy value of a given tree for Closed Geodetic Game. In order to do this, they use dynamic programing with subtrees that decrease in size at each step: for any given selected vertex u, each of the subtrees rooted at a neighbour of u are completely independent in Closed Geodetic Game, since by Lemma 2 selecting a vertex in a subtree will never impact any of the other subtrees. The base case happens when the subtree is either empty or a single vertex. Hence, it is possible to define a recursive algorithm computing, for each vertex of the tree, the Sprague-Grundy value of the subtrees rooted at each of its neighbours, and then apply the mex operation to the resulting set. Since each subtree is independent, and the intermediate values can be stored, the algorithm returns the result in linear-time.

In this section, we extend this idea to other tree-like graph classes: block graphs and cacti. Indeed, the building blocks of those graphs are simple graphs for which we can easily compute the Sprague-Grundy values: complete graphs and cycles. We will present polynomial-time algorithms for those classes, adapted from the trees' algorithm in [2]. Note that the building blocks have specific properties, and are close to the class of *geodetic graphs*, that is, graphs for which there is a unique shortest path between any pair of vertices (this class includes trees, block graphs, and cacti with only odd cycles); the only non-geodetic class is cacti with even cycles, for which the shortest paths between two vertices will share common articulation points and thus are almost as constrained.

5.1 Block Graphs

A graph is a *block graph* if every biconnected component is a clique. Informally, we can see a block graph as a tree-like structure in which cliques are connected to each other by one of their vertices.

Theorem 5. *There is a linear-time algorithm computing the Sprague-Grundy value of a block graph for* Closed Geodetic Game.

Proof. The algorithm is based on dynamic programing. First, for a given block graph G and one of its vertices u, we present how to recursively compute $\mathcal{G}(G, \{u\})$, that is, we consider that the first player selects u. We will show that some clique vertices are equivalent for this first move, and that we can store the intermediary results to reuse them for other vertices (thus avoiding polynomial-time when applying the algorithm for all the non-equivalent vertices of G). Note that the base case is when the graph is a complete graph, since we can compute its Sprague-Grundy value using Proposition 1 (each selected vertex within a complete graphs flips the Sprague-Grundy value between 0 and 1).

There are two types of vertices in G: those that are articulation points between several cliques, and those that are not. First, assume that the selected vertex u is an articulation point, then we can compute $\mathcal{G}(G, \{u\})$ by using Lemma 2 and store the intermediate values for future computations.

Assume now that the selected vertex u is not an articulation point, hence u is a vertex in a given clique A of G. Note that selecting any other non-articulation point of A is equivalent to selecting u. Now, consider the components that are attached to A in G (that is, they share one vertex with A), denote them by $B_1, \ldots, B_k$ and the articulation point between A and B_i by v_i. Note that selecting any other non-articulation point in A will not affect any of the B_i's, and that selecting v_i can only affect B_i. Hence, we can decompose the game into almost-independent subgraphs: $A \setminus \{v_1, \ldots, v_k\}$ with the vertex u selected, and each component containing B_i with a leaf attached to v_i, which is selected too. This is depicted on Fig. 5. Note that this decomposition works since A is a clique, and so all its vertices are at distance 1 from each other.

Now, we need to see how we can decompose the graph with one selected vertex u when another vertex v is selected. Again, if v is an articulation point, we can use Lemma 2. If v is also in A, then we do not need to decompose further. So, assume that v is in another clique B and is not an articulation point. We can apply the same principle of decomposition, after covering the vertices in the shortest paths between u and v. Note that if A and B are not adjacent cliques, then let C be a clique with two vertices on the shortest path between u and v, two vertices of C are covered and the other ones will have no impact on A or B for the rest of the game. This decomposition is depicted on Fig. 6.

Note that both decompositions have the property that a subgraph that is not a clique has at most one selected vertex. This covers all possible cases, and thus the algorithm will further decompose the input block graph until it is a disjoint sum of cliques, for which the Sprague-Grundy value can be computed in constant time. By using the nim-sum and the mex, we can then go back and compute

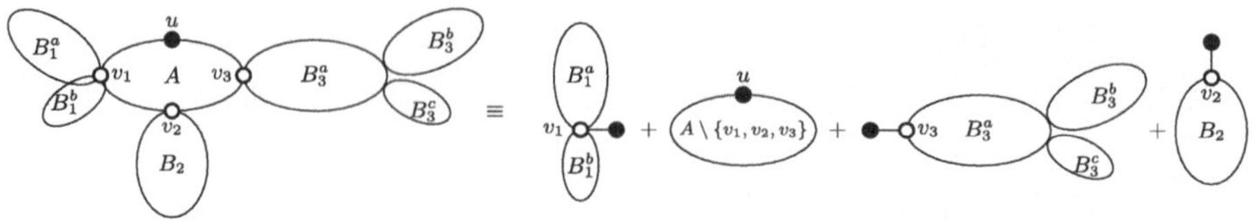

Fig. 5. Selecting a vertex in a clique in a block graph allows us to decompose the graph as follows (here, each B_i^x is a clique).

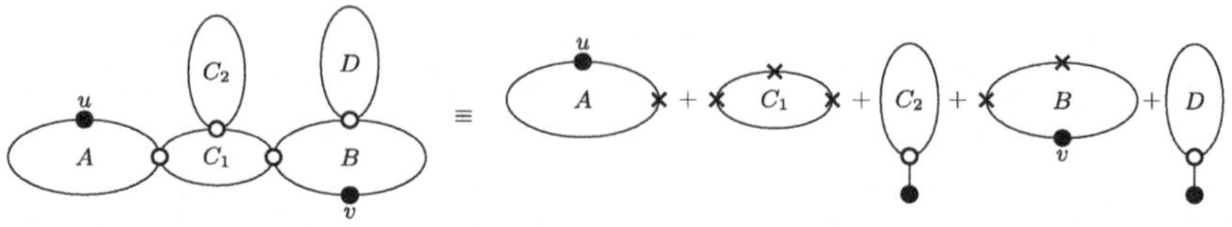

Fig. 6. Selecting a second vertex in a block graph allows us to decompose it further.

the Sprague-Grundy value of the block graph. The algorithm ends in linear time for a given vertex, and allows us to store the intermediary results for future computations. In fact, the number of block components being linear, any future computation will see similar decompositions, and thus reuse the intermediary results, hence allowing the algorithm to remain linear-time when considering all possible first moves. □

5.2 Cacti

A *cactus* (plural: cacti) is a graph where every edge is in at most one simple cycle. Alternately, cacti have a tree-like structure in which cycles and trees are connected to each other by one of their vertices.

Theorem 6 (∗). *There is a quadratic-time algorithm computing the Sprague-Grundy value of a cactus for* Closed Geodetic Game.

Proof (Sketch). The algorithm is also based on dynamic programing. We identify three types of cacti with either one or two selected vertices, and show that any move in either type yields a graph that can be decomposed (thanks to Lemma 2) into a disjoint sum of subgraphs, each of one of those three types. We use the algorithm in [2], Theorems 2 and 4 and Proposition 4 as base cases. □

6 Future Research

In this paper, we explored Closed Geodetic Game in three directions: general results such as graph operations (Cartesian product) and decomposition through articulation points, characterization of outcomes and Sprague-Grundy values, and polynomial-time algorithms based on dynamic programming for constrained graph classes. However, many more general graph classes remain open, algorithmically as well as structurally.

A first interesting research direction would be to fully characterize the graphs for which Closed Geodetic Game is a She Loves Me-She Loves Me Not game, or those that are close to such a situation. We already exhibited complete graphs and stars as She Loves Me-She Loves Me Not games in Sect. 3, but there might be other very structured classes with many simplicial vertices that are in the same situation. Note that the unicity of minimum-size geodetic sets and the structural backbone of the graph are not sufficient conditions for this property.

Another idea would be to further the work initiated in Sect. 4 and to define a type of graph symmetry adapted to Closed Geodetic Game, that would allow us to obtain the outcomes of other structured graph classes. Expansions of cycles could be good candidates.

As explained in Sect. 5, the dynamic programing framework we used for block graphs and cacti requires strong constraints for shortest paths between vertices, which do not hold for more general graph classes such as outerplanar graphs. A research direction would be to try to express weaker constraints allowing a dynamic programing framework based on decomposition into independent subcomponents to apply for other classes. Outerplanar graphs are a natural candidate, since the geodetic set problem is polynomial for them but NP-hard for planar graphs.

Finally, other graph products than Cartesian (studied in Sect. 2) could be explored, such as the tensor, strong, lexicographical, modular or rooted products. However, those products may be more involved: for example, the outcome of the tensor product of two graphs is not necessarily a function of the outcomes of the two graphs. Some further conditions on the structure of the graphs or their product may be needed to obtain similar results.

Acknowledgements. The authors would like to thank the universities' libraries, without which the seminal paper [7] wouldn't be obtainable; as well as Sagnik Sen for his advice and tremendous help.

References

1. Albert, M., Nowakowski, R., Wolfe, D.: Lessons in Play: An Introduction to Combinatorial Game Theory. AK Peters/CRC Press (2019)
2. Araújo, S.N., Brito, J.M., Folz, R., de Freitas, R., Sampaio, R.M.: Graph convexity impartial games: complexity and winning strategies. Theoret. Comput. Sci. **998**, 114534 (2024)
3. Atici, M.: Computational complexity of geodetic set. Int. J. Comput. Math. **79**(5), 587–591 (2002)
4. Benesh, B.J., Ernst, D.C., Meyer, M., Salmon, S.K., Sieben, N.: Impartial geodetic building games on graphs. Int. J. Game Theory 1–34 (2025)

5. Berlekamp, E.R., Conway, J.H., Guy, R.K.: Winning Ways for Your Mathematical Plays. AK Peters/CRC Press (2001–2004)
6. Buckley, F., Harary, F., Quintas, L.: Extremal results on the geodetic number of a graph. Scientia **2A**, 17–26 (1988)
7. Buckley, F., Harary, F.: Closed geodetic games for graphs. Congr. Numer. **47**, 131–138 (1985)
8. Buckley, F., Harary, F.: Geodetic games for graphs. Quaest. Math. **8**(4), 321–334 (1985)
9. Buckley, F., Harary, F.: Distance in Graphs. Addison and Wesley, New York (1990)
10. Chakraborty, D., Das, S., Foucaud, F., Gahlawat, H., Lajou, D., Roy, B.: Algorithms and complexity for geodetic sets on planar and chordal graphs. In: 31st International Symposium on Algorithms and Computation (ISAAC 2020). Schloss Dagstuhl-Leibniz-Zentrum für Informatik (2020)
11. Chakraborty, D., Gahlawat, H., Roy, B.: Algorithms and complexity for geodetic sets on partial grids. Theoret. Comput. Sci. **979**, 114217 (2023)
12. Conway, J.H.: On Numbers and Games. CRC Press (2000)
13. Dailly, A., Gahlawat, H., Myint, Z.M.: The closed geodetic game: algorithms and strategies. arXiv preprint arXiv:2409.20505 (2024)
14. Ekim, T., Erey, A., Heggernes, P., van't Hof, P., Meister, D.: Computing minimum geodetic sets of proper interval graphs. In: Latin American Symposium on Theoretical Informatics, pp. 279–290. Springer (2012)
15. Fraenkel, A.S., Harary, F.: Geodetic contraction games on graphs. Internat. J. Game Theory **18**, 327–338 (1989)
16. Grundy, P.M.: Mathematics and games. Eureka **2**, 6–8 (1939)
17. Harary, F.: Convexity in graphs: achievement and avoidance games. In: North-Holland Mathematics Studies, vol. 87, p. 323. Elsevier (1984)
18. Harary, F., Loukakis, E., Tsouros, C.: The geodetic number of a graph. Math. Comput. Model. **17**(11), 89–95 (1993)
19. Harary, F., Nieminen, J.: Convexity in graphs. J. Differ. Geom. **16**(2), 185–190 (1981)
20. Haynes, T.W., Henning, M., Tiller, C.A.: Geodetic achievement and avoidance games for graphs. Quaest. Math. **26**(4), 389–397 (2003)
21. Mezzini, M.: Polynomial time algorithm for computing a minimum geodetic set in outerplanar graphs. Theoret. Comput. Sci. **745**, 63–74 (2018)
22. Nečásková, M.: A note on the achievement geodetic games. Quaestiones Math. **12**, 115–119 (1988)
23. Sieben, N.: Impartial hypergraph games. Electron. J. Comb. P2–13 (2023)
24. Siegel, A.N.: Combinatorial Game Theory, vol. 146. American Mathematical Society (2013)
25. Sprague, R.: Über mathematische kampfspiele. Tohoku Math. J. First Ser. **41**, 438–444 (1935)

ETH Lower Bounds for n-Queens: Time Waits for Nobody

MIT Hardness Group, Josh Brunner[1], Erik D. Demaine[1], Timothy Gomez[1](✉), Markus Hecher[2], and Meryl Zhang[3]

[1] MIT Computer Science and Artificial Intelligence Laboratory, 32 Vassar St., Cambridge, MA 02139, USA
{brunnerj,edemaine,tagomez7}@mit.edu
[2] Univ. Artois, CRIL, CNRS, UMR8188, Lens, France
hecher@cril.fr
[3] Plano Senior High School, Plano, TX, USA
merylz@mit.edu

Abstract. We develop lower bounds with often-matching exponential algorithms for several variants of the n-queens problem. In particular, we prove that placing n queens onto an $n \times n$ board with holes requires $2^{\Theta(n)}$ time for both decision and counting, assuming the Exponential Time Hypothesis (ETH) and #ETH respectively. The same result extends to more general manifolds. If the $n \times n$ board has no holes, but some of the queens are already placed, then completing the placement of n queens is known to be NP-complete and #P-complete; we show that both versions require between $2^{\Omega(\sqrt{n})}$ and $2^{O(n)}$ time, again assuming (#)ETH.

1 Introduction

The classic $\boldsymbol{n-queens}$ problem asks to place n queens onto an empty $n \times n$ chessboard such that no two queens can capture each other, i.e., no two queens share the same row, column, or diagonal. The $n = 8$ case was first posed in 1848 [2], and was subsequently studied by such luminaries as Gauss and Dijkstra [3]. The *decision version* of the problem is trivial: the answer is YES for all $n > 3$. *Counting* the number of solutions, even for a torus, is "Beyond #P" [8] because the output number requires $\Theta(n \log n)$ bits while the input is merely $\log n$ bits (the binary encoding of n).

To make the n-queens decision problem interesting from a complexity perspective, we can suppose that some k of the n queens are already placed on the board. The ***completion*** problem asks to place the remaining $n - k$ queens to form an n-queens placement. This problem is known to be NP-complete [6]. The known reduction is ***parsimonious*** (preserves the number of solutions), so counting completions is also #P-complete.

Artificial first author to highlight that the other authors (in alphabetical order) worked as an equal group. Please include all authors (including this one) in your bibliography, and refer to the authors as "MIT Hardness Group" (without "et al.").

H. Fernau and B. Zhu (Eds.): IWOCA 2025, LNCS 15885, pp. 287–301, 2025.
https://doi.org/10.1007/978-3-031-98740-3_21

While these n-queens problems are famous for being a benchmark for many AI models and computing formalisms [3,5,6,10,12], the *algorithmic complexity* of the problem remains unresolved. Assuming P $\neq$ NP, n-queens completion requires superpolynomial time, but how superpolynomial?

The most common approach to answering this type of question is to assume the ***Exponential Time Hypothesis (ETH)*** of Impagliazzo and Paturi [9]: solving 3SAT requires $2^{\varepsilon n}$ time for some $\varepsilon > 0$. As essentially all NP-hardness reductions can be traced back from SAT, we only need to track the polynomial growth of the problem size. If a reduction from 3SAT to a problem X increases the size of the instance from n to $O(n^c)$, then X requires $2^{\Omega(n^{1/c})}$ time assuming ETH. In particular, if we can find a reduction that expands the problem size only linearly, we obtain an $2^{\Omega(n)}$ lower bound. There is also a counting complexity analog of ETH called ***#ETH*** [4]: solving #3SAT requires $2^{\varepsilon n}$ time for some $\varepsilon > 0$. Further techniques such as sparsification [9] allow us to assume restrictions on 3SAT such as only a linear number of clauses, or stronger that each variable appears a constant number of times.

So how fast can we solve n-queens completion? Do we obtain ETH-tight bounds? What about counting solutions? Are there other board shapes where even n-queens decision becomes hard?

1.1 Our Results

Table 1 summarizes our results (and related known results) for n-queens. We consider both the decision and counting problems on several geometries, forming a hierarchy of natural generalizations of the $n \times n$ board. Beyond an $n \times n$ board, we consider allowing already placed queens (completion), or ***holes***—cells where queens cannot be placed but attacks can still go through. More generally, we consider ***manifolds*** made from gluing together a oriented unit squares to form l horizontal/vertical/diagonal lines.

Our lower bounds start in Sect. 2 by analyzing a chain of reductions to show that ETH and #ETH implies $2^{\Omega(n)}$ lower bounds for many variants of SAT such as 3SAT-3, 1-in-3SAT-3, and a recently introduced variant called Generalized #SAT [7] which allows assignments to be weighted. Next, in Theorem 3.1, we compute the ETH bound given by the chain of reductions in [6] to n-queens, where the last step of the reduction blows up the size by a quadratic factor, so we achieve a weaker bound of $2^{\Omega(\sqrt{n})}$ for both the decision and counting version with preplaced queens. Then, in the rest of Sect. 3, we improve this result to a singly exponential lower bound of $2^{\Omega(n)}$ when we have cells that are not allowed to have queens ($n \times n$ board with holes). Along the way, we build a reduction to the more general case of a manifold. In Sect. 4, we prove the same set of lower bounds for the counting versions of these problems.

Our algorithms in Sect. 5 generalize the $O^*(8^n)$-time[1] dynamic program from [13] to handle boards with holes, establishing that our lower bounds are tight

[1] We use $O^*(f(n))$ to represent $O(f(n) \cdot n^{O(1)})$. Hiding such polynomial factors is reasonable when f is superpolynomial.

up to constant factors assuming (#)ETH. We use the same technique to get a $O^*(2^l)$ algorithm on a manifold, which is tight according to this parameter.

Table 1. Complexity of n-queens placement/completion on the stated geometry. Lower bounds assume ETH, or #ETH for counting. Upper bounds use O^* to omit polynomial factors.

Type	Geometry	Lower Bound	Upper Bound
Decision	$n \times n$ grid completion	NP-complete [6]	$O^*(8^n)$ [Theorem 5.1]
Decision	$n \times n$ grid completion	$2^{\Omega(\sqrt{n})}$ (ETH) [Theorem 3.1]	$O^*(8^n)$ [Theorem 5.1]
Decision	$n \times n$ grid with holes	$2^{\Omega(n)}$ (ETH) [Theorem 3.3]	$O^*(8^n)$ [Theorem 5.1]
Decision	l-line manifold	$2^{\Omega(l)}$ (ETH) [Corollary 3.1]	$O^*(2^l)$ [Theorem 5.2]
Decision	a-area manifold	$2^{\Omega(a)}$ (ETH) [Theorem 3.2]	$O^*(2^a)$ [Theorem 5.3]
Counting	$n \times n$ grid	Beyond #P [8]	$O^*(8^n)$ [13]
Counting	$n \times n$ grid completion	#P-complete [6]	$O^*(8^n)$ [Theorem 5.1]
Counting	$n \times n$ grid completion	$2^{\Omega(\sqrt{n})}$ (#ETH) [Theorem 3.1]	$O^*(8^n)$ [Theorem 5.1]
Counting	$n \times n$ grid with holes	$2^{\Omega(n)}$ (#ETH) [Theorem 4.2]	$O^*(8^n)$ [Theorem 5.1]
Counting	l-line manifold	$2^{\Omega(l)}$ (#ETH) [Corollary 4.1]	$O^*(2^l)$ [Theorem 5.2]
Counting	a-area manifold	$2^{\Omega(a)}$ (#ETH) [Theorem 4.1]	$O^*(2^a)$ [Theorem 5.3]

2 Preliminaries

These problems all form a hierarchy. We start with the canonical version of this problem which is on a $n \times n$ chess board which we will call a *grid*. Each of these problems has two versions, *decision* which asks if a solution exists, and *counting* which asks how many solutions exist.

Definition 2.1 (n-Queens Placement).
INSTANCE: *$\langle n \rangle$ where n is an integer ≥ 1.*
SOLUTION: *A set Q of queens which all fit on board size n such that: $|Q| = n$; and for any two distinct queens $\mathbf{q}_1, \mathbf{q}_2 \in Q$, $\mathbf{q}_1$ does not attack $\mathbf{q}_2$.*

Definition 2.2 (n-Queens Completion).
INSTANCE: *$\langle n, P \rangle$ where n is an integer and P is a set of queens and their locations on board size n.*
SOLUTION: *A set S_2 of queens which is a solution to n-Queens Placement for n and such that $P \subseteq S_2$.*

Various variants were given as intermediate problems in [6] with given disallowed rows, columns, and diagonals which can simulate preplaced queens. Most of these reduce to n-queens completion with only a linear blow up. The next generalization we investigate is a *grid with holes*[2], where we have a $n \times n$ grid with certain squares where we cannot place queens but which do not prevent attacking over.

Definition 2.3 (n-Queens Placement with holes).
INSTANCE: *$\langle n, H\rangle$ where n is an integer, H is a set of locations where queens cannot be placed.*
SOLUTION: *A set S_2 of queens which is a solution to n-Queens Placement for n and such that for all $S_2 \bigcap H = \emptyset$.*

We refer to rows, columns, and diagonals as *lines*. More generally a line is a set of colocated points. A generalization of all these is n-queens on a *manifold*. The area a of a manifold is the number of cells on the manifold, or the total number of cells. Here each square has up to 8 neighbors each assigned a unique direction from the set north, south, east, west, north-east, north-west, south-east, and south-west. Each direction has an opposite in the obvious way. Here lines are formed by picking a cell and building a set by picking a direction recursively adding the next and previous cell to the set. Each direction is paired with the opposite side. We refer to the total number of lines as l. It is easy to see that grid and torus are a special case of a manifold. For a grid with holes we could not include those cells and connect the adjacent cells. More specifically if there is a hole at position (x, y) we do skip adding a cell for that point and instead setting $(x-1, y)$ and $(x+1, y)$ as east/west neighbors and $(x, y-1)$ and $(x, y+1)$ as north/south neighbors. We do similar for the diagonals. Using these observations we can describe a manifold as a collection of grids with holes which are interconnected in some way.

Definition 2.4 (n-Queens Placement on a manifold).
INSTANCE: *$\langle n, M\rangle$ where n is an integer ≥ 1 and a graph M which is a manifold.*
SOLUTION: *A set Q of queens which all fit on the manifold such that: $|Q| = n$; and for any two distinct queens $\mathbf{q}_1, \mathbf{q}_2 \in Q$, $\mathbf{q}_1$ is not on the same line as q_2.*

2.1 Assumptions and Satisfiability

The canonical problem we will use for showing hardness is Satisfiability (SAT). This decision problem asks given a v-variable CNF formula does there exists an assignment to the variables which satisfies all the clauses. kSAT restricts each class to have size k. The corresponding counting problem #3SAT asks to count the number of satisfying assignments. 3SAT is NP-complete and #3SAT is #P-complete, so a polynomial-time algorithm is likely to not exist. However in order to get stronger lower bounds ruling out subexponential time algorithms we use assumptions such as the Exponential Time Hypothesis.

Conjecture 2.1 (Exponential Time Hypothesis (ETH)[9]*).* There exists a constant ε such that 3SAT with v variables requires $2^{\varepsilon v}$ time.

---

[2] In [2], this version is referred to as "blocked n-queens".

For counting we refer to the analogous assumption as #ETH, which is implied by ETH. Let $\#(x)$ for an instance x of a problem X denote the number of its solutions. A ***c* − *monious reduction*** f is a function from problem X to problem Y such that for any instance x of X, $\#(x) = c \cdot \#(f(x))$. We only use polynomial-time computable c-monious functions; we say f is ***parsimonious*** if $c = 1$.[3]

Conjecture 2.2 (#ETH [4]*).* There exists a constant ε such that #3SAT with v variables requires $2^{\varepsilon v}$ time.

Let 3SAT-c be the version of 3SAT where each variable appears only in up to c clauses, note that when c is constant all instances are sparse. For many reductions we require 3SAT-3 which has been shown to be NP-hard [15]. It is not immediately clear that this version has the same lower bound assuming ETH.

Lemma 2.1. *Assuming ETH, there exists a constant* ε_1 *such that 3SAT-3 with* v *variables requires* $2^{\varepsilon_1 v}$ *time. Assuming #ETH, there exists a constant* ε_2 *such that #3SAT-3 with* v *variables requires* $2^{\varepsilon_2 v}$ *time.*

Proof. To start we use "sparsification" [9] which reduces any SAT instance to $2^{\varepsilon_1 v}$ many ***sparse*** instances with $m = O(v)$ clauses. Surprisingly this reduction also results in formulas where each variable only appears in poly$(\frac{1}{\varepsilon_1})$ clauses. The reduction in [15] from 3SAT-c to 3SAT-3 (shown in Fig. 1) introduces cv new variables. Because ε_1 is a constant, this reduction preserves linear size.

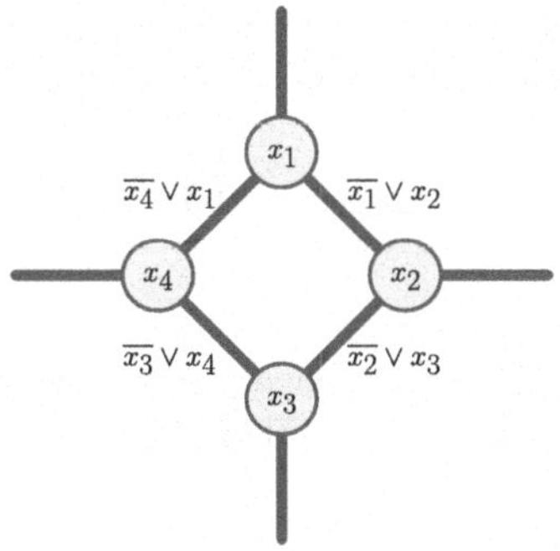

Fig. 1. Converts a degree-4 variable x into four degree-3 variables x_1, x_2, x_3, x_4 connected by clauses of size 2.

The sparsification lemma for counting in [4, Thm 1.1, A.1] has $O(v)$ clauses which for kSAT has $f(k, \frac{1}{\varepsilon_2})$ appearances. Thus for #3SAT we get a constant #3SAT-c. The reduction in [15] turns a single variable into a chain of variables (Fig. 1) that are all forced to be equal, so the reduction is parsimonious. □

[3] Such a strong reduction is not required as obtaining an #ETH lower bound can be done with only a c-monious subexponential time reduction.

A variant of 3SAT we use is 1-in-3SAT. Here we still have CNF clauses but each clause must have exactly 1 literal true. There exists a reduction that is parsimonious, preserves #ETH, and preserves constant variable appearance [1].

Lemma 2.2. *Assuming ETH there exists a constant ε_1 such that 1-in-3SAT-3 with v variables requires $2^{\varepsilon_1 v}$ time. Assuming #ETH there exists a constant ε_2 #1-in-3SAT-3 with v variables requires $2^{\varepsilon_2 v}$ time.*

Proof. We know that #1-in-3SAT-3 is NP-hard and #P-hard, even under parsimonious reductions from #3SAT-3 [1, Proposition 3; Appendix B: Figs. 21,22,23]. The number of appearances of the original variables remains the same during the reduction, and any new variables are used max 3 times. This reduction linearly preserves the number of variables, so under ETH and #ETH we can not solve either problem in time $2^{o(n)} \cdot n^c$ for some constant c. □

2.2 Generalized #SAT

Generalized #SAT has been defined in [7]. Each version of Generalized #SAT is specified by a set of ***clause types***, non-negative integer functions $f\colon \{0,1\}^k \to \mathbb{Z}_{\geq 0}$, which describes the number of ways a given assignment of k literals satisfies the clause. The two clause types we use in this paper are $\boldsymbol{f(x+y+z)}$ and $\boldsymbol{f(x+y)}$. We allow negations of variables, denoted with a bar (like $\bar{x}$), to be used freely in clauses. An input to Generalized #SAT consists of a number n, the number of variables (denoted $x = (x_1, \ldots, x_n)$), and a set of m clauses $C = \{\phi_1, \ldots, \phi_m\}$. Each ϕ_i is a k-tuple of literals, like $(x_{i_1}, \bar{x}_{i_2}, \ldots, x_{i_k})$. Let x_ϕ denote the restriction of a variable assignment x to the variables in a clause ϕ. The goal is to compute the number of ways to satisfy all clauses:

$$\sum_{(x_1,\ldots,x_n)\in\{0,1\}^n} \prod_{\phi\in C} f(x_\phi). \tag{1}$$

Previous work [7] showed that with just clause type $f(x+y)$ was #P-complete via reduction from #2SAT. Here we prove a $2^{\Omega(n)}$ lower bound when using both $f(x+y)$ and $f(x+y+z)$ assuming #ETH via reduction #1-in-3SAT-3.

Theorem 2.1. *Assuming #ETH, there exists a ε such that Generalized #SAT with clause types $f(x+y+z)$ and $f(x+y)$ with v variables requires $2^{\varepsilon v}$ when each variable appears in up to three clauses.*

Proof. By Lemma 2.2, it suffices to reduce from #1-in-3SAT-3. The goal of this reduction is for a given 1-in-3SAT clause (x, y, z) to create a set of Generalized #SAT clauses, which multiply to 1 when exactly 1 literal is true and 0 otherwise. We must preserve the number of variables and clauses linearly as well.

$$(x+y+z)(\overline{x}+\overline{y})(\overline{y}+\overline{z})(\overline{x}+\overline{z})$$

The first term will be 0 if there are no true variables. The next clauses all are 0 when any two variables are both true. If exactly 1 variable is true then the values of this expression is 2. Each satisfying assignment has a weight of 2^m, thus this is a 2^m-monious reduction. We do not introduce any new variables and increase clauses by a factor of 4. For appearances so far we have increase by a factor of 3. We can use the same degree reduction technique shown in Fig. 1. For each variable x which appears 9 times, we add variables $\{x_1 \ldots x_9\}$ and clauses $(\overline{x_i} + x_{(i+1)\%9})$ for $1 \leq i \leq 9$. When all these variables are equal, all the clauses are weight 1. When any two clauses are different some clause will equal 0. Thus this last step is parsimonious. Now our whole chain is 2^m-monious, preserves linear size, and has up to 3 appearances of each variable. □

3 Decision Hardness

First we point out the lower bound given by the reduction in [2] for the grid, given an n variable 3SAT instance this creates a instance of n^2-queens, however this lower bound is not tight with the best algorithms. Next we show the cases where there does exist an ETH tight lower bound. Even though our definition of manifold was very general, we still attempt to describe the manifold reduction as a collection of $O(1) \times O(1)$ grids, or gadgets, which are then linked together. For the manifold case we show that we create only a linear number of gadgets and thus we get singly exponential lower bounds in Theorem 3.2. We do this to make it easier to describe how to linearly embed all the gadgets into a $n \times n$ grid with holes to get a $2^{\Omega(n)}$ lower bound for a grid with holes.

Theorem 3.1. *n-queens completion on a grid requires $2^{\Omega(\sqrt{n})}$ time assuming ETH. #n-queens completion on a grid requires $2^{\Omega(\sqrt{n})}$ time assuming #ETH.*

Proof. We follow the chain of reductions performed in [6] from 1-in-3SAT-3[4] to n-queens completion. These reductions are between intermediate variants of n-queens which have constraints on (1) Allowed Rows and Columns and (2) Allowed Rows and Columns, Disallowed Diagonals. The reductions from (2) to (1) to the grid case are all linear blow up and involve preplacing queens to simulate constraints. However the reduction, which shows (2) is NP-hard, has super-linear blow up. This is from a variant consisting of M instances of n-queens completion, and a solution must be found for each instance such that none of the queens capture each other along one of the diagonals, but can along all other lines. The reduction, summarized in Fig. 2, embeds multiple chessboards spaced apart, thus the parameter board size n increases by a factor of M. Proving this last variant hard reduces from 1-in-3SAT-3 and creates $M = O(n)$ subinstances, a constant number for each variable and clause. The reduction in [6] is in fact parsimonious so it preserves #ETH as well. □

[4] This problem is called "Restricted 1-in-3SAT" in [6].

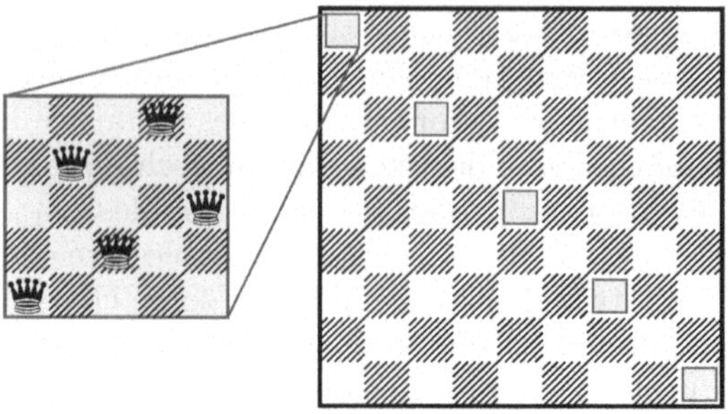

Fig. 2. Summary of the reductions in [6].

The bottleneck here is the intermediate problem which has $O(n)$ subinstances with limited interactions. Breaking it up into these gadgets made arrangement easier as the authors did not have to worry about many of the interactations between them. However in order to reduce this problem back to n-queens completion each subproblem needs its own set of rows, columns, and opposite diagonals.

3.1 Manifold

We use the gadgets almost identical to those used in [6], however by linking them on a manifold we can avoid the blow up given by the previous reduction. Each gadget consists of a $O(1) \times O(1)$ grid, the neighbors internally are given by adjacency in this grid. The variable gadget is a 4×4 grid shown in Fig. 3. There exists two ways to satisfy this gadget shown in each figures. The variable gadget has holes filling the second row from the bottom and the second column from the right. The clause gadget is a 3×7 grid with the three possible configurations shown in Fig. 4. Each column where a queen is not placed is filled with holes. Each column with a queen represents a literal of the clause. The queen being in the bottom cell of column indicates that a clause is satisfied by that literal.

(a) True variable gadget

(b) False variable gadget

Fig. 3. Two possible arrangements of the variable gadget.

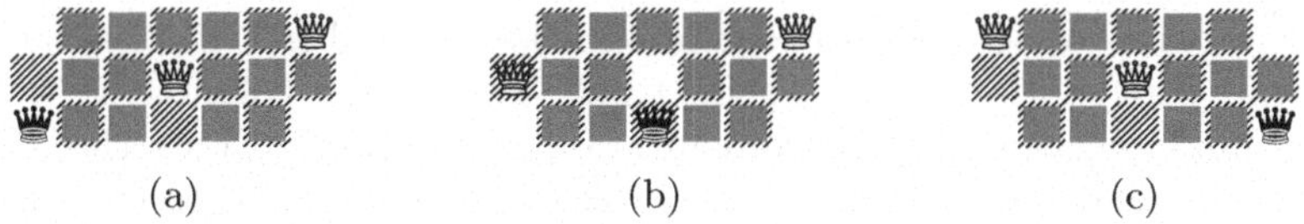

Fig. 4. The three possible arrangements of the clause gadget. The queen in the bottom row forces the truth value of a corresponding variable.

Gadgets are connected by diagonals, each variable gadget has up to three outgoing diagonals each connecting to the bottom cell of a clause gadget. This is shown in Fig. 5.

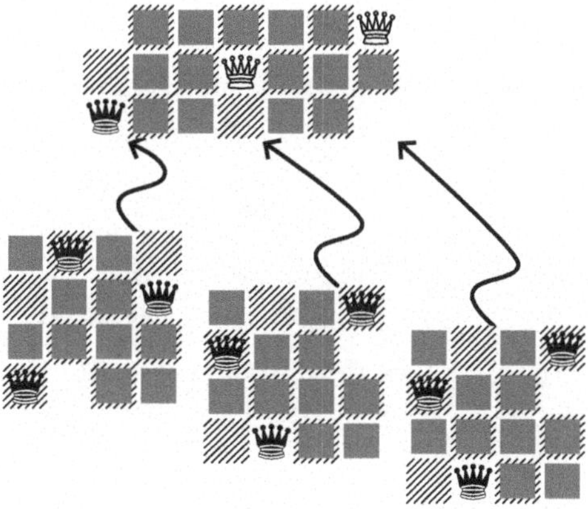

Fig. 5. How to link variable gadgets to clauses. The clause is $(x_1 \vee x_2 \vee \overline{x_3})$ and the variables are assigned 100. The configuration of the clause uses x_1 to satisfy it thus the queen in the bottom row is placed in the first column.

Theorem 3.2. *n-queens completion on a manifold with area a requires* $2^{\Omega(a)}$ *time assuming ETH.*

Proof. We reduce from 3SAT-3 which has a lower bound of $2^{\Omega(v)}$ from Lemma 2.1 for v variables. The reduction we use $O(1)$ cells for each variable and clause. Because the number of clauses is also $O(v)$, the total number of cells in this reduction $a = O(v)$. Thus we achieve a lower bound of $2^{\Omega(a)}$. Here $n = 3(v + c)$.

For correctness assume that there exists a satisfying assignment for 3SAT-3, then we may place the queens in the variable gadgets each corresponding to the assignment of that variable. Each clause is satisfied so we may select some literal which satisfies it and place the queen in that column. This cell is legal to place a queen because the configuration of the variable gadget represents the assignment of the variables.

If there exists a valid n-queens configuration then there exists a satisfying assignment. First note that each gadget only has 3 valid columns so they can only contain 3 queens each. The configurations shown in Fig. 3 are the only two configurations that are valid thus can represent an assignment to the variables. Each clause gadget needs to place a queen in the bottom row since the gadget only contains 3 rows. The bottom row cells each connect to a variable gadget so the clause must be satisfied by that variable. □

Corollary 3.1. *n-queens completion on a manifold with area l requires $2^{\Omega(l)}$ time assuming ETH.*

Proof. Each cell can only be in up to 4 lines which we use 8 directions. Thus $l \leq O(a)$ and the lower bound holds for l as well. □

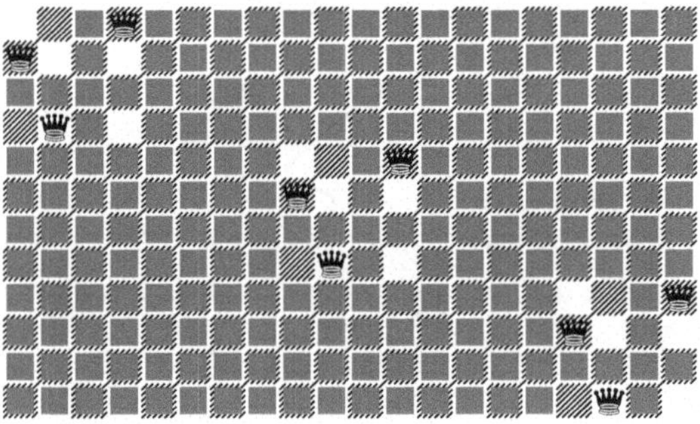

Fig. 6. How to arrange the variable gadgets with holes so that they don't unintentionally interact.

3.2 Grid with Holes

Theorem 3.3. *n-queens completion on a $n \times n$ grid with holes requires $2^{\Omega(n)}$ time assuming ETH.*

Proof. We will use the same variable and clause gadgets that are used in the proof of Theorem 3.2 above. The difference is that now, we need to describe an embedding of these gadgets in a $O(n) \times O(n)$ grid such that the only rows, columns and diagonals that are shared between gadgets are the diagonals that variables use to interact with clauses. When placing the gadgets, we refer to two squares of the chessboard as *interacting* if they are on the same row, column, or diagonal (i.e. a queen on one of the squares attacks the other, and thus constrains queens from being on both squares). For the clause gadget, we note that we can stretch the gadget by adding additional rows or columns between the queens as long as the nine non-hole squares where queens can go do not interact with each other along any diagonals. Thus, a clause gadget is really just nine squares, located in 3 (not necessarily adjacent) rows and 3 (not necessarily adjacent) columns, which do not interact with each other on any diagonals.

To lay out the variables, we place each variable along a 2:1 slope line, as shown in Fig. 6. This ensures that all of the variable gadgets are on distinct rows and columns, and the diagonals they will use to interact with clauses are all parallel diagonals.

To the upper left of the variable line, we will place our clauses. We'll place the rows of the clauses one at a time, greedily picking the lowest row which does not create any undesired interactions. For each clause, we place its bottom row on the lowest row which only interacts with the three variables it contains. On that row, we fill the entire row with holes except for the 3 squares which interact on the diagonals with the three variables present in the clause. Such a placement is *valid* if those 3 squares do not interact with any other clauses. The other two rows will be placed similarly on the lowest rows such that the three squares that are part of that clause do not interact with any other clause or variable.

Now we need to show that this greedy placement of clauses will guarantee that all of the clauses fit within $O(n)$ rows. We use induction on the number of clauses. Suppose we can fit the first n clauses in $O(n)$ rows. When placing the last clause, we place it in the lowest row which has no unintended interactions. For each square from a previous clause, there are up to 6 rows which contain a square that interacts with both that square and one of the three desired variables, preventing the clause from being placed on that row. These 6 possible interactions come from the intersection of the column and upward diagonal of the square in question intersecting each of the three variable diagonals. Since there are $9n$ squares in previous clauses, there are at most $54n = O(n)$ rows which have an unintended interaction with a previous clause. Thus, the lowest row which has no such interaction must be at a height of $O(n)$, so this clause can be placed (and along with all of the other clauses) can fit within $O(n)$ rows. □

4 Counting Hardness

We now utilize the lower bound proved in Theorem 2.1 for generalized #SAT. First note that the reduction from 3SAT-3 in Theorem 3.2 is not parsimonious. The clause gadget has exactly 3 legal configurations each corresponding to a literal. When a literal is satisfied then that configuration is possible. Thus a clause satisfied by multiple literals can have multiple configurations. The original #P-completeness in [2] overcame this by using a larger clause gadget and reducing from #1-in-3SAT-3. However we handle this "at the SAT" level and reduce from generalized #SAT.

Theorem 4.1. *#n-queens completion on a manifold with area a requires* $2^{\Omega(a)}$ *time assuming #ETH.*

Proof. Consider the reduction from Theorem 3.2, connect clause and variables in the same way. We must also create clauses of size 2 which we may do by create a slightly smaller gadget, dropping the rightmost and topmost column and row.

Let $X = (x_1, \ldots, x_n)$ be some assignment to the formula, define $Q(X)$ to to be the number of valid configurations with variable gadgets in states representing the bits of X. We then wish to show that for each X,

$$Q(X) = \prod_{\phi \in C} f(x_\phi)$$

Since we are essentially fixing the variable gadgets to represent X, the only degree of freedom is the clause gadgets. Since no two clause gadgets share a line they are independent thus we multiply the numbers of configurations of each. The number of possible configurations of a clause gadget is the number of true literals, in other words $f(x+y+z)$. Thus the reduction is parsimonious and still preserves linear size. □

The result extends to lines in the same way as Corollary 3.1. Finally changing the problem we're reducing from does not change the embedding size so we get our holes lower bound as well.

Corollary 4.1. *#n-queens completion on a manifold with l lines requires $2^{\Omega(l)}$ time assuming #ETH.*

Theorem 4.2. *#n-queens completion on an $n \times n$ grid with holes require $2^{\Omega(n)}$ time assuming #ETH.*

5 Algorithm

Previous work [13] has given a $O^*(8^n)$ algorithm for n-queens counting on both the grid and the torus. It easily generalizes to when we have a grid with holes, and thus for completion. We first prove this generalization showing that our lower bound is tight up to ETH for grids with holes. We further generalize the algorithm to a manifold and show our lower bounds are tight when parameterized by both l and a.

Theorem 5.1. *n-queens and #n-queens on a $n \times n$ grid even with holes can be solved in $O^*(8^n)$.*

Proof. n-queens counting can be solved using dynamic programming in $O^*(8^n)$ time [13]. The key to this dynamic program over the set of currently blocked lines, or lines which already contain a queen which we will refer to as line configurations. The algorithms stores line configurations and the number of ways to place queens to reach that it. Then they loop through each available position, attempting to place a queen there, then for each placement incrementing the count of line configurations or adding new ones to the queue. When starting with holes we can just skip these in the placement step. For the decision version we just check whether the output is greater than 0. □

Next we point out that the general form of the algorithm in [13] achieves $O^*(2^l)$ run time. We note naively running this version on an $n \times n$ grid gives a worse run time. In total an $n \times n$ grid contains $6n - 2$ lines however by merging partial configurations in [13, Theorem 3.2] the authors are able to decrease this search space to only these semi-exhausted lines when processing in row-major order. However this doesn't immediately extend to more general manifolds to decrease the run time.

Theorem 5.2. *n-queens and #n-queens on a manifold on l-line manifold can be solved in* $O^*(2^l)$.

Proof. As stated in [13] the run time is exponential in the number of lines. The dynamic program represents configurations uses currently blocked lines, thus the number of objects that can be pushed in the queue is $O(2^l)$. On a manifold we can run the same algorithm selecting each cell and placing a queen there and adding the new configurations to the queue. □

Finally we give a singly exponential upper bound even parameterized by area. Here this algorithm would be much worse on a grid as its area is n^2.

Theorem 5.3. *n-queens and #n-queens on a manifold with area a can be solved in* $O^*(2^a)$.

Proof. There are up to 2^a possible configurations as each cell either contains a queen or does not. We can check each one to see if it is a solution. □

6 Conclusion and Open Problems

In this paper we gave ETH bounds for generalized version of the n-queens problem. We first give the $2^{\Omega(\sqrt{n})}$ lower bound proven by previous reductions. We then provide singly exponential lower bounds starting at a grid with holes. We finish by analyzing the best deterministic upper bounds.

We believe that n-queens and its variants are a great candidate to better study the complexity of NP-complete and #P-complete problems. While we make progress toward this we leave a number of open problems.

- Can we get a $2^{\Omega(n)}$ lower bound for n-queens completion, i.e. can we eliminate the quadratic blow up?
- Can we decrease the base to get a $O^*(2^n)$ or smaller algorithm for counting? Can we further get a tighter bound on the number of lines we must process?
- Can we get any $\Omega(2^n)$ lower bounds using the Strong Exponential Time Hypothesis (SETH) on a manifold?
- On an $n \times n$ torus is the problem still hard?
- Are there any manifolds which are easier, perhaps parameterized?

In more detail we wish to close the gap for completion on a $n \times n$ board. While we do not believe a subexponential time algorithm exists, we proved that such an algorithm must take advantage of the board being a perfect grid, not even missing any holes. The botttleneck of our reduction is embedding the manifold into the grid which requires arbiratry placement of holes. If we could instead only block complete rows, columns, and diagonals, we could use the reductions from [6] to get a tight bound. While there exists techniques to handle some of the holes we include[5] the problematic ones are those that allow us to arrange variables and gadgets together as in Fig. 6. Currently the fastest algorithms on grids [13] decrease the naive base of 64 to 8 by merging configurations to bound the total number of lines we must process. Are there any other insights we can use to get a smaller bound? What is the smallest the base can be? ETH alone is not powerful enough, however using SETH we can get lower bounds of the form $\Omega(2^n)$ by reducing from kSAT. However in this case we also are still missing tools such as sparsification in this setting. To our knowledge there are no hardness results or lower bounds for the torus, can any of these reductions be extended? Finally we can further analyze these algorithms by studying simple manifolds such as those having bounded treewidth.

Acknowledgements. We found this open problem in thanks to MIT Future Tech's series of algorithmic surveys [11,14]. We would like to thank Jayson Lynch for helpful discussion and guiding early steps of the project and Jenny Diomidova for help with Generalized #SAT. This collaboration originated through the Research Science Institute (RSI), jointly sponsored by the Center for Excellence in Education and MIT. We thank the program for enabling this research partnership.

References

1. Barbanchon, R.: On unique graph 3-colorability and parsimonious reductions in the plane. Theoret. Comput. Sci. **319**(1–3), 455–482 (2004)
2. Bezzel, M.: Schachfreund. Berliner Schachzeitung **3**, 363 (1848)
3. Dahl, O-J., Dijkstra, E.W., Hoare, C.A.R.: Structured Programming, volume 8 of A.P.I.C. Studies in Data Processing. Academic Press (1972)
4. Dell, H., Husfeldt, T., Marx, D., Taslaman, N., Wahlen, M.: Exponential time complexity of the permanent and the Tutte polynomial. ACM Transactions on Algorithms (TALG) **10**(4), 1–32 (2014)
5. Gebser, M., Maratea, M., Ricca, F.: The seventh answer set programming competition: design and results. Theory Pract. Logic Program. **20**(2), 176–204 (2020)
6. Gent, I.P., Jefferson, C., Nightingale, P.: Complexity of n-queens completion. J. Artif. Intell. Res. **59**, 815–848 (2017)
7. MIT Hardness Group, Brunner, J., Demaine, E.D., Diomidova, J., Gomez, T., Hecher, M., Stock, F., Zhou, Z.: Easier ways to prove counting hard: a dichotomy for generalized #SAT, applied to constraint graphs. In: Mestre, J., Wirth, A., (eds.) 35th International Symposium on Algorithms and Computation (ISAAC 2024). LIPIcs, vol. 322, pp 51:1–51:14. Dagstuhl, Germany. Schloss Dagstuhl – Leibniz-Zentrum für Informatik (2024)

[5] such as reducing from 1-in-3SAT and having larger clause gadgets as in [6].

8. Hsiang, J., Hsu, D.F., Shieh, Y-P.: On the hardness of counting problems of complete mappings. Discrete Math. 277(1–3), 87–100 (2004)
9. Impagliazzo, R., Paturi, R.: On the complexity of k-SAT. J. Comput. Syst. Sci. **62**(2), 367–375 (2001)
10. Knuth, D.E.: The Art of Computer Programming, Combinatorial Algorithms, vol. 4B. Pearson Education (2022)
11. Liu, E.: A metastudy of algorithm lower bounds. PhD thesis, Massachusetts Institute of Technology (2021)
12. Google OR-Tools. The N-queens problem (2025). https://developers.google.com/optimization/cp/queens
13. Rivin, I., Zabih, R.: A dynamic programming solution to the n-queens problem. Inf. Process. Lett. **41**(5), 253–256 (1992)
14. Sherry, Y., Thompson, N.C.: How fast do algorithms improve? Proc. IEEE **109**(11), 1768–1777 (2021)
15. Tovey, C.A.: A simplified NP-complete satisfiability problem. Discret. Appl. Math. **8**(1), 85–89 (1984)

On Solving Simple Curved Nonograms

Maarten Löffler[1], Günter Rote[2], Soeren Terziadis[3(✉)], and Alexandra Weinberger[4]

[1] Utrecht University, Utrecht, The Netherlands
m.loffler@uu.nl
[2] Freie Universität Berlin, Berlin, Germany
rote@inf.fu-berlin.de
[3] TU Eindhoven, Eindhoven, The Netherlands
s.d.terziadis@tue.nl
[4] FernUniversität in Hagen, Hagen, Germany
alexandra.weinberger@fernuni-hagen.de

Abstract. Nonograms are a popular type of puzzle, where an arrangement of curves in the plane (in the classic version, a rectangular grid) is given together with a series of hints, indicating which cells of the subdivision are to be colored. The colored cells yield an image. Curved nonograms use a curve arrangement rather than a grid, leading to a closer approximation of an arbitrary solution image. While there is a considerable amount of previous work on the natural question of the hardness of solving a classic nonogram, research on curved nonograms has so far focused on their creation, which is already highly non-trivial. We address this gap by providing algorithmic and hardness results for curved nonograms of varying complexity.

Keywords: Nonogram · Arrangement · Puzzle · Dynamic Programming

1 Introduction

Nonograms, also known as *Japanese puzzles*, *paint-by-numbers*, or *griddlers*, are a popular puzzle type where one is given an empty grid in which some grid cells are to be colored (filled); the remaining cells remain empty (unfilled). For every row and column, there is a *description* that constrains the set of colored grid cells in this row or column. The description specifies how many consecutive blocks of cells should be filled and how large these blocks are. Two filled blocks need to be separated by one or more unfilled cells. A solved nonogram typically results in a picture (see Fig. 1).

Nonograms provide an accessible and contained environment for logical deduction. They have been used successfully to teach logical thinking [5,16], and have been shown to stimulate brain activity to prevent dementia [8].

Batenburg et al. [1] introduce the notion of a *simple* nonogram, which can be solved efficiently. A nonogram is *simple* when it can be solved by only looking at a single row or column at a time. More precisely, they consider a nonogram

H. Fernau and B. Zhu (Eds.): IWOCA 2025, LNCS 15885, pp. 302–315, 2025.
https://doi.org/10.1007/978-3-031-98740-3_22

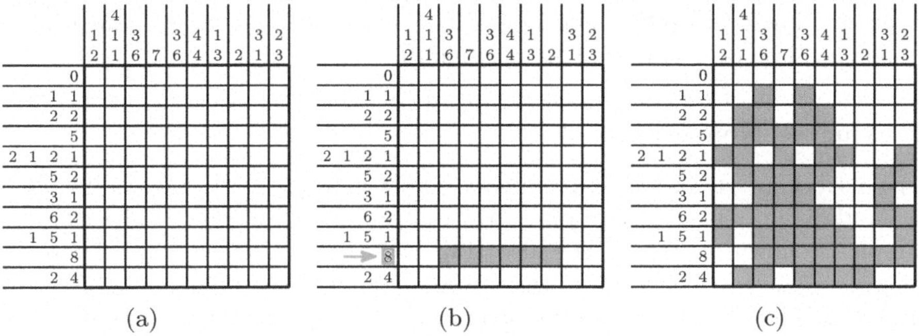

Fig. 1. A classic nonogram puzzle. An inference based on the highlighted clue. The solved nonogram.

simple if it can be solved by repeatedly considering a row or column, enumerating all possible solutions for it that are consistent with the fixed cells determined so far, and fixing all cells which have the same value in any possible solution. This procedure is called *settling a row/column* (or simply SETTLE) and will be considered in more detail in the preliminaries (Sect. 2). Note that since repeated application of settling a row or column is a deterministic process, the existence of multiple solutions for a nonogram immediately implies that it cannot be simple; however, the converse is not true, i.e., there are uniquely solvable nonograms that are not simple. In fact, Batenburg and Kosters introduce a whole hierarchy of complexity for nonograms, depending on the number of rows and columns which have to be considered simultaneously (by a specific solver) in order to definitively identify a cell whose status can be settled; puzzles with unique solutions can be found at all levels of this hierarchy.

Nonogram puzzles that appear in newspapers or similar platforms tend to be of this simple type [2], which contradicts to some extent the popular opinion that all interesting games and puzzles are NP-hard [6,14].

1.1 Solving Nonograms

Beside Jan Wolter's online survey [15], there has been substantial academic interest in nonograms. The natural question is to come up with an algorithm to decide whether a given nonogram can be solved. A great variety of approaches have been considered, ranging from heuristics to exact algorithms, see the full version [10] for more references.

The computational problem of deciding if a nonogram has a solution is NP-complete, as was first shown by Uada and Nagao [11]. This of course implies that computing this solution is also at least NP-hard. Uada and Nagao additionally prove that, given a nonogram and a solution, deciding if this solution is unique is also NP-hard, via a parsimonious reduction from three-dimensional matching.

In contrast, Batenburg and Kosters [2] gave a polynomial-time algorithm that, given a sequence of partially settled cells and a corresponding description, finds a cell that is either filled or unfilled in every possible solution, if such a cell exists. Their procedure can be used to either solve a given nonogram in polynomial time or decide that it is not simple.

1.2 Curved Nonograms

Van de Kerkhof et al. [12] introduced *curved* nonograms, a variant in which the puzzle is no longer played on a grid but on any arrangement of curves (an example is shown in Fig. 2); see also [13]. For distinction, we will refer to the nonograms played on a grid as *classic nonograms*. In curved nonograms, the numbers of filled faces of the arrangement in the sequence of faces that appear along a side of a curve, are specified by a description (one on each side). Curved nonograms allow cells with more organic shapes than classic nonograms, and thus lead to clearer or more specific pictures. Van de Kerkhof et al. focus on heuristics to automatically generate such puzzles from a desired solution picture by extending curve segments to a complete curve arrangement.

Additionally, they define three different levels of complexity of curved nonograms—not in terms of how hard it is to *solve* a puzzle, but how hard it is to understand the rules (see Fig. 2). It turns out that these difficulty levels nicely correspond with properties of the underlying curve arrangement as observed by De Nooijer et al. [7]. Specifically, *basic* curved nonograms are exactly the puzzles in which each description corresponds to a sequence of *distinct* faces. The analogy with descriptions in classic nonograms is straightforward. In an *advanced* curved nonogram, a face may be incident to the same curve multiple times, but only on the same side, and therefore a face can appear more than once in a sequence. If such a face is filled it is also counted multiple times when checking consistency with a description; in particular, it is no longer true that the sum of the numbers in a description is equal to the total number of filled faces incident to the curve. *Expert* curved nonograms may have descriptions in which a single face is incident to the same curve on *both* sides (which corresponds to the presence of a self-intersecting curve in the arrangement).

Research on curved nonograms has so far focused on their production. Klute, Löffler and Nöllenburg [9] investigate the geometric problem of adding descriptions to the ends of curves and provide polynomial-time algorithms for restricted cases and hardness results for the general problem. De Noojier et al. [7] aim to eliminate all faces with multiple incidences to the same curve (so-called *popular faces*) from a nonogram by adding one additional curve to the arrangement. The same goal was recently pursued by reconfiguring the curve arrangement through local crossing resolution [4].

1.3 Contribution

In this paper, we investigate for the first time the computational problem of *solving* curved nonograms. In particular, we investigate how the concept of *simple* nonograms translates to curved nonograms. After some preliminaries in Sect. 2, we present in Sect. 3 a dynamic program which leverages the nested structure of popular faces in advanced nonograms to check for a given sequence of faces along a curve, some of which are already filled or unfilled, if it can still be extended to

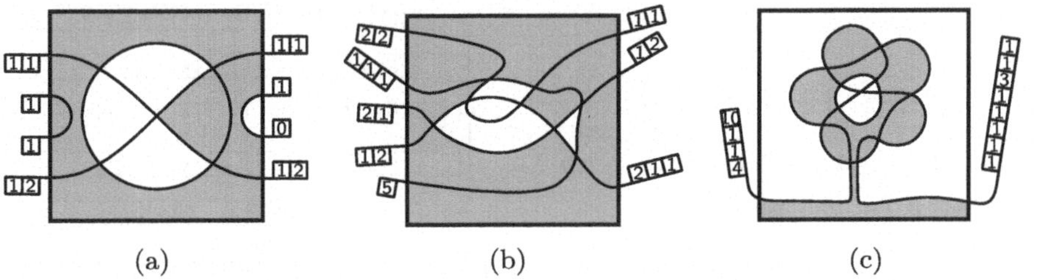

Fig. 2. Types of curved nonograms of increasing complexity [12], with solutions. (a) *Basic* puzzles have no popular faces. (b) *Advanced* puzzles may have popular faces, but no self-intersections. (c) *Expert* puzzles have self-intersecting curves.

a solution that is consistent with a given description. This implies a procedure solving simple advanced curved nonograms in $O(l^7)$ time, where l is the length of the longest description. This runtime can be improved to $O(l^6)$ by using an additional top-down phase of the dynamic program. In the case the nonogram is basic, the dynamic program coincides with a special case of the one presented by Batenburg and Kosters [2], showing that simple basic curved nonograms can be solved in the same way as simple classic nonograms. Then Sect. 4 shows that self-intersecting curves likely make curved nonograms significantly harder to solve, since even simple curved expert nonograms are at least as hard to solve as classic nonograms with a guaranteed unique solution. We close with some further research questions in Sect. 5.

2 Preliminaries

Nonograms. Let $\mathcal{A}$ be a curve arrangement consisting of h curves $A_1, \ldots, A_h$ all contained in and starting and ending at a rectangle called the *frame*. Every piece of a curve A between two consecutive intersections (or the start or end of A) is a curve segment of A. A face of $\mathcal{A}$ (also called cell) is *popular* if two or more curve segments incident to the face belong to the same curve. Every cell initially has the value *unsettled* which we denote with ?. If a value is assigned one of the two values *empty* (0) or *filled* (1) we say that the cell is *settled*. Here we follow the notation of Batenburg and Kosters [2].

We choose an arbitrarily orientation for each curve; accordingly, a face f incident to a curve segment s is said to be *on the left* or *on the right* side of s. Let $s_1, s_2, \ldots, s_k$ be the curve segments of a curve ℓ in $\mathcal{A}$. We call the list of faces $f_1, \ldots, f_k$, s.t. f_i is on the right (left) of s_i the *right (left) sequence* S^r_ℓ (S^l_ℓ) of ℓ. Popular faces can appear multiple times in the same sequence, and if ℓ is a self-intersecting curve, faces can appear in both sequences.

A specification for a sequence S is a string $\Psi^S = \psi_1\psi_2\ldots\psi_k \in \{0, 1, ?\}^k$, and it encodes the current state of knowledge about the faces in the sequence. If Ψ^S contains no ?, then it is a *fix*. If the sequence in question is clear from context, we may omit the superscript. If for two specifications Ψ and Ψ' of the same sequence S it holds that either $\psi_i = ?$ or $\psi_i = \psi'_i$ for all i , we say that Ψ' *refines* Ψ.

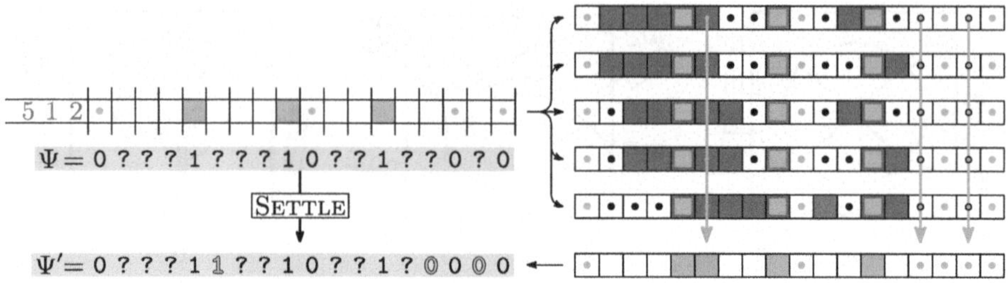

Fig. 3. An example of the SETTLE function. Filled cells are shown with a filled square, empty cells with a dot; all other cells are unsettled.For example, in a row with specification Ψ and description $D = 5\text{-}1\text{-}2$ (shown in the top left) there are five possible fixes of Ψ consistent with D shown in the top right (cells already settled in Ψ are colored gray, the remaining ones black). The blocks corresponding to the three parts of D are indicated in corresponding colors. One cell is filled and two are empty for all five fixes. This is indicated with yellow entries. (Color figure online)

A description $D = d_1, \ldots, d_t$ is a list of t numbers. One such number d_i will be called a *clue* of D. A fix Ψ of S is *consistent with* D if and only if Ψ contains exactly t maximal blocks of consecutive 1s and the i-th block consists of exactly d_i 1s. Since these blocks are maximal, consecutive blocks are separated by one or more 0s. A specification Ψ of S is *consistent with* D if there exists a fix that is consistent with D and refines Ψ.

In a curved nonogram, a face can appear more than once along a curve. This leads to additional constraints in the form of equations $\psi_i = \psi_j$. We encode this by a sequence of letters $f_1 \ldots f_l$, like abcdefdbgbh, where repeated letters indicate positions that belong to the same face. For example, the 2nd, 8th, and 10th edges lie on a common face, marked b. We call this the *letter description* of the sequence. We will only consider specifications Ψ that fulfill all equality constraints.

A *curved nonogram* C consists of a curve arrangement together with a set of descriptions and specifications (one for each sequence in C respectively). If all specifications are fixes, we say the nonogram is solved and conversely *solving* a given nonogram means obtaining a fix for every specification that is consistent with its description. For any $i \leq j$ we write $i \mathinner{..} j$ for the list of numbers between i and j (including both).

Settling and Nonogram Complexity. Given a specification Ψ, which is consistent with a description D, obtaining a specification Ψ' that refines Ψ and is still consistent with D is called *making progress on* Ψ. The procedure $\mathit{Settle}(\Psi, D)$ takes a specification and a description and (if possible) returns a specification Ψ', which refines Ψ and is consistent with D. It does so by settling any unsettled cells to be filled (or empty) if they have the same value in all possible fixes which refine Ψ and are consistent with D. This procedure is illustrated in Fig. 3.

Note that there can be exponentially many such fixes. However, while SETTLE is defined via equality of the value of a cell over all possible fixes, an implementation of SETTLE does not necessarily need to enumerate all possible fixes to find such a cell. For example in a classic nonogram, the dynamic program of Batenburg and Kosters [2] finds such a cell in polynomial time or decides that no such cell exists. Applying SETTLE to all rows and columns of a nonogram until no progress can be made is called a FULLSETTLE.

If every specification of a nonogram has a fix consistent with its description it is *solvable* and correspondingly we will call a nonogram in which every specification is a fix the *solution* of the nonogram. If a nonogram is solvable via a FULLSETTLE it is called *simple*. We remark that this definition is in line with [2], whose algorithm can solve simple classic nonograms in polynomial time.

3 Solving Simple Advanced Curved Nonograms

In this section we present a dynamic program which, given a sequence S together with a specification Ψ and a description D decides in polynomial time if there exists a fix Ψ' consistent with D that refines Ψ. This is analogous to the existing dynamic program by Batenburg and Kosters for classic nonogram [2]. Readers familiar with their work will easily spot the parallels; however, the presence of popular faces requires the maintenance of an additional data structure. The application of our algorithm to simple basic curved nonograms (i.e., those without popular faces) is discussed at the end of the section.

The property of advanced nonograms that is crucial for us is that the equality constraints are properly nested:

Observation 1. *Let i, j, k and l be four indices of letters in the letter description of a sequence S belonging to a curve A in an advanced curved nonogram, such that $f_i = f_j \neq f_k = f_l$. W.l.o.g. assume $\min(i, j, k, l) = i$ and $k < l$. Then either (i) $j < k \wedge j < l$ or (ii) $j > k \wedge j > l$.*

Proof. We only have to exclude the order $i < k < j < l$. Assume w.l.o.g. that we are considering the left side of A. Let a and b be points on the i-th and j-th segment of A, respectively. Since these segments lie on a common face, we can connect a and b by a curve B in that face, on the left side of A. Let $A[a, b]$ be the subcurve of A between a and b. Then $C = A[a, b] \cup B$ is a Jordan curve (simple and closed). If $i < k < j < l$, C encloses the face on the left side of the k-th segment, but it does not enclose the face on the left side of the l-th segment. Hence, these faces cannot be identical, contrary to our assumption $f_k = f_l$, and therefore, the order $i < k < j < l$ is impossible. □

We note that the nesting property of Observation 1 is the only property on which our algorithm relies. If a curve A has self-intersections but there are no faces that lead to a violation of the nesting property, our algorithm can be applied.

Theorem 1. *Consistency of a sequence S of length l with a description D with $\sum_{d \in D} d = k$ can be decided in time $O(k^3 l) = O(l^4)$.*

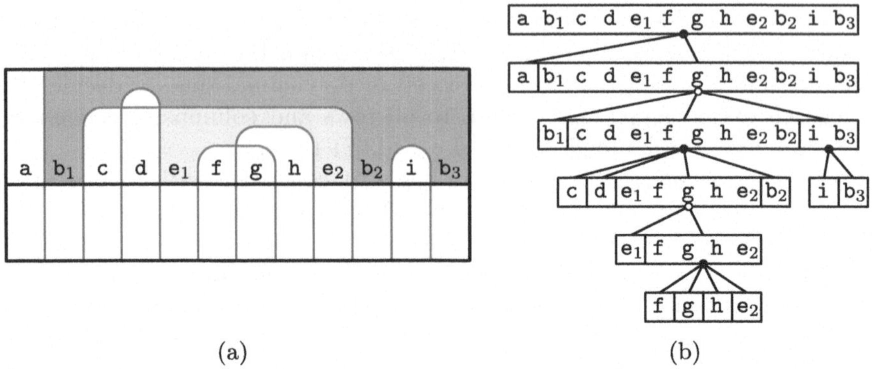

Fig. 4. (a) A schematic representation of a curve arrangement indicating the face incidences above the horizontal line and (b) its hierarchical decomposition into subintervals. Multiple occurrences of the same face (such as $\mathtt{b}_1, \mathtt{b}_2, \mathtt{b}_3$) are distinguished by indices. A white node denotes a decomposition of a complete group into brackets; a black node denotes decomposition of a bracket into complete groups. Black and white nodes occur in alternate levels of the tree.

Proof. In a bottom-up phase of the dynamic program we try to match larger and larger intervals of the specification with larger and larger parts of the description. In a subsequent top-down phase we will discover which assignments are consistent with an overall solution.

We translate $D = d_1 d_2 \dots d_t$ to a list $a_1 a_2 \dots a_k$ of `0`s and `1`, by creating blocks of d_i `1`s for every $1 \le i \le t$ and concatenating them with one `0` between consecutive blocks. Additionally we artificially pad the list by an extra `0` at the beginning and at the end. This assumption implies that every row and column starts and ends with an empty cell. Every nonogram can obviously be padded with empty rows and columns to achieve this. For example, `5-1-2` is translated to `011111010110`, with the understanding that a `0` has the potential to stretch to an arbitrary larger number of unfilled cells. A specification is consistent with this `0`-`1`-string variant of a description if one can create two equal strings by replacing every `?` in the specification with either `0` or `1` and replacing any `0` in the description with one or more `0`. This example shows this for $D =$ 5-1-2:

$$\text{0-1-string of } D\text{: } a_1 a_2 \dots a_i \dots a_k = \texttt{011111010110}$$

$$\text{Specification } \Psi\text{: } \psi_1 \psi_2 \psi_3 \dots \psi_j \dots \psi_{l-1} \psi_l = \texttt{0???1???10??1??0?0}$$

In the finished nonogram, all ?'s should be turned into 0's or 1's, subject to the requirement that the resulting sequence fits the specification. We want to know whether a particular `?` can be turned only into `0` or only into `1` in *all* possible solutions, because then this `?` can be fixed to this value. In other words we aim to implement the SETTLE procedure.

Recall that the finished solution must satisfy certain equations $\psi_i = \psi_j$ when two edges are incident to a common face, which we have encoded by a letter description $f_1 \dots f_l$, like **abcdefdbgbh**, where repeated letters indicate

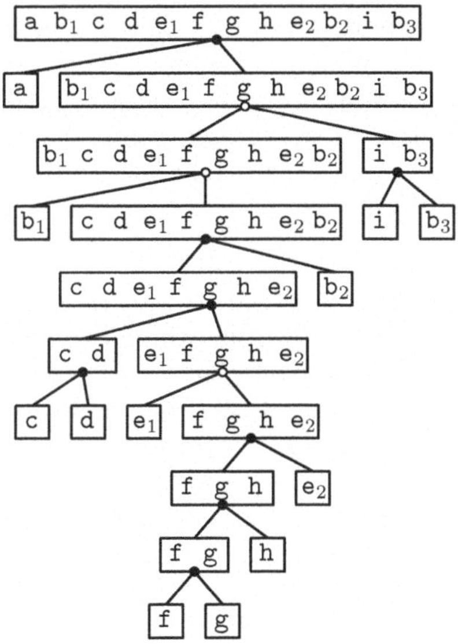

Fig. 5. The binary composition tree $\mathcal{T}$ corresponding to the tree of Fig. 4b, in which nodes of higher degree have been replaced by sequences of binary nodes. This is a binary tree whose leaves are the singleton intervals.

that edges belong to the same face. According to Observation 1 these repeated occurrences are *nested*: The pattern $\ldots x \ldots y \ldots x \ldots y \ldots$ cannot occur in the sequence.

We solve the following subproblems $\mathit{Match}_{j\,..\,j'}^{i\,..\,i'}$, for every $1 \le i \le i' \le k$ and for a certain set J of $2l - 1$ selected intervals $j\,..\,j'$ with $1 \le j \le j' \le l$:

> Can the ?'s in $\psi_j\,..\,\psi_{j'}$ be turned into 0's or 1's such that the resulting string is consistent with the description $a_i\,..\,a_{i'}$?

The subproblem $\mathit{Match}_{j\,..\,j'}^{i\,..\,i'}$ results in a Boolean value *true* or *false*. Accordingly, we will say that a subproblem is consistent or inconsistent. The set J of intervals $j\,..\,j'$ of the curve that we consider is defined as follows.

Suppose that the cells $j_1 < \cdots < j_m$ are the cells belonging to some common face: $f_{j_1} = \cdots = f_{j_m}$. We call the interval $j_1\,..\,j_m$ a *complete group*, and we call the intervals $j_1\,..\,j_p$, for $p = 1, \ldots, m$ the *progressive sleuths*. The first progressive sleuth is the singleton interval $j_1\,..\,j_1$. If a face occurs only once along the curve, at position j, then the singleton interval $j\,..\,j$ forms a complete group.

An interval $j_p + 1\,..\,j_{p+1}$ between two successive occurrences of the same face, including the second occurrence but not the first, is called a *bracket*. A bracket consists of a (possibly empty) succession of complete groups, followed by of the face of the enclosing group at position j_{p+1}. We consider the interval $1\,..\,l$ a bracket although it lacks the final element of the enclosing group.

We will build up the whole curve $1 \,..\, l$, starting from singleton intervals $j \,..\, j$. These can be seen as the leaves of a binary composition tree $\mathcal{T}$ (see Fig. 5) This binary tree represents how we will combine certain pairs of consecutive intervals $j \,..\, j'$ and $j'+1 \,..\, j''$ into larger intervals $j \,..\, j''$, as follows. Every complete group is built up from left to right by successive addition of brackets:

$$[j_1 \,..\, j_p] \cup [j_p + 1 \,..\, j_{p+1}] = [j_1 \,..\, j_{p+1}]$$

Similarly, every bracket is built up from left to right by successive addition of the complete groups that make it up (plus the final cell of the enclosing group). In total, a set J of $2l - 1 = O(l)$ intervals $j \,..\, j''$ are considered. Each such interval with $j < j''$ – an internal node in $\mathcal{T}$ – is built in a unique way from two disjoint subintervals in J (see also Fig. 5 for an example):

$$[j \,..\, j''] = [j \,..\, j'] \cup [j' + 1 \,..\, j''] \tag{1}$$

The singleton subproblems of the form $Match_{j..j}$ are trivial to solve: For a description of length 1, we have $Match_{j..j}^{i..i} \iff \psi_j = ? \vee \psi_j = a_i$, while $Match_{j..j}^{i..i'}$ is trivially inconsistent for $i < i'$, since a single cell can never be consistent with a description of length larger than 1.

Following the decomposition (1), each subproblem of the type $Match_{j..j''}$ with $j'' > j$ is associated to two families of smaller subproblems $Match_{j..j'}$ and $Match_{j'+1..j''}$, for some fixed j'. We solve these subproblems by this recursion:

$$\begin{aligned} Match_{j..j''}^{i..i''} \iff & \bigvee_{i':i \le i' \le i''-1} \left(Match_{j..j'}^{i..i'} \wedge Match_{j'+1..j''}^{i'+1..i''} \wedge a_{i'} = a_{i''} \right) \\ \vee & \bigvee_{i':i \le i' \le i''} \left(a_{i'} = 0 \wedge Match_{j..j'}^{i..i'} \wedge Match_{j'+1..j''}^{i'..i''} \wedge a_{i'} = a_{i''} \right) \end{aligned} \tag{2}$$

where the final condition $a_{i'} = a_{i''}$ is present only in case of composing a progressive sleuth with a bracket. It ensures that occurrences of the same face have the same color 0 or 1; when combining two complete groups, there are no shared faces that need to be considered, and the condition $a_{i'} = a_{i''}$ is omitted.

The first clause considers all possibilities of splitting the interval $i \,..\, i''$ into two disjoint parts $i \,..\, i'$ and $i' + 1 \,..\, i''$. The second clause considers in addition the possibility that the two parts of the curve can use overlapping parts of the description if the overlap is a 0. This completes the description of the bottom-up phase. The *target problem* $Match_{1..l}^{1..k}$ describes the original problem: consistency of the whole specification Ψ with the complete description D.

In total, there are $O(k^2 l)$ subproblems, and each subproblem can be evaluated by trying $O(k)$ choices for i', for a total running time of $O(k^3 l)$. □

3.1 Making Progress

Having solved the consistency problem, we immediately get a polynomial-time solution algorithm for making progress.

Theorem 2. *Given a single sequence S of length l and a description D with $\sum_{d \in D} d = k$ we can make progress or decide that no progress can be made in time $O(k^3 l^2) = O(l^5)$.*

Proof. We tentatively set a `?` to `0` or `1` and check consistency again. If one of the options is inconsistent, then we know that `?` must be replaced by the other letter, thus making progress. This is repeated most l times, for each occurrence of `?`. □

However, we can solve this more efficiently and avoid the additional factor l by a top-down phase, in which we mark certain subproblems as *extensible*. This is explained in detail in the full version [10, Section 3.1, Theorem 3].

Theorem 2 (or Theorem 3 of [10]) can be used to obtain the following corollary by the simple fact that there are only a linear number of rows and columns and cells in the nonogram. After applying the dynamic program once to every row and column, we must have made progress on at least one sequence, so there is at most an overhead of $O(l^2)$.

Corollary 1. *Simple advanced curved nonograms can naïvely be solved in time $O(l^7)$, or in time $O(l^6)$ using the top-down phase of [10, Theorem 3].*

3.2 Back to Basic Nonograms

When our algorithm is applied to a curve in a basic nonogram, there are no groups, and the whole sequence is just one bracket whose sequence of "complete groups" consists of singletons. The decomposition tree degenerates, and the algorithm simply grows the intervals $1 \mathinner{..} i$ and $1 \mathinner{..} j$ by adding one symbol at a time. Here our dynamic program reduces to the seminal algorithm of Batenburg and Kosters [2] (which is actually more general because it can deal with a specified *range* of lengths for each `1`-block instead of a fixed length).

4 Solving Simple Expert Curved Nonograms

In this section we show that solving a simple curved expert nonogram is at least as hard as finding the solution to a not necessarily simple classic nonogram provided that the classic nonogram has a unique solution.

Unique Solution Nonogram (1-SN). *Given a classic nonogram N with the guarantee it has a unique solution, find the solution for the nonogram N.*

Note that testing if a classic nonogram has a solution is NP-hard in general [11]. This of course implies that finding such a solution is also NP-hard. Ueda and Nagao [11] also show that testing if a given nonogram has more than one solution, even if we are given a solution, is NP-hard.

However this does not directly imply that finding a solution for a classic nonogram is still NP-hard if we are guaranteed that it has a unique solution. By providing a reduction from 1-SN to SIMPLE CURVED EXPERT NONOGRAM

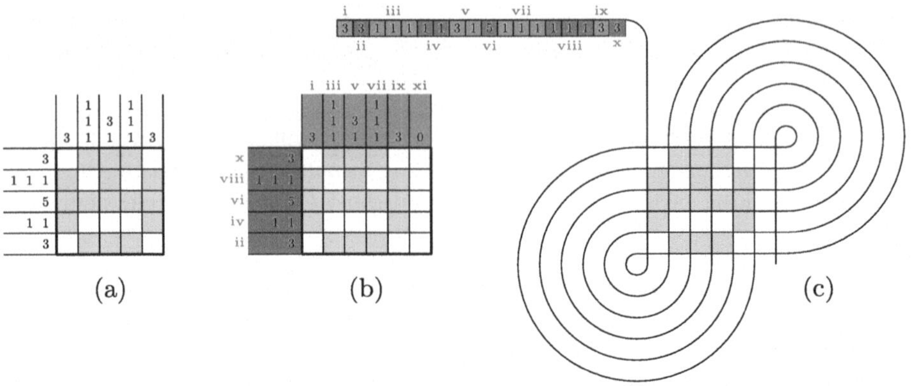

Fig. 6. (a) A classic nonogram with $w = 5$ columns and $h = 5$ rows. (b) In the reduction we pad a given nonogram to guarantee it has one more column than rows. The result here is a padded nonogram with $w = 6$ columns and $h = 5$ rows. The hints are annotated with the order in which they are collected into (c) a single self-intersecting *vital* curve that contains all grid lines of the classic nonogram and all descriptions in one *vital* description.

we still show that finding the solution to a simple curved expert nonogram is at least as hard as solving 1-SN.

The high-level overview of our reduction is as follows. We describe the construction of a curved expert nonogram C based on a given not-necessarily-simple classic nonogram N. The existing structure of N (after some additional padding to obtain a classic nonogram of desired proportions) can be replicated using a single *vital* curve, which is built by alternatingly connecting row and column boundaries of N with circular arcs. This is illustrated in Fig. 6.

If N has a unique solution, then C will equally have a unique solution. One can show that C is simple if an only if N has a unique solution. To guarantee this equivalence we have to deal with the fact that unifying all rows and columns into one big sequence risks that parts of the filled cells which satisfy a clue of the description can be placed outside the original row or column. To avoid this, the construction is padded with a sufficient number of necessarily filled cells around and intersecting the vital curve, s.t., large sequences of necessarily filled cells are inserted into the sequence restricting the remaining clues (which correspond to the original descriptions of N) to their correct rows and columns (see Fig. 7).

Now, given C, filling the trivially filled cells in the outer parts of the nonogram (the filled orange cells on the right in Fig. 7) yields a partially filled simple expert curved nonogram, in which all cells, which are not yet colored in, are part of the right sequence S_ℓ^r along a single curve ℓ; in fact they are part of both the right and the left sequence, but one such sequence is sufficient. Moreover the specification Ψ of this sequence includes already filled chains of cells (orange hints in the vital description on the right in Fig. 7) and there is a one-to-one correspondence between any chain of unsettled cells to a row or column of the

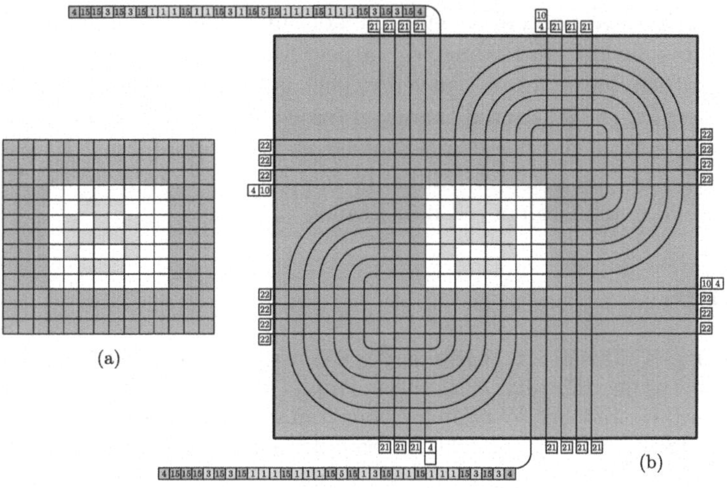

Fig. 7. (a) An even more padded version of the nonogram from Fig. 6. (b) The final construction including $k = 3$ additional rows and columns of filled cells on all sides. Original filled cells are yellow; padding cells are orange. In the vital description, numbers are orange if they are at least $2k$ and yellow otherwise. (Color figure online)

input classic nonogram (recall also Figs. 6b and c). Given any fix which refines Ψ and is consistent with the description of S_ℓ^r will fill in all remaining cells of C and immediately yields a solution for N.

This implies that any polynomial time algorithm that would compute progress on a single sequence of a simple curved expert nonogram – such an algorithm would again imply a polynomial-time scheme to solve these nonograms – would immediately imply a polynomial-time algorithm for 1-SN.

Containment in NP-is rather straight-forward since consistency of a fix with a description can be checked in polynomial time and therefore correctness of a given solution can be verified efficiently. The result is summarized in the following theorem.

Theorem 3. *Solving* Simple Curved Expert Nonogram *is (a) at least as hard as 1-SN and (b) in NP.*

Note that the construction of our hardness proof produces a *simple curve arrangement*: No three curves intersect in the same point, no two curves touch without crossing, and no curves locally overlap in more than a single point.

5 Conclusions

We have shown that the concept of *simple* nonograms extends to curved nonograms to some extent. In general, simple curved nonograms are not necessarily easy to solve: even the problem of testing for progress on a single description is already as hard as solving a classic nonogram under the assumption that it has a unique solution, and while we suspect this problem to be NP-hard, the complexity of solving a simple curved expert nonogram remains an open question. However, for the restricted classes of *basic* and *advanced* curved nonograms, we show that simple puzzles can be solved in polynomial time. It would be of interest how other measures of difficulty like the ones proposed by Batenburg and Kosters [3] extend to curved nonograms.

References

1. Batenburg, K., Henstra, S., Kosters, W., Palenstijn, W.: Constructing simple nonograms of varying difficulty. Pure Math. Appl. (PU. M. A.), **20**, 1–15 (2009)
2. Batenburg, K., Kosters, W.: Solving nonograms by combining relaxations. Pattern Recogn. **42**(8), 1672–1683 (2009). https://doi.org/10.1016/j.patcog.2008.12.003
3. Batenburg, K.J., Kosters, W.A.: On the difficulty of nonograms. J. Int. Comput. Games Assoc. **35**(4), 195–205 (2012). https://doi.org/10.3233/ICG-2012-35402
4. Brunck, F., Chang, H.-C., Löffler, M., Ophelders, T., Schlipf, L.: Reconfiguring popular faces. Comput. Reconfiguration Low-Dimensional Topological Spaces (Dagstuhl Sem. 22062), pp. 24–34 (2022). https://doi.org/10.4230/DagRep.12.2.17
5. Cigas, J., Hsin, W.-J.: Teaching proofs and algorithms in discrete mathematics with online visual logic puzzles. J. Educ. Resour. Comput. **5**(2) (2005). https://doi.org/10.1145/1141904.1141906
6. Cormode, G.: The hardness of the Lemmings game, or "Oh no, more NP-completeness proofs. In: Proceedings of 3rd International Conference on Fun with Algorithms, pp. 65–76 (2004)
7. de Nooijer, P., et al.: Removing popular faces in curve arrangements. J. Graph Algorithms Appl. **28**(2), 47–82 (2024). https://doi.org/10.7155/jgaa.v28i2.2988
8. Kasinathan, V., Mandhana, A., Mustapha, A., Adam, N.: How logical and mathematical games treat patients and adults suffering from dementia. Int. J. Curr. Res. **12**(18), 9–13 (2020). https://doi.org/10.31782/IJCRR.2020.121829
9. Klute, F., Löffler, M., Nöllenburg, M.: Labeling nonograms: boundary labeling for curve arrangements. Comput. Geom. Theory Appl. **98**, 101791 (2021). https://doi.org/10.1016/j.comgeo.2021.101791
10. Löffler, M., Rote, G., Terziadis, S., Weinberger, A.: On solving simple curved nonograms (2025). arXiv:2505.01554
11. Ueda, N., Nagao, T.: NP-completeness results for NONOGRAM via parsimonious reductions. Technical report TR96-0008, Department of Computer Science, Tokyo Institute of Technology. CiteSeerX (1996). 10.1.1.57.5277
12. van de Kerkhof, M., de Jong, T., Parment, R., Löffler, M., Vaxman, A., van Kreveld, M.: Design and automated generation of Japanese picture puzzles. In: 40th Annual Conference on European Association Computer Graphics (2019). https://doi.org/10.1111/cgf.13642
13. van Kreveld, M. J.: On nonogram and graph planarity puzzle generation. In: 30th Canadian Conference on Computational Geometry (CCCG), pp. 326–327 (2018). http://www.cs.umanitoba.ca/~cccg2018/papers/invited_paper_3.pdf
14. Viglietta, G.: Gaming is a hard job, but someone has to do it! Theory Comput. Syst. **54**(4), 595–621 (2013). https://doi.org/10.1007/s00224-013-9497-5

15. Wolter, J.: Survey of paint-by-number puzzle solvers. https://webpbn.com/survey/. Accessed 22 Feb 2025
16. Yang, Y., Zhang, D., Ji, T., Li, L., He, Y.: Designing educational games based on intangible cultural heritage for rural children: a case study on "Logic Huayao. In: Advances in Human Factors in Wearable Technologies and Game Design (AHFE), pp. 378–389 (2019). https://doi.org/10.1007/978-3-319-94619-1_38

Tile-Based Knot Assembly with Celtic!

Divya Bajaj, Ryan Knobel, Juan Manuel Perez, Rene Reyes, Ramiro Santos, and Tim Wylie(✉)

University of Texas Rio Grande Valley, Edinburg, TX, USA
timothy.wylie@utrgv.edu

Abstract. In this paper we focus on the intersection of tile assembling systems, edge-matching puzzles, combinatorial games, and knot construction and identity. As a basis, we utilize the game Celtic!, which is a 2-Player board game where the goal of the game is to construct knots where one knot uses more of a player's pieces than the other player over all knots. All pieces must build off an existing knot and a valid knot must be closed. We consider three variations: a 0-player self-assembly variation that deterministically places pieces to form a closed knot of some length, a 1-player puzzle variation where the goal is to form a closed knot of some length, and the original 2-player game with restricted pieces. We show these are P-complete, NP-complete (depending on the pieces), and PSPACE-complete (for a first-player win), respectively. We nearly fully characterize the hardness of the 1-player puzzle based on the pieces.

1 Introduction

Square edge-matching puzzles generally employ the use of (possibly rotatable) tiles, with each side given a specific label that dictates how they may be used in conjunction with other tiles. These types of puzzles have existed for centuries, and are often deceptively difficult to solve. For instance, the Eternity II puzzle [19], introduced in 2007, offered $2,000,000 for the first complete solution. While no restrictions were placed on the methods used, no complete solution has yet to be found, with the closest solutions falling short a few pieces. The work of [6] shows that edge-matching puzzles, including MacMahon Squares [12], Scramble Squares, and TetraVex [17], are NP-complete and are equivalent to other types of puzzles such as jigsaw and polyomino packing puzzles.

The puzzles often remain NP-hard even for small instances. In [7], the authors show that even for a 1-by-n puzzle, unsigned edge-matching with rotation is still NP-hard and fixed parameter tractable in the number of unique labels. This work was later improved in [3], where it was shown that for both signed and unsigned edges, 1-by-n puzzles are NP-hard, even to approximate. Naturally, variations of these types of puzzles have been considered, such as having additional inequality constraints between adjacent tiles, triangular edge matchings, and when no target shape is specified [2].

This research was supported in part by National Science Foundation Grant CCF-2329918.

H. Fernau and B. Zhu (Eds.): IWOCA 2025, LNCS 15885, pp. 316–331, 2025.
https://doi.org/10.1007/978-3-031-98740-3_23

(a) Completed Game (b) Completed Knots (d) Bounded 2CL

Fig. 1. (a) Final Celtic! board where all knots are closed and no valid moves exist. (b) Celtic! board displaying completed knots. The knot score for Red (R) and Blue (B) players based on the knot. Red knot: R(9) - B(2), Blue knot: R(1) - B(5), Yellow knot: R(4) - B(1), and White knot: R(0) - B(1). Given these knots, the best score for Red is 9 and 5 points for Blue. (c) Gadgets for 0-player non-planar Bounded Deterministic CL. (d) Gadgets for Bounded 2-player CL. (Color figure online)

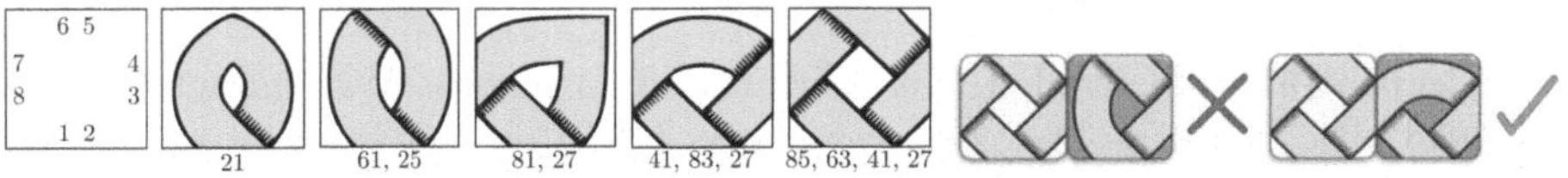

(a) Piece paths (b) Invalid vs. Valid

Fig. 2. (a) The 5 types of pieces in the game as well as their path connections (in this orientation). Each player (red and blue) has two copies of each type and there is a shared set of white-backed ones. (b) Invalid vs. valid placement. (Color figure online)

In general, edge-matching puzzles are undecidable with unlimited pieces [1]. More abstractly, generalized edge-matching of squares (without rotation) is closely related to computational complexity as exhibited through Wang tiles [18], and more recently, self-assembly models such as the abstract Tile Assembly Model (aTAM) [20], which are both Turing Universal even with small tile sets.

Our work focuses on a game with similar edge-matching mechanics: Celtic! [5]. Celtic! is a 2-Player (Blue/Red) tile placement game where each player attempts to complete knots by alternatively placing pieces onto a fixed size board. Each piece can be thought of as a rotatable square with a 'closed' or 'open' label on each side, similar to that of unsigned edge-matching with two labels. The goal of the game is alternate playing pieces and completing knots. The player who has the most number of pieces of their color in a knot (over all knots) wins the game. See Fig. 1 for an example of a completed game with scoring.

The pieces of Celtic! use crossing paths, which makes it a type of connection game where the goal is to build connecting paths. Path based pieces and mechanics show up in several games such as Tsuro [14], Squiggle [13], Travel [8],

Table 1. Summary of results for different Celtic! variations. All reductions use a $O(n) \times O(n)$ size board.

Version	Player Pieces	Board Pieces	Complexity	Theorem
0-Player		all	P-complete	Thm. 8
1-Player		all	$O(n^2)$	Thm. 1
		all	$O(n^2)$	Thm. 2
		all	$O(n^2)$	Thm. 3
			NP-complete	Thm. 4
		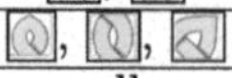	NP-complete	Thm. 5
		all	OPEN	-
		all	OPEN	-
			NP-complete	Thm. 6
2-Player			PSPACE-complete	Thm. 7

Kaliko/Psyche-Paths/Cram [16], and many others. Please see [4] for an overview of connection-based (and edge-matching) games.

Our Contributions. Table 1 overviews the main results. We characterize the game of Celtic! for 0, 1, and 2 player games based on the types of pieces. We consider the complexity of three variations of the board game. To start, we look at a 1-player variation of Celtic! in which some pieces initially exist on the board and the goal of the player is to form a single closed knot of some length in k moves. We show that for some pieces, this is NP-complete in Sect. 3. We show that it is PSPACE-complete to determine if there is a 1st-player win in a 2-player game (Sect. 4). In Sect. 5, we analyze a 0-player version with deterministic placement and show that making a closed knot of some length is P-complete.

2 Preliminaries

We first detail some game mechanics from the original game, as well as provide formal definitions for the general game and problems afterwards.

2.1 Game Mechanics

Pieces. The game pieces consist of 5 distinct types of pieces, with 5 copies of each type (25 pieces in total), with path segments (Fig. 2a). The Blue and Red players each own 10 pieces (2 of each type with their background color), and they share 5 pieces with a **white** background.

Initial Configuration and Rules. The  piece is placed to begin. Players take turns placing a piece adjacent to already existing pieces to form knots; each piece must be placed within the playing area such that no knot is prevented from closing (Fig. 2b). In the standard game, it must remain in a 5×5 square.

2.2 Definitions

Let $D(p_1, p_2)$ be the ℓ_1-norm (Manhattan distance) between points p_1, p_2. We say 2 cells $[x_1, y_1], [x_2, y_2]$ are **adjacent** if $D((x_1, y_1), (x_2, y_2)) = 1$.

Definition 1 (Board and cells). *A* ***board*** *B is a fixed $w \times h$ lattice surface composed of* ***cells****, with each cell $[x, y] \in B$ corresponding to a location in row x and column y, and $[0, 0]$ corresponding to the bottom-leftmost cell.*

Definition 2 (Path-segment). *A path-segment is a pair of connected edge points on a piece that represent a continuous path from one point to the other.*

Definition 3 (Piece). *A* ***piece*** $p \in \{$ , , , ,  $\}$ *is a rotatable 4-sided unit square, where each side is labeled* ***open*** *or* ***close*** *denoting the type of side that may or may not attach on this side. Each open side has at most two path-segment points (over and under). Label two points on each side $\langle 1, 2, 3, 4, 5, 6, 7, 8 \rangle$ going counterclockwise around the square (Fig. 2a) as all path-segment points, Celtic! uses pieces (ignoring rotation) with path-segments (21)* *, (25, 61)* *, (27, 81)* *, (27, 41, 83)* *, (27, 41, 63, 85)* *. We use the convention that each path-segment has the over number followed by the under number, and we use this rotation as the canonical orientation.*

Definition 4 (Matching and Paths). *Two pieces p_1, p_2 are* ***matching*** *if they are in adjacent cells and the touching sides are both labeled open. A* ***path*** *is made of a sequence of path-segments from pieces. A path between p_1, p_2 conceptually connects the internal path-segment $i_1 o_1$ from p_1 to path-segment $i_2 o_2$ on p_2, and is defined as $P = \langle \ldots, i_1 o_1, i_2 o_2, \ldots \rangle$, where o_1, i_2 lie on the touching sides of p_1, p_2, and $o_1 \mod 2 = i_2 \mod 2$.*

Definition 5 (Valid piece). *A piece in cell $[x_1, y_1]$ is said to be* ***valid*** *if:*

1. *$\exists [x_i, y_j]$ for $(i, j) \in \{(2, 1), (1, 2), (0, 1), (1, 0)\}$ such that $[x_1, y_1], [x_i, y_j]$ contain matching pieces, and*
2. *All other adjacent cells $[x_i, y_j]$ for $(i, j) \in \{(2, 1), (1, 2), (0, 1), (1, 0)\}$ satisfy (a) $[x_1, y_1], [x_i, y_j]$ contain matching pieces, (b) the touching sides of the pieces in $[x_1, y_1]$ and $[x_i, y_j]$ are both labeled closed, or (c) $[x_i, y_j]$ is empty.*

Definition 6 (Knot). *A* ***knot*** *is a continuous path $P = \langle i_1 o_1, \ldots, i_k o_k \rangle$ through path-segments on valid pieces $\langle p_1, \ldots, p_k \rangle$ where each path piece p_a identifies a corresponding input and output path-segment $i_a o_a$ on a piece such that*

1. *each p_a matches with p_{a+1} for $1 \leq a \leq k - 1$ and p_k matches with p_1, and*
2. *each o_a continues through i_{a+1} for $1 \leq a \leq k - 1$ and o_k continues to i_1.*

Note that multiple path-segments through the same piece may exist. The length (score) of a knot is not the number of path-segments in pieces it crosses, but the number of ***distinct*** *pieces (or grid locations) the total path crosses. This definition ignores the topological properties of a knot in relation to other knots.*

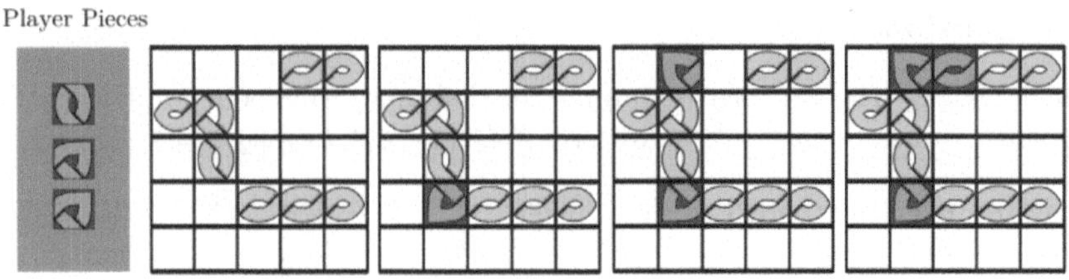

Fig. 3. Example of an initial game board configuration for generalized 1-Player Celtic! and a sequence of valid moves that form a closed knot of length $|K| = 11$ in $k = 3$ moves given 2 pieces and 1 piece.

Problem Definitions. This paper looks at 3 variations of Celtic!: a 0-Player simulation (Fig. 13), a 1-Player puzzle (Fig. 3), and a 2-Player game (Fig. 1).

Definition 7 (0-Player Celtic!). *A* 0*-player Celtic! simulation is defined as:*

INPUT: *A $w \times h$ board partially filled with pieces* $\in \{$, , , , $\}$*, a piece as a starter piece and a positive integer L.*

OUTPUT: *Is it possible to place the starter piece in a cell* $[0, i]$ *rotated* 180 *degrees for some integer $0 < i < w$ and create a knot of length $\geq L$?*

Definition 8 (1-Player Celtic!). *A* 1*-Player Celtic! puzzle is defined as:*

INPUT: *A $w \times h$ board partially filled with pieces, a multi-set S of pieces (with $k = |S|$) and positive integer L where $k \leq L \leq k + L \leq wh$.*

OUTPUT: *Can k valid pieces (from S) be placed such that a closed knot of length $\geq L$ is formed?*

Definition 9 (2-Player Celtic!). *A* 2*-Player Celtic! game is defined as:*

INPUT: *A $w \times h$ board with a white piece in cell $[i, j]$, where $1 \leq i \leq w$ and $1 \leq j \leq h$, multi-sets S_R, S_B, and S_W of red, blue, and white pieces, respectively.*

MOVES: *Players take turns placing a valid move using one of their colored pieces or a white piece. Once no valid moves exist, each player chooses a completed knot and receives a score. The score M_R, M_B for each player is a positive integer denoting the number of pieces of their color from their chosen knot.*

OUTPUT: *Whether there is a Blue player win, a Red player win, or a draw based on* $\max(M_R, M_B)$.

2.3 Constraint Logic

The 0-player and 2-player reductions make use of constraint logic (CL) [9]. Problems in CL are based on a constraint graph where each weighted directed edge has value 2 (blue) or 1 (red), and every vertex has a minimum inflow constraint of 2. The game is based on flipping edges of the graph, where an edge can be flipped only if the minimum inflow constraints are maintained. In [9], they show that all graphs in CL can be reduced to a few simple gadgets, each of degree 3, that are equivalent for the games.

The 0-player gadgets for Bounded Deterministic Constraint Logic (BDCL), Fig. 1c, require that the graph is non-planar and every edge may only be flipped once. They consist of an AND, FANOUT, and OR. BDCL is P-complete.

The Bounded 2-player Constraint Logic (B2CL) gadgets (Fig. 1d) have assigned edges (white is 1st player and black is 2nd) and may only be flipped once by the owner. They consist of an AND, FANOUT, OR, CHOICE, and VARIABLE. B2CL is PSPACE-complete even for planar graphs. Note that the 2nd player can only move in the variable gadget.

3 Generalized 1-Player Celtic!

This section looks at generalized 1-Player Celtic! played on an $O(n) \times O(n)$ board, where the goal is to build a closed knot of length $\geq L$ in k moves. We start with polynomial time algorithms when restricted to placing , , or  pieces. We then show a general framework for the NP-hardness reductions, which is used in the final section to show NP-completeness when restricted to either placing  or , as well as for other piece combinations.

3.1 Polynomial Cases

Theorem 1. *Building a closed knot of length $\geq L$ with k moves by placing*  *pieces in Generalized 1-Player Celtic!, where the initial board configuration can contain* , , , , *and*  *pieces, is solvable in $O(n^2)$ time.*

Theorem 2. *Building a closed knot of length $\geq L$ with k moves by placing*  *pieces in Generalized 1-Player Celtic!, where the initial board configuration can contain* , , , , *and*  *pieces, is solvable in $O(n^2)$ time.*

Theorem 3. *Building a closed knot of length $\geq L$ with k moves by placing*  *pieces in Generalized 1-Player Celtic!, where the initial board configuration can contain* , , , , *and*  *pieces, is solvable in $O(n^2)$ time.*

3.2 General Framework for Hardness Reductions

We outline the framework used for the hardness reductions in Sect. 3.3. For each result, we reduce from directed, planar Hamiltonian cycle with max degree 3 [15]. At a high-level, given a directed, planar graph, we transform it into an equivalent rectilinearly embedded graph G_e [10]. We space out the vertices in G_e so each edge may be transformed to have roughly the same length (Fig. 4). Denote this new graph G. We then construct the Celtic! board used for the reduction. We replace edges with  and  pieces, leaving a 7×7 empty region centered at each vertex, which is where the vertex gadget will be placed. The gadget consists of a  piece with other pieces placed depending on the allowable pieces, and the main objective being that all incoming edges lie either to the left or right of the central vertex piece, and all outgoing edges lie on the other side. This forces the

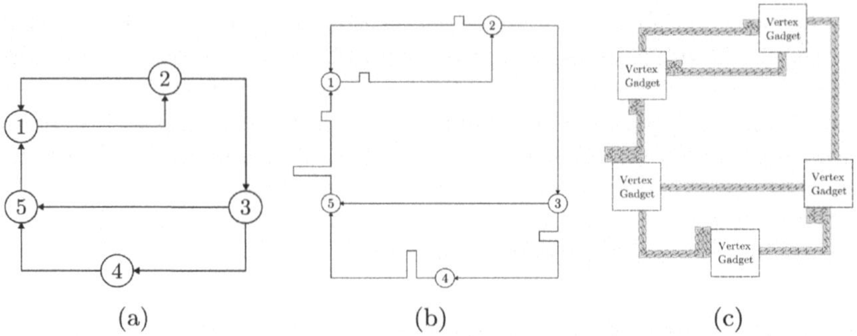

Fig. 4. The 3-step process for creating the initial board for the NP-complete reductions. (a) Example directed, planar graph with max degree 3. (b) The graph is scaled by a factor of $|V|^2$, then transformed to have edges of roughly equal lengths using the free space around each vertex. The edges within 3 units of the vertex are left untouched. (c) The initial board created for the graph. The vertex gadget changes per reduction depending on the pieces used.

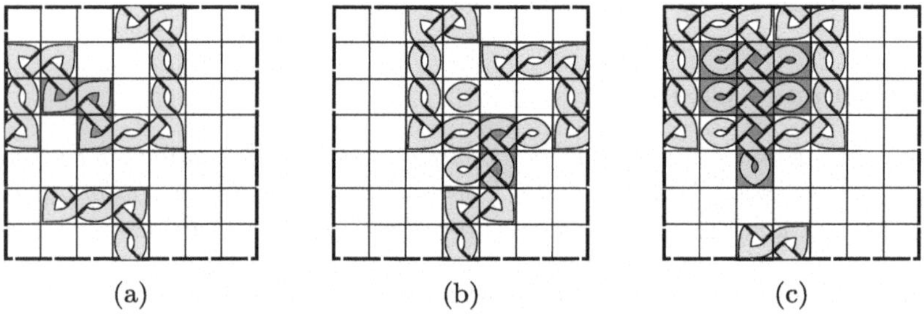

Fig. 5. Vertex gadgets for the (a)  piece selecting the western edge, (b)  selecting the southern edge, and (c)  and  selecting the western edge.

player to place pieces that choose exactly one incoming edge and one outgoing edge per vertex gadget. If these choices are consistent across all vertices, then a knot of length L is formed corresponding to a Hamiltonian cycle in the original graph G_e, otherwise, the knot is not formed.

For each reduction, we give a vertex gadget with the different combinations of input and output on different sides for the construction using the pieces for that problem. Depending on the reduction, some additional pieces are needed to aid with making a choice between the incoming and outgoing edges at each vertex piece. Although not all cases are given, a few examples are shown in Fig. 5.

3.3 Hardness Results

Using the framework discussed in Sect. 3.2, we now show NP-completeness for different piece combinations.

Lemma 1. *Generalized 1-Player Celtic! is in NP.*

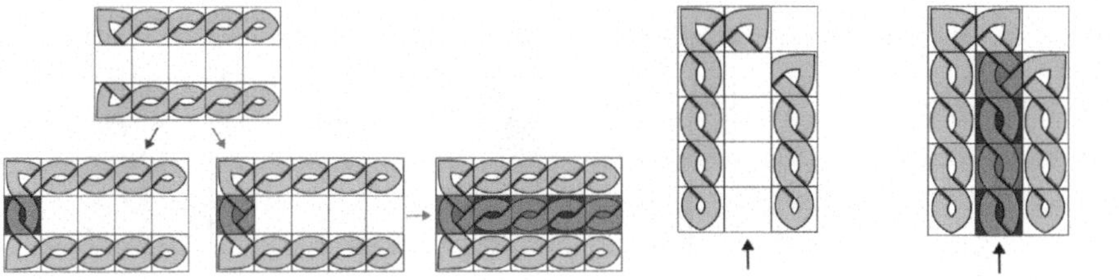

(a) Variable Gadget (b) Win Gadget (c) Win Example

Fig. 6. (a) VARIABLE Gadget in the 2-player game representing the VARIABLE vertex in a Bounded 2CL. If Red player moves first, the knot is closed with [tile], preventing Blue player from continuing an edge to other gadgets. If Blue player moves first, [tile] is placed, which restricts the red player to only continue the knot to other gadgets. (b) In the WINNING Gadget, only the Blue player is allowed to finally close the knot. Once the blue player closes the knot, the knot can no longer be extended. (c) Example moves in the WINNING gadget. The red player can only continue in the gadget, while blue player makes the last move. (Color figure online)

Theorem 4. *Building a closed knot of length $\geq L$ with k moves by placing [tile] pieces in Generalized 1-Player Celtic!, where the initial board configuration can contain [tile] and [tile] pieces, is NP-complete.*

Theorem 5. *Building a closed knot of length $\geq L$ with k moves by placing [tile] pieces in Generalized 1-Player Celtic!, where the initial board configuration can contain [tile], [tile], and [tile] pieces, is NP-complete.*

Theorem 6. *Building a closed knot of length $\geq L$ with k moves by placing [tile] and [tile] pieces in Generalized 1-Player Celtic!, where the initial board configuration can contain [tile], [tile], and [tile] pieces, is NP-complete.*

4 Constraint-Graph Reduction for 2-Player

We now analyze generalized 2-player Celtic! played on an $O(n) \times O(n)$ board where both Blue and Red players have a multi-set of pieces.

Lemma 2. *Given an $O(n) \times O(n)$ board configuration where both Blue and Red player have a multi-set of pieces containing pieces [[tile], [tile], [tile], [tile]]. Deciding whether there is a sequence of moves to force a Blue player (player 1) win starting at a given board position is PSPACE-hard.*

Proof. To show that determining if Blue player has a forced win is PSPACE-hard, we reduce Bounded 2-player Constraint Logic, where a sequence of moves in the 2-Player Celtic! game represents the flipping of an edge in the Constraint Graph whose configuration represents the constraint-satisfaction problem of 2CL. The problem is defined as: Does the Blue player have a forced win?

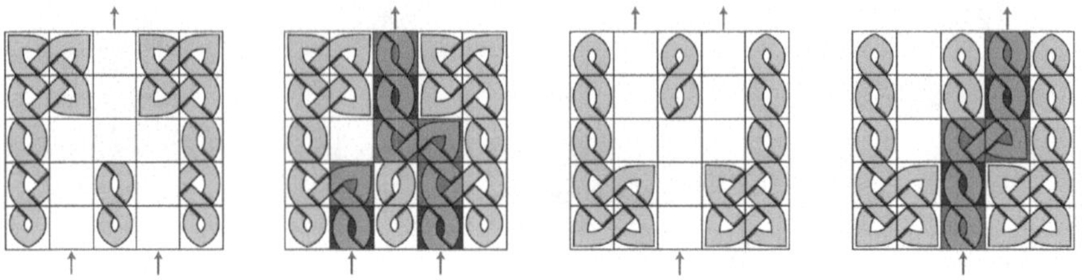

(a) OR Gadget (b) OR Example (c) CHOICE (d) CHOICE Ex.

Fig. 7. (a) OR Gadget in the 2-player game representing the OR vertex in a Bounded 2CL. For the player to continue with a valid knot, the player can connect a knot from the right side using the or from the left side using the . The player can then complete the knot from the opposite side as well. (b) An example of a sequence of moves in the OR gadget when the knot from the right side is continued. (c) CHOICE Gadget in the 2-player game representing the CHOICE vertex in a Bounded 2CL. For the player to continue a valid knot, the player can use either or . Depending on the piece, the knot continues right or left. No knot can be completed from the opposite side. (d) Moves in the CHOICE gadget continuing the knot to the right side.

We reduce Bounded 2CL by designing the VARIABLE, AND, OR, CHOICE and FANOUT gadgets for our game that are joined to create a constraint graph representing the sequence of moves. There is a Winning gadget that the Blue player can use to finish the knot. We show that the constraint graph accepts the configuration when the Blue player wins.

VARIABLE Gadget. (Figure 6a) We need to show that this gadget satisfies the same constraints as that of a variable vertex in the Bounded 2CL constraint graph. If the Red player moves first, it closes the knot using the piece preventing Blue player from continuing an edge to other gadgets. But if the Blue player moves first, it can place a piece that restricts the red player to only continue the knot to other gadgets.

WIRE Gadget. A wire is a walled path of width 1 that connect gadgets.

WINNING Gadget. (Figure 6b) The winning gadget is used by the blue player to close the knot once they reach it. In all other gadgets, the players continue with the knot, and they finally lead into the winning gadget. Here only the blue player is able to close the knot using the as shown in Fig. 6c.

OR Gadget. (Figures 7a, 7b) The construction of this OR gadget in our game satisfies the same constraints as a Bounded 2CL OR vertex. For the player to continue with a valid knot, the player can connect a knot from the right side using the or from the left side using the . Another knot can later be used to close the opposite Input side using .

CHOICE Gadget. (Figure 7c) The construction of this gadget satisfies the same constraints as that of a CHOICE vertex in a Bounded 2CL where a sequence of moves mark whether the knot can be extended towards left or right. For the player to continue with a valid knot, the player can use either or . Depending on the piece used, the knot will then continue right or left. No knot can be completed from the opposite side.

AND Gadget. (Figure 8a) We now show that the construction of this gadget satisfies the same constraints as that of the AND vertex in a Bounded 2CL constraint. For the player to continue with a valid knot, the player can use both and . If the player only connects a knot from one side, the knot will not be closed. Hence, the pieces in the knot will not count towards the players.

FANOUT Gadget. (Figure 8c) This gadget satisfies the same constraints as that of the FANOUT vertex in a Bounded 2CL constraint graph. The construction is quite similar to the AND gadget where the Output edge is now the Input edge. If the player is able to connect to the Input edge, then the player can complete the knot by connecting on both sides using and .

Since the Red player only has , they can use it in the Variable gadget to block, or continue with existing knots in all the gadgets. The Blue player makes moves in the Variable gadgets. If the Blue player is able to assign the Variables to true (place in the Variable gadget), they can then satisfy other constraints by continuing knots in other gadgets. Once the knot reaches the Winning gadget, the Blue player can close it, thus ending and winning the game. If both players make alternate moves and Blue player makes the first move, the winning gadget

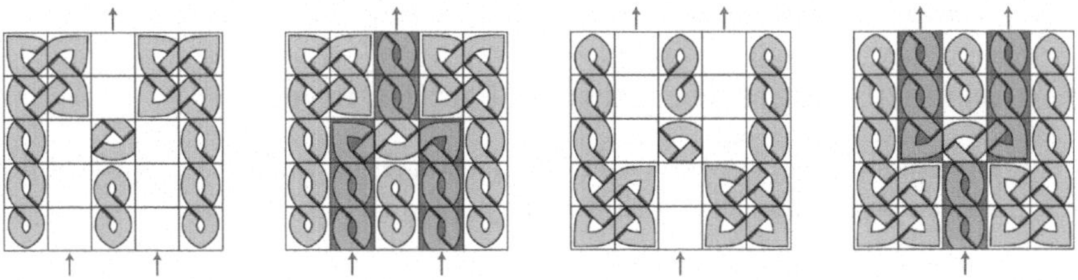

(a) AND Gadget (b) AND Example (c) FANOUT (d) FANOUT Ex.

Fig. 8. (a) AND Gadget in the 2-player game representing the AND vertex in a Bounded 2CL. For the player to continue with a valid knot, the player can use both and . If the player only connects a knot from one side, the knot will not be closed. Hence, the pieces in the knot will not count towards the players. (b) An example of moves in the AND gadget. To complete the knot the player has to connect from both sides. (c) FANOUT Gadget in the 2-player game representing the FANOUT vertex in a Bounded 2CL. The player can complete the knot by connecting on both sides using and . (d) An example of moves in the FANOUT gadget. The player is able to continue knots on both sides.

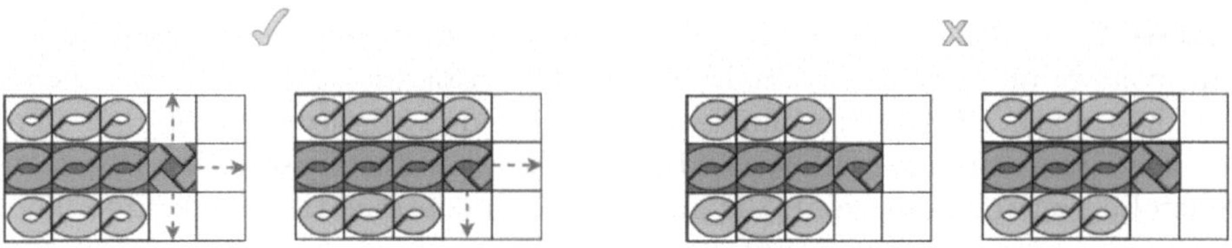

Fig. 9. Valid (left) and invalid (right) moves. In the invalid moves, the left has an empty cell on the north side, so the player must place a piece. In the other example, there is an adjacent piece to the north with a closed side, thus the played piece needs a closed side on its north.

will ensure that the Blue player is the last player to make a move. Thus, Blue has at least one piece more than Red, making them the winner.

The construction of these gadgets satisfies all constraints as that of vertices in a constraint graph of a Bounded 2CL. We know that finding the variables assignments for the constraint logic problem for a Bounded 2CL is PSPACE-complete. Such variable assignments correspond to the Blue and Red player making moves in the variable gadgets, setting them to true or false. Therefore deciding if the Blue player can make certain moves in the variable gadgets such that the Blue player always wins is PSPACE-hard. □

Lemma 3. *Given an $O(n) \times O(n)$ board configuration where both Blue and Red player have a multi-set of pieces containing pieces [, , ,]. Deciding whether there is a sequence of moves the Blue player can make that results in a forced win starting at a given board position is in PSPACE (Fig. 9).*

Theorem 7. *Given an $O(n) \times O(n)$ board configuration where both Blue and Red player have a multi-set of pieces containing pieces [, , ,]. Deciding if the given board configuration is a 1st player win is PSPACE-complete.*

5 Constraint Logic Reduction for 0-Player

This section considers a 0-player variation of Celtic! with rule constraints to allow for automation, and shows that 0-player Celtic! is P-complete.

5.1 0-Player Decisions

We define placement rules for an automated 0-player to make decisions.

- Structural Pieces - Existing white pieces used as walls or requiring specific pieces to traverse.
- Signal Pieces - The automated blue pieces that propagate through the board.
- Close South Rule - A piece is always placed to close a southward path unless there is another adjacent cell that requires a different piece (Fig. 10a).
- 3-Open Piece Rule - A piece is placed when 3 sides are possible (Fig. 10b).
- 4-Open Piece Rule - A is placed when 4 open sides are possible, which also allows for crossovers in the graph (Fig. 10c) (Fig. 11).

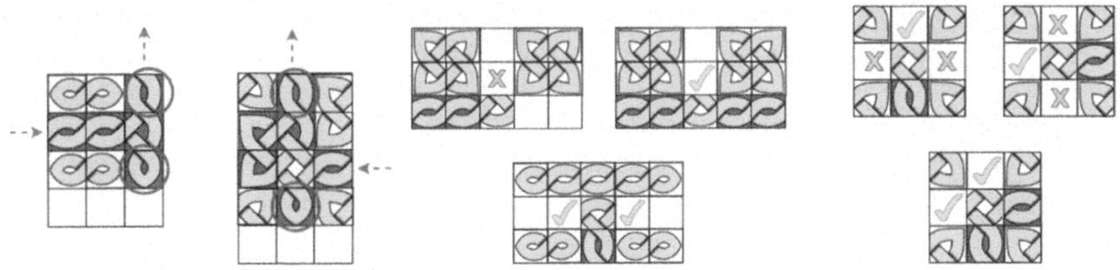

(a) (b) (c)

Fig. 10. (a) An example of the Close South Movements rule. The red arrows show the direction that the player is placing the pieces. As it continues north, any movement south will result in a piece to close that path. (b) The conditions needed in order to continue through a piece. (c) The conditions necessary to continue through a piece. (Color figure online)

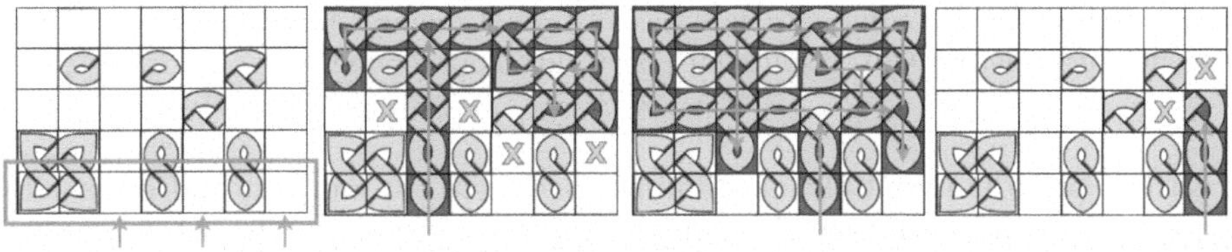

Fig. 11. Example 0-player simulations of a board, depending on where the starter piece is placed. Here, the longest knot is created by placing it in the middle. If the goal is to get a knot ≥ 5, the player can place the starter piece in the left or middle starting location. The green arrows show the path the simulation takes and the red X's are the directions it can not continue through. (Color figure online)

5.2 0-Player Bounded Deterministic Constraint Logic Gadgets

Bounded DCL is P-complete [9] since we can use it to simulate monotone boolean circuits, which only requires us to simulate simple AND, OR and FANOUT gadgets. We reduce from Bounded DCL to 0-Player Celtic! simulation. We show the construction of the AND, OR and FANOUT in Fig. 12.

WIRE Gadget. A wire is a walled (by structural pieces) path of width 1 that connects gadgets.

OR Gadget. The OR gadget has two inputs and one output. If a signal piece propagates through the left input, by the **4-Open Sided Rule**, it will propagate vertically through the 4 open-sided structural piece (Fig. 12). As it propagates north, it splits into 2 signals: one of them continues as the output of the gadget, the other takes a turn left. By the **Close South Movement Rule** and **4-Open Sided Rule** it will then connect with the 4 open-sided piece and continue horizontally until it closes going southward through the other input path. Similarly, if the signal piece propagates on the right input, the signal traverses the

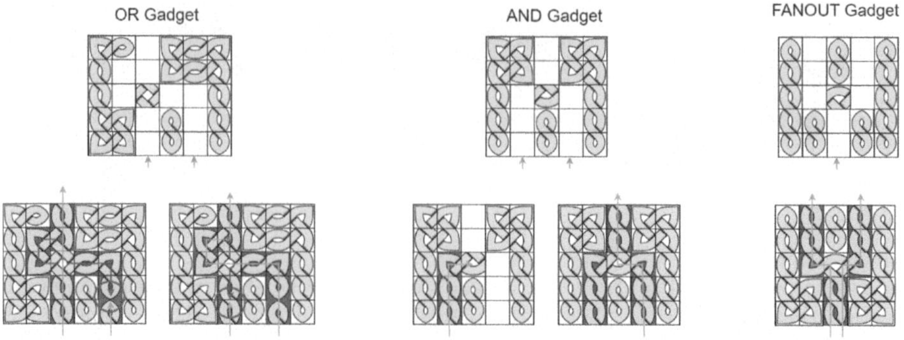

Fig. 12. Examples of an OR, AND, and FANOUT gadget, and how they are traversed. The green arrows show inputs propagating inward, red arrows depict southward propagation getting closed. In the OR gadget, one input causes the other input path to close, and any other signal coming in will close. The AND requires both inputs in order to continue outwards. FANOUT splits into 2. (Color figure online)

structural piece horizontally, continuing north with its output, and eventually propagating south vertically through the piece and closes the other input. If another signal tries to propagate through the opposite side it will immediately close, but the signal has already propagated from the first input.

AND Gadget. By the **3-Open Sided Rule**, a signal piece coming from either input cannot continue through the output unless the automated player has connected signal pieces to both parallel sides of the structural piece (Fig. 12).

FANOUT Gadget. By the **3-Open Sided Rule**, a signal piece propagating through the perpendicular open side of the piece allows the automated player to split the signal and propagate through both outputs (Fig. 12).

Simulating DCL with Gadgets. Figure 13 demonstrates how the gadgets on a board can simulate bounded DCL. The goal is to attempt to reach the output from the top AND gadget. Notice it can only reach the output by playing in the middle starting position where the signal splits in the FANOUT gadget, through the 2 OR gadgets, and then into the final AND gadget. The goal of a 0-Player Celtic! simulation is to generate a knot of size $\geq L$, so you can extend the output of the last top gadget to be longer than any possible knot before that, and let L equal to that. Thus, the only knot that can achieve this is the knot that can output through the last gadget.

Theorem 8. *0-Player Celtic! is P-complete.*

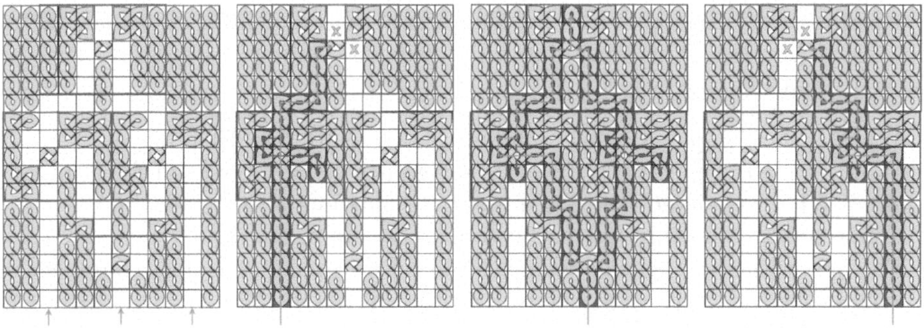

Fig. 13. Example of the 0-Player Celtic! reduction from DCL. The red squares highlight the gadgets locations. On top we have an AND with inputs coming from the outputs of the middle OR gadgets. At the bottom, we can place our starter piece as input into one of the ORs, or we can input it into the FANOUT. Here, only the middle starting position will reach the opposite side of the board. (Color figure online)

6 Generalized Celtic! with Boards of Odd Dimension

This section considers generalized Celtic! played on boards of odd dimension, where the count of each , , ,  and  white pieces are even and the set of pieces each player gets is equal. This includes when the white pieces are excluded altogether, or if the number of white pieces are assumed infinite. We show that through a *strategy-stealing* argument, where player 2 takes advantage of the games symmetry to prevent player 1 from having a winning strategy, the game is guaranteed to be a draw. Let S_B, S_R and S_W be multi-sets of blue, red and white pieces, respectively.

Theorem 9. *Celtic! played on an $n \times n$ board, where n is odd, is a guaranteed draw when $S_B = S_R$ and the count of each* , , , ,  $\in S_W$ *is even.*

Proof. Although no formal proof is provided due to space, the basic idea is strategy stealing based on board symmetry. If the board size is odd, and there are an even number of each neutral white piece, the second player can always mirror the play of the first player on the other side of the board. This would ensure that all knots end up with an equal number of each color or that there is an equivalent mirrored knot with the player colors inverted.

7 Conclusion

This work characterizes and analyzes the complexity of 3 variations of Celtic!: The 0, 1, and 2 player variants depending on the pieces allowed. In the 1-Player version, we show NP-completeness or membership in P depending on the pieces allowed. For the 2-player variation, deciding if there is a first-player win is PSPACE-complete. We show that a 0-Player variation with forced play, determining if a closed knot of some length can be formed after an initial starter piece is P-complete. This work naturally leads to some open questions, such as those in the table and below.

- What is the complexity with different pieces for Red or Blue player?
- What is the complexity of the different games given a board of fixed dimension? What restricted board instances are polynomial time solvable?
- What is the complexity of the two open cases in Table 1?
- We show that for any general board of odd dimension with even numbers of each piece, the game always ends in a draw. Without these restrictions, is the standard game a first-player win?
- What if we look at pattern complexity based on the knots? General patterns could be answered similar to the PATS (patterned self-assembly tile set synthesis) problem [11], but if we consider the topology of the knots, the pattern is no longer based solely on the tile-type, or even color, at some location.

References

1. Berger, R.L.: The undecidability of the domino problem. Mem. Am. Math. Soc. (1966)
2. Bosboom, J., et al.: Edge matching with inequalities, triangles, unknown shape, and two players (2020). arXiv:2002.03887
3. Bosboom, J., Demaine, E.D., Demaine, M.L., Hesterberg, A., Manurangsi, P., Yodpinyanee, A.: Even $1 \times n$ edge-matching and jigsaw puzzles are really hard (2016). arXiv:1701.00146
4. Browne, C.: Connection Games: Variations on a Theme. A K Peters, Ltd. (2005)
5. Browne, C.: Celtic! [Board Game] (2009)
6. Demaine, E.D., Demaine, M.L.: Jigsaw puzzles, edge matching, and polyomino packing: connections and complexity. Graphs Comb. **23**, 195–208 (2007)
7. Ebbesen, M., Fischer, P., Witt, C.: Edge-matching problems with rotations (2017). arXiv:1703.09421
8. Hardesty, F.W., Schenck, I.W.: Competitive road building and travel game. US grant 3309092A (1967)
9. Hearn, R.A., Demaine, E.D.: Games, Puzzles, and Computation. CRC Press, Boca Raton (2009)
10. Liu, Y., Morgana, A., Simeone, B.: A linear algorithm for 2-bend embeddings of planar graphs in the two-dimensional grid. Disc. Appl. Math. **81**(1–3), 69–91 (1998)
11. Ma, X., Lombardi, F.: Synthesis of tile sets for dna self-assembly. IEEE Trans. Comput. Aided Des. Integr. Circuits Syst. **27**(5), 963–967 (2008)
12. MacMahon, P.A.: New Mathematical Pastimes. Cambridge University Press, Cambridge (1921)
13. McMurchie, T.: Squiggle Game. US grant 4180271A (1979)
14. McMurchie, T.: Tsuro. Calliope Games (2005)
15. Plesńik, J.: The np-completeness of the hamiltonian cycle problem in planar diagraphs with degree bound two. Inf. Process. Lett. **8**(4), 199–201 (1979)
16. Schensted, C., Titus, C.: Kalico (Psyche-Paths). Funtastic, Kadon (1968)
17. Takenaga, Y., Walsh, T.: Tetravex is np-complete. Inf. Process. Lett. **99**(5), 171–174 (2006)

18. Wang, H.: Proving theorems by pattern recognition – ii. Bell Syst. Techn. J. **40**(1), 1–41 (1961)
19. Wikipedia: Eternity ii, en.wikipedia.org/wiki/Eternity_II_puzzle (2024)
20. Winfree, E.: Algorithmic Self-Assembly of DNA. Ph.D. thesis, California Institute of Technology (1998)

On the Existence of a Subgroup Magic Rectangle

S. Karthik(✉), Aruna Venkatesan, and Krishnan Paramasivam

Department of Mathematics, National Institute of Technology Calicut, Kozhikode 673601, India
{karthik_p210136ma,aruna_p220282ma,sivam}@nitc.ac.in
https://nitc.ac.in/department/mathematics

Abstract. Let Γ be a finite additive Abelian group. A subgroup magic rectangle is an $m \times n$ array $\mathbb{S} = (\zeta_{ij})$, where $\zeta_{ij} \in \Gamma$ for all i, j and $|\Gamma| = mn$, each element appears exactly once in such a way that the set of all row-sums forms a subgroup of order m and the set of all column-sums forms a subgroup of order n. This combinatorial object is a generalization of the constant sum partition of a finite Abelian group and group magic rectangle. In this article, we prove the existence of a subgroup magic rectangle over Abelian group Γ of order p^3, where p is an odd prime and a characterization for the existence of a subgroup magic rectangle over $\mathbb{Z}_{mn}$, where $\gcd(m, n)$ is odd.

Keywords: Constant sum partition · Kotzig array · group magic rectangle · Abelian group · coset

2020 Mathematics Subject Classification: 05A18 · 05B05 · 05B15 · 20K27

1 Introduction and Preliminaries

Magic square is a fascinating mathematical object that has captivated humans for centuries. A magic square is an $n \times n$ array, whose elements are 1 to n^2 such that each element appears exactly once; each row-sum and column-sum and the main diagonal and anti-diagonal sum are the same constant. A magic rectangle is a natural generalization of the magic square, which has long intrigued mathematicians, theoretical computer scientists and the general public. A magic rectangle with m rows and n columns is an $m \times n$ is an $m \times n$ array, whose entries are 1 to mn in such a way that the sum of the integers in each row is a constant, and the sum of the integers in each column is a constant. A lot of work is done on various construction of magic rectangles [1]. Now, a natural question arises, what happens if the positive integers 1 to n^2, are replaced by elements of the finite additive Abelian group of order n^2, in which the binary operation of group plays a vital role in such combinatorial object. In 1997, Wen et al. [14] introduced group magic rectangles, whose entries are from an Abelian group of

H. Fernau and B. Zhu (Eds.): IWOCA 2025, LNCS 15885, pp. 332–346, 2025.
https://doi.org/10.1007/978-3-031-98740-3_24

order n^2 and later Cichacz [3] studied the existence of a $m \times n$ magic rectangle whose entries are from an Abelian group of order mn. In 2022, Morini et al. [11] introduced the concept of magic partially filled arrays on Abelian groups as the generalization of the Heffter array [2]. In 2013, Wallis [10] introduced the Kotzig array based on an unpublished paper by Kotzig and a Kotzig array is an $m \times n$ matrix $\mathcal{KA}(m,n)$, each row being a permutation of $\{0,1,\ldots,n-1\}$ and each column having a constant sum $\frac{m(n+1)}{2}$. Kaplan et al. [9] and Freyberg [6] generalized the concept of integer partitions to constant sum partition for subsets in a finite Abelian group. For more details, one can refer to [4,7].

For two integers i and j with $i<j$, $[i,j]$ denotes the set of integers $i, i+1, \ldots, j$. If $\Gamma = \{\zeta_{ij} : i \in [1,m], j \in [1,n]\}$ is an additive Abelian group of order mn, then $\alpha_i[k,\ell:\zeta_{ij}]$ denotes the partial sum $\zeta_{k,j} + \zeta_{k+1,j} + \cdots + \zeta_{\ell,j}$ and $\alpha_j[k,\ell:\zeta_{ij}]$ denotes the partial sum $\zeta_{i,k} + \zeta_{i,k+1} + \cdots + \zeta_{i,\ell}$. Throughout this article, Γ denotes a finite additive Abelian group with identity 0, $\sum_{g\in\Gamma} g$ is the sum of all elements in Γ, and $\mathcal{C}_n$ is the cyclic group of order n. If $g \in \Gamma$ is an element of order k, then the cyclic subgroup generated by g is $\langle g\rangle = \{0, g, 2g, \ldots, (k-1)g\}$, and is isomorphic to $\mathbb{Z}_n$. If Γ is a group of order pq, where p and q are two distinct primes with $p < q$, then Γ is cyclic, or Γ has two generators a and b such that $pa = e$, $qb = e$, and $(-a)ba = rb$, where $r \neq 1 \pmod{q}$, $pr \equiv 1 \pmod{q}$, and p divides $q-1$. If $p \nmid q-1$, then Γ is cyclic and is isomorphic to $\mathbb{Z}_{pq}$. However, if $p \mid q-1$, then $\mathbb{Z}_{pq}$ and a non-Abelian group are two distinct groups of order pq (up to isomorphism). If $\mathbb{A} = (a_{i,j})$ is an $m \times n$ matrix, define the permutation π such that it rearranges the rows of $\mathbb{A}$ by fixing the first row, then systematically swapping the second row with the last row, the third row with the second-to-last row, etc. Explicitly, this permutation can be written as

$$\pi(\mathbb{A})_{i,j} = \begin{cases} a_{i,j} & \text{if } i = 1, \\ a_{m-i+2,j} & \text{if } i \in [2,m]. \end{cases}$$

This defines the permutation matrix $\pi(\mathbb{A})$ applied to the matrix $\mathbb{A}$. Now, we recall the necessary definitions and theorems that are used in our main results.

Definition 1 ([12]). *If H is a subgroup of a group Γ, then a left coset of H is a set $xH = \{xh : h \in H\}$, where $x \in \Gamma$ is some fixed element.*

Proposition 1 ([12]). *Let H be a non-trivial subgroup of a group Γ.*

1. *The union of all the left cosets of H equals Γ.*
2. *All left cosets of H have the same size as H.*
3. *The set of all the left cosets of H forms a partition of Γ.*

Definition 2 ([13]). *Let Γ be a group with identity e and let $\{\mathcal{H}_n\}$ be a sequence of subgroups of Γ. The subgroups $\mathcal{H}_1, \mathcal{H}_2, \ldots, \mathcal{H}_n$ are independent if and only if $\prod_{i=1}^{n} h_i = e \Leftrightarrow \forall i \in [1,n] : h_i = e$, where $h_k \in \mathcal{H}_k$.*

Theorem 1 (Structure theorem for finite Abelian groups [12]). *Let Γ be a finite Abelian group. Then there are integers $m_1, m_2, \dots, m_r$ such that $\Gamma \cong \prod_{i=1}^{r} \mathbb{Z}_{m_i}$. Further, we may take $m_1, m_2, \dots, m_r$ as the prime powers.*

Corollary 1 ([12]). *Let Γ be an Abelian group of order n. If a positive integer m divides n, then Γ has a subgroup of order m.*

Theorem 2 ([12]). *The group $\prod_{i=1}^{n} \mathbb{Z}_{m_i}$ is cyclic and isomorphic to $\mathbb{Z}_{m_1 m_2 \dots m_n}$ if and only if the numbers m_i are pairwise co-prime.*

Recall that if $|\Gamma| = p^\alpha$, where p is prime and $\alpha \geq 1$, then Γ is called a *p-group*. The *Sylow p-subgroup* of Γ is the maximal subgroup L of Γ, the order of which is a power of p. For example, by the Fundamental Theorem of Abelian groups, one can see that any Abelian group Γ can be factorized as $\Gamma \cong L \times H$, where L is the Sylow 2-group of Γ and the order of H is odd. A non-zero element g of Γ is an involution if $g = -g$, where $-g$ is the inverse of g in Γ and an element g of Γ is an involution if and only if $o(g) = 2$. We use $I(\Gamma)$ to denote the set of all involutions in Γ. Since any cyclic finite group of even order has exactly one involution, $|I(\Gamma)| = 2^e - 1$, where e is the number of cyclic components in the factorization of even ordered group Γ. Note that the sum of all the elements of the group Γ equals the sum of its involutions and the identity element. The following lemma is well known.

Lemma 1 ([5]). *If $|I(\Gamma)| = 1$, then $\sum_{g \in \Gamma} g = \tau$, otherwise $\sum_{g \in \Gamma} g = 0$, where τ is an involution in Γ.*

For more group theory notation and terminology, one can refer to Silverman [12], Warner [13], and Holt et al., [8].

2 $\mathcal{G}$-Coset Array and $\mathcal{H}$-Kotzig Array

Throughout this article, $\zeta_{i,j}$, ΣR_i and ΣC_j denote the group element assigned to the (i, j)-th cell, the sum of all group elements in the i-th row and the sum of all group elements in the j-th column of an $m \times n$ array, with $m \leq n$, respectively. When we say *perform* $\zeta_{i,j} \rightleftharpoons \zeta_{i',j'}$', the group elements in the (i, j)-th cell and (i', j')-th cell are interchanged. Let $A = (\zeta_{i,j})$ and $B = (\zeta_{i',j'})$ be two $m \times n$ arrays and we say *perform* $A \rightleftharpoons B$', we mean *perform* $\zeta_{i,j} \rightleftharpoons \zeta_{i',j'}$' for all i and j.

Definition 3. *Let Γ be a group of order mn and $\mathcal{G}$ be a subgroup of order n. A $\mathcal{G}$-coset array is an $m \times n$ array, whose entries are all distinct elements of Γ in such a way that any two rows are distinct left cosets of $\mathcal{G}$ and it is denoted by $\mathcal{CA}_\Gamma(m, n : \mathcal{G})$.*

Note that m is the index of $\mathcal{G}$ in Γ and n is the order of $\mathcal{G}$ in a $\mathcal{G}$-coset array $\mathcal{CA}_\Gamma(m, n : \mathcal{G})$.

Proposition 2. *Let Γ be a group of order mn, and let $\mathcal{G}$ be any n-order subgroup of Γ. Then there exists an $m \times n$ $\mathcal{G}$-coset array.*

Proof. The proof follows from Proposition 1. □

Motivated by the combinatorial objects discussed earlier, where the entries consist of elements from an Abelian group, we introduce the $\mathcal{H}$-Kotzig array as a generalization of a constant-sum partition of an additive Abelian group.

Definition 4. *Let Γ be an Abelian group of order mn. Then a $\mathcal{H}$-Kotzig array $\mathcal{KA}_\Gamma(m,n:\mathcal{H})$ is an $m \times n$ array $\mathbb{S} = (\zeta_{ij})$, where $\zeta_{ij} \in \Gamma$ for all i, j, each appearing exactly once such that whenever $\mathcal{H} = \{h_1, h_2, \ldots, h_m\}$ is a set of m elements, where $h_i = \sum_{j=1}^{n} \zeta_{ij}$, for any $i \in [1, m]$, then $\mathcal{H}$ satisfies following condition:*

$$h_k, h_\ell \in \mathcal{H}, \text{ implies } h_k - h_\ell \in \mathcal{H}.$$

Note that $\mathcal{H}$ is an m order subgroup of Γ and the index of $\mathcal{H}$ in Γ is n.

Example 1. A $\mathcal{H}$-Kotzig array over $\mathbb{Z}_{27}$ with $m = 3$ and $n = 9$ is given in Fig. 1, where $\mathcal{H} = \{0, 9, 18\}$. Note that the elements in each row of a 3×9 array are the distinct left cosets of another subgroup of $\mathbb{Z}_{27}$.

0	3	6	9	12	15	18	21	24
1	4	7	10	13	16	19	22	25
2	5	8	11	14	17	20	23	26

$\mathcal{H}$
0
9
18

Fig. 1. An example of $\mathcal{H}$-Kotzig array

Proposition 3. *Suppose Γ is an Abelian group of order mn. If there exists a $\mathcal{H}$-Kotzig array, then $\sum_{h \in \mathcal{H}} h = \sum_{g \in \Gamma} g$.*

Proof. Let Γ be an Abelian group of $|\Gamma| = mn$ and let $\mathcal{H}$ be a subgroup of Γ with $|\mathcal{H}| = m$. If there exists an $\mathcal{H}$-Kotzig array $\mathcal{KA}(m, n : \mathcal{H})$, then

$$\sum_{h \in \mathcal{H}} h = \alpha_j[1, n : \zeta_{1,j}] + \alpha_j[1, n : \zeta_{2,j}] + \cdots + \alpha_j[1, n : \zeta_{m,j}] = \sum_{g \in \Gamma} g.$$

□

Lemma 2. *Suppose Γ is a cyclic group of order mn. Then a $\mathcal{G}$-coset array $\mathcal{CA}_\Gamma(m,n:\mathcal{G})$ is a $\mathcal{H}$-Kotzig array $\mathcal{KA}_\Gamma(m,n:\mathcal{H})$ if and only if either m is even or both m and n are odd.*

Proof. Let $\mathcal{G}$ be an n order subgroup of Γ. Suppose that $\mathcal{CA}_\Gamma(m, n : \mathcal{G})$ exists. Let the set of all row-sums of $\mathcal{CA}_\Gamma(m, n : \mathcal{G})$ be

$$\mathcal{H} = \{0 + \sum_{g \in \mathcal{G}} g, nh + \sum_{g \in \mathcal{G}} g, 2nh + \sum_{g \in \mathcal{G}} g, \ldots, (m-1)nh + \sum_{g \in \mathcal{G}} g\}.$$

Claim: $\mathcal{H}$ is a m order subgroup of Γ.
Case 1. Suppose m is even.
If n is odd, then $\sum_{g\in\mathcal{G}} g = 0$. Thus $\mathcal{H} = \{0, nh, 2nh, \dots, (m-1)nh\}$ is an m order subgroup of Γ. Since $\mathcal{H}$ is an even order subgroup, $\mathcal{H}$ has an involution (say) τ. On the other hand, if n is even, then $\sum_{g\in\mathcal{G}} g = \tau$. Since $\tau \in \mathcal{H}$, we have $\tau + \mathcal{H} = \mathcal{H} = \{0, nh, 2nh, \dots, (m-1)nh\}$, which is a subgroup of order m.

Case 2. Suppose m is odd and n is odd.
Since $\mathcal{G}$ is a subgroup of odd order n, we have $\sum_{g\in\mathcal{G}} g = 0$. Consequently, the set $\mathcal{H} = \{0 + \sum_{g\in\mathcal{G}} g, nh + \sum_{g\in\mathcal{G}} g, 2nh + \sum_{g\in\mathcal{G}} g, \dots, (m-1)nh + \sum_{g\in\mathcal{G}} g\} = \{0, nh, 2nh, \dots, (m-1)nh\}$ is an m order subgroup of Γ.
Case 3. Suppose m is odd and n is even.
Assume that the $\mathcal{G}$-coset array is a $\mathcal{H}$-Kotzig array. Since $\mathcal{H}$ is an m order subgroup, we have $\sum_{h\in\mathcal{H}} h = 0$. By Proposition 3, we have

$$\begin{aligned}
\sum_{h\in\mathcal{H}} h &= 0 + \sum_{g\in\mathcal{G}} g + nh + \sum_{g\in\mathcal{G}} g + 2nh + \sum_{g\in\mathcal{G}} g + \dots + (m-1)nh + \sum_{g\in\mathcal{G}} g \\
&= nh + 2nh + \dots + (m-1)nh + m\sum_{g\in\mathcal{G}} g \\
&= nh + 2nh + \dots + (m-2)nh + (m-1)nh + m\sum_{g\in\mathcal{G}} g \\
&= \begin{cases} m\sum_{g\in\mathcal{G}} g & \text{if } m \text{ isodd,} \\ \left(\frac{m}{2}\right) nh + m\sum_{g\in\mathcal{G}} g & \text{if } m \text{ iseven.} \end{cases}
\end{aligned} \tag{1}$$

Now from Eq. 1, we have

$$m\sum_{g\in\mathcal{G}} g = \sum_{h\in\mathcal{H}} h$$
$$m\sum_{g\in\mathcal{G}} g = 0.$$

Thus either $\sum_{g\in\mathcal{G}} g$ is 0 or $o(\sum_{g\in\mathcal{G}} g)$ divides m. Clearly, $\mathcal{G}$ is a subgroup of order n. Then $\sum_{g\in\mathcal{G}} g = 0$ is not possible. On the other hand, suppose $\sum_{g\in\mathcal{G}} g = \tau$. Then $o(\sum_{g\in\mathcal{G}} g) = o(\tau) = 2$, which is again not possible. □

Corollary 2. *Suppose Γ is a cyclic group of order mn. Then a $\mathcal{G}$-coset array $\mathcal{CA}_\Gamma(m, n : \mathcal{G})$ is a $\mathcal{G}$-Kotzig array $\mathcal{KA}_\Gamma(m, n : \mathcal{G})$ if and only if $m = n$.*

Proof. The proof follows from Proposition 3 and Lemma 2. □

3 Subgroup Magic Rectangles

In this section, we introduce an $m \times n$ subgroup magic rectangle over an Abelian group Γ as a generalization of the $\mathcal{H}$-Kotzig array by taking the set of all the sums of each row as a subgroup and the set of all the sums of each column as a subgroup. A subgroup magic rectangle over Γ is shortly termed as subgroup magic rectangle when Γ under consideration is clear from the context.

Definition 5. *Let Γ be an Abelian group of order mn. A subgroup magic rectangle over Γ, $\mathrm{SMR}_\Gamma(m,n:\mathcal{H}_1,\mathcal{H}_2)$, is an $m\times n$ array (ζ_{ij}), whose entries $\zeta_{ij}\in\Gamma$ for all i and j, each appearing exactly once in such that whenever the sets $\mathcal{H}_1=\{h_1,h_2,\ldots,h_m\}$ and $\mathcal{H}_2=\{h'_1,h'_2,\ldots,h'_n\}$, where $h_i=\sum_{j=1}^m\zeta_{ij}$ and $h'_j=\sum_{i=1}^n\zeta_{ij}$, for all i and j, satisfy the following conditions:*

(i) $h_k-h_\ell\in\mathcal{H}_1$ *whenever* $h_k,h_\ell\in\mathcal{H}_1$,
(ii) $h'_k-h'_\ell\in\mathcal{H}_2$ *whenever* $h'_k,h'_\ell\in\mathcal{H}_2$.

Note 1. Clearly, $\mathcal{H}_1$ is a m order subgroup of Γ and index of $\mathcal{H}_1$ in Γ is n and $\mathcal{H}_2$ is a n order subgroup of Γ and index of $\mathcal{H}_2$ in Γ is m.

Note 2. The subgroups $\mathcal{H}_1$ and $\mathcal{H}_2$ in Γ are the row-subgroup and the column-subgroup of $\mathrm{SMR}_\Gamma(m,n:\mathcal{H}_1,\mathcal{H}_2)$. From Definition 5, any $\mathrm{SMR}_\Gamma(m,n:\mathcal{H}_1,\mathcal{H}_2)$ is a $\mathcal{KA}_\Gamma(m,n:\mathcal{H}_1)$.

Example 2. Suppose $\Gamma=\mathbb{Z}_3\times\mathbb{Z}_3\times\mathbb{Z}_5$ of order 45 and $m=9$ and $n=5$. A subgroup magic rectangle $\mathrm{SMR}\Gamma(9,5:\mathcal{H}_1,\mathcal{H}_2)$ is given below (Fig. 2).

						$\mathcal{H}_1$
	(0,0,0)	(0,0,1)	(0,0,2)	(0,0,3)	(0,0,4)	(0,0,0)
	(0,1,0)	(0,1,1)	(0,1,2)	(0,1,3)	(0,1,4)	(0,2,0)
	(0,2,0)	(0,2,1)	(0,2,2)	(0,2,3)	(0,2,4)	(0,1,0)
	(1,0,0)	(1,0,1)	(1,0,2)	(1,0,3)	(1,0,4)	(2,0,0)
	(1,1,0)	(1,1,1)	(1,1,2)	(1,1,3)	(1,1,4)	(2,2,0)
	(1,2,0)	(1,2,1)	(1,2,2)	(1,2,3)	(1,2,4)	(2,1,0)
	(2,0,0)	(2,0,1)	(2,0,2)	(2,0,3)	(2,0,4)	(1,0,0)
	(2,1,0)	(2,1,1)	(2,1,2)	(2,1,3)	(2,1,4)	(1,2,0)
	(2,2,0)	(2,2,1)	(2,2,2)	(2,2,3)	(2,2,4)	(1,1,0)
$\mathcal{H}_2$	(0,0,1)	(0,0,2)	(0,0,3)	(0,0,4)	(0,0,0)	

Fig. 2. 9×5 subgroup magic rectangle

Note that there is a trivial 1×1 subgroup magic rectangle over a trivial group by taking $\mathcal{G}=\mathcal{H}=\{0\}$ and hence we assume $m\geq2$ or $n\geq2$ and $m\leq n$.

Lemma 3. *Let Γ be an Abelian group of order mn and $\mathcal{H}_1$ and $\mathcal{H}_2$ be two subgroups of Γ of orders m and n, respectively. If there exists a subgroup magic rectangle, then*

$$\sum_{h\in\mathcal{H}_1}h=\sum_{g\in\Gamma}g=\sum_{h'\in\mathcal{H}_2}h'$$

Proof. Assume that $\mathrm{SMR}_\Gamma(m,n:\mathcal{H}_1,\mathcal{H}_2)$ exists. Then, $\sum\limits_{h\in\mathcal{H}} h = \alpha_j[1,n:\zeta_{1,j}] + \cdots + \alpha_j[1,n:\zeta_{m,j}] = \sum\limits_{g\in\Gamma} g = \alpha_j[1,n:\zeta_{i,1}] + \cdots + \alpha_j[1,n:\zeta_{i,m}] = \sum\limits_{h'\in\mathcal{H}_2} h'$. □

From Lemma 3, we conclude that if there exists an $m\times n$ subgroup magic rectangle over a finite cyclic group Γ, then m and n must be same parity.

Observation 3. *If Γ is a cyclic group of order $2\prod_{i=1}^{k} p_i^{\alpha_i}$, where p_i's are all odd primes, then there does not exist a subgroup magic rectangle.*

Proof. Suppose that $\mathcal{H}_1$ and $\mathcal{H}_2$ are a row-subgroup and a column-subgroup of a $\mathrm{SMR}_\Gamma(m,n:\mathcal{H}_1,\mathcal{H}_2)$. Then $2\prod_{i=1}^{k} p_i^{\alpha_i} = mn$. Hence, m and n must be of different parity, which contradicts Lemma 3. □

Observation 4. *Let Γ be an Abelian group of order mn, which contains a unique involution τ. If $\mathrm{SMR}_\Gamma(m,n:\mathcal{H}_1,\mathcal{H}_2)$ exists, then both subgroups $\mathcal{H}_1$ and $\mathcal{H}_2$ contain τ.*

Proof. The proof directly follows from Lemma 1 and Lemma 3. □

Suppose Γ is an Abelian group of order mn with exactly one involution element. Then from Observation 4 and Lemma 3, it is clear that if there exists a $\mathrm{SMR}_\Gamma(m,n:\mathcal{H}_1,\mathcal{H}_2)$, then both m and n are even.

Proposition 4. *Let Γ be an Abelian group of order n. Then a $1\times n$ subgroup magic rectangle exists if and only if $n\geq 3$ and $\sum_{g\in\Gamma} g = 0$.*

Proof. Assume that there exists a $\mathrm{SMR}_\Gamma(1,n:\mathcal{H}_1,\mathcal{H}_2)$. Thus, the subgroup $\mathcal{H}_1 = \{0\}$. Therefore, by Lemma 3, $\sum_{g\in\Gamma} g = 0$. It enforces that the order of Γ must be greater than 2. For the converse, let $\Gamma = \{g_0 = 0, g_1, g_2, \ldots, g_{n-1}\}$, and assign the group element $g_{j-1}, j\in[1,n]$ to the cells in an $1\times n$ array, so that $\zeta_{1,j} = g_{j-1}, j\in[1,n]$. Then there exists a row-subgroup $\mathcal{H}_1 = \{0\}$ and column-subgroup $\mathcal{H}_2 = \Gamma$. □

Theorem 5. *Let Γ be an Abelian group of order $2n$ and $n\not\equiv 0 \pmod 4$. Then there does not exist a $2\times n$ subgroup magic rectangle.*

Proof. Suppose n is odd. By Observation 4, a $2\times n$ subgroup magic rectangle does not exist. When $n\equiv 2 \pmod 4$, assume that there exists a $\mathrm{SMR}_\Gamma(2,n:\mathcal{H}_1,\mathcal{H}_2)$. Then, $\mathcal{H}_1 = \{0,\tau\}$, where $\tau\in I(\Gamma)$. Therefore, $\sum_{h\in\mathcal{H}_1} h = \tau$. By Lemma 3, $\sum_{h'\in\mathcal{H}_2} h' = \tau$. Consequently, $\tau\in\mathcal{H}_2$. Therefore τ is the unique involution is in both $\mathcal{H}_2$ and Γ and hence Γ is isomorphic to $\mathbb{Z}_{2n}$. Since $\Gamma\cong\mathbb{Z}_{2n}$, and $\mathcal{H}_2\cong\langle 2\rangle$, we have $\mathcal{H}_2\cong\langle 2\rangle = \{0,2,4,\ldots,2n-2\}$, the subgroup of even elements in $\mathbb{Z}_{2n}$. Let us consider the parity of positions. Since all elements of $\mathcal{H}_2$ are even, and the number of columns is n, it follows that each row contains n even elements. But when $n\equiv 2 \pmod 4$, n is not divisible by 4. Hence, it is impossible to construct a $2\times n$ subgroup magic rectangle. □

Theorem 6. *Let Γ be a cyclic group of odd order $|\Gamma| = mn$. If there exists a $\mathcal{G}$-coset array, then there exists a subgroup magic rectangle.*

Proof. Let $|\Gamma| = mn$. Assume that $\mathcal{CA}_\Gamma(m,n:\mathcal{G})$ exists, where $\mathcal{G}$ is an n order subgroup of Γ. Then $\Lambda = \{0+\mathcal{G}, 1+\mathcal{G}, -1+\mathcal{G}, 2+\mathcal{G}, -2+\mathcal{G}, \dots, \frac{m-1}{2}+\mathcal{G}, \frac{m+1}{2}+\mathcal{G}\}$ is the set of all the distinct rows of $\mathcal{CA}_\Gamma(m,n:\mathcal{G})$. Therefore for any $j \in [1,n]$, the column-sum in $\mathcal{CA}_\Gamma(m,n:\mathcal{G})$ can be uniquely written as, $\alpha_i[1,m:\zeta_{i,j}] = \sum_{i=1}^{m} \zeta_{ij} = g + (1+g) + (-1+g) + \cdots + ((m-1)/2+g) + ((m+1)/2+g) = g$ for all $g \in \mathcal{G}$. Hence, the set of all column-sums form a subgroup of order n, which is equal to $\mathcal{G}$. Since Γ is cyclic, and take $\mathcal{H}_2 = \mathcal{G}$. The existence of the row subgroup follows from Lemma 2. Therefore, $\mathrm{SMR}_\Gamma(m,n:\mathcal{H}_1,\mathcal{H}_2)$ exists. □

Corollary 3. *Let Γ be an Abelian group of $|\Gamma| = \prod_{i=1}^{k} p_i$, where all p_i's are distinct odd primes. Then there exists $\mathrm{SMR}_\Gamma(m,n:\mathcal{H}_1,\mathcal{H}_2)$, for any m and n.*

Proof. The proof follows from Theorem 6 and Γ is a cyclic group of order $\prod_{i=1}^{k} p_i$. □

Theorem 7. *Let Γ be an odd order cyclic group. If $\mathcal{H}_1$ and $\mathcal{H}_2$ are independent subgroups of Γ with orders m and n, respectively, then there exists a $m \times n$ subgroup magic rectangle.*

Proof. Let Γ be a cyclic group of order mn, and let $\mathcal{H}_1 = \{0, h_1, h_2, h_3, \dots, h_{m-1}\}$ and $\mathcal{H}_2 = \{0, h'_1, h'_2, \dots, h'_{n-1}\}$ are independent subgroups. Then $\gcd(m,n) = 1$. Since n is odd, we have

$$\sum_{h \in \mathcal{H}_1} h = \sum_{h' \in \mathcal{H}_2} h' = 0.$$

Also, $\Gamma \cong \mathcal{H}_1 \otimes \mathcal{H}_2$ and any $g \in \Gamma$ is uniquely written as $g = h_k + h'_\ell$, where $h_k \in \mathcal{H}_1$, and $h'_\ell \in \mathcal{H}_2$. Now, we assign the group elements from Γ to the $m \times n$ array $\mathbb{S}$ in such a way that, $\zeta_{i,j} = h_{i-1} + h'_{j-1}$, where $i \in [1,m], j \in [1,n]$. Hence, the row-sums in each row are given by,

$$\begin{aligned}
\alpha_j[1,n:\zeta_{1,j}] &= n \cdot 0 + \sum_{h' \in \mathcal{H}_2} h', \\
\alpha_j[1,n:\zeta_{2,j}] &= nh_1 + \sum_{h' \in \mathcal{H}_2} h', \\
&\vdots \qquad\qquad \vdots \\
\alpha_j[1,n:\zeta_{m,j}] &= nh_{m-1} + \sum_{h' \in \mathcal{H}_2} h'.
\end{aligned}$$

Let Λ be the set of all row-sums. Then,

$$\begin{aligned}
\Lambda &= \{n \cdot 0 + \sum_{h' \in \mathcal{H}_2} h', nh_1 + \sum_{h' \in \mathcal{H}_2} h', nh_2 + \sum_{h' \in \mathcal{H}_2} h', \dots, nh_{m-1} + \sum_{h' \in \mathcal{H}_2} h'\} \\
&= \{0, nh_1, nh_2, \dots, nh_{m-1}\}.
\end{aligned}$$

Similarly, suppose Λ' is the set of all column-sums. Then,

$$\begin{aligned}\Lambda' &= \{m.0 + \sum_{h\in\mathcal{H}_1} h, mh'_1 + \sum_{h\in\mathcal{H}_1} h, mh'_2 + \sum_{h\in\mathcal{H}_1} h, \ldots, mh'_{n-1} + \sum_{h\in\mathcal{H}_1} h\}\\ &= \{0, mh'_1, mh'_2, \ldots, mh'_{n-1}\}.\end{aligned}$$

It is clear that Λ and Λ' are subgroups of Γ of order m and n, respectively. Since Γ is finite cyclic group, $\Lambda \cong \mathcal{H}_1 = \langle n\rangle$ and $\Lambda' \cong \mathcal{H}_2 = \langle m\rangle$. □

Theorem 8. *Let Γ be a cyclic group of order mn and $\gcd(m,n) = 1$. Then a subgroup magic rectangle exists if and only if m and n are odd.*

Proof. Let Γ be an Abelian group of order mn. Assume that there exists a $\text{SMR}_\Gamma(m, n : \mathcal{H}_1, \mathcal{H}_2)$. Then, by Lemma 3, both m and n are of same parity. Since $\gcd(m,n) = 1$, m and n are odd. The converse part follows directly from Theorem 7. □

From Theorem 7 and Theorem 8, we see that if the order of Γ is even and $\mathcal{H}_1$ and $\mathcal{H}_2$ are two independent subgroups of Γ, then the subgroup magic rectangle over Γ does not exist with a row-subgroup $\mathcal{H}_1$ and column subgroup $\mathcal{H}_2$.

Corollary 4. *Let p be an odd prime and Γ be an Abelian group of order $2p^t$, $t \geq 1$. Then there does not exist a $2p^m \times p^n$ subgroup magic rectangle, where $m + n = t$.*

Proof. Since Γ is a group of order $2p^t$, Γ has a unique involution (say) τ. Now, by Lemma 1, $\sum_{g\in\Gamma} g = \tau$. Let $\mathcal{H}_2$ be a subgroup of order p^n. Then $\sum_{g\in\mathcal{H}_2} g = 0$, which contradicts Lemma 3. □

4 Existence of a Subgroup Magic Rectangle over an Abelian Group of Product of Prime Powers Order

In this section, we prove the existence of a subgroup magic rectangle over an Abelian group Γ whose orders are p^k, where p is an odd prime and $k > 1$.

Theorem 9. *If Γ is an Abelian group of order p^2, where p is an odd prime, then there exists a $p \times p$ subgroup magic rectangle.*

Proof. Let Γ be a cyclic group of order p^2. By Theorem 6, $\text{SMR}_\Gamma(p, p : \mathcal{H}_1, \mathcal{H}_2)$ exists. Suppose Γ is a non-cyclic Abelian group of order p^2. Then $\Gamma \cong \mathbb{Z}_p \times \mathbb{Z}_p$. Let $\mathbb{S}$ be an $p \times p$ array. Now, assign the group elements from Γ to the cells in $\mathbb{S}$ in such a way that,

$$\zeta_{ij} = ((i-1) \bmod p, (j-1) \bmod p), 1 \leq i \leq j \leq p.$$

Now, for all $i, j \in [1, p]$, $\alpha_i[1, p : \zeta_{i,j}] = \alpha_j[1, p : \zeta_{i,j}] = (0 \bmod p, 0 \bmod p)$.
In the next step, perform $\zeta_{1,j} \rightleftharpoons \zeta_{j,1}$ for all $j \in [1, p]$, Then all row-sums are,

$$\begin{aligned}
\Sigma R_1 &= \alpha_j[1, p : \zeta_{1,j}] = (\alpha_j[1, p : j - 1] \bmod p, \alpha_j[1, p : i - 1] \bmod p)\,, \\
\Sigma R_2 &= \alpha_j[1, p : \zeta_{2,j}] = (0, 1) + (\alpha_j[2, p : i - 1] \bmod p, \alpha_j[2, p : j - 1] \bmod p)\,, \\
&\vdots \qquad\qquad \vdots \qquad\qquad \vdots \\
\Sigma R_p &= \alpha_j[1, p : \zeta_{p,j}] = (0, p - 1) + (\alpha_j[2, p : i - 1] \bmod p, \alpha_j[2, p : \zeta_{p,j}](j - 1) \bmod p)\,.
\end{aligned}$$

Suppose $\mathcal{H}_1$ is the set of all row-sums. Then,

$$\mathcal{H}_1 = \{(0, 0), (p - 1, 1), \ldots, ((p - 1)(p - 1) \bmod p, (p - 1) \bmod p).$$

Similarly, by considering the column-sums in each column of $\mathbb{S}$, we have

$$\begin{aligned}
\Sigma C_1 &= \alpha_i[1, p : \zeta_{i,1}] = (\alpha_i[1, p : j - 1] \bmod p, \alpha_i[1, p : i - 1] \bmod p) \\
\Sigma C_2 &= \alpha_i[1, p : \zeta_{i,2}] = (1, 0) + (\alpha_i[2, p : i - 1] \bmod p, \alpha_i[2, p : j - 1) \bmod p)\,, \\
&\vdots \qquad\qquad \vdots \qquad\qquad \vdots \\
\Sigma C_p &= \alpha_i[1, p : \zeta_{i,p}] = (p - 1, 0) + (\alpha_i[2, p : i - 1] \bmod p, \alpha_i[2, p : j - 1] \bmod p)\,.
\end{aligned}$$

Suppose the set of all column-sums is $\mathcal{H}_2$. Then,

$$\mathcal{H}_2 = \{(0, 0), (1, p - 1), \ldots, ((p - 1) \bmod p, (p - 1)(p - 1) \bmod p).$$

It is not hard to verify that $\mathcal{H}_1$ and $\mathcal{H}_2$ are subgroups of order p.
Therefore, $\mathrm{SMR}_\Gamma(p, p : \mathcal{H}_1, \mathcal{H}_2)$ exists. □

Lemma 4. *If $\Gamma = \mathbb{Z}_{p^2} \times \mathbb{Z}_p$, then there exists a $p^2 \times p$ subgroup magic rectangle.*

Proof. Let $\Gamma = \mathbb{Z}_{p^2} \times \mathbb{Z}_p$ and $\mathbb{S}$ be an $p^2 \times p$ array. Now, assign the group elements from Γ to the cells in $\mathbb{S}$ in such a way that

$$\zeta_{i,j} = ((i - 1) \bmod p^2, (j - 1) \bmod p).$$

Now we consider the array $\mathbb{S}(\zeta_{i,j})$ as an $p^2 \times p$ matrix, which can be partitioned into p block matrices, each of order p given by

$$\mathbb{S}(\zeta_{i,j}) = \begin{pmatrix} \mathbb{S}_1 \\ \mathbb{S}_2 \\ \vdots \\ \mathbb{S}_p \end{pmatrix}$$

where,

$$\mathbb{S}_1 = \begin{pmatrix} \zeta_{1,1} & \zeta_{1,2} & \cdots & \zeta_{1,p} \\ \zeta_{2,1} & \zeta_{2,2} & \cdots & \zeta_{2,p} \\ \vdots & \vdots & \vdots & \vdots \\ \zeta_{p,1} & \zeta_{p,2} & \cdots & \zeta_{p,p} \end{pmatrix}, \quad \mathbb{S}_2 = \begin{pmatrix} \zeta_{p+1,1} & \zeta_{p+1,2} & \cdots & \zeta_{p+1,p} \\ \zeta_{p+2,1} & \zeta_{p+2,2} & \cdots & \zeta_{p+2,p} \\ \vdots & \vdots & \vdots & \vdots \\ \zeta_{2p,1} & \zeta_{2p,2} & \cdots & \zeta_{2p,p} \end{pmatrix},$$

$$\ldots, \mathbb{S}_p = \begin{pmatrix} \zeta_{p^2-p+1,1} & \zeta_{p(p-1)+1,2} & \cdots & \zeta_{p(p-1)+1,p} \\ \zeta_{p^2-p+2,1} & \zeta_{p(p-1)+2,2} & \cdots & \zeta_{p(p-1)+2,p} \\ \vdots & \vdots & \vdots & \vdots \\ \zeta_{p^2,1} & \zeta_{p^2,2} & \cdots & \zeta_{p^2,p} \end{pmatrix}.$$

Define $\mathbb{S}'(\zeta_{i,j})$ from $\mathbb{S}(\zeta_{i,j})$ by replacing the transpose of each of its blocks. That is, $\mathbb{S}'(\zeta_{i,j}) = \{\mathbb{S}_1^T, \mathbb{S}_2^T, \ldots, \mathbb{S}_p^T\}$. Now, for $k \in [1,p]$, each block matrix $\mathbb{S}_k^{\mathrm{T}}$ is partitioned into three submatrices such as $p \times 1$ block matrix $\mathbb{S}_k^1$ corresponding to first column of $\mathbb{S}_k^{\mathrm{T}}$ and two $p \times \frac{p-1}{2}$ submatrices $\mathbb{S}_k^2$ and $\mathbb{S}_k^3$, corresponding to the remaining columns of $\mathbb{S}_k^{\mathrm{T}}$. Thus,

$$\mathbb{S}'(\zeta_{i,j}) = \begin{pmatrix} \mathbb{S}_1^T \\ \mathbb{S}_2^T \\ \vdots \\ \mathbb{S}_p^T \end{pmatrix} = \begin{pmatrix} \mathbb{S}_1^1 & \mathbb{S}_1^2 & \mathbb{S}_1^3 \\ \mathbb{S}_2^1 & \mathbb{S}_2^2 & \mathbb{S}_2^3 \\ \vdots & \vdots & \vdots \\ \mathbb{S}_p^1 & \mathbb{S}_p^2 & \mathbb{S}_p^3 \end{pmatrix}.$$

Define $\mathbb{S}''(\zeta_{i,j})$ from $\mathbb{S}'(\zeta_{i,j})$ by replacing each $\mathbb{S}_k^3$ by $\pi(\mathbb{S}_\ell^3)$, where $\ell \equiv (p-k+1) \pmod p$. Then, the matrix

$$\mathbb{S}''(\zeta_{i,j}) = \begin{pmatrix} \mathbb{S}_1^1 & \mathbb{S}_1^2 & \pi(\mathbb{S}_p^3) \\ \mathbb{S}_2^1 & \mathbb{S}_2^2 & \pi(\mathbb{S}_{p-1}^3) \\ \vdots & \vdots & \vdots \\ \mathbb{S}_p^1 & \mathbb{S}_p^2 & \pi(\mathbb{S}_1^3) \end{pmatrix}.$$

In the next step, $\zeta_{1,j'} \rightleftharpoons \zeta_{1,\ell'}$ in $\mathbb{S}''$, where $\ell' \equiv (p-j'+1) \pmod p$. Now, let $\mathbb{S}''\ (\zeta_{i'j'})$. Then, the row-sum of the i'-th row in $\mathbb{S}''$ is,

$$\alpha_{j'}[1, p^2 : \zeta_{i',j'}] = \zeta_{i',1}.$$

Let $\mathcal{H}_1$ be the set of all row-sums,

$$\mathcal{H}_1 = \{\zeta_{i',1} : 1 \le i' \le p^2\} = \{(ap \bmod p^2, b \bmod p) : a \in [1,p] \text{ and } b \in [1,p]\}.$$

Similarly, the column-sum of the j'-th column in $\mathbb{S}''$ is,

$$\alpha_{i'}[1, p^2 : \zeta_{i',j'}] = -\zeta_{1,j'} + \zeta_{1,\ell'}.$$

Let $\mathcal{H}_2$ be the set of all column-sums,

$$\mathcal{H}_2 = \{-\zeta_{1,j'} + \zeta_{1,\ell'} : 1 \le j', \ell \le p\} = \{(ap, 0) : a \in [1,p]\}.$$

Hence, $\mathcal{H}_1$ and $\mathcal{H}_2$ are subgroups of Γ. □

Lemma 5. *If $\Gamma = \mathbb{Z}_p \times \mathbb{Z}_p \times \mathbb{Z}_p$, then there exists a $p^2 \times p$ subgroup magic rectangle.*

Proof. Let $\Gamma = \mathbb{Z}_p \times \mathbb{Z}_p \times \mathbb{Z}_p$ and $\mathbb{S}$ be a $p^2 \times p$ array. Assign the group elements from Γ to the cells in $\mathbb{S}$ by

$$\zeta_{i,j} = \big((i-1) \bmod p, (j-1) \bmod p, \lfloor (i-1)/p \rfloor \bmod p\big)$$

Similar to Lemma 4, let $\mathbb{S}(\zeta_{ij})$ be a $p^2 \times p$ matrix, whose entries are all group elements from Γ and $\mathbb{S}(\zeta_{ij}) = \{\mathbb{S}_1, \mathbb{S}_2, \ldots, \mathbb{S}_p\}$, where $\mathbb{S}_\ell$'s are block matrices of order p. In the next step, each block matrix $\mathbb{S}_\ell$ is partitioned into two subblock matrices $\mathbb{A}_\ell$ and $\mathbb{B}_\ell$ of orders $p\times(p-1)$ and $p\times 1$, respectively. Now we define $\mathbb{S}'(\zeta_{ij})$ as follows:

$$\mathbb{S}(\zeta_{ij}) = \begin{pmatrix} \mathbb{A}_1 & \mathbb{B}_1 \\ \vdots & \vdots \\ \mathbb{A}_{\frac{p+1}{2}} & \mathbb{B}_{\frac{p+1}{2}} \\ \vdots & \vdots \\ \mathbb{A}_p & \mathbb{B}_p \end{pmatrix} \xrightarrow[\pi(\mathbb{B}_k)\rightleftharpoons\pi(\mathbb{B}_{k'}),\,k'\equiv(p-k+1) \bmod p]{\text{Define } \mathbb{S}'(\zeta_{ij})} \mathbb{S}'(\zeta_{ij}) = \begin{pmatrix} \mathbb{A}_1 & \pi(\mathbb{B}_p) \\ \vdots & \vdots \\ \mathbb{A}_{\frac{p+1}{2}} & \pi(\mathbb{B}_{\frac{p+1}{2}}) \\ \vdots & \vdots \\ \mathbb{A}_p & \pi(\mathbb{B}_1) \end{pmatrix}$$

In the next step, perform $\mathbb{S}'(\zeta_{1,j'}) \rightleftharpoons \mathbb{S}'(\zeta_{1,\ell'})$, where $\ell' \equiv p - j' + 1 \pmod p$ and $j' \in [1, p]$. Let $\mathbb{S}'(\zeta_{ij}) = (\zeta_{i'j'})$. Then, the row-sum of the i-th row in $\mathbb{S}'(\zeta_{ij})$ is given below.
Case 1. If $i' \equiv 1 \bmod p$ and $i' \in [1, p^2]$, then $i' \to p(p-1) - i' + 2$ in $\pi(\mathbb{B}_\ell)$. Therefore,

$$\begin{aligned} \alpha_{j'}[1, p : \zeta_{i',j'}] &= \big(\alpha_{j'}[1, p-1 : i'-1] \bmod p, \alpha_{j'}[1, p-1 : j'-1] \bmod p, \\ &\quad \alpha_{j'}[1, p-1 : \lfloor (i'-1)/p \rfloor] \bmod p\big) + \big((1-i') \bmod p, (p-1) \bmod p, \\ &\quad \lfloor (p^2 - p - i' + 1)/p \rfloor \bmod p\big), \\ &= \big(2(1-i') \bmod p, p-1 \bmod p, \big(\lfloor (p^2-p-i'+1)/p \rfloor - \lfloor (i'-1)/p \rfloor\big) \bmod p\big). \end{aligned}$$

Case 2. If $i' \not\equiv 1 \bmod p$ and $i' \in [1, p^2]$, then $i' \to p^2 - (\lceil i'/p \rceil - 1)\, p + i'$ in $\pi(\mathbb{B}_\ell)$. Therefore,

$$\begin{aligned} \alpha_{j'}[1, p : \zeta_{i',j'}] &= \big(\alpha_{j'}[1, p-1 : i'-1] \bmod p, \alpha_{j'}[1, p-1 : j'-1] \bmod p, \\ &\quad \alpha_{j'}[1, p-1 : \lfloor (i'-1)/p \rfloor] \bmod p\big) \\ &\quad + \big((i'-1) \bmod p, -1 \bmod p, \lfloor p^2 - (\lceil i'/p \rceil - 1)\, p + i' + 1)/p \rfloor \bmod p\big), \\ &= \big(0, -1 \bmod p, \big(\lfloor p^2 - (\lceil i'/p \rceil - 1)\, p + i' + 1)/p \rfloor - \lfloor (i'-1)/p \rfloor\big) \bmod p\big). \end{aligned}$$

Let $\mathcal{H}_1$ be the set of all row-sums of $\mathbb{S}'(\zeta_{ij})$. Then one can write $\mathcal{H}_1 = A \cup B$, with $A \cap B = \varnothing$.

Let $\mathcal{H}_1$ be the set of all row-sums of $\mathbb{S}'(\zeta_{ij})$. Then $\mathcal{H}_1 = A \cup B$, with $A \cap B = \varnothing$, where

$$A = \left\{ (a,b,c) \;\middle|\; \begin{array}{l} a = 2(1-i') \bmod p, \\ b = p-1, \\ c = \left(\lfloor (p^2 - p - i' + 1)/p \rfloor - \lfloor (i'-1)/p \rfloor\right) \bmod p, \\ i' \equiv 1 \bmod p,\ i' \in [1,p^2] \end{array} \right\},$$

$$B = \left\{ (0,b,c) \;\middle|\; \begin{array}{l} b = p-1, \\ c = \left(\lfloor (p^2 - (\lceil i'/p \rceil - 1)p + i' + 1)/p \rfloor - \lfloor (i'-1)/p \rfloor\right) \bmod p, \\ i' \not\equiv 1 \bmod p,\ i' \in [1,p^2] \end{array} \right\}.$$

Since all elements in $\mathcal{H}_1$ are distinct, $|\mathcal{H}_1| = p^2$ and hence $\mathcal{H}_1$ is a subgroup of order p^2.
Further, the column-sum of the j'-th column in $\mathbb{S}'(\zeta_{ij})$ is given below.
Case 1. If $j' = 1$, then

$$\begin{aligned} \alpha_{i'}[1,p^2 : \zeta_{i',1}] &= (0,p-1,0) + \big(\alpha_{i'}[2,p^2 : i'-1] \bmod p,\ 0, \\ &\qquad \alpha_{i'}\left[2,p^2 : \lfloor (i'-1)/p \rfloor\right] \bmod p\big) \\ &= (0,p-1,0). \end{aligned}$$

Case 2. If $j' \in [2,p-1]$, then

$$\begin{aligned} \alpha_{i'}[1,p^2 : \zeta_{i',j'}] &= (0,p-j',0) + \big(\alpha_{i'}[2,p^2 : i'-1] \bmod p, \alpha_{i'}[2,p^2 : j'-1] \bmod p, \\ &\qquad \alpha_{i'}\left[2,p^2 : \lfloor (i'-1)/p \rfloor\right] \bmod p\big) \\ &= (0,\ (p-j') \bmod p,\ 0) + (0,\ (1-j') \bmod p,\ 0) \\ &= (0,\ (1-2j') \bmod p,\ 0)\,. \end{aligned}$$

Case 3. If $j' = p$, then

$$\begin{aligned} \alpha_{i'}[1,p^2 : \zeta_{i',p}] &= (0,0,0) + \big(\alpha_{i'}[2,p^2 : i'-1] \bmod p,\ \alpha_{i'}[2,p^2 : p-1] \bmod p, \\ &\qquad \alpha_{i'}\left[2,p^2 : \lfloor (i'-1)/p \rfloor\right] \bmod p\big) \\ &= (0,1,0). \end{aligned}$$

Let $\mathcal{H}_2$ be the set of all column-sums of $\mathbb{S}'(\zeta_{ij})$. Then $\mathcal{H}_2 = A \cup B \cup C$, where $A = \{(0,p-1,0)\}$, $B = \{(0,b,0) : b \equiv 1-2j' \pmod p,\ j' \in [2,p-1]\}$, $C = \{(0,1,0)\}$. Thus, $\mathcal{H}_2 = \{(0,b,0) : b \in \mathbb{Z}_p\}$, which is a subgroup of order p. □

Example 3. The construction of a 9×3 subgroup magic rectangle over $\mathbb{Z}_3 \times \mathbb{Z}_3 \times \mathbb{Z}_3$ is given below. In Step 1, assign all group elements ζ_{ij}'s to all cells (i,j) in the 9×3 array $\mathbb{S}$ by

$$\zeta_{i,j} = ((i-1) \bmod p, (j-1) \bmod p, \lfloor (i-1)/p \rfloor \bmod p)\,.$$

In Step 2, partition the array $\mathbb{S}$ into three distinct 3×3 sub-arrays, and then partition each sub-array into a 3×2 sub-array $\mathbb{A}$ and a 3×1 sub-array $\mathbb{B}$. Finally, apply the permutation π on $\mathbb{B}$. In step 3, perform $\pi(\mathbb{B}_1) \rightleftharpoons \pi(\mathbb{B}_3)$ and fix $\pi(\mathbb{B}_2)$. In Step 4, perform $\zeta_{1,1} \rightleftharpoons \zeta_{1,3}$ and fix $\zeta_{1,2}$ (Fig. 3).

(0,0,0)	(0,1,0)	(0,2,0)
(1,0,0)	(1,1,0)	(1,2,0)
(2,0,0)	(2,1,0)	(2,2,0)
(0,0,1)	(0,1,1)	(0,2,1)
(1,0,1)	(1,1,1)	(1,2,1)
(2,0,1)	(2,1,1)	(2,2,1)
(0,0,2)	(0,1,2)	(0,2,2)
(1,0,2)	(1,1,2)	(1,2,2)
(2,0,2)	(2,1,2)	(2,2,2)

Step 1

(0,0,0)	(0,1,0)	(0,2,0)
(1,0,0)	(1,1,0)	(2,2,0)
(2,0,0)	(2,1,0)	(1,2,0)
(0,0,1)	(0,1,1)	(0,2,1)
(1,0,1)	(1,1,1)	(2,2,1)
(2,0,1)	(2,1,1)	(1,2,1)
(0,0,2)	(0,1,2)	(0,2,2)
(1,0,2)	(1,1,2)	(2,2,2)
(2,0,2)	(2,1,2)	(1,2,2)

Step 2

(0,0,0)	(0,1,0)	(0,2,2)
(1,0,0)	(1,1,0)	(2,2,2)
(2,0,0)	(2,1,0)	(1,2,2)
(0,0,1)	(0,1,1)	(0,2,1)
(1,0,1)	(1,1,1)	(2,2,1)
(2,0,1)	(2,1,1)	(1,2,1)
(0,0,2)	(0,1,2)	(0,2,0)
(1,0,2)	(1,1,2)	(2,2,0)
(2,0,2)	(2,1,2)	(1,2,0)

Step 3

(0,2,2)	(0,1,0)	(0,0,0)
(1,0,0)	(1,1,0)	(2,2,2)
(2,0,0)	(2,1,0)	(1,2,2)
(0,0,1)	(0,1,1)	(0,2,1)
(1,0,1)	(1,1,1)	(2,2,1)
(2,0,1)	(2,1,1)	(1,2,1)
(0,0,2)	(0,1,2)	(0,2,0)
(1,0,2)	(1,1,2)	(2,2,0)
(2,0,2)	(2,1,2)	(1,2,0)

Step 4

Fig. 3. A step-by-step illustration of Example 3

Suppose $\mathcal{H}_1 = \{\Sigma R_1, \Sigma R_2, \ldots, \Sigma_9\}$ and $\mathcal{H}_2 = \{\Sigma C_1, \Sigma C_2, \Sigma C_3\}$. Then, we get $\mathcal{H}_1 = \{(a, 0, c) : a, c \in \mathbb{Z}_3\}$ and $\mathcal{H}_2 = \{(0, b, b) : b \in \mathbb{Z}_3\}$ as subgroups of order 9 and 3, respectively.

Theorem 10. *If Γ is an Abelian group of order p^3, where p is an odd prime, then there exists a $p^2 \times p$ subgroup magic rectangle.*

Proof. The proof follows from Theorem 6, Lemma 4, and Lemma 5. □

Theorem 11. *If Γ is an Abelian group of order p^2q, where p and q are two odd primes, then there exists a $p^2 \times q$ subgroup magic rectangle.*

Proof. If $p = q$, then the proof follows from Theorem 10. If $p \neq q$, then $\Gamma \cong \mathbb{Z}_{p^2q}$ and by Theorem 6, $p^2 \times q$ subgroup magic rectangle exists. □

Theorem 12. *If $\Gamma = \mathbb{Z}_p \times \mathbb{Z}_p \times \cdots \times \mathbb{Z}_p$ (n times), then there exists a $p^{n-1} \times p$ subgroup magic rectangle.*

Proof. The proof is analogous to Lemma 5, by assigning the group elements to the $p^{n-1} \times p$ array in such a way that,

$$\zeta_{i,j} = \left\{ (a_1, a_2, \ldots, a_p) \;\middle|\; \begin{array}{l} a_1 = (i-1) \bmod p, \\ a_2 = (j-1) \bmod p, \\ a_k = \lfloor (i-1)/p \rfloor \bmod p \;, k \in [3, p] \end{array} \right\}.$$

□

5 Conclusion and Scope

The motivation of this article is to find certain combinatorial object as close to group magic rectangle as possible, when magic rectangle mainly uses the structural properties of an Abelian groups. First $\mathcal{G}$-coset arrays and $\mathcal{H}$-Kotzig arrays are defined and constructed. Then, using such combinatorial objects, we characterize the existence of the subgroup magic rectangle over $\mathbb{Z}_{mn}$ where m or n is odd. However, when both m and n are even, the existence of a subgroup magic rectangle over $\mathbb{Z}_{mn}$ is still open. Further, when p is an odd prime, we prove the existence of a subgroup magic rectangle for all Abelian groups of order up to p^3. The following conjectures are still open.

Problem 1. Suppose $\mathcal{G}$ and $\mathcal{H}$ are two subgroups of an Abelian group Γ of order mn. Then a $\mathcal{G}$-coset array is a $\mathcal{H}$-Kotzig array if and only if $\gcd(m, n) = 1$.

Problem 2. Does a subgroup magic rectangle exist for all Abelian p-groups, where p is odd?

References

1. Andrews, W.S.: Magic squares and cubes. Open Court Publishing Company (1917)
2. Archdeacon, D.: Heffter arrays and biembedding graphs on surfaces. Electron. J. Comb. **22**(1), 1.74, 14 (2015)
3. Cichacz, S.: A Γ-magic rectangle set and group distance magic labeling. In: Kratochvíl, J., Miller, M., Froncek, D. (eds.) IWOCA 2014. LNCS, vol. 8986, pp. 122–127. Springer, Cham (2015). https://doi.org/10.1007/978-3-319-19315-1_11
4. Cichacz, S.: On zero sum-partition of Abelian groups into three sets and group distance magic labeling. Ars Math. Contemp. **13**(2), 417–425 (2017)
5. Combe, D., Nelson, A.M.: Magic labellings of infinite graphs over infinite groups. Aust. J. Combin. **35**, 193–210 (2006)
6. Freyberg, B.: On constant sum partitions and applications to distance magic-type graphs. AKCE Int. J. Graphs Comb. **17**(1), 373–379 (2020)
7. Gao, W., Geroldinger, A.: Zero-sum problems in finite abelian groups: a survey. Expo. Math. **24**(4), 337–369 (2006)
8. Holt, D.F., Eick, B., O'Brien, E.A.: Handbook of Computational Group Theory. Chapman & Hall/CRC Press, Boca Raton (2005)
9. Kaplan, G., Lev, A., Roditty, Y.: On zero-sum partitions and anti-magic trees. Disc. Math. **309**(8), 2010–2014 (2009)
10. Marr, A.M., Wallis, W.D.: Magic Graphs, 2nd edn. Birkhäuser/Springer, New York (2013)
11. Morini, F., Pellegrini, M.A.: Magic partially filled arrays on abelian groups. J. Comb. Des. **31**(8), 347–367 (2023)
12. Silverman, J.H.: Abstract algebra—an integrated approach, vol. 55 of Pure and Applied Undergraduate Texts. American Mathematical Society, Providence (2022)
13. Warner, S.: Modern algebra. Dover Books on Advanced Mathematics, 2nd edn. Dover Publications Inc., New York (1990)
14. Wen, Y., Sun, H.: Note on magic squares and magic cubes on abelian groups. J. Math. Res. Exposit. **17**(2), 176–178 (1997)

FPT Algorithms and Other Approaches to Optimization

Parameterized Algorithms for Power Edge Set and Zero Forcing Set

Sriram Bhyravarapu[1], Lawqueen Kanesh[3], Madhumita Kundu[4](✉), Daniel Lokshtanov[5], and Saket Saurabh[2,4]

[1] Indian Institute of Technology Guwahati, Guwahati, India
sriramb@iitg.ac.in
[2] The Institute of Mathematical Sciences, HBNI, Chennai, India
saket@imsc.res.in
[3] Indian Institute of Technology Indore, Indore, India
[4] University of Bergen, Bergen, Norway
{Madhumita.Kundu,Saket.Saurabh}@uib.no
[5] University of California Santa Barbara, Santa Barbara, USA
daniello@ucsb.edu

Abstract. In this article, we study the parameterized complexity of the POWER EDGE SET problem (abbreviated as PES). In PES, we are given a graph G and an integer k, and the goal is to find a set $S \subseteq E(G)$ of size at most k such that S *monitors* all the vertices of G. An edge set S is said to monitor the vertex set $V(G)$ if, starting with the endpoints of the edges in S (which are initially considered monitored), the entire vertex set can be monitored by repeatedly applying the following rule: if there exists a monitored vertex with exactly one unmonitored neighbor, then that neighbor becomes monitored.

The parameterized complexity of this problem was initiated by Darties et al. (Journal of Discrete Algorithms, 2018), who explicitly posed the question of whether PES is fixed-parameter tractable (FPT)—that is, whether it admits an algorithm with running time $f(k) \cdot n^{\mathcal{O}(1)}$. Cazals et al. (IWOCA, 2019) subsequently showed that a precolored variant of the problem is W[2]-hard. While their introduction mentions that PES is W[2]-hard, the reduction only applies to the precolored version.

In this paper, we clarify the complexity of PES and show that both PES and its natural variant, ZERO FORCING SET, are indeed fixed-parameter tractable. Our results include an FPT algorithm for PES parameterized by the treewidth of the input graph, thereby improving upon a previously known XP algorithm for this parameter. Furthermore, we present efficient FPT algorithms for PES when parameterized by vertex cover number and by neighborhood diversity.

Keywords: Parameterized Algorithms · Power edge set · Zero forcing set · PMU placement problem · Tree-width · Neighborhood diversity

H. Fernau and B. Zhu (Eds.): IWOCA 2025, LNCS 15885, pp. 349–361, 2025.
https://doi.org/10.1007/978-3-031-98740-3_25

1 Introduction

Phasor Measurement Unit (PMU) is an expensive device that is used in power networks to continuously collect the voltage and phase angle. Thus, the problem of minimizing the number of PMUs to place on a network for complete network monitoring naturally arises (see [1,5,9,13] and references therein for further details). When abstracted out this has led to the study of "propagation variants" of well-studied Dominating Set and Edge Cover problems [1,5,9,13]. In this paper we study one such problem namely PES from the perspective of parameterized complexity.

We start by defining two propagation rules that govern PES. Let $S \subseteq E(G)$.

Rule 1 The endpoints of each edge in S are monitored.
Rule 2 If there exists a monitored vertex v such that all but one neighbor $w \in N(v)$ are monitored, then monitor w.

Given a graph G, the *power edge set* $S \subseteq E(G)$ is a set of edges that monitors all the vertices of G after the application of Rule 1 and repeated applications of Rule 2. An illustration of a power edge set is given in Fig. 1. In this paper we study the following decision version of PES.

Power Edge Set (PES)
Input: An undirected graph G, and an integer k.
Question: Does there exist a power edge set $S \subseteq E(G)$ of size at most k?

1.1 Background and Motivation

PES is a variant of Power Dominating Set (in short PDS). PDS has been around for more than a decade and is an extremely well understood problem [7]. In this problem, we have to find a k sized vertex set $D \subseteq V(G)$ that monitors $V(G)$ with the initial set $N[D]$ as monitored, and then Rule 2 is applied repeatedly. PDS turned out to be a highly intractable problem both within the classical complexity and parameterized complexity. In order to circumvent the intractability of PDS, PES was introduced by Toubaline at al. [12]. Even before PES was introduced, a vertex variant of PES, called Power Vertex Set (in short PVS), existed [12]. It is similar to Power Dominating Set with the only change that the initial set of monitored vertices is just the set D. This is also known as Zero Forcing Set (in short ZFS) [6]. Thus, in all the three problems, PES, PDS and PVS, the propagation rule remains the same, what differs is the set of vertices that are considered monitored initially. Our main conceptual message is that, in the realm of parameterized complexity, *the set of vertices that are considered monitored initially determines the parameterized complexity.*

Toubaline et al. [12] show that PES is polynomial time solvable on trees and grids. They also show that it is NP-hard to approximate within a factor $(1.12-\epsilon)$, for any $\epsilon > 0$. Darties et al. [5] continue this study and show that PES is NP-hard on planar graphs of degree at most five, bipartite graphs with bounded degree,

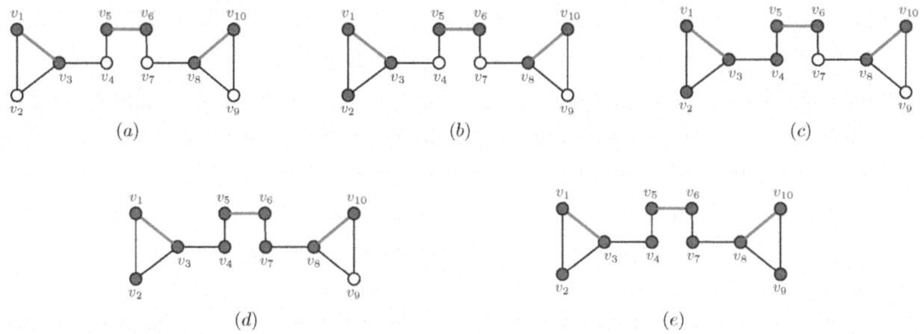

Fig. 1. An illustration of a propagation of monitored vertices. The vertices colored green are the monitored vertices at the current stage. The red colored edges are the ones that constitute the power edge set. (Color figure online)

unit disk graphs, and grids with maximum degree three. They also initiate the study of parameterized complexity of the problem and give an XP-algorithm parameterized by the tree-width of the input graph. This implies polynomial time algorithms for series-parallel and outerplanar graphs. Darties et al. [5] also show that PES does not admit an $2^{o(k)}n^{\mathcal{O}(1)}$ algorithm when parameterized by the solution size k. As follow up open problems they ask:

> Is PES FPT parameterized by (a) k – the solution size; and (b) the tree-width of the input graph?

Cazals et al. [2] study PES and ZFS on cubic graphs and among other results show that a precolored version of PES and ZFS is W[2]-hard. However, in the contribution part of their introduction (as well in the first paragraph of Section 3) they say: "We show that Power Edge Set and Zero Forcing Set are W[2]-hard parameterized by the size of the solution in Sect. 3." We clear this confusing message about PES and ZFS and answer in affirmative, both the open questions posed in [5] (stated above).

1.2 Our Results

We show that PES and ZFS are FPT with respect to either of the parameters the solution size k (the number of edges/vertices) and the tree-width (tw) of the input graph. To design an FPT algorithm for PES (ZFS) we first show that if the input is a yes-instance then the path-width of the input graph is bounded as a function of solution size k. That is, if G has a power edge set of size k then the path-width of G, $\mathsf{pw}(G) \leq 2k$ (Lemma 1). Then, we show that the PES can be formulated in MSO_2. Thus, using Courcelle's Theorem [3] we get that PES is FPT with the combined parameters tree-width and k. Since $\mathsf{tw}(G) \leq \mathsf{pw}(G)$ and $\mathsf{pw}(G) \leq 2k$ (Lemma 1), we get that PES is FPT with k alone. The same strategy (or algorithm) shows that ZFS is FPT parameterized by the solution size (Sect. 5). These results are primarily classification results.

Towards achieving an explicit FPT algorithm for PES with k alone, it is sufficient if we give an explicit FPT algorithm for PES with tw alone (because of Lemma 1). Darties et al. in [5] gave a $2^{\mathcal{O}(\mathsf{tw}^2)}n^{\mathcal{O}(\mathsf{tw})}$ time algorithm. Using the ideas from FPT algorithm for PDS parameterized by the tree-width [7], given by Guo, Niedermeier and Raible, we design an explicit FPT algorithm for PES running in time $2^{\mathcal{O}(\mathsf{tw}^2)}n^{\mathcal{O}(1)}$. Thus, we show that PES is FPT parameterized by the treewidth of the input graph. We also consider the parameter vertex cover number and obtain an algorithm running in time $2^{\mathcal{O}(\mathsf{vc}\log \mathsf{vc})}n^{\mathcal{O}(1)}$ better than $2^{\mathcal{O}(\mathsf{vc}^2)}n^{\mathcal{O}(1)}$ (that follows from the tree-width algorithm), in Sect. 6. In Sect. 7, we show that PES is FPT parameterized by the neighborhood diversity of the graph. More precisely it can be solved in $2^t n^{\mathcal{O}(1)}$ time, where t is the neighborhood diversity of the graph on n vertices.

2 Preliminary Notations

For a graph G, we use n and m to denote the number of vertices and edges of G. We use $V(G)$ and $E(G)$ to denote the vertex set and the edge set of G. Let $X \subseteq E(G)$ be a subset of edges. We use $V_e(X)$ to denote the set of endpoints of all edges in X. Let $f : X \to \{1, 2, \ldots, \ell\}$. We denote $f(X) = \bigcup_{x \in X} \{f(x)\}$. We denote $[\ell]$ to denote the set $\{1, 2, \ldots, \ell\}$.

Definition 1 (Tree decomposition [4]). *A* tree decomposition *of a graph $G = (V, E)$ is a pair $\mathcal{T} = (T, \{X_t : t \in V(T)\})$, where T is a tree where every node t is assigned a vertex subset $X_t \subseteq V(G)$, called a* bag, *such that the following three conditions hold:*

(i) *Every vertex of G is in some bag. That is, $\bigcup_{t \in V(T)} X_t = V(G)$.*
(ii) *For each $uv \in E(G)$, there exists a node t of T such that $\{u, v\} \subseteq X_t$.*
(iii) *For every $u \in V(G)$, the set $T_u = \{t \in V(T) : u \in X_t\}$, i.e., the set of nodes whose corresponding bags contain u, induces a connected subtree of T.*

The *width* of a tree decomposition $\mathcal{T} = (T, \{X_t \mid t \in V(T)\})$ equals $\max_{t \in V(T)} |X_t| - 1$. The *tree-width* of G, denoted by $\mathsf{tw}(G)$, is the minimum possible width of a tree decomposition of G. If the tree T in the pair $\mathcal{T}$ is replaced by a path, then we get the concept of path decomposition. The *path-width* of a graph G, denoted $\mathsf{pw}(G)$, is the minimum width over all path decompositions of G.

Definition 2 (Neighborhood Diversity [10]). *Let G be a graph. Two vertices $u, v \in V(G)$ are said to be* equivalent *(or same type) if and only if $N(u) \setminus \{v\} = N(v) \setminus \{u\}$. A graph G has neighborhood diversity at most t, if there exists a partition of $V(G)$ into at most t sets $V_1, V_2, \ldots, V_t$ such that for each $i \in [t]$, vertices in V_i are pairwise equivalent (same type).*

Notice that, for all $i \in [t]$, V_i is a clique or an independent type in G. That is, V_i is a *clique type* (resp. *independent type*) if $G[V_i]$ is a clique (resp. independent). For each $i, j \in [t]$, $i \neq j$, either every vertex in V_i is adjacent to every vertex in V_j or no vertex in V_i is adjacent to any vertex in V_j.

3 Parameterization by the Solution Size

In this section we give an FPT algorithm for PES parameterized by the solution size. To this end, in the next subsection we show that if a graph G has a PES of size at most k, then its path-width is bounded by $2k$. Then, we provide an MSO_2 formulation for PES in the full version of the paper.. This combined with Courcelle's Theorem [3] shows the desired result.

3.1 Upper Bound on Treewidth as a Function of Solution Size

Lemma 1. *If G has a power edge set of size k then* $\mathsf{pw}(G) \leq 2k$.

Proof. Let $S \subseteq E(G)$, of size k, be a power edge set of G. This implies that the number of vertices that are incident to the edges in S is at most $2k$. We now construct a series of bags $\mathcal{B}$ in an iterative fashion and argue that it is a path decomposition of G. Let $B_j \in \mathcal{B}$ represent the jth bag of vertices constructed in this process. We initialize B_0 to be the set of vertices that are the endpoints of S. We explain the construction of $\mathcal{B}$ below.

Construction of $\mathcal{B}$: For each $i \geq 1$, we construct the bags B_{2i-1} and B_{2i} from B_{2i-2}. To this end, first we check for the existence of a vertex $u \in B_{2i-2}$ that satisfies the following premise:
Premise: There exists a vertex $u \in B_{2i-2}$ such that there is a vertex $v \in N(u)$ where $N(u) \setminus \bigcup_{j=0}^{2i-2} B_j = \{v\}$.

If there exists a vertex u satisfying the above premise, then we define $B_{2i-1} := B_{2i-2} \cup \{v\}$ and $B_{2i} := B_{2i-1} \setminus \{u\}$. It is easy to see that $|B_{2i-1}| \leq 2k+1$ and $|B_{2i}| \leq 2k$. Else if there is no vertex u satisfying the above premise, then the following two cases arise: either $\bigcup_{j=0}^{2i-2} B_j = V(G)$ in which case we stop the process, or we have that for every vertex $u \in B_{2i-2}$, there are two vertices $p, q \in N(u)$ such that $\{p, q\} \subseteq N(u) \setminus \bigcup_{j=0}^{2i-2} B_j$ which contradicts that S is a valid power edge set of G. Let B_ℓ be the last bag constructed in this process. We now show that the bags $B_0, B_1, \ldots, B_\ell$ form a path decomposition of G. That is, we need to show that $\mathcal{B}$ satisfies the properties (i), (ii) and (iii) in Definition 1 with respect to path decomposition.

Since $\bigcup_{j=0}^{\ell} B_j = V(G)$, every vertex of G is in some bag B_j and the property (i) is satisfied. Consider an edge $xy \in E(G)$. If $x, y \in B_0$ then the property (ii) trivially holds. We now argue for the other case depending on which vertex is added first to the bags. Without loss of generality, let the first occurrences of x and y be in the bags B_i and $B_{i'}$ respectively, for some $0 \leq i < i'$. Notice that in the construction of the bags, a vertex $w \in B_{j-1}$ is not included in B_j only when w is responsible for adding its lone neighbor in $N(w) \setminus \bigcup_{j'=0}^{j-2} B_{j'}$ to the bag B_{j-1}.

Since x is introduced before y whether or not x is responsible for the inclusion of y in the bag $B_{i'}$, both x and y belong to $B_{i'}$. Hence the property (ii) is satisfied.

Consider a vertex w. Let the first and the last occurrences of w be in the bags B_i and $B_{i'}$ respectively, $0 \leq i < i' \leq \ell$. We claim that $w \in B_j$, for all $i \leq j \leq i'$. Suppose not. Let $B_{j'}$, $i < j' < i'$, be the first bag that does not contain w. To add w to some bag $B_{j''}$, $j' < j'' \leq i'$, there should exist a vertex $u \in B_{j''-1}$ such that $N(u) \setminus \bigcup_{\widehat{j}=0}^{j''-1} B_{\widehat{j}} = \{w\}$, which is not true as $w \notin B_{j'}$ where $j' < j''$. Hence $w \in B_j$ for all $i \leq j \leq i'$ and thus satisfying the property (iii).

The size of the bag B_0 is at most $2k$. The size of the bag B_{2j-1} is at most $2k+1$ and the size of the bag B_{2j} is at most $2k$, for all j. Hence we get that the path-width of G is at most $2k$. □

For any graph G, $\mathsf{tw}(G) \leq \mathsf{pw}(G)$. Thus, we get the following.

Corollary 1. *If G has a power edge set of size k then* $\mathsf{tw}(G) \leq 2k$.

3.2 FPT Parameterized by Solution Size

First we show that PES can be expressed in MSO_2. Due to space constraints, we provide the details in the full version of the paper. Further we show that the problem is FPT parameterized by solution size.

Theorem 1. PES *is FPT when parameterized by solution size.*

Proof. For each value of q in $[k, 2k]$, we formulate the PES problem in MSO_2 as described above. From Courcelle's Theorem [3], we get that PES is FPT when parameterized by the tree-width and q. Since the tree-width of a graph is at most twice the solution size, from Corollary 1, and $q \leq 2k$, we get that PES is FPT when parameterized by k, the size of the solution. □

The running time is of the form tower$(h(k))$, where the height of the tower depends on the size of the MSO_2 formula and the parameter k appears at the topmost exponent. Hence, we do not state the bound explicitly. Therefore, in Sect. 4, we present an explicit FPT algorithm, which also allows us to demonstrate that the problem is FPT when parameterized by tree-width.

4 Valid Edge Orientation with Minimum Origin

In this section we give the following two results.

Theorem 2. *There is an algorithm that solves* PES *in* $2^{\mathcal{O}(k^2)} n^{\mathcal{O}(1)}$ *time.*

Theorem 3. *There is an algorithm that solves* PES *in* $2^{\mathcal{O}(\mathsf{tw}^2)} n^{\mathcal{O}(1)}$ *time.*

As observed before, Theorem 2 follows from Theorem 3 and Corollary 1. In what follows, we focus on Theorem 3. We first introduce a new problem called VALID EDGE ORIENTATION WITH MINIMUM ORIGIN (in short VEOMO) and give a polynomial time parameter preserving reduction from PES to VEOMO. Next, we obtain an FPT algorithm for VEOMO parameterized by tree-width. This gives us an FPT algorithm for PES when parameterized by tree-width. We now define modified versions of edge orientation and dependency path, adapted to suit our purposes based on the algorithm by Guo, Niedermeier, and Raible [7].

Each vertex not incident to a solution edge is monitored by the application of Rule 2. That is, at each stage we want a monitored vertex to monitor its only unmonitored neighbor. Further the newly monitored vertex can monitor an unmonitored neighbor. In some sense, it resembles like a path of monitored vertices where a vertex on the path is monitored and it further monitors its adjacent vertex on the path. We encode these properties on a graph with an orientation on the edges and define paths satisfying these properties.

Definition 3 (Edge Orientation). *An* edge orientation, *or simply* orientation, *of an undirected graph* $G = (V, E)$ *is a graph* $D = (V, E_1 \cup E_2 \cup E_3)$ *such that, for each* $\{u, v\} \in E(G)$, *there is either a directed edge* (u, v) *or* (v, u) *in* E_1 *or an undirected edge* $\{u, v\}$ *in* E_2 *or a* special edge $\{u, v\} \in E_3$.

The indegree *of a vertex* v *in* D, *denoted by* $d^-(v)$, *is defined as* $|\{(u, v) \mid (u, v) \in E_1\}|$ *and the* outdegree *of* v, *denoted by* $d^+(v)$, *is defined as* $|\{(v, u) \mid (v, u) \in E_1\}|$. *If the special edge* $\{u, v\} \in E_3$ *then* $d^-(u) = d^-(v) = 0$. *The subgraph* $D[V']$ *induced by* $V' \subseteq V$ *in* D *is called a* suborientation.

Definition 4 (Dependency Path). *A* dependency path *in an edge orientation* $D = (V, E_1 \cup E_2 \cup E_3)$ *is a sequence of vertices and edges* $v_1, e_1, v_2, e_2, \ldots, e_{i-1}, v_i$, *where* $i \geq 3$, *satisfying:*

1. *for all* $1 \leq j, k \leq i$, $v_j = v_k \iff j = k$,
2. *for all* $1 \leq j \leq i - 1$, *either* $e_j = (v_j, v_{j+1}) \in E_1$, $e_j = \{v_j, v_{j+1}\} \in E_2$ *or* $e_j = \{v_j, v_{j+1}\} \in E_3$, *and*
3. *for all* $1 \leq j \leq i - 1$, *at least one of* e_j *and* e_{j+1} *belongs to* E_1.

The vertices v_1 *and* v_i *are called the* tail endpoint *and the* head endpoint *of the dependency path, respectively. A* dependency cycle *in an orientation is a sequence of vertices and edges* $v_1, e_1, v_2, e_2, \ldots, e_{i-1}, v_i$, *such that* $v_1 = v_i$ *and* $v_1, e_1, v_2, e_2, \ldots, v_{i-1}$ *is a dependency path and the edge between* v_1 *and* v_i *can be in* E_2 *or* E_3 *only if* $(v_1, v_2) \in E_1$ *and* $(v_{i-2}, v_{i-1}) \in E_1$; *otherwise, it is directed. A* directed path *is a dependency path with only directed edges.*

Definition 5 (Valid Edge Orientation). *A* valid edge orientation $D = (V, E_1 \cup E_2 \cup E_3)$ *of an undirected graph* $G = (V, E)$ *is an edge orientation such that* $\forall v \in V : d^-(v) \leq 1$ *and* $d^+(v) \leq 1$, *and there is no dependency cycle in* D. *The* origin O *of* D *is defined as the set of vertices* $v \in V$ *with* $d^-(v) = 0$.

Remark: Let $D = (V, E_1 \cup E_2 \cup E_3)$ be a valid edge orientation of G. Then from Definitions 3 and 5, for all edges $\{u, v\} \in E_3$ we have that $u, v \in O$.

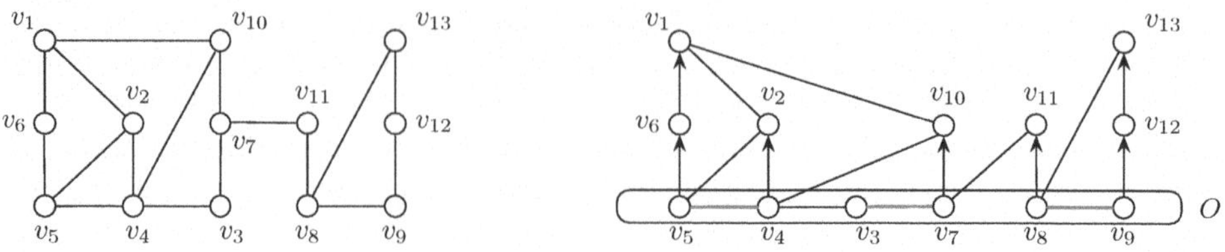

Fig. 2. (right) Illustration of a valid edge orientation of the graph (left). The special edges $\{v_5v_4, v_3v_7, v_8v_9\}$ are marked in red color. The vertex set $\{v_3, v_4, v_5, v_7, v_8, v_9\}$ corresponding to the special edges forms the origin O. (Color figure online)

Lemma 2. *Let $D = (V, E_1 \cup E_2 \cup E_3)$ denote a valid edge orientation with origin $O \subseteq V$. Then,*

1. *For each $v \in V \setminus O$, there exists a unique vertex $u \in O$ such that there is a directed path from u to v, and*
2. *Any two directed paths starting from vertices in O to two vertices in $V \setminus O$ are vertex disjoint.*

Proof. From Definition 5, for each $v \in V \setminus O$, we have that $d^-(v) = 1$. Let the corresponding edge be $(u, v) \in E_1$. If $u \in O$, then the property (1) is trivially satisfied. Hence let us assume that $u \notin O$. Again from the definition, we have $d^-(u) = 1$ and we apply the same arguments as above to u. Since the number of vertices in D is finite there has to be a vertex $x \in O$ with a directed path from x to v. Since $d^+(x) = 1$ the path thus constructed is the unique directed path from the vertices in O to v.

Since each vertex v in V has $d^+(v) \leq 1$ and $d^-(v) \leq 1$, the property (2) follows. □

An illustration of a valid edge orientation of a graph is given in Fig. 2. We are now ready to define the problem. The VALID EDGE ORIENTATION WITH MINIMUM ORIGIN problem asks to find a minimum sized set $X \subseteq E(G)$ such that there exists a valid edge orientation $D = (V, E_1 \cup E_2 \cup X)$ of G with the special edge set $E_3 = X$ and the origin of D as $V_e(X)$. The decision version of the problem is formulated as follows:

> VALID EDGE ORIENTATION WITH MINIMUM ORIGIN (VEOMO)
> **Input:** An undirected graph $G = (V, E)$ and a positive integer k.
> **Question:** Does there exist a set $X \subseteq E(G)$ of size k such that G has a valid edge orientation $D = (V, E_1 \cup E_2 \cup X)$ with $V_e(X)$ as the origin?

Theorem 4. *G has a power edge set X of size k if and only if G has a valid edge orientation $D = (V, E_1 \cup E_2 \cup X)$ with $|X| = k$ and $V_e(X)$ as the origin.*

Proof. (⇒) Let $X = \{e_1, e_2, \ldots, e_k\} \subseteq E(G)$ be a power edge set of G and $V_e(X) = \{v_1, v_2, \ldots, v_\ell\}$, where $\ell \leq 2k$. We construct a valid edge orientation D of G (in rounds) with $O = V_e(X)$ as the origin. We first add the vertices of $V_e(X)$ in some arbitrary order. Let D_i, $1 \leq i \leq n$, be the orientation of the subgraph on the vertex set $V(D_{i-1}) \cup \{w\}$ where w is the vertex that is added in the ith round, with $D_0 = (\emptyset, \emptyset)$. We also associate a timestamp t_v, for each vertex $v \in V(G)$, that is equal to the round in which the vertex is added to the orientation. Since we add exactly one new vertex in each round, for any two distinct vertices u and v, $t_u \neq t_v$. After the ℓth round, $D_\ell = (V_e(X), E_1^\ell \cup E_2^\ell \cup X)$ is a suborientation of G with $E_1^\ell = \emptyset$ and E_2^ℓ being the set of edges between the vertices of $V_e(X)$ in G excluding X. That is, $E_2^\ell = E(G[V_e(X)]) \setminus X$. We construct a suborientation D_{i+1}, where $\ell \leq i \leq n$ according to the following rule.

Rule: If there exists a vertex v in the orientation D_i such that there is a vertex $u \in N(v)$ where $N(v) \setminus V(D_i) = \{u\}$, then assign $t_u = i + 1$, add a directed edge (v, u) to E_1^{i+1}, and add $\{u, w\}$ to E_2^{i+1}, for all $w \in (N(u) \setminus \{v\}) \cap V(D_i)$. Note that, if there are more than one vertices that satisfy the aforementioned **Rule**, then choose one of them arbitrarily to be v and proceed.

That is, D_{i+1} is a suborientation with $V(D_{i+1}) = V(D_i) \cup \{u\}$, $E_1^{i+1} = E_1^i \cup \{(v, u)\}$, $E_2^{i+1} = E_2^i \cup \{\{u, w\} \mid w \in (N(u) \setminus \{v\}) \cap V(D_i)\}$ and $E_3^{i+1} = E_3^i$. Since the number of vertices of G is finite, the above process stops after n rounds. We now have to show that $D = D_n = (V, E_1^n \cup E_2^n \cup X)$ is a valid edge orientation. Observe that since X is a power edge set of G which monitors all the vertices of G, by our construction of D, all vertices of G are captured in D and each edge of G belongs to one of the partitions E_1^n, E_2^n and $E_3^n = X$. Thus D is an orientation of G. All the vertices in $V_e(X)$ have indegree 0 and outdegree at most one. Also from the construction, each vertex in $V \setminus O$ has indegree one and outdegree at most one. Hence D satisfies the first part of the Definition 5. We will now show that there is no dependency cycle in D. We prove this by contradiction. Suppose that there is a dependency cycle in D with the vertex ordering $u_0, u_1, \ldots, u_i$, with $i > 2$, and $u_i = u_0$. Let $P_1, P_2, \ldots, P_\ell$ be the directed paths constructed in the orientation starting from $v_1, v_2, \ldots, v_\ell$ respectively. Consider the paths that intersect with the cycle. WLOG let $P_1, P_2, \ldots, P_s$, $s \leq \ell$, be paths that have vertices from the cycle and also let it be the order in which the vertices of the cycle appear. Let x_i and y_i be the first vertex and the last vertex of the path P_i that intersects with the cycle respectively. Notice that $\{x_1, y_s\} \in E_2$. It is easy to see that $t_{x_j} < t_{y_j}$, for all $j \in [s]$. From the **Rule** we get that $t_{y_j} < t_{y_{j+1}}$ for all j. This is because of the fact that the directed edge in P_{j+1} from x_{j+1} to its neighbor is added after all the neighbors of x_{j+1} are added in the orientation. Thus $t_{x_1} < t_{y_s}$. This is a contradiction because to have a directed edge from x_1 to its neighbor of P_1, all of its neighbors should be in the orientation, implying $t_{y_s} < t_{x_1}$.

(⇐) Let $D = (V, E_1 \cup E_2 \cup X)$ be a valid edge orientation with $|X| = k$ and $V_e(X)$ as the origin. Let $X = \{e_1, e_2, \ldots, e_k\}$. We claim that X is a power edge set of G. Suppose not. Then there exists an unmonitored vertex $v \in V \setminus V_e(X)$. Since D is a valid orientation of G there is a directed path, say P_1, from a vertex in $V_e(X)$, due to Lemma 2. WLOG let v be the first unmonitored vertex on P_1. Consider the predecessor u of v on P_1. The vertex u is monitored because of the assumption that v is the first unmonitored vertex on P_1. The reason for u not being able to monitor v must be that there exists another neighbor w of u that is not monitored yet. Let w be on the directed path P_2. Again from Lemma 2, notice that two directed paths starting from origin are vertex disjoint. Consider the first unmonitored vertex on P_2. Let it be x. By a similar argument, there exists a predecessor y of x on P_2 for which there exists an unmonitored vertex z in the neighborhood of y. Let z be on the path P_3. Since the graph is finite, and all the paths are pairwise vertex disjoint, we will find a vertex on one of the already considered paths and this leads to a dependency cycle which is a contradiction to the fact that D is a valid edge orientation. □

Observe that the reduction takes polynomial time and since the graph is not modified, this preserves tree-width.

Proof (Proof of Theorem 2). From Corollary 1, we get that $\mathsf{tw}(G) \leq 2k$. From Theorem 3, we obtain a $2^{\mathcal{O}(\mathsf{tw}^2)} n^{\mathcal{O}(1)}$ time algorithm for PES. Combining the above two results, we obtain a $2^{\mathcal{O}(k^2)} n^{\mathcal{O}(1)}$ time algorithm for PES.

Proof (Proof of Theorem 3). Theorem 4 gives a polynomial time parameter preserving reduction from PES to VEOMO. Now it remains to give an FPT algorithm for VEOMO parameterized by tree-width. Once we get an instance of VEOMO, we can verbatim use the dynamic programming-based FPT algorithm for a variant of VEOMO from [7], that runs in $2^{\mathcal{O}(\mathsf{tw}^2)} n^{\mathcal{O}(1)}$ time. Using this, we obtain an FPT algorithm for VEOMO parameterized by tree width. Since this is a verbatim adaptation, we omit the proof. □

5 Zero Forcing Set (ZFS)

In this section, we show that ZFS (or PVS) is FPT when parameterized by the solution size or the tree-width of the input graph. It is easy to get an analogous result of Lemma 1 for ZFS. That is, if G has a power vertex set of size k, then $\mathsf{pw}(G) \leq k$. The proof is identical to the proof of Lemma 1. This gives us the following corollary.

Corollary 2. *If G has a power vertex set of size k, then* $\mathsf{tw}(G) \leq k$.

Similar to PES, we can modify the dynamic programming algorithm from [7] to get an explicit FPT algorithm for ZFS parameterized by tree-width. From Corollary 2 and the explicit FPT algorithm for ZFS, we get the following FPT result.

Theorem 5. *There is an algorithm that solves* ZFS *in* $2^{\mathcal{O}(k^2)} n^{\mathcal{O}(1)}$ *time.*

6 PES Parameterized by Vertex Cover

Since $\mathsf{tw(G)} \leq \mathsf{vc(G)}$ for any graph G, from Theorem 3 we obtain an algorithm for PES running in time $2^{\mathcal{O}(\mathsf{vc}^2)}n^{\mathcal{O}(1)}$, where vc is the vertex cover number of the graph G. In this section, we present a more efficient algorithm for the problem parameterized by vertex cover number.

Theorem 6 (⋆).[1] PES *can be solved in time* $2^{\mathcal{O}(\mathsf{vc}\log \mathsf{vc})}n^{\mathcal{O}(1)}$.

Let G be a graph and $X \subseteq V(G)$ of size vc, be a vertex cover of G. We assume that X is given as input. Otherwise we can compute such a set in time $1.25284^{\mathsf{vc}}n^{\mathcal{O}(1)}$ [8]. Let (G, vc, k) be an instance of PES, where $k \in \mathbb{N}$. We partition the vertices of $V(G) \setminus X$ based on their neighborhood in X. For each $A \subseteq X$, let $T_A = \{v \in V(G) \setminus X \mid N(v) \cap X = A\}$. We first observe the following.

Observation 1 (⋆). *Let* $S \subseteq E(G)$ *be a solution of* (G, vc, k). *Then* $|V(G) \setminus V_e(S)| \leq 2\mathsf{vc}$.

Observation 2. *Let* $S \subseteq E(G)$ *be a solution of* (G, vc, k). *Consider a type* T_A, *for some* $A \subseteq X$. *Then* $|V_e(S) \cap T_A| \geq |T_A| - 1$.

Moreover, there is a solution $\widehat{S} \subseteq E(G)$, *of size* $|S|$, *of* (G, vc, k) *such that either* $|V_e(\widehat{S}) \cap T_A| = |T_A|$ *or,* $|V_e(\widehat{S}) \cap T_A| = |T_A| - 1$ *and for each vertex* $v \in T_A$ *the number of edges of* $\widehat{S}$ *incident to* v *is at most one.*

Proof. Suppose that $|V_e(S) \cap T_A| = |T_A| - 2$. Then there are two vertices $u, v \in T_A$ such that u and v are not incident to any edge of S. Since u and v are false twins, it is not possible to monitor both u and v.

From (1), we have that $|V_e(S) \cap T_A| = |T_A| - 1$ or $|V_e(S) \cap T_A| = |T_A|$. For the latter case, $\widehat{S} = S$. Suppose that $|V_e(S) \cap T_A| = |T_A| - 1$, and there exists a vertex $w \in T_A$ such that number of edges of S incident to w is at least two. Let $T_A \setminus V_e(S) = \{v\}$ and $wu_1, wu_2 \in S$. Then $\widehat{S} = S \cup \{vu_2\} \setminus \{wu_2\}$ is also a solution of (G, vc, k). Therefore we have $|V_e(\widehat{S}) \cap T_A| = |T_A|$. □

Using Observation 2 we design a dynamic programming algorithm for PES parameterized by vertex cover in the full version of the paper.

7 Neighborhood Diversity

In this section, we show that PES is FPT when parameterized by the neighborhood diversity.

Theorem 7. PES *can be solved in* $2^t n^{\mathcal{O}(1)}$ *time, where* t *is the neighborhood diversity of the graph on* n *vertices.*

[1] Results marked with ⋆ will be given in the full version of the paper.

We say that a set $S \subseteq E(G)$ *maximizes* a type V_i if either (a) $|V_i \cap V_e(S)| = |V_i|$, or (b) for each $v \in V_i$ the number of edges of S incident with v is at most 1. Let (G, t, k) be an instance of PES, where t is the neighborhood diversity of the graph G and $k \in \mathbb{N}$.

Lemma 3 (⋆). *Let $S \subseteq E(G)$ be a solution of (G, t, k). Then,*

1. *for all $i \in [t]$, we have that $|V_i \cap V_e(S)| \in \{|V_i|, |V_i| - 1\}$.*
2. *There exists a solution S' such that $|S| = |S'|$ and S' maximizes V_i for all i.*

Proof (Proof of Theorem 7). Let $S \subseteq E(G)$ be a solution of (G, t, k) of size at most k. Moreover, from Lemma 3 we can assume that S maximizes V_i, for all $i \in [t]$, and $|V_i \cap V_e(S)| \in \{|V_i|, |V_i| - 1\}$, for each $i \in [t]$. For each V_i, we guess whether $|V_i \cap V_e(S)| = |V_i|$ or $|V_i \cap V_e(S)| = |V_i| - 1$. For the former case, we pick all vertices from V_i. For the latter case, we arbitrarily pick $|V_i| - 1$ many vertices from V_i, as each pair of vertices in V_i are twins. Let X be the set of vertices of G that are picked from each V_i, $i \in [t]$. We can check in polynomial time if the set of vertices X monitors $V(G)$ using the propagation Rule 2. Now all that remains is to find a minimum set of edges $S \subseteq E(G[X])$ such that $V_e(S) = X$. We say that a set of edges $\widehat{S} \subseteq E(H)$ is an *edge cover* of H if every vertex in $V(H)$ is incident to at least one of the edges in $\widehat{S}$. The input to Edge Cover is a graph H and an integer ℓ. The objective is to check if there exists an edge cover of size at most ℓ. It is easy to see that (G, t, k) has a power edge set of size k if and only if there is a guess X such that $G[X]$ has an edge cover of size k. A minimum sized edge cover can be found in polynomial time [11]. We consider the minimum over all the values returned for all the 2^t instances by the Edge Cover problem and check if that minimum is at most k. □

8 Conclusion

In this paper, we showed that PES can be solved in time $2^{\mathcal{O}(k^2)} n^{\mathcal{O}(1)}$. Darties et al. [5] showed that PES does not admit an $2^{o(k)} n^{\mathcal{O}(1)}$ algorithm when parameterized by k. It will be interesting to obtain an algorithm that runs in time $c^k n^{O(1)}$. We give an efficient FPT algorithm for vertex cover number parameter and show that the problem is FPT parameterized by neighborhood diversity or tree-width. Another question is whether the running time of our vertex cover algorithm is optimal? Similar questions can be asked for ZFS.

References

1. Brueni, D.J., Heath, L.S.: The pmu placement problem. SIAM J. Disc. Math. **19**(3), 744–761 (2005). https://doi.org/10.1137/S0895480103432556
2. Cazals, P., Darties, B., Chateau, A., Giroudeau, R., Weller, M.: Power edge set and zero forcing set remain difficult in cubic graphs. In: Colbourn, C.J., Grossi, R., Pisanti, N. (eds.) IWOCA 2019. LNCS, vol. 11638, pp. 122–135. Springer, Cham (2019). https://doi.org/10.1007/978-3-030-25005-8_11

3. Courcelle, B.: The monadic second-order logic of graphs. i. recognizable sets of finite graphs. Inf. Comput. **85**(1), 12–75 (1990). https://doi.org/10.1016/0890-5401(90)90043-H
4. Cygan, M., et al.: Parameterized Algorithms, vol. 5. Springer, Heidelberg (2015)
5. Darties, B., Champseix, N., Chateau, A., Giroudeau, R., Weller, M.: Complexity and lowers bounds for power edge set problem. J. Disc. Algor. **52-53**, 70–91 (2018). https://doi.org/10.1016/j.jda.2018.11.006. https://www.sciencedirect.com/science/article/pii/S1570866718301667
6. Group, A.M.R.S.G.W., et al.: Zero forcing sets and the minimum rank of graphs. Linear Algebra Appl. **428**(7), 1628–1648 (2008)
7. Guo, J., Niedermeier, R., Raible, D.: Improved algorithms and complexity results for power domination in graphs. Algorithmica **52**(2), 177–202 (2008). https://doi.org/10.1007/s00453-007-9147-x
8. Harris, D.G., Narayanaswamy, N.S.: A faster algorithm for vertex cover parameterized by solution size (2024)
9. Haynes, T.W., Hedetniemi, S.M., Hedetniemi, S.T., Henning, M.A.: Domination in graphs applied to electric power networks. SIAM J. Disc. Math. **15**(4), 519–529 (2002). https://doi.org/10.1137/S0895480100375831
10. Lampis, M.: Algorithmic meta-theorems for restrictions of treewidth. Algorithmica **64**(1), 19–37 (2012)
11. Micali, S., Vazirani, V.V.: An o(sqrt(|v|) |e|) algorithm for finding maximum matching in general graphs. In: 21st Annual Symposium on Foundations of Computer Science, Syracuse, New York, USA, 13–15 October 1980, pp. 17–27. IEEE Computer Society (1980). https://doi.org/10.1109/SFCS.1980.12
12. Toubaline, S., D'Ambrosio, C., Liberti, L., Poirion, P.-L., Schieber, B., Shachnai, H.: Complexity and inapproximability results for the Power Edge Set problem. J. Comb. Optim. **35**(3), 895–905 (2017). https://doi.org/10.1007/s10878-017-0241-y
13. Toubaline, S., Poirion, P.-L., D'Ambrosio, C., Liberti, L.: Observing the state of a smart grid using bilevel programming. In: Lu, Z., Kim, D., Wu, W., Li, W., Du, D.-Z. (eds.) COCOA 2015. LNCS, vol. 9486, pp. 364–376. Springer, Cham (2015). https://doi.org/10.1007/978-3-319-26626-8_27

Minimizing ℓ_2 Norm of Flow Time by Starvation Mitigation

Tung-Wei Kuo(✉)

Department of Computer Science, National Chengchi University, Taipei, Taiwan
twkuo@cs.nccu.edu.tw

Abstract. The assessment of a job's Quality of Service (QoS) often revolves around its flow time, also referred to as response time. This study delves into two fundamental objectives for scheduling jobs: the average flow time and the maximum flow time. While the Shortest Remaining Processing Time (SRPT) algorithm minimizes average flow time, it can result in job starvation, causing certain jobs to experience disproportionately long and unfair flow times. In contrast, the First-Come-First-Served (FCFS) algorithm minimizes the maximum flow time but may compromise the average flow time.

To strike a balance between these two objectives, a common approach is to minimize the ℓ_2 norm of flow time. SRPT and FCFS are $O(n^{\frac{1}{2}})$-competitive for this problem, where n is the number of jobs. Prior to this work, no algorithm is known to achieve a competitive ratio better than SRPT and FCFS. In this paper, we use FCFS to mitigate the starvation caused by SRPT. Given a good estimate of n, we prove that this approach achieves a much better competitive ratio of $O(n^{\frac{1}{3}})$. Our results provide the first theoretical evidence that mitigating starvation in SRPT leads to a provable improvement in scheduling performance.

Keywords: ℓ_2 Norm of Flow Time · SRPT · FCFS

1 Introduction

In a server system, the Quality of Service (QoS) of a job is normally measured by its response time, which is defined as the amount of time between job release and job completion. In job scheduling, the response time is also called the flow time. Specifically, for any schedule $\mathcal{S}$, let $c_i(\mathcal{S})$ be the completion time of job J_i under $\mathcal{S}$ and r_i be the release time of J_i. Then the flow time of a job J_i under $\mathcal{S}$, denoted by $f_i(\mathcal{S})$, is defined as

$$f_i(\mathcal{S}) = c_i(\mathcal{S}) - r_i.$$

Two natural QoS objectives for a server system are the average, or equivalently, ℓ_1 norm, of job flow times, and the maximum, or ℓ_∞ norm, of job flow times. It

A full version of this paper can be found in [16].

H. Fernau and B. Zhu (Eds.): IWOCA 2025, LNCS 15885, pp. 362–375, 2025.
https://doi.org/10.1007/978-3-031-98740-3_26

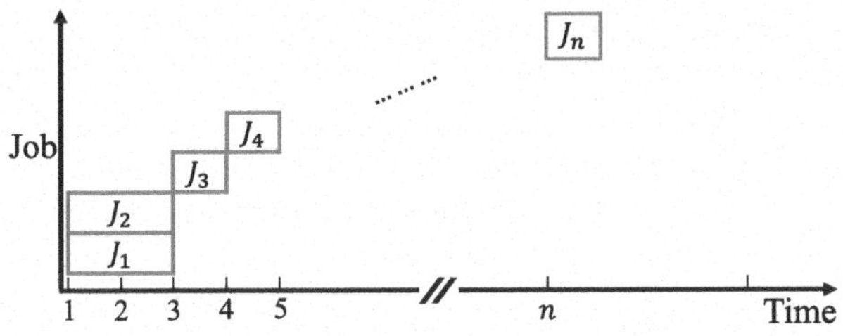

Fig. 1. A bad instance for SRPT.

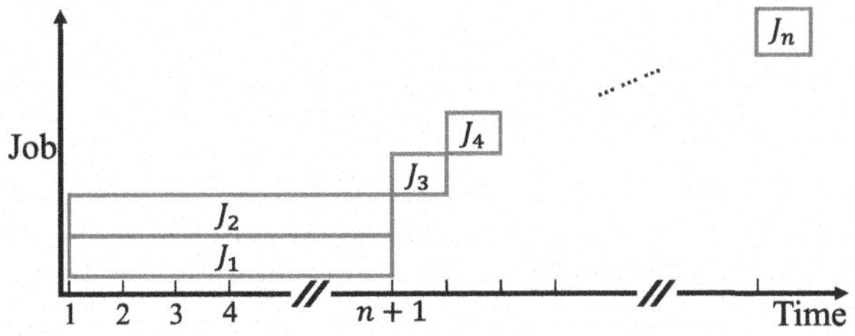

Fig. 2. A bad instance for FCFS.

is well-known that Shortest Remaining Processing Time (SRPT), which minimizes average flow time, may cause some jobs to have long and inequitable flow times (i.e., job starvation). In contrast, First-Come-First-Served (FCFS), which minimizes the maximum flow time [7], may deteriorate the average flow time.

Example 1. In Fig. 1, jobs J_1 and J_2 are released at time 1, and both have processing times of 2. All the other jobs J_i are released at time i and have unit processing times. Assume that under SRPT, J_1 is executed first. Thus, J_2 is starving in the sense that it has to wait until all the other jobs are completed. Under FCFS, however, the flow time of every job is only $O(1)$.

Example 2. Consider Fig. 2. Compared with Fig. 1, the processing times of J_1 and J_2 in Fig. 2 are increased to n. Under FCFS, the flow time of every job is $\Theta(n)$, and thus FCFS's average flow time is $\Theta(n)$. Under SRPT, however, the average flow time is only $O(1)$.

The standard approach to balance the ℓ_1 norm and the ℓ_∞ norm of flow time is minimizing the ℓ_2 norm of flow time [5,13]. Specifically, the ℓ_2 norm of flow time under schedule $\mathcal{S}$ is

$$\sqrt{f_1(\mathcal{S})^2 + f_2(\mathcal{S})^2 + \cdots + f_n(\mathcal{S})^2},$$

where n is the number of jobs. In this paper, we aim to minimize the ℓ_2 norm of flow time online on a single machine.

For the online problem of minimizing the ℓ_2 norm of flow time, only standard algorithms, including SRPT, Shortest Job First (SJF), Shortest Elapsed Time

First (SETF), and Round Robin (RR), have been analyzed, assuming that the machine is gifted extra speed [5,13]. However, SRPT, SJF, and SETF may cause job starvation, and FCFS and RR may deteriorate average flow time. These drawbacks worsen the ℓ_2 norm of flow time (see [16]). For instance, the examples illustrated in Figs. 1 and 2 demonstrate that FCFS and SRPT are $\Omega(n^{1/2})$-competitive for minimizing the ℓ_2 norm of flow time.

In this paper, we use FCFS to mitigate the starvation caused by SRPT. Our algorithm follows SRPT initially. If a job has been in the system for too long and its remaining processing time is small enough, then the job becomes a starving job. Specifically, a job J_i becomes starving at time t if

$$\frac{t - r_i}{\text{remaining processing time of } J_i} \geq \theta,$$

where θ is called the *starvation threshold.* Whenever some jobs are starving, the algorithm processes the job that becomes starving first. We use $BAL(\theta)$ to denote the algorithm when the starvation threshold is set to θ. For example, $BAL(0)$ is equivalent to FCFS and $BAL(\infty)$ is equivalent to SRPT.

1.1 Our Result

Our analysis on $BAL(\theta)$'s competitive ratio suggests that $\theta = n^{\frac{2}{3}}$ is the best starvation threshold. Among all the aforementioned standard scheduling algorithms, SRPT and FCFS achieve the best known competitive ratio of $O(n^{\frac{1}{2}})$ for minimizing the ℓ_2 norm of flow time (see [16]). We prove that $BAL(n^{\frac{2}{3}})$, a combination of SRPT and FCFS, achieves a significantly better competitive ratio of $O(n^{\frac{1}{3}})$.

In practice, we can only estimate n. For example, usually job scheduling is only critical during peak hours, and normally the job arrival process is modeled as a Poisson process. Thus, the product of the job arrival rate and the duration of peak hours would be a good estimate of n. Nonetheless, in this study, we do not make any assumptions about the job arrival process or the estimation method. Instead, we assume that an estimate $\tilde{n}$ of n with bounded error is given. The following theorem gives the competitive ratio of $BAL(\tilde{n}^{\frac{2}{3}})$. Specifically, if $\tilde{n} = \Theta(n)$, then $BAL(\tilde{n}^{\frac{2}{3}})$ is $O(n^{\frac{1}{3}})$-competitive.

Theorem 1. *Let $\tilde{n}$ be an estimate of n such that $\beta n \leq \tilde{n} \leq \alpha n$ for some $\frac{1}{n} \leq \beta \leq 1$ and $\alpha \geq 1$. Then the competitive ratio of $BAL(\tilde{n}^{\frac{2}{3}})$ for minimizing the ℓ_2 norm of flow time is $O\left(\tilde{n}^{\frac{1}{3}} + n^{\frac{1}{2}}/\tilde{n}^{\frac{1}{6}}\right) = O\left(n^{\frac{1}{3}}\left(\alpha^{\frac{1}{3}} + \beta^{\frac{-1}{6}}\right)\right)$.*

Observe that if the predicted number of jobs $\tilde{n}$ overestimates the true number n, $BAL(\tilde{n}^{\frac{2}{3}})$ achieves an $O(\tilde{n}^{\frac{1}{3}})$-competitive ratio, which corresponds to the usual competitive ratio of the algorithm, but expressed in terms of the predicted number of jobs rather than the actual count. Conversely, if the prediction underestimates the true count, the competitive ratio is $O(n^{\frac{1}{2}}/\tilde{n}^{\frac{1}{6}})$, which is no worse than that of standard algorithms without prediction.

1.2 Related Work

In their seminal work, Bansal and Pruhs introduced the online problem of minimizing the ℓ_p norm of flow time [5]. They proved that SRPT and SJF are $(1+\epsilon)$-speed $O(\frac{1}{\epsilon})$-competitive and that SETF is $(1+\epsilon)$-speed $O(\frac{1}{\epsilon^{2+2/p}})$-competitive. The results have been extended to all symmetric norms of flow time [12] and identical machines [8,11]. Moreover, Im et al. showed that RR is $O(1)$-speed $O(1)$-competitive for the ℓ_2 norm of flow time [13]. In [1,15], more general objective functions were considered. Specifically, for a job J_i with flow time f_i, a cost $g_i(f_i)$ is incurred. The only restriction on g_i is that g_i must be non-decreasing. For this general cost minimization problem, there are $O(1)$-speed $O(1)$-competitive online algorithms [1,15].

For the offline setting, Bansal and Pruhs first proposed an $O((\log\log P)^{1/p})$-approximation algorithm using linear programming rounding, where P is the ratio of the maximum to minimum job size [4]. Subsequently, several $O(1)$-approximation algorithms were developed, leveraging linear programming rounding, dynamic programming, or a combination of both [2,6,14].

2 Definitions and the Algorithm

We consider n jobs, $J_1, J_2, \cdots, J_n$, and one machine. Each job J_i has a processing time p_i and a release time r_i. As in [4,14], we assume that p_i and r_i are integers. We allow job preemption and consider clairvoyant scheduling. Define $F(\mathcal{S}) = \sum_{i=1}^{n} f_i(\mathcal{S})^2$. The goal is to compute online a schedule $\mathcal{S}$ that minimizes the ℓ_2 norm of flow time, i.e., $\sqrt{F(\mathcal{S})}$.

For every $t \in \mathbb{N}$[1], the **time slot** $[t]$ is defined as the time interval between time t and time $t+1$. Thus, we can divide time into time slots $[0], [1], [2], \cdots$. We can view each job J_i as a chain of tasks $J_{i,1}, J_{i,2}, \cdots, J_{i,p_i}$, where each task has a unit processing time. Because all the processing times and release times are integers, by a simple exchange argument, we can assume that under an optimal schedule, the machine never executes more than one task in time slot $[t]$ for any $t \in \mathbb{N}$ (i.e., the machine is either idle or executing the same task throughout the entire time slot $[t]$). Thus, for every time slot $[t]$, a schedule assigns a (possibly empty) task to be executed in $[t]$. If a task $J_{i,k}$ is executed in time slot $[t]$ under schedule $\mathcal{S}$, then its completion time, denoted by $c_{i,k}(\mathcal{S})$, is $t+1$. Throughout this paper, we use $\mathcal{BAL}(\theta), \mathcal{SRPT}, \mathcal{FCFS}, \mathcal{RR}$, and $\mathcal{OPT}$ to denote the schedule obtained by $BAL(\theta)$, SRPT, FCFS, RR, and an optimal schedule, respectively. If θ is clear from the context, we simply write $\mathcal{BAL}$ instead of $\mathcal{BAL}(\theta)$.

A job is said to be **active** at time t under schedule $\mathcal{S}$ if it is released by time t but has not yet been completed by time t under $\mathcal{S}$. We use $A(t, \mathcal{S})$ to denote the index set of the active jobs under $\mathcal{S}$ at time t. In this paper, for any $a, b \in \mathbb{N}$ with $a \le b$, $[a, b]$ is defined as $\{i | i \in \mathbb{N}, a \le i \le b\}$. If $a > b$, then $[a, b] = \varnothing$. Moreover, we define a **map** as a non-negative function with a finite domain. In this paper, we use the terms "remaining processing time" and "number of remaining tasks" interchangeably. We then introduce the most important map in this paper.

[1] In this paper, we assume $\mathbb{N}$ contains 0.

Definition 1. *For any schedule $\mathcal{S}$, any time t, and any $i \in A(t, \mathcal{S})$, define $q_{t,\mathcal{S}}(i)$ as the number of remaining tasks of J_i at time t under $\mathcal{S}$; if $i \in [1, n] \setminus A(t, \mathcal{S})$, define $q_{t,\mathcal{S}}(i) = 0$.*

Example 3. Consider the schedule obtained by SRPT for the instance shown in Fig. 3. Assume that SRPT first executes J_1. Then, at time 3, the number of remaining tasks of J_1 and J_2 are 4 and 6, respectively. Thus, $q_{3,\mathcal{SRPT}}(1) = 4$ and $q_{3,\mathcal{SRPT}}(2) = 6$. Because jobs J_3, J_4, J_5 are not active at time 3, we have $q_{3,\mathcal{SRPT}}(i) = 0$ for all $3 \leq i \leq 5$.

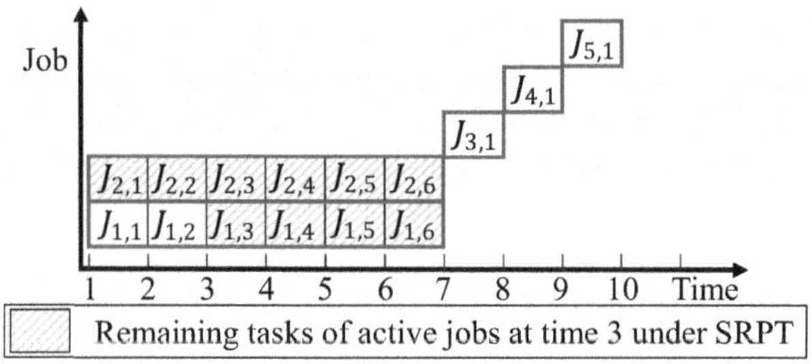

Fig. 3. An example of $q_{3,\mathcal{SRPT}}(i)$.

2.1 The Algorithm

Next, we formalize the algorithm $BAL(\theta)$ introduced in Sect. 1. For brevity, define $q_t(i)$ as $q_{t,\mathcal{BAL}}(i)$. In other words, $q_t(i)$ is the number of remaining tasks of J_i at time t under $BAL(\theta)$. $BAL(\theta)$ categorizes active jobs into two types, starving and normal, according to the given starvation threshold θ. Every job is **normal** initially, and tasks of normal jobs are called normal tasks. An active job J_i is said to be **starving** at time t if

$$\frac{t - r_i}{q_t(i)} \geq \theta. \tag{1}$$

Moreover, once J_i becomes starving, every remaining task of J_i is said to be starving as well. For any starving job J_i, define t_i as the time at which J_i becomes starving (i.e., t_i is the smallest $t \in \mathbb{N}$ that satisfies Eq. (1)).

When there are starving jobs, $BAL(\theta)$ executes the starving job that has the smallest t_i. In other words, $BAL(\theta)$ executes the job that becomes starving first. In Algorithm 1, ST denotes the set of the starving job indices.

When there are no starving jobs, $BAL(\theta)$ compensates normal jobs for their lost execution time (as compared to SRPT). More precisely, during the execution of a starving job by $BAL(\theta)$, there is an opportunity for a normal job to be executed under SRPT, resulting in a delay compared to SRPT for this particular

Algorithm 1. $BAL(\theta)$

$ST \leftarrow \varnothing$
for $t \leftarrow 0$ to ∞ **do**
  for every active job J_i **do**
    if $i \notin ST \wedge \frac{t-r_i}{q_t(i)} \geq \theta$ **then**
      $t_i \leftarrow t$
      Add i to ST
  if $ST \neq \varnothing$ **then**
    $i^* = \arg\min_{i \in ST} t_i$
    Execute job J_{i^*} in time slot $[t]$
    Remove i^* from ST if J_{i^*} is completed at time $t+1$
  else if $M_{SRPT}(t) \setminus M_{BAL}(t-1) \neq \varnothing$ **then**
    Among all the tasks in $M_{SRPT}(t) \setminus M_{BAL}(t-1)$, execute the one that has the smallest completion time under SRPT

job. To address this delay, when there are no starving jobs, $BAL(\theta)$ gives priority to these delayed normal jobs. In particular, among all the delayed normal tasks, $BAL(\theta)$ executes the task that has the earliest completion time under SRPT, as it has experienced the longest delay. Specifically, let $M_{SRPT}(t)$ and $M_{BAL}(t)$ be the sets of tasks that are executed in time slots $[0], [1], [2], \cdots, [t]$ under SRPT and $BAL(\theta)$, respectively. Define $M_{BAL}(-1) = \varnothing$. At time t, if there are no starving jobs, then among the tasks in $M_{SRPT}(t) \setminus M_{BAL}(t-1)$, $BAL(\theta)$ executes the one that is completed the earliest under SRPT.[2]

3 Searching for the Best θ: Proof of Theorem 1

To find the best starvation threshold θ, we partition the lifetime of J_i into two periods, normal and starving. Specifically, if a job never becomes starving under $BAL(\theta)$, it is termed a **Finished-as-Normal (FaN) job**; otherwise, it is termed a **Finished-as-Starving (FaS) job**. We further extend the definition of t_i to FaN jobs as follows: If J_i is an FaN job, t_i is defined as its completion time (i.e., $t_i = c_i(\mathcal{BAL})$). For any job J_i, we then define

$$norm_i(\theta) = t_i - r_i$$

and

$$starv_i(\theta) = c_i(\mathcal{BAL}) - t_i.$$

Thus, $f_i(\mathcal{BAL}) = norm_i(\theta) + starv_i(\theta)$ and

$$F(\mathcal{BAL}) = \Theta\left(\sum_{i=1}^{n} norm_i(\theta)^2 + \sum_{i=1}^{n} starv_i(\theta)^2\right).$$

[2] All the variables associated with $BAL(\theta)$ depend on θ (e.g., t_i, ST, and $M_{BAL}(t)$). For simplicity, we do not explicitly show their dependency on θ in our notation.

To minimize the ℓ_2 norm of flow time, we balance $\sum_{i=1}^{n} norm_i(\theta)^2$ and $\sum_{i=1}^{n} starv_i(\theta)^2$. Intuitively, $norm_i(\theta)$ decreases as the starvation threshold θ decreases. In contrast, $starv_i(\theta)$ decreases as the starvation threshold θ increases. The following two theorems, whose proofs are postponed to Sects. 4 and 5, relate $\sum_{i=1}^{n} norm_i(\theta)^2$ and $\sum_{i=1}^{n} starv_i(\theta)^2$ to $F(\mathcal{OPT})$.

Theorem 2. *For any starvation threshold* $\theta \geq 0$,

$$\sum_{i=1}^{n} norm_i(\theta)^2 = O(\theta)F(\mathcal{OPT}).$$

Theorem 3. *For any starvation threshold* $\theta > 0$,

$$\sum_{i=1}^{n} starv_i(\theta)^2 = O\left(\frac{n}{\sqrt{\theta}}\right) F(\mathcal{OPT}).$$

Solving $\theta = \frac{n}{\sqrt{\theta}}$ yields $\theta = n^{\frac{2}{3}}$. Thus, Theorems 2 and 3 suggest that, to minimize the ℓ_2 norm of flow time, the best starvation threshold is $\theta = n^{\frac{2}{3}}$, leading to a competitive ratio of $O(n^{\frac{1}{3}})$. In addition, Theorem 1 is a direct result of Theorems 2 and 3.

3.1 Proof of Theorem 1

$$\begin{aligned} F(\mathcal{BAL}(\tilde{n}^{\frac{2}{3}})) &= \sum_{i=1}^{n} \left(norm_i(\tilde{n}^{\frac{2}{3}}) + starv_i(\tilde{n}^{\frac{2}{3}})\right)^2 \\ &= O\left(\sum_{i=1}^{n} norm_i(\tilde{n}^{\frac{2}{3}})^2 + \sum_{i=1}^{n} starv_i(\tilde{n}^{\frac{2}{3}})^2\right) \\ &\overset{\text{by Theorems 2 and 3}}{=} O\left(\tilde{n}^{\frac{2}{3}} + \frac{n}{\tilde{n}^{\frac{1}{3}}}\right) F(\mathcal{OPT}) \end{aligned}$$

Thus, the competitive ratio of $BAL(\tilde{n}^{\frac{2}{3}})$ for minimizing the ℓ_2 norm of flow time is $O\left(\tilde{n}^{\frac{1}{3}} + n^{\frac{1}{2}}/\tilde{n}^{\frac{1}{6}}\right) = O\left(n^{\frac{1}{3}}\left(\alpha^{\frac{1}{3}} + \beta^{\frac{-1}{6}}\right)\right)$.

4 Proof of Theorem 2

To minimize the ℓ_2 norm of flow time, we should not waste any time slot. Specifically, at every time $t \in \mathbb{N}$, if some active jobs are not completed, then one of them must be executed in time slot $[t]$. A schedule that satisfies the above property is called a **work-conserving** schedule. Clearly, for any time t, all work-conserving schedules have the same total number of remaining tasks over all active jobs. Thus, we define $q(t)$ as the total number of remaining tasks over all active jobs at time t under any work-conserving schedule. For example, consider the instance shown in Fig. 3 at time $t = 3$. We then have $q(t) = 10$. The following lemma gives the relationship between $q(t)$ and $F(\mathcal{OPT})$.

Lemma 1. *For any work-conserving schedule* $\mathcal{S}$,

$$F(\mathcal{OPT}) \geq \sum_{t \in \mathbb{N}} q(t) = \sum_{i=1}^{n} \sum_{t=r_i}^{c_i(\mathcal{S})} q_{t,\mathcal{S}}(i).$$

Proof. Observe that for any job J_i and any time t, we have $f_i(\mathcal{OPT}) \geq p_i \geq q_{t,\mathcal{OPT}}(i)$. Thus, for any work-conserving schedule $\mathcal{S}$, we have

$$\begin{aligned} F(\mathcal{OPT}) &= \sum_{i=1}^{n} \sum_{t=r_i}^{c_i(\mathcal{OPT})-1} f_i(\mathcal{OPT}) \\ &\geq \sum_{i=1}^{n} \sum_{t=r_i}^{c_i(\mathcal{OPT})-1} q_{t,\mathcal{OPT}}(i) \\ &= \sum_{t \in \mathbb{N}} q(t) = \sum_{i=1}^{n} \sum_{t=r_i}^{c_i(\mathcal{S})} q_{t,\mathcal{S}}(i). \end{aligned}$$

□

To prove Theorem 2, we use Lemma 1 to upper bound $\sum_{i=1}^{n} norm_i(\theta)^2$. Let $I = \{i | norm_i(\theta) = 1, i \in [1,n]\}$. Obviously, $\sum_{i \in I} norm_i(\theta)^2 \leq F(\mathcal{OPT})$. Thus, we only need to consider job indices that are not in I. Let $i \in [1,n] \setminus I$. Define $T_i = [r_i, t_i - 1] = \{r_i, r_i+1, r_i+2, \cdots, t_i - 1\}$. Because $norm_i(\theta) = t_i - r_i = |T_i|$, we have

$$norm_i(\theta)^2 = |T_i|^2 = \Theta\left(\sum_{t \in T_i} (t - r_i)\right).$$

When $t \in T_i$, $\frac{t-r_i}{q_t(i)} < \theta$, and thus $t - r_i < \theta q_t(i)$. As a result, $norm_i(\theta)^2 = O(\theta) \sum_{t \in T_i} q_t(i)$. We then have

$$\sum_{i \in [1,n] \setminus I} norm_i(\theta)^2 = O(\theta) \sum_{i \in [1,n] \setminus I} \sum_{t \in T_i} q_t(i) \overset{\text{by Lemma 1}}{=} O(\theta) F(\mathcal{OPT}).$$

5 Proof of Theorem 3

Let $u(t)$ be the number of starving tasks at time t. Thus, by the design of $BAL(\theta)$, $c_i(\mathcal{BAL}) \leq t_i + u(t_i)$. Therefore, $starv_i(\theta) = c_i(\mathcal{BAL}) - t_i \leq u(t_i)$. Define $t^* = \arg\max_{t \in \mathbb{N}} u(t)$. We then have $starv_i(\theta) \leq u(t^*)$ and thus $\sum_{i=1}^{n} starv_i(\theta)^2 = O(n \cdot u(t^*)^2)$. As a result, to prove Theorem 3, it is sufficient to show

$$F(\mathcal{OPT}) = \Omega\left(\sqrt{\theta} \cdot u(t^*)^2\right). \qquad (*)$$

5.1 The First Lower Bound of $F(\mathcal{OPT})$

By Lemma 1, we have

$$F(\mathcal{OPT}) \geq \sum_{i=1}^{n} \sum_{t=r_i}^{t_i} q_t(i) \geq \sum_{i=1}^{n} \sum_{t=r_i}^{t_i} q_{t_i}(i) \geq \theta \sum_{i=1}^{n} q_{t_i}(i)^2, \quad (2)$$

where the last inequality holds because by the design of $BAL(\theta)$, $t_i - r_i \geq \theta q_{t_i}(i)$ for any FaS job J_i. For any FaN job J_i, we have $q_{t_i}(i) = 0$.

5.2 The Second Lower Bound of $F(\mathcal{OPT})$

Observe that to minimize $F(\mathcal{S})$ when all jobs are released at the same time, jobs should be executed in increasing order of their processing times. To derive the second lower bound of $F(\mathcal{OPT})$ by this idea, jobs that are not active at time t^* under $\mathcal{OPT}$ are ignored (and thus their flow times are not counted in $F(\mathcal{OPT})$). For every job J_i that is active at time t^* under $\mathcal{OPT}$, we gift $F(\mathcal{OPT})$ by setting $r_i = t^*$ and $p_i = q_{t^*,\mathcal{OPT}}(i)$. We then execute jobs in increasing order of their new processing times. The above idea will be used in Lemma 3.

5.3 Proof of Eq. (*)

For any two maps f and f', we write f **dominates** f', or f' is dominated by f, if $\operatorname{dom} f' = \operatorname{dom} f$ and $f'(i) \leq f(i), \forall i \in \operatorname{dom} f$, where $\operatorname{dom} f$ denotes the domain of f. We prove the following lemma in Sect. 6.

Lemma 2. *There is a map $m_{\mathcal{OPT}}$ dominated by $q_{t^*,\mathcal{OPT}}$ such that*

P1: $\sum_{i=1}^{n} m_{\mathcal{OPT}}(i)^2 \leq \sum_{i=1}^{n} q_{t_i}(i)^2$, *and*
P2: $\sum_{i=1}^{n} m_{\mathcal{OPT}}(i) \geq \frac{1}{4} \cdot u(t^*)$.

By Eq. (2) and P1, we have

$$F(\mathcal{OPT}) \geq \theta \sum_{i=1}^{n} m_{\mathcal{OPT}}(i)^2. \quad (3)$$

Because $m_{\mathcal{OPT}}$ is dominated by $q_{t^*,\mathcal{OPT}}$, we can use the technique developed for the second lower bound to prove the following lemma.

Lemma 3. *Let $m_{\mathcal{OPT}}$ be the map defined in Lemma 2. Reindex jobs so that*

$$m_{\mathcal{OPT}}(1) \leq m_{\mathcal{OPT}}(2) \leq \cdots \leq m_{\mathcal{OPT}}(n). \quad (4)$$

We then have $F(\mathcal{OPT}) \geq \sum_{i=1}^{n} \left(\sum_{h=1}^{i} m_{\mathcal{OPT}}(h) \right)^2$.

Proof. To derive a lower bound of $F(\mathcal{OPT})$, we remove some tasks from the instance at time t^* so that under $\mathcal{OPT}$, every job J_i has exactly $m_{\mathcal{OPT}}(i)$ remaining tasks at time t^*. The above modification is achievable because $q_{t^*,\mathcal{OPT}}$ dominates $m_{\mathcal{OPT}}$. Let i^* be the smallest integer such that $m_{\mathcal{OPT}}(i^*) > 0$. Further assume that starting from time t^*, the goal of $\mathcal{OPT}$ becomes to minimize

$$\sum_{i=i^*}^{n} (c'_i(\mathcal{OPT}) - t^*)^2, \tag{5}$$

where $c'_i(\mathcal{OPT})$ is the completion time of J_i under $\mathcal{OPT}$ in the modified instance. Observe that for all $i \geq i^*$, because $q_{t^*,\mathcal{OPT}}(i) \geq m_{\mathcal{OPT}}(i) > 0$, J_i's original release time is at most t^*. Thus, Eq. (5) is a lower bound of $F(\mathcal{OPT})$. By Eq. (4), to minimize Eq. (5), $\mathcal{OPT}$ should execute $J_{i^*}, J_{i^*+1}, \cdots, J_n$ in increasing order of their job indices. Thus, for any $i \geq i^*$, $c'_i(\mathcal{OPT}) - t^* = \sum_{h=i^*}^{i} m_{\mathcal{OPT}}(h)$, which implies Lemma 3.

We are now ready to prove Eq. (∗). Like Lemma 3, we reindex jobs so that Eq. (4) holds. We then have

$$\begin{aligned} F(\mathcal{OPT}) &= \Omega\left(\sum_{i=1}^{n}\left(\sum_{h=1}^{i} m_{\mathcal{OPT}}(h)\right)^2 + \theta\sum_{i=1}^{n} m_{\mathcal{OPT}}(i)^2\right) \\ &= \Omega\left(\sqrt{\theta}\right)\sqrt{\left(\sum_{i=1}^{n}\left(\sum_{h=1}^{i} m_{\mathcal{OPT}}(h)\right)^2\right)\left(\sum_{i=1}^{n} m_{\mathcal{OPT}}(i)^2\right)} \\ &= \Omega\left(\sqrt{\theta}\right)\sum_{i=1}^{n}\left(\sum_{h=1}^{i} m_{\mathcal{OPT}}(h)\right) m_{\mathcal{OPT}}(i) \\ &= \Omega\left(\sqrt{\theta}\right)\left(\sum_{i=1}^{n} m_{\mathcal{OPT}}(i)\right)^2 \\ &= \Omega\left(\sqrt{\theta}\right) u(t^*)^2, \end{aligned}$$

where the first equality follows from Lemma 3 and Eq. (3), the second equality follows from the AM–GM inequality, the third equality follows from the Cauchy-Schwarz inequality, and the last equality follows from P2 in Lemma 2.

6 Proof of Lemma 2

Because Lemma 2 is trivial when $u(t^*) = 0$, we assume $u(t^*) \geq 1$ in the following proof. We first introduce the following shorthand notations.

Definition 2. *For any map f, define $S(f) = \sum_{i \in \mathrm{dom}\, f} f(i)$, and define f^2 as a map such that $\mathrm{dom}\, f^2 = \mathrm{dom}\, f$ and $f^2(x) = f(x)^2$ for any $x \in \mathrm{dom}\, f^2$.*

Informally, to prove Lemma 2, we have to construct a map $m_{\mathcal{OPT}}$ dominated by $q_{t^*,\mathcal{OPT}}$ such that $S({m_{\mathcal{OPT}}}^2)$ is sufficiently small but $S(m_{\mathcal{OPT}})$ is sufficiently large. Thus, for any $c > 0$, we say that a map f is a c-**proper** map if f satisfies the following two constraints:

1. $S(f^2) \leq \sum_{i=1}^{n} q_{t_i}(i)^2$,
2. $S(f) \geq c \cdot u(t^*)$.

To prove Lemma 2, it is sufficient to construct a $\frac{1}{4}$-proper map dominated by $q_{t^*,\mathcal{OPT}}$. To this end, we first analyze the relationship between $q_{t^*,\mathcal{OPT}}$ and $q_{t^*,\mathcal{SRPT}}$. In particular, we show that $q_{t^*,\mathcal{SRPT}}$ *majorizes* $q_{t^*,\mathcal{OPT}}$. We then construct another map h based on q_{t^*} such that h majorizes $q_{t^*,\mathcal{SRPT}}$. Because majorization is transitive, h majorizes $q_{t^*,\mathcal{OPT}}$. We then reduce the task of constructing a $\frac{1}{4}$-proper map dominated by $q_{t^*,\mathcal{OPT}}$ to the task of constructing a 1-proper map dominated by h.

6.1 The Relationship Between $q_{t^*,\mathcal{OPT}}$ and $q_{t^*,\mathcal{SRPT}}$

Recall that SRPT always executes the active job J_i that has the smallest $q_{t,\mathcal{SRPT}}(i)$. Thus, SRPT avoids executing the active job that has the largest number of remaining tasks. It is not difficult to show that, for any positive integer k, among all work-conserving schedules, SRPT maximizes the sum of the top-k largest numbers of remaining tasks at any time. Lemma 4 formalizes the above statement using majorization. Roughly speaking, if map f majorizes map g, then for any positive integer k, the sum of the top-k largest outputs of f is greater than or equal to that of g.

To define majorization, we first introduce the following definition, which sorts the domain of a map in decreasing order of their outputs.

Definition 3. *Let f be any map. Define π_f as a function that maps any $k \in [1, |\operatorname{dom} f|]$ to the element in $\operatorname{dom} f$ that has the kth largest output of f (ties can be broken arbitrarily). Thus, $\operatorname{dom} f = \{\pi_f(1), \pi_f(2), \cdots, \pi_f(|\operatorname{dom} f|)\}$ and $f(\pi_f(1)) \geq f(\pi_f(2)) \geq \cdots \geq f(\pi_f(|\operatorname{dom} f|))$.*

Next, we define the sum of the top-k outputs of a map f, denoted by $S_k(f)$.

Definition 4. *Let f be any map. Define*

$$S_k(f) = \begin{cases} 0 & \text{if } k = 0 \\ \sum_{j=1}^{k} f(\pi_f(j)) & \text{if } k \in [1, |\operatorname{dom} f|] \\ S(f) & \text{if } k > |\operatorname{dom} f| \end{cases}$$

Definition 5. *For any two maps f and g, f **majorizes** g if the following two conditions are met:*

1. $S(f) = S(g)$,
2. $S_k(f) \geq S_k(g), \forall k \in \mathbb{N}$.

The next lemma formalizes the previous discussion on SRPT. The proof is based on the simple property that SRPT avoids executing the active jobs that have the most remaining tasks and can be found in [16].

Lemma 4. *Let $\mathcal{S}$ be any work-conserving schedule. For any $t \in \mathbb{N}$, $q_{t,\mathcal{SRPT}}$ majorizes $q_{t,\mathcal{S}}$.*

6.2 A Map h that Majorizes $q_{t^*,\mathcal{SRPT}}$ and a 1-Proper Map Dominated by h

We begin by defining a map $\hat{q}_{t^*}$ that *freezes* the number of remaining tasks for a job once it becomes starving. For an FaN job (or an FaS job with $t^* \leq t_i$) we set

$$\hat{q}_{t^*}(i) = q_{t^*}(i),$$

while for an FaS job with $t^* > t_i$ we set

$$\hat{q}_{t^*}(i) = q_{t_i}(i).$$

By Line 12 of Algorithm 1, this ensures that for every job i,

$$\hat{q}_{t^*}(i) \geq q_{t^*,\mathcal{SRPT}}(i),$$

so that $\hat{q}_{t^*}$ dominates $q_{t^*,\mathcal{SRPT}}$.

Although $\hat{q}_{t^*}$ dominates $q_{t^*,\mathcal{SRPT}}$, its sum $S(\hat{q}_{t^*})$ may exceed $q(t^*) = S(q_{t^*,\mathcal{SRPT}})$. To address this, we truncate $\hat{q}_{t^*}$ by reducing the smallest nonzero entries of $\hat{q}_{t^*}$ so that $S(\hat{q}_{t^*}) = q(t^*)$ while preserving the majorization relation. The resulting truncated map is h. The detailed construction of h can be found in [16].

To obtain a 1-proper map h' dominated by h, we set the values of h to zero for all FaN jobs, ensuring that h' satisfies the required properness conditions. The detailed construction of h' can be found in [16].

6.3 The Reduction

The next lemma shows that to construct a $\frac{1}{4}$-proper map dominated by $q_{t^*,\mathcal{OPT}}$, it suffices to construct a 1-proper map dominated by h (e.g., h').

Lemma 5. *Let f and g be such that f majorizes g. For any $c > 0$, if there is a c-proper map dominated by f, then there is a $\frac{c}{4}$-proper map dominated by g.*

We only give a proof sketch of Lemma 5 here. The complete proof can be found in [16]. To prove Lemma 5, it suffices to prove that for any map f' dominated by f, there is a map g' dominated by g such that $S(g'^2) \leq S(f'^2)$ and $S(g') \geq \frac{1}{4}S(f')$. Thus, if f' is a c-proper map, then g' is a $\frac{c}{4}$-proper map. The construction of g' is purely combinatorial. Initially, $g' = g$ and we find a set

$I \subseteq \operatorname{dom} f'$ such that $\sum_{i\in I} f'(i) \geq \frac{1}{2}S(f')$. We then construct a family $\{B_i\}_{i\in I}$ of mutually disjoint subsets of $\operatorname{dom} g'$, such that for each $i \in I$,

$$\frac{f'(i)}{2} \leq \sum_{b\in B_i} g'(b) \leq f'(i).$$

To this end, we may decrease g'. For any $b \notin \bigcup_{i\in I} B_i$, we set $g'(b) = 0$. g' is then the desired map.

6.4 Proof of Lemma 2

Because majorization is transitive, h majorizes $q_{t^*,\mathcal{OPT}}$. Because there is a 1-proper map dominated by h, Lemma 5 implies that there is a $\frac{1}{4}$-proper map dominated by $q_{t^*,\mathcal{OPT}}$. This completes the proof of Lemma 2.

7 Concluding Remark

While numerous studies have explored strategies to mitigate the starvation issue inherent in SRPT scheduling [3,9,10,17–21], no theoretical work has rigorously established whether such mitigation efforts lead to meaningful improvements. In this study, we address this gap by analyzing the competitive ratio of the ℓ_2 norm of flow time. Our findings provide the first theoretical evidence that starvation mitigation significantly enhances SRPT performance.

Acknowledgments. This work was supported in part by the Ministry of Science and Technology of Taiwan (MOST 111-2221-E-004-003-MY2) and the National Science and Technology Council of Taiwan (NSTC 113-2221-E-004-011-MY2). The author is grateful to the anonymous reviewers for their valuable suggestions, particularly regarding the refinement of Lemma 2 and its proof.

References

1. Angelopoulos, S., Lucarelli, G., Nguyen Kim, T.: Primal–dual and dual-fitting analysis of online scheduling algorithms for generalized flow-time problems. Algorithmica **81**(9), 3391–3421 (2019). https://doi.org/10.1007/s00453-019-00583-8
2. Armbruste, A., Rohwedder, L., Wiese, A.: Simpler constant factor approximation algorithms for weighted flow time - now for any p-norm. In: SOSA, pp. 63–81 (2024). https://doi.org/10.1137/1.9781611977936.7
3. Bai, W., Chen, L., Chen, K., Han, D., Tian, C., Wang, H.: PIAS: practical information-agnostic flow scheduling for commodity data centers. IEEE/ACM Trans. Network. **25**(4), 1954–1967 (2017). https://doi.org/10.1109/TNET.2017.2669216
4. Bansal, N., Pruhs, K.: The geometry of scheduling. SIAM J. Comput. **43**(5), 1684–1698 (2014). https://doi.org/10.1137/130911317

5. Bansal, N., Pruhs, K.R.: Server scheduling to balance priorities, fairness, and average quality of service. SIAM J. Comput. **39**(7), 3311–3335 (2010). https://doi.org/10.1137/090772228
6. Batra, J., Garg, N., Kumar, A.: Constant factor approximation algorithm for weighted flow-time on a single machine in pseudopolynomial time. SIAM J. Comput., FOCS18–158–FOCS18–188 (2020). https://doi.org/10.1137/19M1244512
7. Bender, M.A., Chakrabarti, S., Muthukrishnan, S.: Flow and stretch metrics for scheduling continuous job streams. In: ACM-SIAM SODA (1998). https://doi.org/10.5555/314613.314715
8. Chekuri, C., Goel, A., Khanna, S., Kumar, A.: Multi-processor scheduling to minimize flow time with ϵ resource augmentation. In: ACM STOC (2004). https://doi.org/10.1145/1007352.1007411
9. Cherkasova, L.: Scheduling strategy to improve response time for web applications. In: Sloot, P., Bubak, M., Hertzberger, B. (eds.) HPCN-Europe 1998. LNCS, vol. 1401, pp. 305–314. Springer, Heidelberg (1998). https://doi.org/10.1007/BFb0037157
10. Delgado, P., Didona, D., Dinu, F., Zwaenepoel, W.: Job-aware scheduling in eagle: divide and stick to your probes. In: ACM SoCC (2016). https://doi.org/10.1145/2987550.2987563
11. Fox, K., Moseley, B.: Online scheduling on identical machines using SRPT. In: ACM-SIAM SODA (2011). https://doi.org/10.5555/2133036.2133046
12. Golovin, D., Gupta, A., Kumar, A., Tangwongsan, K.: all-norms and all-l_p-norms approximation algorithms. In: FSTTCS (2008). https://doi.org/10.4230/LIPIcs.FSTTCS.2008.1753
13. Im, S., Kulkarni, J., Moseley, B.: Temporal fairness of round robin: competitive analysis for l_k-norms of flow time. In: ACM SPAA (2015). https://doi.org/10.1145/2755573.2755581
14. Im, S., Moseley, B.: Fair scheduling via iterative quasi-uniform sampling. SIAM J. Comput. **49**(3), 658–680 (2020). https://doi.org/10.1137/18M1202451
15. Im, S., Moseley, B., Pruhs, K.: Online scheduling with general cost functions. SIAM J. Comput. **43**(1), 126–143 (2014). https://doi.org/10.1137/120902288
16. Kuo, T.W.: Minimizing ℓ_2 norm of flow time by starvation mitigation (2025). https://arxiv.org/abs/2112.14403
17. Li, W., Chen, S., Li, K., Qi, H., Xu, R., Zhang, S.: Efficient online scheduling for coflow-aware machine learning clusters. IEEE Trans. Cloud Comput. (2020). https://doi.org/10.1109/TCC.2020.3040312
18. Mangharam, R., Demirhan, M., Rajkumar, R., Raychaudhuri, D.: Size matters: size-based scheduling for MPEG-4 over wireless channels. In: SPIE Multimedia Computing and Networking (2003). https://doi.org/10.1117/12.538820
19. Modiano, E.: Scheduling algorithms for message transmission over a satellite broadcast system. In: IEEE MILCOM (1997). https://doi.org/10.1109/MILCOM.1997.646697
20. Pastorelli, M., Carra, D., Dell'Amico, M., Michiardi, P.: HFSP: bringing size-based scheduling to hadoop. IEEE Trans. Cloud Comput. **5**(1), 43–56 (2017). https://doi.org/10.1109/TCC.2015.2396056
21. Wang, Z., Ye, J., Lin, D., Chen, Y., Lui, J.C.: Designing approximate and deployable SRPT scheduler: a unified framework. In: IEEE/ACM IWQOS (2021). https://doi.org/10.1109/IWQOS52092.2021.9521259

Recoverable Robust Cardinality Constrained Maximization with Commitment of a Submodular Function

Sabine Münch(✉), Stephen Raach, and Sven de Vries

Trier University, 54286 Trier, Germany
{muench,raach,devries}@uni-trier.de

Abstract. We consider a game-theoretic variant of maximizing a monotone increasing, submodular function under a cardinality constraint. Initially, a solution to this classical problem is determined. Subsequently, a predetermined number of elements from the ground set, not necessarily contained in the initial solution, are deleted, potentially reducing the solution's cardinality. If any deleted elements were part of the initial solution, they are replaced with a set of at most equal cardinality.

The objective is to maximize the value of the ultimate solution, with the deletion being maximally disadvantageous to the ultimate solution.

When the submodular function is $M^\natural$-concave, we prove that a simple greedy algorithm computes an optimal solution.

When only one element may be deleted, we propose a polynomial running time algorithm with an approximation factor of at least $\frac{1}{3}$.

Furthermore, assuming that the number of allowed deletions is upper bounded by a term of the order of $\frac{k}{\log_2^2(k)}$, where k is the cardinality parameter, we adapt an algorithm from Bogunovic et al. and show that its approximation factor is at least 0.108.

Keywords: Recoverable robust optimization · Submodular maximization · Cardinality constraint

1 Introduction

Submodular maximization is an important problem in combinatorial optimization, covering a range of classical problems such as graph cuts (see e.g., [11]), facility location (see e.g., [1,6]), generalized assignment (see e.g., [9,10]), and set cover (see e.g., [8,16]). In recent years, submodular maximization has gained attention due to its appearance in various application domains, including sensor placement (see e.g., [16]), data subset selection (see e.g., [25]), influence maximization in social networks (see e.g., [15]), extractive document and image summarization (see e.g., [18,24]).

H. Fernau and B. Zhu (Eds.): IWOCA 2025, LNCS 15885, pp. 376–390, 2025.
https://doi.org/10.1007/978-3-031-98740-3_27

Definition 1. *In the following, let I be a finite set of cardinality $n \in \mathbb{N}$ and $f : 2^I \to \mathbb{R}_{\geq 0}$ be a* ***submodular*** *($f(X \cup \{i\}) - f(X) \geq f(Y \cup \{i\}) - f(Y)$ for $X \subseteq Y \subseteq I$, $i \in I \setminus Y$),* ***monotone increasing*** *($f(X) \leq f(Y)$ for any $X \subseteq Y \subseteq I$), and* ***normalized*** *($f(\emptyset) = 0$) function.*

We encode the classic problem of *maximizing a submodular function under a cardinality constraint*, denoted as SMC, through instances (I, f, k), where I is a finite set, $f\colon 2^I \to \mathbb{R}_{\geq 0}$ is a monotone increasing, normalized, submodular function, and $\mathbb{N}_{>0} \ni k \leq |I|$ is the cardinality parameter. The objective is to solve

$$\max\{f(X)\colon |X| \leq k,\ X \subseteq I\}.$$

Definition 2. *Let (I, f, k) be an instance of* SMC. *We call a set $S \subseteq I$ with $|S| \leq k$ an* ***optimal solution*** *to (I, f, k) if $f(S) = \max\{f(X)\colon X \subseteq I, |X| \leq k\}$, and denote some arbitrary optimal solution of an instance (I, f, k) as* $\mathrm{Opt}^{(I,f,k)}$.

For SMC, a simple greedy algorithm, iteratively adding elements that maximize the marginal increase of the objective function until the cardinality parameter is reached, achieves an approximation factor of at least $1 - e^{-1} \approx 0.63$, as shown in [22].

However, in many applications of interest, *robustness* of the solution set against uncertainties about the set of available elements, the cardinality constraint, or the objective function is required, which cannot be guaranteed by the aforementioned greedy algorithm.

In [16], the following *robust version* (RSM) of SMC was introduced:

$$\max_{\substack{S \subseteq I \\ |S| \leq k}} \min_{\substack{D \subseteq S \\ |D| \leq w}} f(S \setminus D).$$

The goal of RSM is to determine a solution to a given SMC-instance (I, f, k) such that, after the deletion of at most w elements from the initial solution, the remaining set has an objective function value as high as possible. Hence, an instance of RSM is defined by a tuple (I, f, k, w), where (I, f, k) is an SMC-instance and $1 \leq w \leq k$ is a deletion parameter.

This paper considers a problem closely related to RSM: the *recoverable robust submodular maximization problem under a cardinality constraint with commitment* (abbreviated as RSMC).

Definition 3. *An instance of* RSMC *is defined by a tuple (I, f, k, w), where (I, f, k) is an* SMC*-instance and $1 \leq w \leq k$ is a* ***deletion parameter****.*

For any RSMC*-instance (I, f, k, w), the* SMC*-instance (I, f, k) is called the (unique)* ***underlying instance*** *of (I, f, k, w). Conversely, for any* SMC*-instance (I, f, k) and any $1 \leq w \leq k$, we call any* RSMC*-instance (I, f, k, w) a* ***robust variant*** *of (I, f, k).*

Given an RSMC-instance, a solution to the underlying SMC-instance is chosen. Next, a predetermined number of elements, less than or equal to the deletion parameter, is deleted from the ground set. Deleted elements that were part of the chosen solution to SMC may be replaced by non-deleted elements.

The deletion is assumed to be adversarial; thus, it is chosen to minimize the value of *our* solution after deletion and replacement.

In contrast, our goal is to select both the initial solution and the replacement elements to maximize the value of the resulting ultimate set; thus, to choose $S \subseteq I$ with $|S| \leq k$, and, after the adversarial choice of $D \subseteq I$ with $|D| \leq w$, to choose $R \subseteq I \setminus (S \cup D)$ with $|R| \leq |S \cap D|$ such that $f((S \setminus D) \cup R)$ approximates

$$\max_{\substack{X \subseteq I \\ |X| \leq k}} \min_{\substack{Y \subseteq I \\ |Y| \leq w}} \max_{\substack{E \subseteq I \setminus (X \cup Y) \\ |E| \leq |X \cap Y|}} f((X \setminus Y) \cup E). \tag{1}$$

In the following, we assume $f(i) > 0$ for all $i \in I$. This assumption is without loss of generality since omitting any chosen element i with $f(i) = 0$ cannot worsen the choice of the selecting party, either the adversary or us.

Definition 4. *Let (I, f, k, w) be an* RSMC*-instance. We call any set $S \subseteq I$ with $|S| \leq k$ an* ***initial solution*** *to* RSMC*, and any set $D \subseteq I$ with $|D| \leq w$ a* ***deletion set****. For an initial solution S and deletion set D, a set $R(S, D) \subseteq I \setminus (S \cup D)$ with $|R(S, D)| \leq |S \cap D|$ is called a* ***second-stage solution*** *of S and D. We call $(S, D, R(S, D))$ a* ***solution tuple****, and $(S \setminus D) \cup R(S, D)$ the corresponding* ***ultimate solution****.*

Given an arbitrary solution tuple $(S, D, R(S, D))$, we call $R(S, D)$ an ***optimal second-stage solution*** *to (S, D) if $R(S, D) \in \arg\max_{E \subseteq I \setminus (S \cup D), |E| \leq |S \cap D|} f((S \setminus D) \cup E)$.*

A solution tuple $(S, D, R(S, D))$ and its corresponding ultimate solution $(S \setminus D) \cup R(S, D)$ are called ***optimal*** *if $f((S \setminus D) \cup R(S, D))$ attains the value of (1), which we refer to as $f(\mathrm{Opt}^{(I,f,k,w)})$, and if D is a minimizer of the ultimate solution given S and an optimal second-stage solution; that is, $D \in \arg\min_{Y \subseteq I, |Y| \leq w} \max_{E \subseteq I \setminus (S \cup Y), |E| \leq |S \cap D|} f((S \setminus Y) \cup E)$.*

For an optimal solution tuple $(S, D, R(S, D))$, we call S an ***optimal initial solution****.*

When the initial solution S and the deletion set D are clear from the context, we simplify the notation by writing (S, D, R) for $(S, D, R(S, D))$.

We aim to develop a polynomial-time algorithm that, given an instance of RSMC, computes an initial solution and, based on this solution and a given deletion set, determines a second-stage solution such that the corresponding ultimate solution approximates an optimal ultimate solution within a constant factor.

Definition 5. *Let (I, f, k, w) be an* RSMC*-instance, and $\mathcal{A}$ be an algorithm that computes an initial solution S, and a second stage solution R, given S and an arbitrary but fixed deletion set D. Then, the approximation factor of $\mathcal{A}$ is defined as*

$$\min_{(I,f,k,w) \text{ is an RSMC-instance}, D \subseteq I, |D| \leq w} \frac{f((S \setminus D) \cup R)}{f(\mathrm{Opt}^{(I,f,k,w)})}.$$

Notice that the approximation factor is well-defined because $k > 1$ and $f(i) > 0$, for every $i \in I$ in each RSMC-instance (I, f, k, w), hence $f(\mathrm{Opt}^{(I,f,k,w)}) > 0$ for every RSMC-instance (I, f, k, w).

1.1 Results

First, we consider a setting where polynomial-time algorithms can return optimal solutions: For instances (I, f, k, w) with $\mathrm{M}^\natural$-concave objective functions, an important natural subset of submodular functions, we show that the initial and second-stage solution of an optimal solution tuple can be computed in polynomial time, by the greedy algorithm analyzed in [22]. This generalizes the result in [12], which focuses on the special case that the $\mathrm{M}^\natural$-concave function is *additionally* the rank function of some matroid.

Next, we examine the special case where the deletion parameter equals one. In this case, we present a polynomial-time algorithm with a constant approximation factor, lying within the interval $\left[\frac{1}{3}, \frac{1-e^{-1}}{2-e^{-1}}\right]$.

Finally, for RSMC-instances (I, f, k, w) with $w \leq \frac{k}{12\left(\lceil \log_2(\frac{k}{12})\rceil^2 + 5\lceil \log_2(\frac{k}{12})\rceil\right)}$, building on [2], we present a polynomial running time algorithm that achieves an approximation factor of at least 0.108.

In the following, we use $(*)$ to indicate results where we omitted the proofs due to space constraints.

1.2 Related Work

In [17], the concept of recoverable robust optimization was introduced in the context of timetabling and linear programming for rail optimization. In this framework, the goal is to determine, given an optimization problem with an uncertainty set and a limited set of recovery actions, a feasible solution to the initial optimization problem that can be transformed into a feasible solution in each scenario realization through a recovery action, such that the objective value of the new solution is maximized. For many classical combinatorial problems like shortest path (see e.g., [4,14]), knapsack (see e.g., [5]), spanning tree (see e.g., [13]) and matching (see e.g., [7]) complexity has been studied in this framework. For some classical optimization problems, adapted to recoverable robust optimization, even approximation results have been achieved (see e.g., [3]).

Recoverable robust optimization *with commitment*, which is closely related to recoverable robust optimization, was introduced in [12]. In this model, one chooses a feasible solution to an underlying optimization problem, defined on a finite ground set of elements. Subsequently, some elements from the ground set are deleted. Then, one is allowed to add new elements to the remaining initial solution, ensuring that the new solution remains feasible and that all elements from the initial solution that were not deleted remain part of the solution (we commit to the initial solution, or to what remains of it).

The objective is to identify an initial solution to the underlying optimization problem, as well as a set of elements to be added post-deletion, so as to maximize the quality of the resulting solution.

In [12], complexity results for several optimization problems adapted to this model are presented. In addition, it was shown that an optimal initial solution to the recoverable robust matroid base problem with commitment can be computed by a simple greedy algorithm.

As already mentioned, RSM, a robust version of SMC, was introduced in [16]. An *asymptotic* constant factor approximation for RSM (with the cardinality constraint growing towards infinity), under the assumption that the number of deletions grows asymptotically as the square root of the cardinality constraint, was presented in [23]. This result was generalized in [2] to the assumption that the number of deletions grows asymptotically as the ratio of the allowed deletions to the cube of the logarithm of the allowed deletions.

RSMC, the problem considered in this paper, is closely related to RSM and a special case of recoverable robust optimization with commitment.

2 Maximizing a Submodular Function Under a Cardinality Constraint

Before addressing RSMC, we consider the simpler problem of maximizing a submodular function under a cardinality constraint. We recall the classic greedy algorithm for this fundamental problem and provide additional insights that will serve as a basis for addressing the more complex problem of RSMC.

2.1 A Greedy Algorithm for SMC

We start by recalling a simple greedy algorithm that gradually chooses the next element according to maximal marginal gain until the cardinality constraint is reached, first analyzed in [22]. To convert this greedy algorithm into a deterministic algorithm, we fix an arbitrary tie-breaking rule for comparing elements by their marginal gain by usually choosing a permutation.

Algorithm 1:

Input: An instance (I, f, k) of SMC.
Output: Approximately optimal solution.

Set $S \leftarrow \emptyset$
while $|S| \leq k$ **do**
  Choose $i^* \in \arg\max \{f(S \cup \{i\}) - f(S) \colon i \in I \setminus S\}$ with tie-breaking according to a fixed rule
  Set $S \leftarrow S \cup \{i^*\}$
return S

For our analysis, it is useful to consider the elements of the set S, returned by Algorithm 1, in the order they were added to S. Thus, we assume that iteratively constructed sets are ordered according to the insertion order of all contained elements.

Definition 6. *Let (I, f, k) be an* SMC*-instance, and let $S \subseteq I$ be any set constructed by iteratively adding elements, starting with $S = \emptyset$, e.g., the set constructed in Line 4 of Algorithm 1. In the following, we assume that S is ordered according to the insertion order of the elements, and for any $1 \leq i \leq |S|$, we denote by S_i the element added to S in the i-th iteration.*

We present some auxiliary results concerning the approximation guarantees of Algorithm 1. These will assist in analyzing the algorithms to determine initial and second-stage solutions to RSMC.

We recall the worst-case performance of Algorithm 1.

Proposition 1 [22, Theorem 4.1]. *For any* SMC*-instance (I, f, k), Algorithm 1 returns a set S with $|S| \leq k$ such that $\frac{f(S)}{f(\mathrm{Opt}^{(I,f,k)})} \geq 1 - e^{-1} = 0.63\ldots$.*

We show that if the value of the element chosen in the first iteration of the while loop in Line 2 of Algorithm 1 is greater than or equal to half the value of the set S returned by Algorithm 1, then $f(S)$ approximates an optimal solution by at least $\frac{2}{3}$.

Lemma 1. *Let (I, f, k) be an* SMC*-instance. Let S be the set resulting from Algorithm 1 with $f(S_1) \geq \frac{f(S)}{2}$. Then, we have*

$$f(S) \geq \frac{2}{3} f(\mathrm{Opt}^{(I,f,k)}).$$

Proof. By submodularity of f, we have

$$\begin{aligned} f(\mathrm{Opt}^{(I,f,k)}) &\leq f(\{S_1, \ldots, S_{k-1}\}) + \\ &\sum_{e \in \mathrm{Opt}^{(I,f,k)} \setminus \{S_1, \ldots, S_{k-1}\}} f(\{S_1, \ldots, S_{k-1}\} \cup \{e\}) - f(\{S_1, \ldots, S_{k-1}\}) \\ &\underset{\text{Line 3 of Algorithm 1}}{\leq} f(\{S_1, \ldots, S_{k-1}\}) + \\ &\sum_{e \in \mathrm{Opt}^{(I,f,k)} \setminus \{S_1, \ldots, S_{k-1}\}} f(\{S_1, \ldots, S_{k-1}\} \cup \{S_k\}) - f(\{S_1, \ldots, S_{k-1}\}) \\ &\leq f(\{S_1, \ldots, S_{k-1}\}) + k\left(f(S) - f(\{S_1, \ldots, S_{k-1}\})\right) \\ &= f(S) + (k-1)\left(f(S) - f(\{S_1, \ldots, S_{k-1}\})\right). \end{aligned} \tag{2}$$

Furthermore, submodularity and Line 3 of Algorithm 1 imply

$$f(S) - f(\{S_1, \ldots, S_{k-1}\}) \leq \frac{f(S) - f(S_1)}{k-1}. \tag{3}$$

By adding $(k-1)$ times Inequality (3) to Inequality (2), we obtain $f(\mathrm{Opt}^{(I,f,k)}) \leq 2f(S) - f(S_1)$. Since $f(S_1) \geq \frac{f(S)}{2}$, it immediately follows that $f(\mathrm{Opt}^{(I,f,k)}) \leq \frac{3}{2} f(S)$. □

We require a classical result from [26] that relates the value of the set returned by Algorithm 1 for an SMC-instance to the optimal solution value of an SMC-instance with the same ground set and objective function, but with a greater cardinality parameter.

Lemma 2 [26, Theorem 2]. *Let (I, f, k) be an* SMC*-instance and S be the set returned by Algorithm 1. Then, $f(S) \geq (1 - e^{-\frac{k}{p}})f(\mathrm{Opt}^{(I,f,p)})$ for any $p \geq k$.*

2.2 Connections Between Optimal Solutions to SMC and RSMC

We analyze the relationship between optimal solutions to SMC and both the optimal initial and optimal ultimate solutions to RSMC.

Note that any solution to an SMC-instance can serve as an initial solution to a robust variant of that instance. However, an optimal solution to SMC is not necessarily an optimal initial solution to a corresponding robust variant.

To provide an example of an initial SMC-solution (together with a deletion set and an optimal second-stage solution) that yields an ultimate solution arbitrarily worse than an optimal ultimate solution, we consider coverage functions, a subset of submodular functions.

Definition 7. *Let $E^1, \dots, E^n$ be a collection of subsets of a finite ground set E. Then, $f\colon 2^{\{1,\dots,n\}} \to \mathbb{R}_{\geq 0}, X \mapsto |\bigcup_{i \in X} E^i|$ is called* ***coverage function.***

Obviously, coverage functions are normalized and monotone increasing, and it is well-known that they are submodular.

The following example exhibits an RSMC-instance, in which the submodular function is coverage. It demonstrates that for some optimal solution of the underlying SMC-instance as the initial solution, the resulting ultimate solution to the original problem may be arbitrarily worse than the optimal ultimate solution.

Example 1. For $n \in \mathbb{N}$, let $E = \{1, \dots, n^2 + n - 1\}$, and let $E^1, \dots, E^{2n}$ be a collection of subsets of E with

$$E^i := \begin{cases} \{(i-1)n + 1, \dots, in\}, & \text{if } i \in \{1, \dots, n\} \\ \{n^2 + i - n\}, & \text{if } i \in \{n+1, \dots, 2n-1\} \\ \{1, \dots, n^2\}, & \text{if } i = 2n. \end{cases}$$

Let $f^n\colon 2^{\{1,\dots,2n\}} \to \mathbb{R}_{\geq 0}$, $X \mapsto |\bigcup_{i \in X} E^i|$. Then, f^n is a coverage function and hence submodular.

For $n \in \mathbb{N}$, consider the RSMC-instance $(\{1, \dots, 2n\}, f^n, n, 1)$, with underlying SMC-instance $(\{1, \dots, 2n\}, f^n, n)$. The set $S = \{n+1, \dots, 2n\}$ is an optimal solution to the SMC-instance (I, f^n, n), because $f^n(S) = |\bigcup_{i \in S} E^i| = |\{1, \dots, n^2, n^2 + 1, \dots, n^2 + n - 1\}| = n^2 + n - 1 = f(E)$, and $(S, \{2n\}, \{1\})$ is a solution tuple of $(\{1, \dots, 2n\}, f^n, n, 1)$, where $\{1\}$ is an optimal second-stage solution to the initial solution S and the deletion set $\{2n\}$.

However, the ultimate solution corresponding to $(S, \{2n\}, \{1\})$ is arbitrarily worse than an optimal ultimate solution since

$$\lim_{n\to\infty} \frac{f^n((S \setminus \{2n\}) \cup \{1\})}{f^n((\{1,\ldots,n\} \setminus \{n\}) \cup \{2n\})} = \lim_{n\to\infty} \frac{2n-1}{n^2} = 0.$$

Although an optimal solution to an SMC-instance generally is not an optimal initial solution of some robust variant, its function value provides a trivial upper bound for the function value of an optimal ultimate solution of any robust variant. To provide a possibly stronger upper bound, we define:

Definition 8. *For any* SMC*-instance* (I, f, k) *and* $T \subseteq I$*, we define the* ***restriction*** *of* (I, f, k) *to* T *as* $(I, f, k)|_T := (T, f|_{2^T}, k)$.

The optimal ultimate solution value of an RSMC-instance (I, f, k, w) is at most the minimum optimal solution value, taken over all subsets $Z \subseteq I$ with $|Z| \leq w$, of the restricted SMC-instances $(I, f, k)|_{I\setminus Z}$.

Theorem 1. *Let* (I, f, k, w) *be an* RSMC*-instance and* (X, D, R) *be an optimal solution tuple to it. Let* $Z \in \arg\min_{Y\subseteq I, |Y|\leq w} \max_{E\subseteq I\setminus Y, |E|\leq k} f(E)$*, and let* $T \in \arg\max_{E\subseteq I\setminus Z, |E|\leq k} f(E)$*. Then,*

$$f(T) \geq f((X \setminus D) \cup R).$$

Proof. By the definitions of T and Z, we have

$$\begin{aligned} f(T) &= \max_{E\subseteq I\setminus Z, |E|\leq k} f(E) \geq \max_{E\subseteq I\setminus(Z\cup X), |E|\leq|X\cap Z|} f((X \setminus Z) \cup E) \\ &\geq \min_{Y\subseteq I, |Y|\leq w} \max_{E\subseteq I\setminus(Y\cup X), |E|\leq|X\cap Y|} f((X \setminus Y) \cup E) = f((X \setminus D) \cup R). \end{aligned}$$

□

Notice that Theorem 1 implies that for any RSMC-instance (I, f, k, w) and any $Z \subseteq I$ with $|Z| \leq w$, the function value of the set returned by Algorithm 1 applied to $(I, f, k)|_{I\setminus Z}$ approximates the value of an optimal ultimate solution to (I, f, k, w) by at least $1 - e^{-1}$.

3 RSMC with M$^\natural$-Concave Objective Functions

Before considering RSMC-instances with general monotone increasing submodular objective functions, we focus on the special case where the objective function is M$^\natural$-concave.

The following definition of M$^\natural$-concavity is due to [21], and is far more convenient than the equivalent original definition in [19].

Definition 9. *Let* I *be a finite set and* $f\colon 2^I \to \mathbb{R}$*. Then,* f *is called* **M$^\natural$-*concave*** *if for all* $S, T \subseteq I$ *with* $|S| < |T|$*, we have*

$$f(S) + f(T) \leq \max_{t\in T\setminus S} \{f(S \cup \{t\}) + f(T \setminus \{t\})\}.$$

From Definition 9, one can directly observe that the rank functions of matroids form a subset of $\mathrm{M}^\natural$-concave functions.

For RSMC-instances with an $\mathrm{M}^\natural$-concave objective function, we demonstrate that any optimal solution to the underlying SMC-instance is an optimal initial solution. Additionally, we show that an optimal second-stage solution, given an initial solution and a deletion set, can be computed in polynomial running time.

Theorem 2. *Let $f\colon 2^I \to \mathbb{R}_{\geq 0}$ be $\mathrm{M}^\natural$-concave, and let (I, f, k, w) be an* RSMC-*instance. Then, for any optimal solution S to the* SMC-*instance (I, f, k), there exist D, R such that (S, D, R) is optimal for (I, f, k, w). Furthermore, for any given D, the optimal second-stage solution R can be computed in polynomial running time.*

To prove Theorem 2, we first demonstrate that Algorithm 1 returns an optimal solution for every SMC-instance with an $\mathrm{M}^\natural$-concave objective function.

Lemma 3 (*). *For any* SMC-*instance (I, f, k) with f $\mathrm{M}^\natural$-concave, Algorithm 1 returns an optimal solution.*

The following lemma is equivalent to [20, Theorem 2.3], where the term 'valuated matroid' is used to denote the restriction of an $\mathrm{M}^\natural$-concave function to all sets of a fixed cardinality.

Lemma 4 [20, Theorem 2.3.]. *Let $f\colon 2^I \to \mathbb{R}$ be an $\mathrm{M}^\natural$-concave function. Then, for any $S, T \subseteq I$ with $|S| = |T|$ and any $D \subseteq S \setminus T$, there exists $R \subseteq T \setminus S$, $|R| = |D|$ such that*

$$f(S) + f(T) \leq f((S \setminus D) \cup R) + f((T \setminus R) \cup D).$$

We are ready to prove Theorem 2.

Proof of Theorem 2. Let S be an optimal solution to the SMC-instance (I, f, k) and let D be a deletion set. Further, let $D^S = D \cap S$, and let T be an optimal solution to the SMC-instance $(I, f, k)|_{I \setminus D}$. By Lemma 4, there exists a set $R \subseteq T \setminus S$ with $|R| = |D^S|$ such that

$$f(S) + f(T) \leq f((S \setminus D^S) \cup R) + f((T \setminus R) \cup D^S). \tag{4}$$

By optimality of S and T, we have

$$f((S \setminus D^S) \cup R) \leq f(T) \text{ and } f((T \setminus R) \cup D^S) \leq f(S). \tag{5}$$

Then, inequalities (4) and (5), and optimality of S and T imply $f((S \setminus D^S) \cup R) = f(T)$ (and $f((T \setminus R) \cup D^S) = f(S)$), and therefore, (S, D, R) is an optimal solution tuple.

It remains to show that R can be determined in polynomial running time. Define $\tilde{f}\colon 2^{I \setminus (S \cup D)} \to \mathbb{R}_{\geq 0}$, $X \mapsto f((S \setminus D) \cup X)$. Then $\tilde{f}$ is $\mathrm{M}^\natural$-concave, since f is $\mathrm{M}^\natural$-concave. Let R be the set returned by Algorithm 1 applied to $(I \setminus$

$(S \cup D), \tilde{f}, |S \cap D|)$. Then, R is computed in polynomial running time, and by Lemma 3, we have

$$R \in \underset{E \subseteq I \setminus (S \cup D),\ |E| \leq |S \cap D|}{\arg\max} \tilde{f}(E) = \underset{E \subseteq I \setminus (S \cup D),\ |E| \leq |S \cap D|}{\arg\max} f((S \setminus D) \cup E).$$

□

Notice that, by Lemma 3 and Theorem 2, for any RSMC-instance with an M$^\natural$-concave objective function, the initial solution and the second-stage solution of an optimal solution tuple can be determined in polynomial running time by using Algorithm 1.

4 RSMC with at Most One Deletion

In this section, we focus on instances $(I, f, k, 1)$ of RSMC. As in Sect. 3, the output of Algorithm 1 for the underlying SMC-instance (I, f, k) is a natural candidate for an initial solution.

Recall that, as shown by Example 1, there exist SMC-instances (I, f, k) with f being a non-M$^\natural$-concave function, such that any optimal solution to (I, f, k), when used as an initial solution to the RSMC-instance $(I, f, k, 1)$, yields an arbitrarily bad ultimate solution. In particular, this may occur when a single element of the optimal solution to SMC contributes the majority of the total value to that solution and, in addition, there does not exist a sufficiently valuable replacement element for it.

The following polynomial-time approximation algorithm avoids this trap by constructing an initial solution such that, for any single element of this kind it contains, either a suitable replacement element exists, which is then returned as the second-stage solution, or the remaining initial solution still retains a sufficiently large value after the element is deleted.

Algorithm 2:

Input: An instance $(I, f, k, 1)$ of RSMC.
Output: Initial solution S and second-stage solution R.

1 Let G be the result of Algorithm 1 applied to (I, f, k)
2 **if** $f(G_1) \leq \frac{f(G)}{2}$ **then**
3   Set $S \leftarrow G$ and $R \leftarrow \emptyset$
4 **else**
5   Let H be the result of Algorithm 1 applied to $(I, f, k)|_{I \setminus \{G_1\}}$
6   Set $S \leftarrow \{H_1, \ldots, H_{k-1}\} \cup \{G_1\}$ and $R \leftarrow \{H_k\}$
7 **return** S, R

Assume that G_1, the first element added by Algorithm 1 to the set G in Line 1 of Algorithm 2, contributes less than or equal to half of the total value of G. Then, Algorithm 2 returns G as an initial solution to the RSMC-instance $(I, f, k, 1)$, since even if the adversary deletes the most valuable element G_1, the remaining initial solution maintains more than half of G's value.

In the case where G_1 is more valuable than half the value of G, Algorithm 2 applies Algorithm 1 to a restriction of the original instance to obtain a set that is, by itself (in the absence of G_1), a good approximation of an optimal ultimate solution to the RSMC-instance.

Notice that the second-stage solution returned by Algorithm 2 is independent of the deletion set chosen by an adversary; it solely depends on the initial solution.

We provide an approximation factor of at least $\frac{1}{3}$ for Algorithm 2.

Theorem 3 (*). *Algorithm 2 has an approximation factor of at least* $\frac{1}{3}$.

Additionally to Theorem 3, we provide an example demonstrating that the approximation factor of Algorithm 2 is upper bounded by $\frac{1-e^{-1}}{2-e^{-1}} = 0.387\ldots$ in the full version of this paper.

5 RSMC with More than One Deletion

Next, we examine RSMC-instances that allow for more than one deletion. To this end, we adapt the following polynomial running time approximation algorithm from [2], originally designed for RSM, to determine initial solutions to RSMC.

Algorithm 3:

Input: An RSM-instance (I, f, k, w), $\eta \in \mathbb{N}_{>0}$, and an algorithm $\mathcal{B}$ with the β-iterative property.
Output: Initial solution to (I, f, k, w).

Set $S^0 \leftarrow \emptyset$ and $S^1 \leftarrow \emptyset$
for $i = 0, \ldots, \lceil \log_2(w) \rceil$ **do**
  for $j = 1, \ldots, \lceil \frac{w}{2^i} \rceil$ **do**
    Let $G^{i,j}$ be the set returned by algorithm $\mathcal{B}$ applied to $(I, f, 2^i \eta)|_{I \setminus S^0}$
    Set $S^0 \leftarrow S^0 \cup G^{i,j}$
Let S^1 be the set returned by algorithm $\mathcal{B}$ applied to $(I, f, k - |S^0|)|_{I \setminus S^0}$
Set $S \leftarrow S^0 \cup S^1$
return S

As part of its input, Algorithm 3 requires an algorithm $\mathcal{B}$ with the β-iterative property:

Definition 10. *Let* $\mathcal{B}$ *be an algorithm that applies to an* SMC*-instance* (I, f, k) *and returns an iteratively constructed set* $G \subseteq I$ *with* $|G| = k$. *For* $\beta \geq 1$, *algorithm* $\mathcal{B}$ *has the* β*-iterative property if for each* $j < k$ *the following Inequality is valid:*

$$f(\{G_1, \ldots, G_{j+1}\}) - f(\{G_1, \ldots, G_j\})$$
$$\geq \frac{1}{\beta} \max \{f(\{G_1, \ldots, G_j\} \cup \{G^*\}) - f(\{G_1, \ldots, G_j\}) \colon G^* \in I \setminus \{G_1, \ldots, G_j\}\}.$$

Informally, the β-iterative property is based on the idea that an algorithm selects elements iteratively, ensuring that each chosen element has a marginal gain of at least $\frac{1}{\beta}$ of the maximum possible marginal gain. Note that Algorithm 1 satisfies the β-iterative property for $\beta = 1$.

In [2], the approximation guarantee of Algorithm 3 for RSM is analyzed. For RSM-instances (I, f, k, w) with $2 \leq w \leq \frac{k}{3\eta(\log_2(k+2))}$, it is proven that the initial solution returned by Algorithm 3, after removing an arbitrary deletion set, approximates an optimal solution of RSM depending on $k, w, |S^0|$ and η. Additionally, for RSM-instances with $w \in o(\frac{k}{\eta \log_2(k)})$ and for $\eta \geq \log_2^2(k)$, an asymptotic approximation factor for Algorithm 3 is provided.

Recall that RSMC-instances are similar to RSM-instances. Consequently, Algorithm 3 can be easily adapted to determine an initial solution, given an RSMC-instance.

However, Algorithm 3 does not calculate a second-stage solution. In order to determine the approximation factor of Algorithm 3 despite this, we set, throughout the following, each second-stage solution to the empty set. Hence, Algorithm 3 has a constant approximation factor when it returns, for each RSMC-instance, an initial solution such that the set remaining from the initial solution after removing any arbitrary deletion set approximates an optimal ultimate solution within a constant factor.

Furthermore, in our analysis of Algorithm 3, we fix the input parameter η as well as the algorithm $\mathcal{B}$ used as a subroutine within Algorithm 3.

Definition 11. *Let (I, f, k, w) be an* RSMC-*instance. In the following, we fix the input parameter η of Algorithm 3 to $\eta := 4\lceil \log_2(w) \rceil$. Additionally, we set the subroutine algorithm $\mathcal{B}$ of Algorithm 3 to Algorithm 1.*

We demonstrate that Algorithm 3 achieves a constant approximation factor for RSMC-instances (I, f, k, w) with

$$\frac{w}{k} \leq \frac{1}{12\left(\lceil \log_2(\frac{k}{12}) \rceil^2 + 5\lceil \log_2(\frac{k}{12}) \rceil\right)}. \tag{6}$$

Theorem 4 (∗). *For any* RSMC-*instance (I, f, k, w) that fulfills Inequality* (6), *Algorithm 3 has an approximation factor of at least*

$$\frac{\frac{16}{35}(1 - e^{-\frac{2}{3}})}{1 + \frac{16}{35}(1 - e^{-1})}(1 - e^{-\frac{71}{72}}) = 0.108\ldots.$$

It remains an interesting open question whether it is possible to obtain a constant factor approximation of an optimal ultimate solution for RSMC-instances (I, f, k, w) where more than $\frac{k}{12(\lceil \log_2(\frac{k}{12}) \rceil^2 + 5\lceil \log_2(\frac{k}{12}) \rceil)}$ deletions are allowed.

5.1 New Approximation Results for RSM

As a byproduct of our analysis of Algorithm 3 for RSMC, we obtain a new approximation result of Algorithm 3 for RSM.

Corollary 1 (∗). *For any* RSM-*instance* (I, f, k, w) *that satisfies Inequality* (6)*, Algorithm 3 has an approximation factor of at least*

$$\frac{\frac{16}{35}(1-e^{-\frac{2}{3}})}{1+\frac{16}{35}(1-e^{-1})} = 0.172\dots.$$

6 Conclusion

In this paper, we considered some special cases of the recoverable robust submodular maximization problem under a cardinality constraint with commitment. However, several questions remain open.

We take a first step toward addressing open questions in the full version of the paper, presenting a polynomial-time algorithm for arbitrary RSMC-instances with an approximation guarantee that depends only on the curvature of the instance's objective function.

Acknowledgments. The authors thank the DFG for their support within RTG 2126 "Algorithmic Optimization".

References

1. Ageev, A.A., Sviridenko, M.I.: An 0.828-approximation algorithm for the uncapacitated facility location problem. Disc. Appl. Math. **93**(2–3), 149–156 (1999). https://doi.org/10.1016/S0166-218X(99)00103-1
2. Bogunovic, I., Mitrović, S., Scarlett, J., Cevher, V.: Robust submodular maximization: a non-uniform partitioning approach. In: Proceedings of the 34th International Conference on Machine Learning. Proceedings of Machine Learning Research, vol. 70, pp. 508–516. PMLR (2017)
3. Bold, M., Goerigk, M.: Investigating the recoverable robust single machine scheduling problem under interval uncertainty. Disc. Appl. Math. **313**, 99–114 (2022). https://doi.org/10.1016/j.dam.2022.02.005
4. Büsing, C.: Recoverable Robustness in Combinatorial Optimization. Cuvillier Verlag (2011)
5. Büsing, C., Goderbauer, S., Koster, A.M.C.A., Kutschka, M.: Formulations and algorithms for the recoverable Γ-robust knapsack problem. EURO J. Comput. Optim. **7**(1), 15–45 (2019). https://doi.org/10.1007/s13675-018-0107-9
6. Cornuejols, G., Fisher, M.L., Nemhauser, G.L.: Exceptional paper - location of bank accounts to optimize float: an analytic study of exact and approximate algorithms. Manag. Sci. **23**(8), 789–810 (1977). https://doi.org/10.1287/mnsc.23.8.789
7. Dourado, M.C., Meierling, D., Penso, L.D., Rautenbach, D., Protti, F., de Almeida, A.R.: Robust recoverable perfect matchings. Networks **66**(3), 210–213 (2015). https://doi.org/10.1002/net.21624
8. Feige, U.: A threshold of ln n for approximating set cover. J. ACM **45**(4), 634–652 (1998). https://doi.org/10.1145/285055.285059

9. Feige, U., Vondrák, J.: Approximation algorithms for allocation problems: improving the factor of 1-1/e. In: 2006 47th Annual IEEE Symposium on Foundations of Computer Science (FOCS'06), pp. 667–676. IEEE (2006). https://doi.org/10.1109/FOCS.2006.14
10. Fleischer, L., Goemans, M.X., Mirrokni, V.S., Sviridenko, M.: Tight approximation algorithms for maximum general assignment problems. In: Proceedings of the Seventeenth Annual ACM-SIAM Symposium on Discrete Algorithm, SODA '06, pp. 611–620. Society for Industrial and Applied Mathematics (2006)
11. Goemans, M.X., Williamson, D.P.: Improved approximation algorithms for maximum cut and satisfiability problems using semidefinite programming. J. ACM **42**(6), 1115–1145 (1995). https://doi.org/10.1145/227683.227684
12. Hommelsheim, F., Megow, N., Muluk, K., Peis, B.: Recoverable robust optimization with commitment. arXiv preprint arXiv:2306.08546 (2023)
13. Hradovich, M., Kasperski, A., Zieliński, P.: The recoverable robust spanning tree problem with interval costs is polynomially solvable. Optim. Lett. **11**(1), 17–30 (2017). https://doi.org/10.1007/s11590-016-1057-x
14. Jackiewicz, M., Kasperski, A., Zieliński, P.: Recoverable robust shortest path problem under interval budgeted uncertainty representations. Networks **85**(1), 127–141 (2025). https://doi.org/10.1002/net.22255
15. Kempe, D., Kleinberg, J., Tardos, E.: Maximizing the spread of influence through a social network. Theory Comput. **11**(4), 105–147 (2015). https://doi.org/10.4086/toc.2015.v011a004
16. Krause, A., McMahan, H.B., Guestrin, C., Gupta, A.: Robust submodular observation selection. J. Mach. Learn. Res. **9**(12), 2761–2801 (2008)
17. Liebchen, C., Lübbecke, M., Möhring, R., Stiller, S.: The concept of recoverable robustness, linear programming recovery, and railway applications. In: Ahuja, R.K., Möhring, R.H., Zaroliagis, C.D. (eds.) Robust and Online Large-Scale Optimization. LNCS, vol. 5868, pp. 1–27. Springer, Heidelberg (2009). https://doi.org/10.1007/978-3-642-05465-5_1
18. Lin, H., Bilmes, J.: A class of submodular functions for document summarization. In: Proceedings of the 49th Annual Meeting of the Association for Computational Linguistics: Human Language Technologies, HLT '11, vol. 1, pp. 510–520. Association for Computational Linguistics (2011)
19. Murota, K.: Discrete Convex Analysis, Monographs on Discrete Mathematics and Applications, vol. 10. Society for Industrial and Applied Mathematics (2003)
20. Murota, K.: A stronger multiple exchange property for M$^\natural$-concave functions. Jpn. J. Ind. Appl. Math. **35**(1), 411–421 (2018). https://doi.org/10.1007/s13160-017-0278-4
21. Murota, K., Shioura, A.: Simpler exchange axioms for M-concave functions on generalized polymatroids. Jpn. J. Ind. Appl. Math. **35**(1), 235–259 (2018). https://doi.org/10.1007/s13160-017-0285-5
22. Nemhauser, G.L., Wolsey, L.A., Fisher, M.L.: An analysis of approximations for maximizing submodular set functions-I. Math. Program. **14**, 265–294 (1978). https://doi.org/10.1007/BF01588971
23. Orlin, J.B., Schulz, A.S., Udwani, R.: Robust monotone submodular function maximization. Math. Program. **172**, 505–537 (2018). https://doi.org/10.1007/s10107-018-1320-2
24. Tschiatschek, S., Iyer, R., Wei, H., Bilmes, J.: Learning mixtures of submodular functions for image collection summarization. In: Proceedings of the 28th International Conference on Neural Information Processing Systems, NIPS'14, vol. 1, pp. 1413–1421 (2014)

25. Wei, K., Iyer, R., Bilmes, J.: Submodularity in data subset selection and active learning. In: Proceedings of the 32nd International Conference on Machine Learning. Proceedings of Machine Learning Research, vol. 37, pp. 1954–1963. PMLR (2015)
26. Wolsey, L.A.: Maximising real-valued submodular functions: primal and dual heuristics for location problems. Math. Oper. Res. **7**(3), 410–425 (1982). https://doi.org/10.1287/moor.7.3.410

Bicriteria FPT-Approximation Algorithms for Vertex Deletion to Bounded Degeneracy Graphs

Tanmay Inamdar[1], Lawqueen Kanesh[2], R. Krithika[3(✉)], Harshil Mittal[4], and Saket Saurabh[5,6]

[1] Indian Institute of Technology Jodhpur, Jodhpur, India
[2] Indian Institute of Technology Indore, Indore, India
[3] Indian Institute of Technology Palakkad, Palakkad, India
krithika@iitpkd.ac.in
[4] Indian Institute of Technology Gandhinagar, Gandhinagar, India
mittal_harshil@iitgn.ac.in
[5] Institute of Mathematical Sciences, Chennai, India
saket@imsc.res.in
[6] University of Bergen, Bergen, Norway

Abstract. In this work, we consider the optimization problem of finding a minimum-weight subset of vertices of a given undirected graph on n vertices whose deletion results in a d-degenerate graph. For $d \geq 2$, this problem is known to be constant-factor inapproximable implying that one cannot hope for anything better than bicriteria approximation algorithms. Towards this end, we give a randomized polynomial-time algorithm that for any value of the bicriteria approximation trade-off parameter $\alpha > 1$ and confidence parameter $\delta \in (0, 1)$, returns a $2\alpha d$-degeneracy modulator whose weight is at most $(1 + \delta) \cdot \frac{2\alpha}{\alpha-1}$ times the weight of an optimum solution with high probability. Then, we move on to the decision problem of determining if a graph G on n vertices has a d-degeneracy modulator of size at most k. For each $d \geq 2$, this problem is known to be W[P]-hard with respect to k and we give three FPT-approximation algorithms for solving it. These algorithms return a $2\alpha d$-degeneracy modulator whose size is at most k (if a k-sized d-degeneracy modulator exists) for any $\alpha > 1$. All our algorithms can be tuned to return a $2d$-degeneracy modulator of size at most k (if a k-sized d-degeneracy modulator exists) by setting α appropriately.

1 Introduction

A graph G is *d-degenerate* if every non-empty subgraph of G has a vertex of degree at most d. The *degeneracy* of a graph G, denoted by $\mathsf{degen}(G)$, is the smallest number d such that G is d-degenerate. Degeneracy is a measure of sparseness of a graph and every hereditary class of graphs with the number of edges upper-bounded linearly in the number of vertices is d-degenerate for some constant d. Degeneracy is within a constant factor of another sparsity measure,

H. Fernau and B. Zhu (Eds.): IWOCA 2025, LNCS 15885, pp. 391–404, 2025.
https://doi.org/10.1007/978-3-031-98740-3_28

namely, arboricity and it is well-known that bounded degeneracy is equivalent to bounded arboricity [8]. An ordering of vertices $v_1, \ldots, v_n$ of a graph is called a *d-degeneracy sequence* if every vertex v_i has at most d neighbors among the vertices $v_1, v_2, \ldots, v_{i-1}$ preceding it. It is well-known that a graph is d-degenerate if and only if there is a d-degenerate ordering of its vertices. Further, a d-degeneracy ordering of a d-degenerate graph can be obtained in linear time [13].

Degenerate graphs are amenable to polynomial-time algorithms for classical problems that are NP-hard in general. Further, many problems that are W[1]-hard in general become fixed-parameter tractable when restricted to degenerate graphs [2,5,10]. This motivates one to study the editing problem to the family of bounded degeneracy graphs.

d-Degeneracy Modulator

Input: A graph G on n vertices and a non-negative integer k.
Objective: Determine if there is a set $S \subseteq V(G)$ with $|S| \leq k$ and $\mathsf{degen}(G - S) \leq d$.

A set $S \subseteq V(G)$ is called a d-degeneracy modulator of G if $\mathsf{degen}(G - S) \leq d$. Observe that a 0-degeneracy modulator is a vertex cover and a 1-degeneracy modulator is a feedback vertex set. Thus, for $d \in \{0, 1\}$, d-Degeneracy Modulator is NP-complete and fixed-parameter tractable (FPT) when parameterized by the solution size k. However, Abrahamson et al. [1] showed that d-Degeneracy Modulator is W[P]-complete for $d = 2$. A decade later, Mathieson [12] showed that for each $d \geq 2$, d-Degeneracy Modulator is NP-complete and W[P]-complete even if the input graph has degeneracy $d + 1$ or maximum degree $2d + 1$ (see [11] for some remarks on this result).

In the realm of exact exponential-time algorithms, Pilipczuk and Pilipczuk [14] showed that for any fixed d one can find a maximum induced d-degenerate subgraph in randomized $(2 - \epsilon_d)^n n^{\mathcal{O}(1)}$ time, for some constant $\epsilon_d > 0$ depending only on d. On the approximation side, Demaine et al. [7] gave a polynomial-time $\mathcal{O}(d \log n)$-approximation algorithm for d-Degeneracy Modulator. They also proved a complementary hardness result stating that this problem is $o(\log(n/d))$-inapproximable. Demaine et al. [7] gave constant-factor bicriteria approximation algorithms for Minimum-Weight d-Degeneracy Modulator using the local ratio technique and linear-programming.

Minimum-Weight *d*-Degeneracy Modulator

Input: A graph G on n vertices, with a weight function $w : V(G) \to \mathbb{R}^{\geq 0}$.
Output: A minimum-weight set $S \subseteq V(G)$ such that $\mathsf{degen}(G - S) \leq d$.

For an instance (G, w) of Minimum-Weight d-Degeneracy Modulator, let $\mathsf{opt}(G, w)$ denote the weight $w(S) = \sum_{v \in S} w(v)$ of a minimum weight d-degeneracy modulator S of (G, w). For real numbers $p, q \geq 1$, a set $S \subseteq V(G)$ is called an (p, q)-approximate d-degeneracy modulator (or simply an (p, q)-approximation) of G if $w(S) \leq p \cdot \mathsf{opt}(G, w)$ and $\mathsf{degen}(G - S) \leq q \cdot d$. An algorithm for Minimum-Weight d-Degeneracy Modulator is called an (p, q)-bicriteria approximation algorithm if it returns an (p, q)-approximate d-degeneracy modulator of G.

Our Results. Our first result is a randomized polynomial-time approximation algorithm for Minimum-Weight d-Degeneracy Modulator.

- (Theorem 1) Let $\delta \in (0,1)$ be a constant. For any $d \in \mathbb{N}$ and $\alpha > 1$, Minimum-Weight d-Degeneracy Modulator admits a randomized algorithm that, with high probability, returns a $((1+\delta)\frac{2\alpha}{\alpha-1}, 2\alpha)$-approximate solution in polynomial time.

The algorithms by Demaine et al. [7] can be tuned to return $(4,4)$- and $(6,6)$-approximations. Comparing this with our result, we observe that our algorithm can be tuned to return $(4,4)$- and $(3,6)$-approximations with constant probability. Therefore, our algorithm has a better approximation ratio for certain values of p (i.e., the first criteria). Then, we describe three FPT-approximation algorithms for d-Degeneracy Modulator two of which are inspired by similar algorithms known for Feedback Vertex Set [4,6]. The first one is randomized while the other two are deterministic. Here, for real numbers $p, q \geq 1$, a set $S \subseteq V(G)$ is called a (p, q)-approximation d-degeneracy modulator of G if $|S| \leq \mathrm{p} \cdot k$ and $\mathsf{degen}(G - S) \leq q \cdot d$. Our FPT algorithms have the following behavior: for $\alpha > 1$, if G has a k-sized d-degeneracy modulator, then they run in time $f(k, d, \alpha) \cdot n^{\mathcal{O}(1)}$, and either return a $(1, 2\alpha)$-approximation; or (correctly) report that there is no k-sized d-degeneracy modulator. Further, by setting $\alpha = 1 + \frac{1}{4d}$, we can convert them into $f(k,d) \cdot n^{\mathcal{O}(1)}$-time algorithms that either return a $(1,2)$-approximation; or (correctly) report that there is no k-sized d-degeneracy modulator. For the sake of brevity, by a slight abuse of notation, we say that these algorithms are $(1, 2\alpha)$- (resp. $(1,2)$-) approximations. We mention the specific properties of our FPT algorithms of such kind below.

- (Theorem 2) A *randomized* algorithm that runs in $\lceil \frac{2\alpha}{\alpha-1} \rceil^k n^{\mathcal{O}(1)}$ time, and is a $(1, 2\alpha)$-approximation with probability at least a positive constant.
- (Theorem 3) A *deterministic* $(1, 2\alpha)$-approximation algorithm that runs in $k^k \lceil \frac{2\alpha}{\alpha-1} \rceil^k n^{\mathcal{O}(1)}$ time.
- (Theorem 4) A *deterministic* $(1, 2\alpha)$-approximation algorithm that runs in $c^{k \lceil \frac{2\alpha}{\alpha-1} \rceil} n^{\mathcal{O}(1)}$ time for some constant $c > 1$.

Note that our $(1,2)$-bicriteria approximations can be viewed as the "dual" version of the approximation question studied by Demaine et al. [7] – while their (polynomial-time) algorithm treats the target degeneracy as a fixed constraint and approximates the size of the modulator, our (FPT) algorithm always returns a k-sized solution, while approximating the target degeneracy[1]. It is an interesting question to design a polynomial-time algorithm that outputs a k-sized modulator, while approximating the target degeneracy within a constant factor.

[1] Here we are ignoring the subtlety that, if there is no k-sized d-degeneracy modulator, then the algorithm may output No.

Finally, we remark that all our FPT algorithms generalize easily when the input graph is vertex-weighted and the objective is to obtain a d-degeneracy modulator of size at most k and weight at most W – our algorithms return a $(2\alpha d)$-degeneracy modulator of size at most k and weight at most W if the required solution exists. We only discuss the weighted case for the polynomial-time approximation (cf. Algorithm 1 and Theorem 1), and for the ease of presentation, we stick to the unweighted version in our remaining FPT algorithms.

Preliminaries. Throughout this paper, we will only work with non-negative weight functions and simple undirected graphs. For brevity, we will omit the qualifiers *non-negative* and *simple undirected.* We refer to the book by Diestel [8] for standard graph-theoretic definitions and terminology not defined here. For a graph G, let $V(G)$ and $E(G)$ denote its vertex set and edge set, respectively. The degree of a vertex v is denoted as $\mathsf{deg}(v)$. The input to all the algorithms consists of an undirected (weighted or unweighted) graph G and parameters d denoting the *target degeneracy* and $\alpha > 1$ that tunes the trade-off in the bicriteria approximation. The algorithms begin with simple preprocessing rules.

Preprocessing Rule 1. *If* $\mathsf{degen}(G) \leq 2\alpha d$*, then return* $\emptyset$ *as the solution.*

As all our algorithms are bicritieria approximation algorithms that focus on returning an $(\beta, 2\alpha d)$-approximation for some $\beta \geq 1$, this rule is safe to apply. The next rule executed by our algorithms is the following.

Preprocessing Rule 2. *If there is a vertex* u *in* G *with degree at most* $2\alpha d$*, then delete* u*.*

This preprocessing is safe due to the fact that if $G - u$ has a $2\alpha d$-degeneracy sequence, then appending u to it results in a $2\alpha d$-degeneracy sequence of G. We end the preliminaries section with an important observation. Since d-degenerate graphs on n vertices are *sparse*, it follows that "most" of the edges must be incident to a (fixed) optimal d-degeneracy modulator. This is formalized as Observation 1.

Observation 1. *Let* G *be a graph on* n *vertices and* m *edges, and let* d *be a non-negative integer. For any set* $Z \subseteq V(G)$ *such that* $\mathsf{degen}(G - Z) \leq d$*, the number of edges of* G *that have at least one endpoint in* Z *is at least* $m - dn$*.*

Due to space constraints, the proofs of some results are omitted. The proof of such results can be found in the full paper.

2 A Randomized Polynomial-Time Algorithm

In Algorithm 1, we give a randomized recursive algorithm ($\mathsf{DegenMod}\,1$) which takes as input a weighted graph G and returns a $2\alpha d$-degeneracy modulator with expected weight $\frac{2\alpha}{\alpha-1}$ times the optimal weight. In other words, in expectation, $\mathsf{DegenMod}\,1$ returns a $(\frac{2\alpha}{\alpha-1}, 2\alpha)$-bicriteria approximation.

Description of Algorithm 1. In the base case, if the degeneracy of the graph is already at most $2\alpha d$, then $\emptyset$ is a modulator, which is what we return in line 2 of Algorithm 1. Otherwise, if there is a vertex u of degree at most $2\alpha d$, then we can find a $2\alpha d$-degeneracy modulator Z' for $G - u$ and attach u at the end of the degeneracy sequence for $(G - u) - Z'$. This case is handled in the **else if** case in lines 3–4. Next, if the graph contains a vertex u with weight 0, then it can be included in the solution for free, hence we solve the problem recursively on the remaining graph, and add u to the modulator without increasing the weight. This is done in the **else if** case in lines 5–6. Finally, in the **else** case (lines 7–11), we are in the situation where all the vertices have degree more than $2\alpha d$ and have positive weights. However, since d-degenerate graphs on n vertices contain at most dn edges, it follows that most of the edges must be incident to a (fixed) minimum-weight d-degeneracy modulator (Observation 1). Thus, if an edge $e = uv$ is picked uniformly at random in line 8, with good probability, either u or v belongs to the optimal solution. However, even if this *good event* happens, we do not know whether u or v is the "correct choice". Since we aim for a polynomial-time algorithm, we cannot branch on both choices. However, conditioned on the good event, we know that, in either case, the "correct" vertex has weight at least $\min\{w(u), w(v)\}$. Thus, we "pay" for this lower bound, and subtract it from both of the weights, leading to the specific choice of the *new* weight function w' as defined in line 10. In line 11, we make a recursive call on the same graph, but with respect to the new weight function w' – note that at least one of the two endpoints of the selected edge will have zero weight according to w', and will be included in the solution returned by the recursive call. The intuition about the weight-contribution can be formalized using the local ratio technique [3] to show a formal guarantee on the expected weight of the solution, which is done in Lemma 1.

Algorithm 1. $\mathsf{DegenMod\,1}(G, w, d, \alpha)$

Input: A graph G associated with a weighting w, a positive integer d and a real $\alpha > 1$.
Output: A set $Z \subseteq V(G)$ such that $\mathsf{degen}(G - Z) \leq 2\alpha d$.

1: **if** $\mathsf{degen}(G) \leq 2\alpha d$ **then**
2:     **return** $\emptyset$
3: **else if** there is a vertex $u \in V(G)$ with degree at most $2\alpha d$ **then**
4:     **return** $\mathsf{DegenMod\,1}(G - u, w', d, \alpha)$ ▷ w' is w restricted to $V - u$
5: **else if** there is a vertex $u \in V(G)$ with $w(u) = 0$ **then**
6:     **return** $\mathsf{DegenMod\,1}(G - u, w', d, \alpha) \cup \{u\}$ ▷ w' is w restricted to $V - u$
7: **else** ▷ $\delta(G) > 2\alpha d$ and $w(u) > 0$ for all $u \in V(G)$
8:     Choose $e \in_R E(G)$ and say $e = uv$ ▷ Choose an edge uniformly at random
9:     $\mathsf{W} := \min\{w(u), w(v)\}$
10:     Define $w' : V(G) \to \mathbb{R}^{\geq 0}$ as follows, $w'(y) = \begin{cases} w(y) & \text{if } y \in V(G) \setminus \{u, v\} \\ w(y) - \mathsf{W} & \text{if } y \in \{u, v\} \end{cases}$
11:     **return** $\mathsf{DegenMod\,1}(G, w', d, \alpha)$

Lemma 1. *Let* $d \in \mathbb{N}^+$, $\alpha > 1$. *Given an undirected graph* G, *and a non-negative weight function* w, $\mathsf{DegenMod\,1}(G, w, d, \alpha)$ *is a randomized algorithm that runs in polynomial time and returns a set* $Z \subseteq V(G)$ *such that* $\mathsf{degen}(G - Z) \leq 2\alpha d$ *with probability* 1, *and* $\mathbb{E}[w(Z)] \leq \frac{2\alpha}{\alpha-1} \cdot \mathsf{opt}(G, w)$.

Proof. We define the size of the support of w as the number of vertices of G assigned positive weights by w, and denote it by $\mathsf{supp}_G(w)$. Note that $\mathsf{supp}_G(w)$ is a non-negative integer. We will prove the claim by induction on $\mu(G, w) := |V(G)| + \mathsf{supp}_G(w)$ (which is a non-negative integer). Let $\beta := \frac{2\alpha}{\alpha-1}$. First, for the base case, suppose $\mu(G, w) \leq 2\alpha d$. Clearly, this implies that $\mathsf{degen}(G) \leq |V(G)| \leq 2\alpha d$. In line 2, we return $\emptyset$. In this case, $w(\emptyset) = 0 \leq \beta \cdot \mathsf{opt}(G, w)$, since w is non-negative. Now suppose the claim is inductively true. That is, for any graph G' and any weight function $w' : V(G') \to \mathbb{R}^{\geq 0}$ with $\mu(G', w') \leq k$, if $Z' \subseteq V(G')$ is a random subset returned by $\mathsf{DegenMod\,1}(G', w', d, \alpha)$, then $\mathbb{E}[w'(Z')] \leq \beta \cdot \mathsf{opt}(G', w')$. Now consider a graph G on n vertices, and a weight function $w : V(G) \to \mathbb{R}^{\geq 0}$, with $\mu(G) = k + 1$. As in the algorithm, we consider different sub-cases to complete the inductive proof.

Consider the **else if** case corresponding to line 3, i.e., there is a vertex $u \in V(G)$ with degree at most $2\alpha d$ in G. Note that $\mathsf{opt}(G', w') \leq \mathsf{opt}(G, w)$ since w' is w restricted to $V(G) - \{u\}$. Note that $\mu(G', w') \leq \mu(G, w) - 1$ and we can use induction since in line 4, we return the output, say Z' of $\mathsf{DegenMod\,1}(G - u, w', d, \alpha)$. Then, $\mathbb{E}[w(Z')] = \mathbb{E}[w'(Z')] \leq \beta \cdot \mathsf{opt}(G - u, w') \leq \beta \cdot \mathsf{opt}(G, w)$, where the first inequality follows from induction. Further, by induction $\mathsf{degen}(G' - Z') \leq 2\alpha d$, and a $(2\alpha d)$-degeneracy sequence of $G' - Z'$ can be extended to $G - Z'$ by attaching u at the end of the sequence. Since the degree of u in G is at most $2\alpha d$, this is a valid $2\alpha d$-degeneracy sequence for $G - Z'$, showing that $\mathsf{degen}(G - Z') \leq 2\alpha d$, completing the induction in this case. Now, consider the **else if** case in line 5. Suppose there is a vertex $u \in V(G)$ with $w(u) = 0$. Then, let $G' = G - u$, and consider the set $Z' \subseteq V(G) - \{u\}$ returned by $\mathsf{DegenMod\,1}(G', w', d, \alpha)$ in line 6. Note that $\mu(G', w') = \mathsf{supp}_{G'}(w') + |V(G')| \leq \mathsf{supp}_G(w) + |V(G)| - 1$, since $\mathsf{supp}_{G'}(w') \leq \mathsf{supp}_G(w)$. Thus, by induction, $\mathsf{degen}(G' - Z') \leq 2\alpha d$, and $\mathbb{E}[w(Z' \cup \{u\})] = \mathbb{E}[w'(Z')] \leq \beta \cdot \mathsf{opt}(G', w') \leq \beta \cdot \mathsf{opt}(G, w)$.

Lastly, consider the **else** case corresponding to line 7. As remarked, every vertex in G has degree greater than $2\alpha d$, which implies that $m = |E(G)| > \alpha dn$. Further, every vertex has strictly positive weight according to w. Now, fix an optimal-weight d-degeneracy modulator $X \subseteq V(G)$, i.e., X satisfying (i) $w(X) = \mathsf{opt}(G, w)$, and (ii) $\mathsf{degen}(G - X) \leq d$. We say that an edge is *good* if it has at least one endpoint in X, and *bad* otherwise. By Observation 1, the number of good edges is at least $m - dn$. Let p denote the probability that an edge $e = uv$ chosen in line 8 uniformly at random from $E(G)$ is a good edge. Then, $p \geq \frac{m-dn}{m} = 1 - \frac{dn}{m} > 1 - \frac{dn}{\alpha dn} = \frac{\alpha-1}{\alpha}$. Let $e = uv$ be the random edge chosen in line 8. Let us decompose w into two weight functions $w_1, w_2 : V(G) \to \mathbb{R}^{\geq 0}$. Let $w_1 := w'$ as defined in line 10, and w_2 is defined as follows. For any $y \in \{u, v\}$, $w_2(y) = \mathsf{W}$, and for any $y \in V(G) \setminus \{u, v\}$, $w_2(y) = 0$. Clearly, w_1 and w_2 are non-negative, and it holds that $w = w_1 + w_2$. Note that w_1 and w_2 are random variables since they are defined via the edge e chosen at random in line 8. Let Z denote the solution returned by $\mathsf{DegenMod\,1}(G, w', d, \alpha)$, which we return as the solution in line 11. Note that $\mathsf{supp}_G(w') = \mathsf{supp}_G(w_1) \leq \mathsf{supp}_G(w) - 1$, since at least one of u or v has weight equal to 0 according to w'; and the rest of the weights are unchanged. Therefore, $\mu(G, w') = \mathsf{supp}_G(w') + |V(G)| \leq \mathsf{supp}_G(w) - 1 + |V(G)|$. Hence, we can apply the induction hypothesis. Let X_1 be an optimal solution w.r.t. w_1 (note that X_1 is a random variable since w_1 is random). Therefore, $\mathbb{E}[w_1(X_1)] \leq \mathbb{E}[w_1(X)]$. Then,

$$\mathbb{E}[w_1(Z)] \leq \beta \cdot \mathbb{E}[w_1(X_1)] \leq \beta \cdot \mathbb{E}[w_1(X)]. \tag{1}$$

By inductive hypothesis, we also obtain that $\mathsf{degen}(G - Z) \leq 2\alpha d$. Now let us analyze the expected weight of X w.r.t. w_2.

$$\begin{aligned}
\mathbb{E}[w_2(X)] &= \Pr(e \text{ is good}) \cdot \mathbb{E}[w_2(X)|e \text{ is good}] + \Pr(e \text{ is bad}) \cdot \mathbb{E}[w_2(X)|e \text{ is bad}] \\
&\geq p \cdot w(x) + (1-p) \cdot 0 \\
&\qquad\qquad (\text{If } e \text{ is good, then } 1 \leq |e \cap X| \leq 2\text{; otherwise, } |e \cap X| = 0.) \\
&\geq \tfrac{\alpha-1}{\alpha} \cdot w(x) \qquad (2)
\end{aligned}$$

Now we can show the required upper bound on the expected weight of Z.

$$\begin{aligned}
\mathbb{E}[w(Z)] &= \mathbb{E}[w_1(Z)] + \mathbb{E}[w_2(Z)] \\
&\qquad (\text{By linearity of expectations and since } w = w_1 + w_2) \\
&\leq \beta \cdot \mathbb{E}[w_1(X)] + \mathbb{E}[w_2(Z)] \qquad (\text{From (1)}) \\
&\leq \beta \cdot \mathbb{E}[w_1(X)] + 2 \cdot w(x) \quad (\mathsf{supp}_G(w_2) = 2 \text{ with } w_2(u) = w_2(v) = \mathsf{W}) \\
&\leq \beta \cdot \mathbb{E}[w_1(X)] + \tfrac{2\alpha}{\alpha-1} \cdot \mathbb{E}[w_2(X)] \qquad (\text{From (2)}) \\
&= \beta \cdot \mathbb{E}[w(X)] \qquad (\text{By linearity of expectations}) \\
&= \beta \cdot w(X) \quad (\text{Since the last expression involves no random variable})
\end{aligned}$$

Finally, we note that, in each recursive call, the size of the support of the weight function decreases by at least 1 (with probability 1), which implies that the depth of the recursion tree is at most n. It follows that the running time of the algorithm is polynomial in n, with probability 1.

From Expected Weight to Good Probability Bound. In Lemma 1, we showed an upper bound on the expected weight of the output of Algorithm 1. In order to obtain an algorithm that outputs a solution that does not deviate too far from expectation, one can repeat the algorithm on the original input multiple times, and select the minimum-cost solution.

Theorem 1. *Let $\delta \in (0,1)$ be a constant. For any $d \in \mathbb{N}^+$ and $\alpha > 1$,* Minimum-Weight d-Degeneracy Modulator *admits a randomized polynomial-time algorithm that, with probability at least $1 - \frac{1}{n}$, returns a $((1+\delta)\frac{2\alpha}{\alpha-1}, 2\alpha)$ bicriteria approximation.*

By plugging in $\alpha = (1 + \frac{1}{4d})$ and using the fact that the largest integer no larger than $2d(1 + \frac{1}{4d})$ is $2d$, we obtain the following corollary.

Corollary 1. *Let $\delta \in (0, 1)$ be a constant. For any $d \in \mathbb{N}^+$,* MINIMUM-WEIGHT *d-*DEGENERACY MODULATOR *admits a $((8d + 2)(1 + \delta), 2)$-bicriteria approximation.*

3 FPT Algorithms

In this section, we describe three FPT algorithms for d-DEGENERACY MODULATOR on unweighted graphs. The algorithms begin with preliminary checks based on Preprocessing Rules 1 and 2. Subsequently, the algorithms assume $\mathsf{degen}(G) > 2\alpha d$ and $m > \alpha d n$.

3.1 A Randomized FPT Algorithm

Suppose G has a set S of size at most k such that $\mathsf{degen}(G - S) \leq d$. Then, by Observation 1, most of the edges must be incident to S. Thus, if an edge $e = uv$ is picked uniformly at random, with good probability, either u or v belongs to S. Further, with probability at least half, a randomly chosen vertex among u and v is in S. This observation is the key to Algorithm 2. This and subsequent algorithms return $\perp$ to denote that there is no d-degeneracy modulator of size at most k.

Algorithm 2. DegenMod 2(G, k, d, α)

Input: An undirected graph G, $k \in \mathbb{N}$, $d \in \mathbb{N}^+$, and a real $\alpha > 1$.
Output: Either "No d-degeneracy modulator of size k" or a set $Z \subseteq V(G)$ of size at most k, such that $\mathsf{degen}(G - Z) \leq 2\alpha d$.

if $k < 0$ **then**
  return $\perp$ ▷ `No d-degeneracy modulator of size k`
else if $|V(G)| \leq 2\alpha d$ **then**
  return $\emptyset$
else if there is a vertex $u \in V(G)$ with degree at most $2\alpha d$ **then**
  return DegenMod 2$(G - u, k, d, \alpha)$
else
  Choose $e \in_R E(G)$ and say $e = uv$ ▷ `Choose a uniformly random edge`
  Choose $x \in_R \{u, v\}$ ▷ `Choose a uniformly random endpoint`
  $Z' \leftarrow$ DegenMod 2$(G - x, k - 1, d, \alpha)$
  if $Z' \neq \perp$ **then**
    return $Z' \cup \{x\}$
  else
    return $\perp$ ▷ `No d-degeneracy modulator of size k`

Lemma 2. *Let $k \geq 0$, $\alpha > 1$ and $d \in \mathbb{N}^+$. For any graph G, if there exists a set $X \subseteq V(G)$ of size at most k, such that* $\mathsf{degen}(G-X) \leq d$*, then with probability at least γ^k* $\mathsf{DegenMod}\,2(G,k,d,\alpha)$ *returns a $(1, 2\alpha)$-approximation, where $\gamma = \frac{\alpha-1}{2\alpha}$. That is, with probability at least γ^k, it returns a set Z of size at most k such that,* $\mathsf{degen}(G-Z) \leq 2\alpha d$.

Proof. We prove the lemma by induction on the number of vertices. In the base case, if $|V(G)| \leq 2\alpha d$, then $\mathsf{degen}(G) \leq |V(G)| \leq 2\alpha d$, and in line 4 the algorithm returns $\emptyset$, showing that the claim holds with probability 1. Now suppose the claim is inductively true for all graphs G' with $|V(G')| \leq n-1$, for some $n \geq 1$, and we want to show the claim for a graph G with $|V(G)| = n$.

First, suppose that there exists a vertex $u \in V(G)$ with degree at most $2\alpha d$. Further, assume that there exists a set $X \subseteq V(G)$ of size at most k such that $\mathsf{degen}(G-X) \leq d$. In line 6 we make a recursive call on the graph $G' := G - u$, and the parameter k remains unchanged. Note that, $|V(G')| = n - 1$. In the graph G', there is a set $X' := X \setminus \{u\} \subseteq V(G')$ of size at most k such that $\mathsf{degen}(G'-X) \leq d$. Therefore, by induction, with probability at least γ^k, $\mathsf{DegenMod}\,2(G-u,k,d,\alpha)$ returns a solution $Z \subseteq V(G) \setminus \{u\}$ such that $\mathsf{degen}(G'-Z) \leq 2\alpha d$, and $|Z| \leq k$. By appending u at the end of the $2\alpha d$-degeneracy ordering of $V(G') \setminus Z$, we obtain a $2\alpha d$-degeneracy ordering of $V(G) \setminus Z$, showing that $\mathsf{degen}(G-Z) \leq 2\alpha d$. Thus, with probability at least γ^k, in line 6, the algorithm returns a set with the claimed properties.

Finally, consider the case when $|V(G)| = n$, and there is no vertex with degree at most $2\alpha d$ in G. Thus, we are in the **else** case of line 7. Suppose there exists a set $X \subseteq V(G)$ of size at most k such that $\mathsf{degen}(G-X) \leq d$. Fix this X, and call any edge with at least one endpoint in X a *good* edge, and *bad* otherwise. By Observation 1, the number of good edges is at least $m - dn \geq \alpha dn - dn = (\alpha-1)dn$. Thus, the probability that the edge e chosen in line 8 is a good edge is at least $\frac{\alpha-1}{\alpha}$. So far, this analysis is similar to the proof of Lemma 1. Conditioned on the event that the edge e chosen in line 8 is a good edge, the probability that a random endpoint x chosen in line 9 belongs to X is at least $1/2$ (note that a good edge may have both of its endpoints in X). Then, conditioned on the event that e is a good edge, with probability at least $\frac{\alpha-1}{\alpha} \cdot \frac{1}{2} = \gamma$, the graph $G' = G - x$ contains a set $X' = X \setminus \{x\}$ of size at most $k-1$, such that $\mathsf{degen}(G'-X') \leq d$. Since $|V(G')| = n-1$, by induction hypothesis, with probability at least γ^{k-1}, the recursive call $\mathsf{DegenMod}\,2(G', k-1)$ returns a solution Z' of size at most $k-1$ such that $\mathsf{degen}(G'-Z') \leq 2\alpha d$. In particular, conditioned on the earlier event, the set Z' is not $\perp$ with probability at least γ^{k-1}. Putting everything together, with probability at least $\gamma \cdot \gamma^{k-1} = \gamma^k$, the algorithm will return in line 12 the set $Z' \cup \{x\}$ satisfying that $\mathsf{degen}(G-Z') = \mathsf{degen}(G'-Z') \leq 2\alpha d$. This completes the proof. □

By repeating the algorithm $\lceil \frac{1}{\gamma} \rceil^k$ times, we can boost the success probability of the algorithm to be at least a positive constant. Thus, we obtain the following theorem.

Theorem 2. *For any $d \in \mathbb{N}^+$ and $\alpha > 1$, there is a randomized algorithm that runs in time $\lceil \frac{2\alpha}{\alpha-1} \rceil^k \cdot n^{\mathcal{O}(1)}$, and with probability at least a positive constant, returns a $(1, 2\alpha)$-approximation. In particular, for $\alpha = 1 + \frac{1}{4d}$, the algorithm runs in $(8d+2)^k \cdot n^{\mathcal{O}(1)}$ time and returns a $(1,2)$-approximate solution with probability at least a positive constant.*

3.2 A Deterministic FPT Algorithm

In this subsection, we give a deterministic FPT algorithm that returns a $(1, 2\alpha)$-approximate degeneracy modulator in $k^k \lceil \frac{2\alpha}{\alpha-1} \rceil^k n^{\mathcal{O}(1)}$ time. After applying Preprocessing Rules 1 and 2, we may assume $\mathsf{degen}(G) > 2\alpha d$. Also, for each $v \in V(G)$, we have $d(v) > 2\alpha d$ and it follows that $m > \alpha dn$. Let $v_1, v_2, \ldots, v_n$ be an ordering of vertices of G such that $d(v_1) \geq d(v_2) \geq \ldots \geq d(v_n)$. Let $\ell = \lceil \frac{2\alpha}{\alpha-1} \rceil$. Observe that $\ell > 2$ when $\alpha > 1$. Let $V_{k\ell} = \{v_1, \ldots, v_{k\ell}\}$.

Claim 1. *Any set $S \subseteq V(G)$ of size at most k such that $\mathsf{degen}(G - S) \leq d$ satisfies $S \cap V_{k\ell} \neq \emptyset$.*

Proof. As $G - S$ has at most dn edges, we have $\sum_{v \in S} d(v) \geq m - dn$. If $S \cap V_{k\ell} = \emptyset$ then, we have $2m \geq \sum_{i=1}^{k\ell} d(v_i) \geq \ell \sum_{v \in S} d(v) \geq \ell(m - dn)$. Hence, $2m \geq \ell(m - dn)$ implying that $m \leq \frac{\ell}{\ell-2} dn$. Since $m > \alpha dn$, we get $\alpha < \frac{\ell}{\ell-2}$. However, as $\ell \geq \frac{2\alpha}{\alpha-1}$, we have $\frac{\ell-2}{\ell} = 1 - \frac{2}{\ell} \geq \frac{1}{\alpha}$, this leads to a contradiction. □

Using Claim 1, we obtain a branching algorithm described as Algorithm 3. The correctness of the algorithm is formalized as Lemma 3.

Algorithm 3. $\mathsf{DegenMod\,3}(G, d, k, \alpha)$

Input: An undirected graph G, $k \in \mathbb{N}$, $d \in \mathbb{N}^+$, and a real $\alpha > 1$.
Output: Either "No d-degeneracy modulator of size k" or a set $Z \subseteq V(G)$ of size at most k, such that $\mathsf{degen}(G - Z) \leq 2\alpha d$.

if $k < 0$ **then**
  return $\bot$ ▷ No d-degeneracy modulator of size k
else if $\mathsf{degen}(G) \leq 2\alpha d$ **then**
  return $\emptyset$
else if there is a vertex $u \in V(G)$ with degree at most $2\alpha d$ **then**
  return $\mathsf{DegenMod\,3}(G - u, k, d, \alpha)$
else
  Let $v_1, v_2, \ldots, v_n$ be the vertices of G such that $d(v_1) \geq d(v_2) \geq \ldots \geq d(v_n)$.
  for each $i = 1$ to $\min\{k\ell, n\}$ **do** ▷ $\ell = \lceil \frac{2\alpha}{\alpha-1} \rceil$
    $Z' = \mathsf{DegenMod\,3}(G - v_i, d, k-1, \alpha)$
    if $Z' \neq \bot$ **then**
      return $Z' \cup \{v_i\}$
  return $\bot$ ▷ No d-degeneracy modulator of size k

Lemma 3. *Let $k \geq 0$, $\alpha > 1$ and $d \in \mathbb{N}$. For any graph G, if there exists a set $S \subseteq V(G)$ of size at most k, such that* $\mathsf{degen}(G - S) \leq d$*, then* DegenMod 3$(G, d, k, \alpha)$ *returns a set Z of size at most k such that* $\mathsf{degen}(G-Z) \leq 2\alpha d$ *in $k^k \lceil \frac{2\alpha}{\alpha-1} \rceil^k n^{\mathcal{O}(1)}$ time.*

Now, we have the following result.

Theorem 3. *For any $d \in \mathbb{N}^+$ and $\alpha > 1$, there is an* FPT *algorithm that runs in $k^k \lceil \frac{2\alpha}{\alpha-1} \rceil^k n^{\mathcal{O}(1)}$ time and returns an $(1, 2\alpha)$-approximation degeneracy modulator (if one exists). In particular, for $\alpha = 1 + \frac{1}{4d}$, the algorithm that runs in $k^k \cdot (8d+2)^k \cdot n^{\mathcal{O}(1)}$ time and returns a $(1, 2)$-bicriteria approximation.*

3.3 A Faster Deterministic FPT Algorithm

Next, we describe a faster deterministic algorithm that returns a degeneracy modulator which is a $(1, 2\alpha)$-approximation in $c^{k \lceil \frac{2\alpha}{\alpha-1} \rceil} n^{\mathcal{O}(1)}$ time where $c > 1$ is a constant. Once again, we first apply Preprocessing Rules 1 and 2. Then, we have $\mathsf{degen}(G) > 2\alpha d$ and for each $v \in V(G)$, $\mathsf{deg}(v) > 2\alpha d$ implying that $m > \alpha d n$. Now, we partition the vertex set of G into $\lceil \log n \rceil$ sets that we refer to as bins. This notion of bins was defined and used by Fomin et al. to solve Planar $\mathcal{F}$-Deletion [9]. The key intention behind this partition is to identify a sufficiently large set of vertices that are guaranteed to be in the solution that we seek (if one exists).

Suppose G has a set S of size at most k such that $\mathsf{degen}(G - S) \leq d$. We argue that there is a large enough bin that contains a large number of vertices that are in S. Then, we obtain a simple branching algorithm. Now, we describe the bins. For each $i \geq 1$, $B_i = \{v \in V(G) \mid \frac{n}{2^i} < \mathsf{deg}(v) \leq \frac{n}{2^{i-1}}\}$. As $G - S$ has at most dn edges, it follows that $\sum_{v \in S} \mathsf{deg}(v) \geq m - dn > \alpha dn - dn = (\alpha - 1)dn$. For $0 < t, r < 1$ to be chosen later, we call a bin B_i *nice* if $|B_i \cap S| \geq t|B_i|$, and a bin B_j *large* if $|B_j| \geq r \cdot j$.

Claim 2. *There is a large bin B_i that is nice.*

Proof. Let J denote the indices of bins that are not large and I denote the indices of bins that are not nice. Suppose no large bin is nice. Then, consider:

$$\sum_{v \in S} \mathsf{deg}(v) = \sum_{i=1}^{\lceil \log n \rceil} \sum_{v \in B_i \cap S} \mathsf{deg}(v) \leq \sum_{i \in I} \sum_{v \in B_i \cap S} \mathsf{deg}(v) + \sum_{i \in J} \sum_{v \in B_i \cap S} \mathsf{deg}(v) \quad (3)$$

Here, the last inequality follows from the fact that, since no large bin is nice, $I \cup J$ is equal to the set of all indices, $\{1, 2, \ldots, \lceil \log n \rceil\}$. Now, for a bin B_i that is not nice, we have that,

$$\sum_{v \in B_i \cap S} \mathsf{deg}(v) < t \cdot |B_i| \cdot \frac{n}{2^{i-1}} \leq 2t \cdot \sum_{v \in B_i} \mathsf{deg}(v). \quad (4)$$

On the other hand, for a bin that is not large, we have,

$$\sum_{v \in B_i \cap S} \mathsf{deg}(v) \leq |B_i| \cdot \frac{n}{2^{i-1}} < r \cdot i \cdot \frac{n}{2^{i-1}}. \tag{5}$$

Using (3), (4), and (5), we obtain

$$\sum_{v \in S} \mathsf{deg}(v) \leq 2t \cdot \sum_{i \in I} \sum_{v \in B_i} \mathsf{deg}(v) + 2rn \cdot \sum_{i \in J} \frac{i}{2^i} \leq 4tm + 2rn \cdot \sum_{i=1}^{\infty} \frac{i}{2^i} \leq 4tm + 4rn$$

Let $\rho = t + r\frac{1}{\alpha d}$. Since $m > \alpha dn$ and $\sum_{v \in S} \mathsf{deg}(v) \geq m - dn > \frac{\alpha - 1}{\alpha} m$, we obtain

$$m \cdot \frac{\alpha - 1}{\alpha} < m - dn \leq \sum_{v \in S} \mathsf{deg}(v) < 2m(2t + 2r\frac{1}{\alpha d}) = 4m\rho \tag{6}$$

If we choose t and r such that $\rho \leq \frac{\alpha-1}{4\alpha}$, we get a contradiction, since (6) will then imply $\frac{\alpha-1}{\alpha} m < \frac{\alpha-1}{\alpha} m$. Thus, our assumption that there is no large bin that is nice must be wrong. Hence, there is a bin that is both large and nice. Setting $r = \frac{1}{100}$ and $t = \frac{\alpha-1}{4\alpha} - \frac{r}{\alpha d}$ (note that $t > 0$) guarantees the properties that we want for $\alpha > 1$ and $d \geq 1$. □

Using Claim 2, we obtain the following branching algorithm.

Algorithm 4. DegenMod 4(G, d, k, α)

Input: An undirected graph G, $k \in \mathbb{N}$, $d \in \mathbb{N}^+$, and a real $\alpha > 1$.
Output: Either "No d-degeneracy modulator of size k" or a set $Z \subseteq V(G)$ of size at most k, such that $\mathsf{degen}(G - Z) \leq 2\alpha d$.

if $k < 0$ **then**
  return $\bot$ ▷ No d-degeneracy modulator of size k
else if $\mathsf{degen}(G) \leq 2\alpha d$ **then**
  return $\emptyset$
else if there is a vertex $u \in V(G)$ with degree at most $2\alpha d$ **then**
  return DegenMod 4$(G - u, k, d, \alpha)$
else
  Partition $V(G)$ into bins $B_1, \ldots, B_{\lceil \log n \rceil}$ where $B_i = \{v \in V(G) \mid \frac{n}{2^i} < \mathsf{deg}(v) \leq \frac{n}{2^{i-1}}\}$.
  for each bin B_i with $|B_i| \geq r \cdot i$ **do**
    for each set $X \subseteq B_i$ with $|X| = \lceil t|B_i| \rceil$ **do**
      $Z' =$ DegenMod 4$(G - X, d, k - |X|, \alpha)$
      if $Z' \neq \bot$ **then**
        return $Z' \cup X$
  return $\bot$ ▷ No d-degeneracy modulator of size k

Lemma 4. *Let $k \geq 0$, $\alpha > 1$ and $d \in \mathbb{N}^+$. For any graph G, if there exists a set $S \subseteq V(G)$ of size at most k, such that* $\mathsf{degen}(G - S) \leq d$, *then* $\mathsf{DegenMod}\,4(G, d, k, \alpha)$ *returns a set Z of size at most k such that* $\mathsf{degen}(G - Z) \leq 2\alpha d$ *in $c^{k\lceil \frac{2\alpha}{\alpha-1} \rceil} n^{\mathcal{O}(1)}$ time where $c > 1$ is an absolute constant.*

This leads to the following result.

Theorem 4. *For any $d \in \mathbb{N}^+$ and $\alpha > 1$, there is an FPT algorithm that returns an $(1, 2\alpha)$-approximation degeneracy modulator (if one exists) in $c^{k\lceil \frac{2\alpha}{\alpha-1} \rceil} n^{\mathcal{O}(1)}$ time where $c > 1$ is an absolute constant. In particular, for $\alpha = 1 + \frac{1}{4d}$, the algorithm runs in $2^{\mathcal{O}(dk)} \cdot n^{\mathcal{O}(1)}$ time and returns a $(1, 2)$-bicriteria approximation.*

4 Concluding Remarks

As the reader would have noted, our deterministic algorithms return an $(1, 2)$-approximation in FPT time. A natural question is whether one can design an FPT $(1, 2 - \epsilon)$-approximation algorithm for any $0 < \epsilon \leq 1$. For $\epsilon = \frac{1}{2}$, such an algorithm will solve 2-Degeneracy Modulator on 3-degenerate graphs in FPT time. However, as mentioned in the introduction, 2-Degeneracy Modulator is known to be W[P]-complete even if the input graph has degeneracy 3. Yet, this does not rule out the existence of an FPT $(1, 1.5 + \epsilon)$-approximation algorithm for some $0 < \epsilon < \frac{1}{2}$. Improving the approximation ratio in this direction or proving tight lower bounds is a natural future direction.

Another natural direction is the edge-deletion version of the d-Degeneracy Modulator. Formally, given a graph G, and two non-negative integers k and d, we want to determine whether one can delete at most k edges such that the resulting graph has degeneracy at most d. This problem is known to be W[P]-hard parameterized by k [12]. We give a $(1, 2 - \frac{1}{d})$-bicriteria approximation via looking at arboricity. The details are deferred to the full version of the paper.

Acknowledgements. T. Inamdar acknowledges support from IITJ Research Initiation Grant (grant number I/RIG/TNI/20240072), and the European Research Council (ERC) under the European Union's Horizon 2020 research and innovation programme (grant agreement No. 819416). S. Saurabh acknowledges the support from the European Research Council (ERC) under the European Union's Horizon 2020 research and innovation programme (grant agreement No. 819416), and Swarnajayanti Fellowship grant DST/SJF/MSA-01/2017-18.

References

1. Abrahamson, K.R., Downey, R.G., Fellows, M.R.: Fixed-parameter tractability and completeness IV: on completeness for W[P] and PSPACE analogues. Ann. Pure Appl. Log. **73**(3), 235–276 (1995). https://doi.org/10.1016/0168-0072(94)00034-Z
2. Alon, N., Gutner, S.: Linear time algorithms for finding a dominating set of fixed size in degenerated graphs. Algorithmica **54**(4), 544–556 (2009). https://doi.org/10.1007/S00453-008-9204-0

3. Bar-Yehuda, R., Bendel, K., Freund, A., Rawitz, D.: Local ratio: a unified framework for approximation algorithms. In memoriam: Shimon even 1935-2004. ACM Comput. Surv. **36**(4), 422–463 (2004). https://doi.org/10.1145/1041680.1041683
4. Becker, A., Bar-Yehuda, R., Geiger, D.: Randomized algorithms for the loop cutset problem. J. Artif. Intell. Res. **12**, 219–234 (2000). https://doi.org/10.1613/JAIR.638
5. Cai, L., Kloks, T.: Parameterized tractability of some (efficient) y-domination variants for planar graphs and t-degenerate graphs. In: Proceedings International Computer Symposium 2000 (2000)
6. Cygan, M., et al.: Parameterized Algorithms. Springer, Cham (2015). https://doi.org/10.1007/978-3-319-21275-3
7. Demaine, E.D., et al.: Structural rounding: approximation algorithms for graphs near an algorithmically tractable class. In: 27th Annual European Symposium on Algorithms, ESA 2019. LIPIcs, vol. 144, pp. 37:1–37:15. Leibniz-Zentrum für Informatik (2019). https://doi.org/10.4230/LIPIcs.ESA.2019.37
8. Diestel, R.: Graph Theory, Graduate Texts in Mathematics, 4th edn., vol. 173. Springer (2012)
9. Fomin, F.V., Lokshtanov, D., Misra, N., Saurabh, S.: Planar f-deletion: approximation, kernelization and optimal FPT algorithms. In: 53rd Annual IEEE Symposium on Foundations of Computer Science, FOCS 2012, pp. 470–479. IEEE Computer Society (2012). https://doi.org/10.1109/FOCS.2012.62
10. Golovach, P.A., Villanger, Y.: Parameterized complexity for domination problems on degenerate graphs. In: Broersma, H., Erlebach, T., Friedetzky, T., Paulusma, D. (eds.) WG 2008. LNCS, vol. 5344, pp. 195–205. Springer, Heidelberg (2008). https://doi.org/10.1007/978-3-540-92248-3_18
11. Luo, J., Molter, H., Suchý, O.: A parameterized complexity view on collapsing k-cores. Theory Comput. Syst. **65**(8), 1243–1282 (2021). https://doi.org/10.1007/S00224-021-10045-W
12. Mathieson, L.: The parameterized complexity of editing graphs for bounded degeneracy. Theor. Comput. Sci. **411**(34–36), 3181–3187 (2010). https://doi.org/10.1016/j.tcs.2010.05.015
13. Matula, D.W., Beck, L.L.: Smallest-last ordering and clustering and graph coloring algorithms. J. ACM **30**(3), 417–427 (1983). https://doi.org/10.1145/2402.322385
14. Pilipczuk, M., Pilipczuk, M.: Finding a maximum induced degenerate subgraph faster than 2^n. In: Thilikos, D.M., Woeginger, G.J. (eds.) IPEC 2012. LNCS, vol. 7535, pp. 3–12. Springer, Heidelberg (2012). https://doi.org/10.1007/978-3-642-33293-7_3

Optimal Random Bit Complexity in Efficient Sampling of Set Partition-Like Structures

Olivier Bodini[1] and Francis Durand[1,2](✉)

[1] EREN, USPN, Bobigny, France
olivier.bodini@univ-paris13.fr, francis.durand@ens-paris-saclay.fr
[2] LIP6, Sorbonne Université, Paris, France

Abstract. In this paper (This paper is eligible for the best student paper award), we present an algorithm designed to generate set partitions efficiently while adhering to the Shannon entropy principle, achieving an optimal time complexity of $O(n \log(n))$. Building on the framework of Stam (1983), for which we propose a combinatorial interpretation, we improve its efficiency by integrating the method developed by Devroye (1987) for generating an appropriate random variable. A significant portion of our algorithm's execution time—up to 80%—is dedicated to random generation, highlighting the importance of considering random bit consumption. Our method extends to other combinatorial structures, such as weak orderings and linearly ordered set partitions. We implemented the algorithm in Python and conducted experiments to evaluate its practical efficiency, confirming its scalability even for large input sizes.

Keywords: Random sampling · Combinatorics · Set partition · Entropy-efficient · Optimal randomness complexity

1 Introduction

Combinatorial structures play a pivotal role across a wide range of fields, from fundamental computer science and mathematics to bioinformatics and network analysis. For instance, among numerous examples, in computer science, they are crucial for data structure design and optimization algorithms [11], while in bioinformatics, they aid in modeling complex molecular structures and genetic sequences [22].

The ability to generate these structures randomly is key to understanding their properties and analyzing real-world phenomena. In network analysis, random generation of graphs helps in studying the robustness of networks against failures [1]. Similarly, in algorithm testing, generating random inputs can provide insights into performance and efficiency under various scenarios [21].

Despite the existence of numerous time-efficient random generation algorithms, optimizing them in terms of randomness complexity remains a significant

H. Fernau and B. Zhu (Eds.): IWOCA 2025, LNCS 15885, pp. 405–417, 2025.
https://doi.org/10.1007/978-3-031-98740-3_29

challenge. In this context, Shannon's entropy principle [24] and its combinatorial interpretation by Knuth [18] dictate that at least $\log_2(K)$ random bits are required to generate uniformly an element from a set of size K, implying a minimum generation time of $\log_2(K)$. Classical methods, such as the recursive and Boltzmann methods [5,10], may be sub-optimal in terms of entropic efficiency. For instance, generating a random partition of a set $[n]$ shows a random bit complexity that exceeds Shannon's theoretical lower bound, indicating a waste of randomness and time. In this particular example, it should be noted that the predominant factor in the algorithm's execution time is the time taken for random bit generation. This underscores the importance of being attentive to the consumption of randomness in algorithm design and optimization.

In this paper, we tackle the challenge of generating combinatorial structures in a more entropy-efficient way, while achieving optimal time complexity. We focus on the random generation of a set partition by implementing Stam's framework [25] in an entropy-efficient manner. Notably, our approach represents the first implemented and efficient version of Stam's algorithm, relying on Devroye's method for generating an integer random variable with a log-concave distribution, as outlined in [6]. Moreover, the algorithms are highly parallelizable, allowing for an execution time of almost $O(\log(n))$, depending on the level of parallelization.

We also introduce a new combinatorial interpretation of Stam's algorithm. This new combinatorial interpretation allows us to extend Stam's framework to two other combinatorial structures: Weak orderings of a set and linearly ordered set partitions.

We detail this technique and its application to other combinatorial structures, linking it to combinatorial quantities that can be precisely approximated using tools of analytic combinatorics, such as the saddle-point method.

Let us conclude with a structured overview of the paper. Section 2 is specifically designed to introduce and thoroughly describe the objects under study. In Sect. 3, we introduce an algorithm for generating set partitions with a runtime of $O(n \log(n))$, aligned with the lower bound dictated by Shannon's entropy principle. The basis of this algorithm is drawn from Stam's work, as detailed in [25].

Section 5 provides a detailed analysis of the performance of our algorithm, focusing on two key metrics: runtime and random bit consumption. The latter, often underestimated, is crucial since generating random bits—whether from pseudo-random generators or true random sources—can be time-intensive. Notably, in certain cases, up to 80% of the total time required to generate a set partition is dedicated to random number generation.

To facilitate the algorithm's functionality, we employ a technique from [6] that enables the generation of a random number from any given distribution. A detailed exposition of this method is presented in Sect. 4.1. Subsequently, in Sect. 6, this technique is applied to generate two other combinatorial structures. This generation process leverages the ability to approximate numerous combinatorial quantities, a capability afforded by the tools of analytic combinatorics, particularly the saddle-point method as discussed in [12]. While we occasionally adopt the symbolic method's syntax to describe combinatorial classes, it serves as a supportive tool rather than the core of our approach.

Section 7 details our Python implementation, which scales well and supports parallelization. The full code is available in our GitHub repository.

2 Overview of Combinatorial Structures

We describe the structures that are going to be generated in this paper. In all that follows, $n \geq 1$ is a fixed integer. We work on subsets of a particular set, without loss of generality we assume that this set is $[n] = \{1, ..., n\}$, it could be replaced with any set of size n without any problem.

Definition 1 (Set partition). *A partition of* $[n]$ *is a set of non-empty subsets of* $[n]$ *such that every element of* $[n]$ *is in exactly one of these subsets. The number of set partitions of* $[n]$ *is the Bell number,* B_n.

Definition 2 (Partition of a set into linearly ordered subsets). *A partition into linearly ordered subsets (also called a fragmented set partition) of* $[n]$ *is a collection of non-empty subsets of* $[n]$*, each equipped with a chosen total order, such that every element of* $[n]$ *appears in exactly one subset. A linearly ordered subset is a subset with an assigned total order; for a subset with* k *elements, there are* $k!$ *possible orderings. We denote by* B_n^L *the number of partitions of* $[n]$ *into linearly ordered subsets.*

Unlike standard set partitions, the order of elements within each subset matters in partitions into linearly ordered subsets. For instance, $\{(1,2),(3)\}$ is a different partition in linearly ordered subsets from $\{(2,1),(3)\}$ but is equivalent to $\{(3),(1,2)\}$.

Definition 3 (Weak ordering of a set). *A weak ordering of* $[n]$ *is a sequence of non-empty subsets of* $[n]$ *such that every element of* $[n]$ *is in exactly one of these subsets. A weak ordering of* $[n]$ *can be viewed as a surjective mapping from* $[n]$ *onto a set of a given cardinality. We let* F_n *denote the number of weak orderings of* $[n]$*, the* F_n*s are called the Fubini numbers (because they count the number of ways to compute an* n*-dimensional integral according to Fubini's theorem). They are sometimes referred to as the ordered Bell numbers or the surjection numbers.*

The difference with the set partition is that the order of the subsets matters. For example $(\{1,3\},\{2\})$ is not the same ordered set partition as $(\{2\},\{1,3\})$ but is the same as $(\{3,1\},\{2\})$.

Here is a summary of the properties of these numbers:

Here, we use $\langle k \rangle_0 = 1$ and, for $n \geq 1$, $\langle k \rangle_n = k(k+1)\dots(k+n-1)$ to denote the rising factorial as defined previously. The Stirling numbers of the second kind, $\left\{ {n \atop k} \right\}$, represent the number of partitions of a set of n elements into k non-empty subsets. Similarly, $\left\lceil {n \atop k} \right\rceil$ and $\left\lfloor {n \atop k} \right\rfloor$ refer to the number of weak orderings and linearly

	Set partition	Linearly ordered	Weak orderings
Counted by	B_n	B_n^L	F_n
OEIS reference	A000110	A000262	A000670
Decomposed in	$\sum_{k=0}^{n} \left\{ {n \atop k} \right\}$	$\sum_{k=0}^{n} \left\lfloor {n \atop k} \right\rfloor$	$\sum_{k=0}^{n} \left\lceil {n \atop k} \right\rceil$
Symbolic method	$\mathcal{S}et(\mathcal{S}et_{\geq 1}(\{1\}))$	$\mathcal{S}et(\mathcal{S}eq_{\geq 1}(\{1\}))$	$\mathcal{S}eq(\mathcal{S}et_{\geq 1}(\{1\}))$
Generating function	e^{e^z-1}	$e^{\frac{z}{1-z}}$	$\frac{1}{2-e^z}$
Exact formula	$e^{-1}\sum_{k\geq 0}\frac{k^n}{k!}$	$e^{-1}\sum_{k\geq 0}\frac{\langle k\rangle_n}{k!}$	$2^{-1}\sum_{k\geq 0}\frac{k^n}{2^k}$

ordered partitions of $[n]$ into k subsets, respectively, as outlined above. The operators $\mathcal{S}et$ and $\mathcal{S}eq$ are classical combinatorial constructs where $\mathcal{S}et$ forms an unordered collection (or *bundle*) of elements, and $\mathcal{S}eq$ forms an ordered collection or *sequence*. These operators are studied in detail in Flajolet [12]. We use the subscript ≥ 1 to indicate that only collections or sequences with elements of size at least one are considered. Thus, $\mathcal{S}et(\mathcal{S}eq_{\geq 1}(\{1\}))$ represents the bundles of sequences containing at least one axiom. Since these axioms are labeled, this construction corresponds to the set of non-empty, linearly ordered subsets of $[n]$.

These three combinatorial families share similarities, allowing techniques for one to be adapted to the others. Their low OEIS reference numbers reflect thorough study and many generalizations; see [4,9,13,17] for more details.

In this paper, we do not consider the family $\mathcal{S}eq(\mathcal{S}eq_{\geq 1}(\{1\}))$ because this family essentially corresponds to colored permutations, which are easy to generate. The operator $\mathcal{C}yc$ is out of the framework of this paper.

3 Implementing Stam's Method for Set Partition Generation

Stam's method for generating set partitions, as presented in [25], serves as the foundation for our study. This method generates a set partition of $[n]$ by first determining the number of *urns* U to use and then assigning elements to these urns. The number of urns U is an integer-valued random variable governed by the probability mass function:

$$p_k := P(U = k) = \frac{k^n}{eB_n k!}, \tag{1}$$

where k is the number of urns, and p_k is the probability of $U = k$. Note that p_k depends on n, but this dependency is omitted for simplicity. Once the number of urns U is drawn, each element of $[n]$ is randomly, uniformly, and independently assigned to one of the U urns. Urns containing at least one element then form the resulting set partition.

By Dobiński's formula [8], which states that B_n can be expressed as an infinite sum

$$B_n = \frac{1}{e}\sum_{k\geq 0}\frac{k^n}{k!}, \tag{2}$$

the probabilities p_k correctly define a probability distribution.

Algorithm 1. Stam's Partitioning Algorithm

function GENERATEPARTITION(n)
  Draw $U \sim$ `p_k_distribution` ▷ Step 1: Draw the number of urns using 1
  Initialize empty urns: **`urns`** $\leftarrow$ [EmptyUrn()] $\times U$
  for each element i in $[1, 2, \ldots, n]$ **do**
    Assign i to one of the U urns uniformly at random.
  `partition` $\leftarrow$ Non-empty **`urns`**
  return `partition`

As shown in [25, Theorem 1], the algorithm generates uniform random set partitions of $[n]$.

Remark 1. The number of discarded empty urns is independent of the generated set partition and follows a Poisson distribution with parameter 1, as shown in [25, Theorem 2].

To comprehend how the algorithm functions, we examine Stam's distribution and introduce a novel combinatorial interpretation of his algorithm. This involves the generating functions of the urns: $U_n(u) = \sum_{k \geq 0} \frac{k^n u^k}{k!}$, representing the probability generating function of U up to a constant.

The function $U_n(u)$ also serves as the exponential generating function for the combinatorial family $\mathcal{C}_n$: functions from $[n]$ to a set X, where the size of a function $f : [n] \mapsto X$ is defined by the cardinality of X being k.

Property 1.

$$U_n(u) = \left(u\frac{d}{du}\right)^n e^u = \left(\sum_{k=0}^{n} {n \brace k} u^k\right) e^u$$

Remark 2. Property 1 is a well-known result, tracing back to 1823 with Scherk's work [23], and serves as a foundational element in the theory of normal ordering, as presented by Blasiak et al. [2].

To further illustrate property 1, we give the case $n = 3$. It states that:

$$U_3(u) = \left(u\frac{d}{du}\right)^3 e^u = \left(u + 3u^2 + u^3\right) e^u = \left(\sum_{k=0}^{3} {3 \brace k} u^k\right) e^u. \tag{3}$$

Additionally, by expanding e^u as a power series, we confirm the general form given in property 1 for n = 3.

From property 1, we have a combinatorial interpretation of the algorithm: We sample a uniform element $f : [n] \to X \in \mathcal{C}_n$. We then decompose f into two components:

- The non-empty urns, which describe a set partition, corresponding to the generating function $\sum_{k=0}^{n} \left\{ {n \atop k} \right\} u^k$, which we return as the output.
- A number of empty urns, corresponding to the generating function e^u, which follows a Poisson distribution with parameter 1.

This decomposition simply expresses the trivial identity: $e^{ue^z} = e^{u(e^z-1)}e^u$. The independence of the number of empty urns from the set partition is reflected in the product structure of the generating functions $\left(\sum_{k=0}^{n} \left\{ {n \atop k} \right\} u^k\right) e^u$.

The first step of Stam's algorithm generates the size U of X. The second step generates f.

Putting the elements of $[n]$ in the urns is straightforward and well understood (it can be done efficiently using the algorithm described in [19]) but generating U is not. In [25], the author does not provide a method for generating U, and the proposed procedure is not fully complete. Therefore, we describe below a method to generate U to ensure a functional algorithm.

A classical way to generate an integer with an arbitrary given distribution is to reverse the distribution function of the random variable. This method requires to compute a good amount of the probability coefficients p_k of the distribution but this is too costly. Indeed, computing B_n takes $\Omega(n^2)$ and even computing $k!$ when k is close to n takes $\Omega(n(\log(n)\log(\log(n)))^2)$ with the best algorithm known to this day (it is done in [3]). In the following, we generate U according to Stam's distribution in a way that does not require us to compute B_n exactly.

4 Optimized Generation of the Stam's Distribution

This section details our approach for generating U, a key aspect not addressed in [25]. We employ the method from [6] for generating integer random variables with log-concave distributions, which uses a Monte-Carlo rejection scheme with an efficient dominating function. While extensions of this method exist, such as in [7], the original formulation suffices for our purposes due to its simplicity and effectiveness.

We adapt Devroye's algorithm to generate U, addressing the specifics of our problem and avoiding the pitfalls of a naive application. A direct implementation of the algorithm is impractical for our scenario, as writing down large numbers would require $O(n\log(n))$ time and space. Instead, we modify the algorithm to function efficiently in our setting.

4.1 Devroye's Algorithm for Generating U

While log-concave sampling methods are well-studied in the continuous domain, discrete log-concave distributions have received less attention. In this subsection, we introduce Devroye's algorithm, an efficient method for generating the discrete random variable U following Stam's distribution. This algorithm is crucial for entropy-efficient set partitioning, as it provides the exact count of urns, required for the process.

Devroye's algorithm, as outlined in [6], is a versatile, black-box method for generating discrete random variables with log-concave distributions, including U as defined by Stam's distribution. Its adaptability and efficiency make it a cornerstone of our approach.

Lemma 1. *The sequence* $p_k = \frac{k^n}{eB_n k!}$ *is log-concave that is to say:* $\forall k > 1, p_k^2 \geq p_{k-1}p_{k+1}$.

Despite the algorithm working in a black-box framework, we describe it briefly to show that despite measuring the execution time of our algorithm instead of the number of arithmetic operations, we are still able to have it run in $O(\log(n)^2)$.

Algorithm 2. Generating U According to Stam's Distribution

1: Find a *mode* m of Stam's distribution, i.e. $m = \arg\max_{k\in\mathbb{N}} \frac{k^n}{k!}$.
2: Compute $\tilde{p}_m = \frac{m^n}{m! e\tilde{B}_n}$.
3: Define $h(x) = \min(\tilde{p}_m, \tilde{p}_m e^{1-\tilde{p}_m(|x|-\frac{1}{2})})$, a function with a central rectangular part and two exponential tails as in [6]. The graph of h is over the plot of the distribution (p_k).
4: Generate a point under the curve of h, naming its abscissa Y.
5: Determine X as the nearest integer to Y.
6: Test if $Rh(Y) \leq \tilde{p}_X$, where R is a uniform random number in $[0, 1]$. If true, return X; otherwise, generate another Y and repeat.

Lemma 2. *A mode* m *of the distribution, defined as* $m = \arg\max_{k\in\mathbb{N}} \frac{k^n}{k!}$, *can be found by determining the smallest integer* m *satisfying the inequality* $n \leq \frac{\ln(1+m)}{\ln(1+m^{-1})}$. *This computation can be performed in* $O(\log(n)^2)$ *time.*

Moreover, to run the algorithm, we use the following classical asymptotic equivalent [20] $B_n \sim \frac{\exp(e^{W(n)}-1)}{W(n)^n\sqrt{2\pi W(n)(W(n)+1)e^{W(n)}}}$ with W the principal branch of the Lambert W function.

4.2 Necessity of Numerical Approximations

To effectively use Devroye's algorithm in our context, we need to adapt it slightly. Our focus is on minimizing the number of random bits and computation time. Devroye's algorithm involves arithmetic operations whose complexity becomes significant when dealing with large numbers. For instance, computing $k!$ for k close to n requires substantial computational resources.

Standard floating-point computations may introduce bias into the distribution. Therefore, our approach involves setting a fixed bit precision for calculations and increasing this precision as needed. This works because our method involves only a few computations: determining the mode of the distribution, m; computing $\tilde{p}_m$, $\tilde{B}_n$; generating a random point Y under h's graph; calculating the nearest integer X and $\tilde{p}_X$; and deciding whether or not to reject, by computing $Rh(Y) \leq \tilde{p}_X$. This procedure is detailed in the next subsection.

Precision-Adjusted Generation Procedure. Our procedure for generating U considers precision requirements. We start with a precision level b (e.g., $b = 10$ bits) for computing $Rh(Y)$ and $\tilde{p}_X$. All necessary values are computed, and if the precision is insufficient to decide whether $Rh(Y) \leq \tilde{p}_X$, the precision is doubled, retaining the significant bits of R to avoid bias. If b suffices and $Rh(Y) \leq \tilde{p}_X$, we set $U = X$. Otherwise, we restart with the initial precision, fully re-rolling R in line with the standard Monte-Carlo rejection scheme.

Analysis of Precision Doubling. The expected computation cost depends on the required precision P to compute $Rh(Y) \leq \tilde{p}_X$. The random variable P follows a geometric law of parameter 1/2 because it is a dichotomy. So, $\mathbf{E}(P) = 2$. On average, we double the precision $O(1)$ time. Since we need P bits of precision, we need $O(\log(P))$ doubling of our precision number b (to have $b \geq P$ which allows us to conclude). The running time of computing $Rh(y) \leq \tilde{p}_X$ therefore is $O(\sum_{b\geq 0} \frac{\ln(b)(b^2+\log(n))}{2^b}) = O(\log(n))$. The cost $\ln(b)(b^2 + \log(n))$ comes from the fact that we can compute the first b bits of precision of our quantities (factorial,...) in $O(\ln(b)(b^2 + \log(n)))$. Consequently, the computation time and random bit requirement are both $O(\log(n))$.

Remark 3. The effectiveness of our algorithm hinges on the ability to compute the most significant b bits of our numbers in $O(\log(n) + Poly(b))$ time. The needed approximations are feasible due to efficient algorithms for factorial and related computations as outlined in [15] and [16].

5 Analysis of the Algorithm's Complexity

5.1 Determining the Number of Urns

The runtime and random bit consumption of our algorithm are primarily influenced by the process of generating the number of urns U and the subsequent allocation of elements to these urns. While the generation of U operates in $O(\log(n))$ time, the Shannon entropy lower bound implies that generating a complete element cannot be achieved faster than $O(n \log(n))$ on average. This result, noted in Knuth's work on sorting and searching complexity [18], arises because entropy determines the minimum amount of information required to encode outcomes.

The second phase of the algorithm, which involves assigning the elements of $[n]$ to the urns, accounts for most of the runtime and underscores the significance of randomness in our approach. To analyze the complexity and random bit consumption, it is useful to estimate the average number of urns, $\mathbf{E}(U)$. This quantity influences the computational cost, as the number of urns determines the choices available for element placement.

Proposition 1 ([25]). *The average number of urns,* $\mathbf{E}(U)$, *satisfies*

$$\mathbf{E}(U) = \frac{B_{n+1}}{B_n} = \frac{n}{W(n)} + o(1),$$

where W is the principal branch of the Lambert W function. Consequently, $\mathbf{E}(U) \sim \frac{n}{\ln(n)}$.

Remark 4. The asymptotic behavior of $\mathbf{E}(U)$ provides insight into the complexity of our algorithm. While the computation of h (as in [6]) requires a one-time cost of $O(1+\max(\frac{B_n}{\tilde{B}_n}, \frac{\tilde{B}_n}{B_n}))$ arithmetic operations, i.e. $O(\log(n))$ time since $B_n \sim \tilde{B}_n$. Similarly, generating a candidate X and determining if it satisfies the conditions for U also takes $O(\log(n))$ time on average, with rejection occurring in $O(1)$. As a result, the total average time for generating U is $O(\log(n))$.

5.2 Allocation of Elements Into Urns

Theorem 1. *Algorithm 1 runs in $O(n \log(n))$ time and requires, on average, $n \log_2(n) - n \log_2(\ln(n)) + O(n)$ random bits. The algorithm is quasi-optimal in terms of both runtime and random bit consumption. Specifically:*

- *The runtime $O(n \log(n))$ is optimal, as there exists a lower bound of $\Omega(n \log(n))$ for this problem.*
- *The random bit consumption $n \log_2(n) - n \log_2(\ln(n)) + O(n)$ is quasi-optimal, as it matches the lower bound of $\log_2(B_n) = n \log_2(n) - n \log_2(\ln(n)) + O(n)$ up to the leading terms.*

Proof. The process of assigning each of the n elements in $[n]$ to one of the U urns requires generating n uniform random integers in $[1, U]$. Using the method of [19], this step requires $n \log_2(U) + 2$ random bits and executes in $O(n \log(U))$ time.

Using the concavity of $x \mapsto n \log_2(x)$, Jensen's inequality implies:

$$\mathbf{E}(n \log_2(U)) \leq n \log_2(\mathbf{E}(U)).$$

Since $\mathbf{E}(U) = \frac{n}{W(n)} + o(1)$ and from the well known fact $W(n) \geq \ln(n) - \ln(\ln(n))$, we can bound $\mathbf{E}(n \log_2(U))$ as follows:

$$\mathbf{E}(n \log_2(U)) \leq n \log_2 \left(\frac{n}{\ln(n) - \ln(\ln(n))} + o(1) \right) = n \log_2(n) - n \log_2(\ln(n)) + O(n).$$

Therefore, the algorithm requires an average of $n \log_2(n) - n \log_2(\ln(n)) + O(n)$ random bits. In addition, generating U is done in $O(\log(n))$ time. The allocation of n elements to U urns requires $O(n \log(U))$, which simplifies to $O(n \log(n))$ using $\mathbf{E}(U) \sim \frac{n}{W(n)}$. Thus, the total runtime is $O(n \log(n))$.

Remark 5. As shown in [14], the Stirling numbers of the second kind, $\left(\left\{ {n \atop k} \right\}\right)_{k \in [n]}$, are asymptotically normal with standard deviation $\sigma \sim \frac{\sqrt{n}}{W(n)}$. Additionally, the distribution of empty urns follows a Poisson distribution with parameter 1, as established in [25]. Combining these results, the distribution of U is asymptotically normal with mean $\mu_n \sim \frac{n}{W(n)}$ and variance $\sigma^2 \sim \frac{n}{W(n)^2}$. Furthermore, the tight concentration of U around its mean ensures that the algorithm's execution time is highly stable, an essential feature for applications requiring predictable performance. This stability is directly tied to the underlying probabilistic characteristics of the structures being generated, reinforcing the robustness of our approach.

6 Generalization to Other Combinatorial Structures

6.1 Generation of Linearly Ordered Subsets

Algorithm 1 can be adapted to generate a uniform partition of a set into linearly ordered subsets.

First, the algorithm generates the number U of urns with probability

$$p_k = \mathbf{P}(U = k) = \frac{1}{e}\frac{\langle k\rangle_n}{k!B_n^L}, \tag{4}$$

using the technique of [6] with the modifications established in Sect. 4.1. The next step differs from Algorithm 1: instead of assigning elements uniformly to urns, the n elements are inserted sequentially into U empty piles. For the ith element, there are $k + i - 1$ positions available: it can be placed on top of any pile or below one of the previous $i - 1$ elements in an existing pile.

Once all n elements are distributed, any empty urns are removed. The remaining urns represent linearly ordered, non-empty subsets of the original set, and thus describe a valid partition into linearly ordered subsets. The algorithm then returns this partition.

Notably, leveraging [17, **Theorem 6**], one can derive a combinatorial interpretation of this algorithm analogous to the interpretation provided for Stam's algorithm.

Theorem 2. *Algorithm 1 with the above-described adaptations generates a uniform partition of a set into linearly ordered subsets. It achieves a runtime of $O(n\log(n))$, which is optimal, and its random bit consumption is quasi-optimal.*

6.2 Weak Orderings

To generate an ordered set partition with exactly $n > 1$ elements, we adapt Algorithm 1, modifying it to handle weak orderings. The algorithm proceeds in two main steps:

- Generating the Number of Urns (U): The first step involves generating an integer U, representing the number of urns, with the probability distribution:

$$p_k = \mathbf{P}(U = uk) = \frac{k^n}{2^{k+1}F_n} \tag{5}$$

  The value of F_n can be determined through contour integration of its generating function, calculated as

$$F_n = \frac{n!}{2}\sum_{k=-\infty}^{+\infty}(\ln(2) + 2\pi ik)^{-(n+1)}. \tag{6}$$

  This makes F_n relatively straightforward to approximate, thereby informing the normalization factor needed for our algorithm. The method for generating U follows the approach in [6], adapted for arbitrary precision as outlined in Sect. 4.2.

- Assigning elements to urns: In the second step, we create U *labeled* urns, ranging from 1 to U. Each element $[n]$ is then placed into one of these urns at random. This process is repeated independently for all n elements. After all elements have been assigned, any empty urns are discarded. The result is an ordered set partition, described by the arrangement of elements within the urns. Unlike the final step in Stam's algorithm, we maintain the ordering of the urns, thus creating a weak ordering rather than a simple set partition.

Theorem 3. *Algorithm 1 with the above-described adaptations generates a uniform weak ordering. It achieves an optimal runtime of* $O(n \log(n))$, *and its random bit consumption is quasi-optimal.*

7 Implementations

We implemented Algorithm 1 in Python and the benchmark was performed on an Apple M3 Pro with 36 GB RAM. For an input size of $n = 10^7$ the algorithms runs in around 3 s, with some variation depending on which structure is generated. Even for larger inputs, such as $n = 10^8$ the algorithm manages to run in around 35 s, demonstrating its scalability despite being implemented in Python. The implementation is available in our GitHub repository. The Fig. 1 shows the average execution time based on 10 runs for different values of n in the range $[10^7, \ldots, 10^8]$.

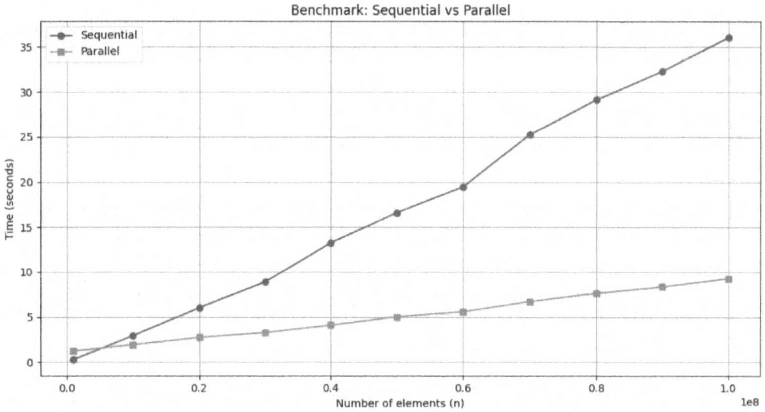

Fig. 1. Average execution time based on 10 runs.

Moreover a major advantage of this Stam-style approach is that filling the urns is very simple and can therefore be easily parallelized. A parallelized version has been produced and is available on the same directory, enabling a billion-size ensemble partition to be drawn uniformly in around 150 s.

8 Future Work and Conclusion

Our ongoing work focuses on extending the generation method of this paper to the normal ordering framework, as explored in [2]. This generalization aims to adapt and refine our techniques to address broader classes of combinatorial structures.

In conclusion, it is important to emphasize that the computational cost of managing a structure is not always the dominant factor. For set partition structures, for instance, the cost of randomness management often outweighs that of structural handling. This highlights the utility of the random bit model as an effective lens for analyzing and optimizing algorithms. By prioritizing randomness efficiency, we uncover valuable insights into computational complexity and the design of more effective methods for addressing intricate combinatorial problems.

Acknowledgments. The authors would like to thank the anonymous referees for their insightful comments and suggestions, which significantly enhanced this work. They also gratefully acknowledge the support of the French National Research Agency under project ANR-23-CE48-0014.

Disclosure of Interests. The authors have no competing interests to declare that are relevant to the content of this article.

References

1. Albert, R., Barabási, A.L.: Statistical mechanics of complex networks. Rev. Mod. Phys. **74**, 47–97 (2002). https://doi.org/10.1103/RevModPhys.74.47
2. Blasiak, P., Flajolet, P.: Combinatorial models of creation-annihilation. Sémin. lothar. comb., Art. B65c (78 pp) (2011)
3. Borwein, P.B.: On the complexity of calculating factorials. J. Algor. **6**(3), 376–380 (1985)
4. Carlitz, L.: Some remarks on the bell numbers. Fib. Quart **18**(1), 66–73 (1980)
5. Denise, A., Zimmermann, P.: Uniform random generation of decomposable structures using floating-point arithmetic. Theor. Comput. Sci. **218**(2), 233–248 (1999)
6. Devroye, L.: A simple generator for discrete log-concave distributions. Computing **39**(1), 87–91 (1987)
7. Devroye, L.: A note on generating random variables with log-concave densities. Stat. Probab. Lett. **82**(5), 1035–1039 (2012)
8. Dobinski, G.: Summirung der reihe sum nm/n! für m= 1, 2, 3, 4, 5. Arch. für Mat. und Physik **61**, 333–336 (1877)
9. Dou, X., Hwang, H.K., Li, C.Y., et al.: Bell numbers in matsunaga's and arima's genjikō combinatorics: modern perspectives and local limit theorems. Electron. J, Comb (2022)
10. Duchon, P., Flajolet, P., Louchard, G., Schaeffer, G.: Boltzmann samplers for the random generation of combinatorial structures. Comb. Probab. Comput. **13**(4–5), 577–625 (2004)

11. Flajolet, P., Fusy, É., Gandouet, O., Meunier, F.: Hyperloglog: the analysis of a near-optimal cardinality estimation algorithm. Disc. Math. Theor. Comput. Sci., 137–156 (2007)
12. Flajolet, P., Sedgewick, R.: Analytic Combinatorics. Cambridge University Press, Cambridge (2009)
13. Good, I.: The number of orderings of n candidates when ties are permitted. Fib. Quart **13**, 11–18 (1975)
14. Hwang, H.K.: On convergence rates in the central limit theorems for combinatorial structures. Eur. J. Comb. **19**(3), 329–343 (1998)
15. Iacono, R., Boyd, J.P.: New approximations to the principal real-valued branch of the lambert w-function. Adv. Comput. Math. **43**, 1403–1436 (2017)
16. Johansson, F.: Arbitrary-precision computation of the gamma function. arXiv preprint arXiv:2109.08392 (2021)
17. Kim, T., Kim, D.: Lah-bell numbers and polynomials. Proc. Jangjeon Math. Soc. **23**, 577–586 (2020). https://doi.org/10.17777/pjms2020.23.4.577
18. Knuth, D.: The complexity of nonuniform random number generation. In: Algorithm and Complexity, New Directions and Results (1976)
19. Lumbroso, J.: Optimal discrete uniform generation from coin flips, and applications. arXiv preprint arXiv:1304.1916 (2013)
20. Moser, L., Wyman, M.: An asymptotic formula for the bell numbers. R. Soc. Canada (1955)
21. Mougenot, A., Darrasse, A., Blanc, X., Soria, M.: Uniform random generation of huge metamodel instances. In: Model Driven Architecture - Foundations and Applications (2009)
22. Ponty, Y.: Efficient sampling of RNA secondary structures from the Boltzmann ensemble of low-energy: the boustrophedon method. J. Math. Biol. **56**(1–2), 107–127 (2008). https://doi.org/10.1007/s00285-007-0137-z
23. Scherk, H.F.: De evolvenda functione yd. yd. yd... ydX/dxn disquisitiones nonnullas analyticas. Bornträger (1824)
24. Shannon, C.E.: A mathematical theory of communication. Bell Syst. Techn. J. **27**(3), 379–423 (1948). https://doi.org/10.1002/j.1538-7305.1948.tb01338.x
25. Stam, A.: Generation of a random partition of a finite set by an urn model. J. Comb. Theory - A **35**(2), 231–240 (1983)

Exact Set Packing in Multimodal Transportation with Ridesharing System for First/Last Mile

Qian-Ping Gu and Jiajian Leo Liang(✉)

School of Computing Science, Simon Fraser University, Burnaby, BC, Canada
{qgu,leo_liang}@sfu.ca

Abstract. We propose a centralized transportation system that integrates public transit with ridesharing to provide multimodal transportation. At each time interval, the system receives a set of personal drivers, designated drivers, and public transit riders. It then assigns all riders to drivers, ensuring that pick-ups and drop-offs occur at designated transit stations. This effectively replaces first-mile/last-mile (FM/LM) segments with a ridesharing alternative, reducing overall commuting time. We study two optimization problems: (1) minimizing the total travel distances of drivers and (2) minimizing the number of designated drivers required to serve all riders. We show the optimization problems are NP-hard and give hypergraph-based integer linear programming exact algorithm and approximation algorithms. To enhance computational efficiency, we introduce a clustering heuristic that utilizes both spatial and temporal aspects of the input data to accelerate rider-to-driver assignments. Finally, we conduct an extensive computational study using real-world datasets and surveys from Chicago to evaluate our model and algorithms at a city-wide scale.

Keywords: Transit and ridesharing · Exact and approximation algorithms · Graph and hypergraph · Clustering · Computational study

1 Introduction

The transportation sector is responsible for about 25% of global greenhouse gas emissions [18,21]. As urban and suburban areas expand, the number of commuters increases, leading to greater traffic congestion and environmental impact. The inconvenience and inflexibility of the first mile (**FM**, commute from one's origin/home to the nearest transit stop or major transportation hub) and last mile (**LM**, commutes from a transit stop or major transportation hub to one's destination/home) transportation, compared to personal vehicles, causes many people to choose to use personal vehicles for commutes [5,8].

Ridesharing and carpooling have been considered effective in reducing traffic congestion and extensively studied. Recent literature reviews on shared mobility

H. Fernau and B. Zhu (Eds.): IWOCA 2025, LNCS 15885, pp. 418–432, 2025.
https://doi.org/10.1007/978-3-031-98740-3_30

can be found in [22,27,28]. An important research direction is to use ridesharing as a *feeder* for public transit to improve the FM/LM efficiency. Several proposed approaches [6,9,17,24] in this direction include using autonomous vehicles to improve bus routes and their flexibility, such as reallocation and on-demand. A recent approach to designing multimodal transportation systems includes ridesharing as an intrinsic part of the transit network [10,11]. Studies in [14,20,26] use a graph matching approach to introduce ridesharing to replace FM/LM transit. Especially, the rider-driver assignment problem is modeled as a graph matching problem in [14].

In this paper, we propose a centralized multimodal transport system (MTRS) that aims to improve the FM/LM transportation to reduce the commuting time and traffic congestion. The system integrates the public transit with ridesharing using *fixed transit stations* which are predefined intermediate stops, such as park-and-ride locations and transportation hubs for pick-ups and drop-offs, where passengers can transfer between ridesharing services and public transit.

Ridesharing providers consist of personal and designated drivers. Personal drivers using their own vehicles for work commute can provide ridesharing service while commuting to/from work. This model offers a more convenient and efficient alternative to traditional FM/LM transit, and multiple riders sharing a single vehicle can lower travel costs and improve access to public transit. As a result, transit ridership may increase and vehicles on the road may decrease. MoD services, such as Uber, Lyft and DiDi, have become popular around the globe for their convenience. As ridesharing and ride-hailing services continue to grow, several studies suggest that integrating MoD with public transportation could enhance urban mobility and increase public transit ridership [2,16,19,23]. By participating in the transportation system, personal drivers can offset their driving costs, while designated drivers can increase their earnings.

The concept of large mobility transportation hubs, which facilitate multimodal travel, has gained popularity in research and policy making as shown in some recent studies (e.g., [4,25]). Fixed drop-off transit stations enhance route visibility and improve schedule predictability for personal drivers. Additionally, park-and-ride facilities and large transportation hubs with parking allow personal drivers to switch to public transit for part of their journey. This hybrid approach can provide a faster commute, particularly when using subways or trains to bypass traffic congestion.

Our system aims at minimizing the total travel distance of drivers and the number of drivers to serve all riders while improving FM/LM efficiency, leading to a more efficient and sustainable urban mobility network. Specifically, we focus on two optimization problems:

(1) Minimizing total travel distance (*MTRS-minDist*): Assign all riders to drivers such that each rider's original commuting time is reduced by a predefined value and the sum of detour distances of personal drivers and travel distances of designated drivers is minimized.

(2) Minimizing the number of designated drivers (*MTRS-minNum*): Assign all riders to drivers such that each rider's original commuting time is reduced by a predefined value and the number of designated drivers used is minimized.

To solve these optimization problems, we adopt a similar model to that in [14,26], providing exact algorithm and approximation algorithms, which are similar to finding a set packing that covers all riders. Our main contributions are summarized as follows:

1. We give a hypergraph-based integer linear program (ILP) exact algorithm approach for solving the MTRS-minDist and MTRS-minNum problems, and we show that both optimization problems are NP-hard.
2. We give a $\frac{\lambda^2\mu+\lambda}{\lambda+1}$-approximation algorithm for MTRS-minDist, where λ is the maximum vehicle capacity and μ is the ratio of the longest to shortest incurred-travel-distance ridesharing routes. For MTRS-minNum, we give a $\frac{\lambda+2}{2}$-approximation algorithm and a local search heuristic.
3. We introduce a clustering heuristic that reduces the computational time needed to compute the hypergraph while preserving solution quality.
4. We conduct an extensive computational study to evaluate the system and optimization algorithms using real-life data from Chicago City. The total travel distance of drivers and number of drivers obtained by our approximation algorithms are close to the optimal solutions while reducing commute time more than the predefined value of 30%.

For the remainder of this paper, Sect. 2 describes our centralized system and defines the MTRS-minDist and MTRS-minNum problems. Then we present the ILP exact algorithm and the NP-hardness results. In Sect. 3, we present approximation algorithms for MTRS-minDist and MTRS-minNum, followed by the clustering heuristic in Sect. 4. We discuss our numerical experiments and results in Sect. 5. Finally, Sect. 6 concludes the paper.

2 Problem Definition and Preliminaries

For each time interval, the multimodal transportation with ridesharing system (MTRS) receives a set $R = \{r_1, \ldots, r_{|R|}\}$ of riders who request ridesharing service and a set $D = \{\eta_1, \ldots, \eta_{|D|}\}$ of drivers who offer ridesharing service. A public transit network with a set TS of *transit stations* (*station* for brevity) and a timetable T is given. Riders of R can be picked up at their origins and dropped off at the stations or picked up at the stations and dropped off at their destinations, by drivers in D. A *public transit route* for rider $r_j \in R$ is a travel plan using only public transportation, whereas a *ridesharing route* for rider r_j is a travel plan using a combination of public transportation and ridesharing to reach r_j's destination. An edge-weighted directed graph $G(V, E, w)$ (road network) is given, which represents locations reachable by both private and public transportation.

The set $D = \Gamma \cup \Delta$ of drivers consists of a set Γ of personal drivers who drive their vehicles for work commute and a set Δ of designated drivers who are employed/self-employed to offer ridesharing service with $\Gamma \cap \Delta = \emptyset$. Personal drivers are on the road independent of ridesharing services, whereas designated drivers are on the road only if necessary (e.g., Uber and Taxi drivers). Each driver $\eta_i \in D$ has travel requirements, including origin o_i, destination d_i, earliest departure time α_i from o_i, latest arrival time β_i at d_i, the vehicle capacity λ_i available for riders, and maximum detour time limit z_i to serve riders. Each rider $r_j \in R$ also has travel requirements with $o_j, d_j, \alpha_j, \beta_j$ as defined in the travel requirements of drivers, and minimum time saved θ_j for r_j to use a ridesharing route (compared to the fastest public transit route). A ridesharing route satisfying a rider r_j's requirements is called a *feasible route* for r_j. Two match types are considered in this paper:

- Match type **FM** (first mile): a driver may make stops to pick up multiple riders but makes only one stop to drop off all riders. The *pick-up locations* are the riders' origins, and the *drop-off location* is a station in TS.
- Match type **LM** (last mile): a driver makes only one stop to pick up riders and may make multiple stops to drop off all riders. The *pick-up location* is a station in TS, and the *drop-off locations* are the riders' destination.

Drivers and riders specify which match type to participate in as inputs to the system. A driver η_i *can serve* rider r_j if a feasible route exists for r_j offered by η_i, η_i and r_j have the same match type (FM or LM), and the requirements of η_i are also satisfied for serving r_j. A driver η_i *can serve* a set $R_i \subseteq R$ of riders means that η_i and all riders of R_i have the same match type, there is a feasible route offered by η_i that satisfies the requirements of η_i and all riders in R_i collectively.

For two locations o and d in road network G, let $FP(o, d)$ be the fastest path from o to d. Let R_i be the set of riders served by driver η_i, and let $\bar{P}(\eta_i, R_i)$ be the ridesharing route in G that η_i uses to serve R_i, which starts at o_i and ends at d_i. For a path P in G, let dist(P) be the distance of P. The *incurred travel distance* of a personal driver $\eta_i \in \Gamma$ is the detour distance η_i needs to travel to serve R_i, which is $TD(\eta_i, R_i) = \text{dist}(\bar{P}(\eta_i, R_i)) - \text{dist}(FP(o_i, d_i))$. The *incurred travel distance* of a designated driver $\eta_i \in \Delta$ is $TD(\eta_i, R_i) = \text{dist}(\bar{P}(\eta_i, R_i))$. The travel distance $TD(\eta_i, R_i)$ is represented by an integer, in meter. We study two optimization problems. For feasibility and solvability, we assume the following.

Assumption 1. *Every rider of R can be assigned a feasible route offered by a distinct driver of Δ.*

The first optimization problem (**MTRS-minDist**):

- Given a set D of drivers and a set R of riders, the system assigns a feasible route for every rider of R offered by some driver of D such that each driver belongs to at most one assigned feasible route and each rider receives exactly one feasible route, and the total travel distance of drivers is minimized.

The second optimization problem (**MTRS-minNum**) focuses on minimizing the number of vehicles on the road. Since personal drivers are on the road regardless, we assume a maximal number of riders has been assigned to personal drivers in Γ by some algorithm, exact or approximate (assigning the maximum number is NP-hard [3,14]). MTRS-minNum is to minimize the number of drivers in Δ needed to serve all remaining riders $R' \subseteq R$. More formally:

- Given a set $D = \Delta$ of designated drivers and a set R' of riders, the system assigns a feasible route for every rider of R' offered by some driver of a set $D' \subseteq D$ such that each driver belongs to at most one assigned feasible route, each rider receives exactly one feasible route, and $|D'|$ is minimized.

2.1 Exact Algorithm and NP-Hard Results

For a driver $\eta_i \in D$ and a group of passengers $R_i \subseteq R$, (η_i, R_i) is a *feasible match* if η_i can serve all riders of R_i by using a feasible route, which is a path in G to serve all riders of R_i. Each feasible match (η_i, R_i) is associated with an integer weight $w(\eta_i, R_i)$, depending on the optimization goal. For MTRS-minDist, $w(\eta_i, R_i) = TD(\eta_i, R_i)$, as defined previously. For MTRS-minNum, $w(\eta_i, R_i) = 1$ (note that $\eta_i \in \Delta$).

The exact algorithm for both problems is as follows: First, we compute all feasible matches for each driver $\eta_i \in D$ as shown in [14,26]. Then, we construct an edge-weighted hypergraph $H(U \cup V, E, w)$, where $U(H) = D$ and $V(H) = R$ are the vertices. For each $\eta_i \in D$ and for every subset R_i of R, create a hyperedge $e = \{\eta_i\} \cup R_i$ in $E(H)$ if (η_i, R_i) is a feasible match. Each edge $e = \{\eta_i\} \cup R_i \in E(H)$ has weight $w(e) = w(\eta_i, R_i)$. For a vertex $u \in U(H) \cup V(H)$, define $E_H(u) = \{e \in \text{E(H)} \mid u \in e\}$ to be the set of edges in $E(H)$ incident to u (containing u); we also use $E(u)$ to denote $E_H(u)$ if H is clear from the context. It was noted in [13] that Observation 1 holds for $H(V \cup U, E, w)$.

Observation 1. *For any edge $\{\eta_i\} \cup R_i \in E(H)$ with $|R_i| \geq 2$, the edge $\{\eta_i\} \cup R_i'$ exists in $E(H)$ for every non-empty subset R_i' of R_i.*

To solve the MTRS-minDist and MTRS-minNum problems, we give an integer linear program (ILP) formulation:

$$
\begin{aligned}
&\text{minimize} && \sum_{e \in E(H)} w(e) \cdot x_e && && \text{(i)} \\
&\text{subject to} && \sum_{e \in E(\eta_i)} x_e \leq 1, && \forall\, \eta_i \in V(H) && \text{(ii)} \\
& && \sum_{e \in E(r_j)} x_e = 1, && \forall\, r_j \in U(H) = R && \text{(iii)} \\
& && x_e \in \{0, 1\}, && \forall\, e \in E(H) && \text{(iv)}
\end{aligned}
$$

For the binary variable x_e corresponding to $e = \{\eta_i\} \cup R_i$, if $x_e = 1$ then all riders of R_i are assigned to η_i, indicating all riders of R_i are delivered by η_i. The objective function (i) is to minimize the total incurred travel distance of the drivers (for MTRS-minDist) or to minimize the number of designated drivers (for MTRS-minNum) needed to serve all riders. Constraint (ii) guarantees that each driver serves at most one feasible set of riders, and each rider is served by exactly one driver by constraint (iii). A feasible solution to MTRS-minDist and MTRS-minNum is a subset $E' \subseteq E(H)$ with $R(E') = R$.

Let $\lambda = \max_{\eta_i \in D} \lambda_i$. If $\lambda = 1$, H is a bipartite graph, and both MTRS-minDist and MTRS-minNum can be solved in polynomial time by finding a minimum weight maximum matching on H [1]. We assume $\lambda > 1$ hereafter. Then our problems are variants of the exact set packing problem. If every edge of H has uniform weight, our problems are similar to exact set packing, except edges of H satisfy Observation 1 and only all riders need to be covered. Our problems are NP-hard, as stated in Theorem 1 (a proof is given in [15]).

Theorem 1. *Both MTRS-minDist and MTRS-minNum are NP-hard.*

Hypergraph $H(U \cup V, E, w)$ and notations related to H below will be used in the rest of the paper. Let $D(H) = U(H) \cap D$, $\Gamma(H) = U(H) \cap \Gamma$ and $\Delta(H) = U(H) \cap \Delta$. For any edge $e \in E(H)$, let $N_H(e)$ be the set of edges in $E(H)$ incident to e ($N(e)$ when H is clear from the context). For any edge $e = \{\eta_i\} \cup R_i$ in H, let $D(e) = \eta_i$ and $R(e) = R_i$. For any subset $E' \subseteq E(H)$, let $R(E') = \bigcup_{e \in E'} R(e)$, $D(E') = \bigcup_{e \in E'} D(e)$ and $w(E') = \sum_{e \in E'} w(e)$. Note that $\Gamma(E') = D(E') \cap \Gamma$, and likewise, $\Delta(E') = D(E') \cap \Delta$.

3 Algorithms for MTRS-MinDist and MTRS-MinNum

Based on Assumption 2, we give algorithms for MTRS-minDist and MTRS-minNum that achieve approximation ratios better than trivial upper bounds. Both problems remain NP-hard even if Assumption 2 is satisfied (as shown in the NP-hardness proof in [15]).

Assumption 2. *$|\Delta| = |R|$ and for every $\eta_i \in \Delta$ and every $r_j \in R$, there is a feasible route for r_j offered by η_i.*

It is worth to mention that our algorithms for MTRS-minDist and MTRS-minNum are modified from algorithms for the maximum (weighted) k-set packing problem [7,12]. Those algorithms do not apply to our optimization problems because not all riders are guaranteed to be served (such as algorithms in [7]) or weight is not considered (e.g., [12]).

Algorithm GreedyMinDist. Let $H(U \cup V, E, w)$ be the hypergraph constructed for an instance of the MTRS-minDist problem. Algorithm GreedyMinDist computes two solutions M_1 and M_2 separately, and then selects either M_1 or M_2 as the solution M with $w(M) = \min\{w(M_1), w(M_2)\}$.

(1) M_1 is computed as follows with the original $H(U \cup V, E, w)$.
Initially, $M_1 = \emptyset$. In each iteration, select an edge e in $E(H)$ with minimum $w(e)$, $M_1 = M_1 \cup \{e\}$, remove all vertices of e from $U(H) \cup V(H)$ and all edges of $\cup_{u \in e} E(u)$ from $E(H)$. Continue to the next iteration until $R(M_1) = R$.

(2) M_2 is computed as follows with the original $H(U \cup V, E, w)$.
For each edge $e \in E(H)$, compute ratio $w'(e) = w(e)/|R(e)|$. Initially, $M_2 = \emptyset$. In each iteration, selects an edge e in $E(H)$ with minimum $w'(e)$, $M_2 = M_2 \cup \{e\}$, and removes all vertices of e from $U(H) \cup V(H)$ and all edges of $\cup_{u \in e} E(u)$ from $E(H)$. Continue to the next iteration until $R(M_2) = R$.

Algorithm GreedyMinDist has at most $|R|$ iterations for each solution computed. Let M be the solution computed by GreedyMinDist and M^* be an optimal solution to the MTRS-minDist problem. A driver η_u or a rider r_v is said to be *covered* by some solution if an edge e containing η_u or r_v is in the solution; otherwise, *uncovered* with respect to (w.r.t.) the solution.

Lemma 1. *Algorithm GreedyMinDist always computes a feasible solution M to the MTRS-minDist problem.*

A proof of Lemma 1 is given in [15]. Recall that $w(e) = TD(\eta_i, R_i)$ is an integer, and in general, a shorter-distance path has a shorter travel time and vice verse, implying $w(e)$ is almost always positive. For analysis purpose, we assume that the more locations a driver needs to visit, the longer the distance and time are required, which implies that $w(e) \geq 1$ for all edges e in H and the the following property.

Property 1. For any pair of edges e and e' in $E(H)$ such that $D(e) = D(e')$ and $R(e') \subseteq R(e)$, $w(e') \leq w(e)$.

By Property 1, we can assume M_1 by GreedyMinDist has property that $|R(e)| = 1$ for every $e \in M_1$ (a proof is given in [15]). Further, M^* can be modified such that every edge e^* of M^* contains exactly λ riders, which is also shown in [15]. Let $\mu = \frac{max_{e \in E(H)} w(e)}{min_{e \in E(H)} w(e)}$, $\hat{M}$ be a non-empty subset of M_1, and a non-empty subset $\hat{M}^* \subseteq M^*$. The proof of Lemma 2 is given in [15].

Lemma 2. *There is a non-empty subset $\hat{M} \subseteq M_1$ and a non-empty subset $\hat{M}^* \subseteq M^*$ such that $|\hat{M}| = |\hat{M}^*|$, $w(\hat{M}) \leq w(\hat{M}^*)$ and $|\hat{M}^*| \geq \lambda |M^*|/(\lambda + 1)$.*

Theorem 2. *Let M be a solution found by Algorithm GreedyMinDist and M^* be an optimal solution to the MTRS-minDist problem. Under Assumption 2 and $w(e) \geq 1$ for all edges e in H, $\frac{w(M)}{w(M^*)} \leq \frac{\lambda^2 \mu + \lambda}{\lambda + 1}$ for $\lambda \geq 2$.*

Proof. Suppose $M = M_1$. From Lemma 2, $w(\hat{M}) \leq w(\hat{M}^*)$ and each rider of $R(M^*) \setminus R(\hat{M})$ is covered by an edge $e \in M \setminus \hat{M}$ with weight at most $max_{e \in E(H)} w(e)$. Hence,

$$\begin{aligned} w(M) &\leq |R(M^*) \setminus R(\hat{M})| \max_{e \in M} w(e) + w(\hat{M}) \\ &\leq |R(M^*) \setminus R(\hat{M})| \max_{e \in E(H)} w(e) + w(\hat{M}^*). \end{aligned} \quad (1)$$

From Lemma 2, $|\hat{M}^*| \geq \lambda|M^*|/(\lambda+1) \geq 1$. Let $x = w(\hat{M}^*)/|\hat{M}^*|$. Then, $1 \leq \min_{e \in E(H)} w(e) \leq \min_{e \in M^*} w(e) \leq x \leq \max_{e \in E(H)} w(e)$. Let $\overline{M^*} = M^* \setminus \hat{M}^*$. From Eq (1),

$$\begin{aligned} \frac{w(M)}{w(M^*)} &\leq \frac{|R(M^*) \setminus R(\hat{M})| \max_{e \in E(H)} w(e) + w(\hat{M}^*)}{w(\overline{M}^*) + w(\hat{M}^*)} \\ &\leq \frac{(\lambda|M^*| - |\hat{M}^*|) \max_{e \in E(H)} w(e) + |\hat{M}^*|x}{|\overline{M^*}| \min_{e \in M^*} w(e) + |\hat{M}^*|x} \\ &\leq \frac{(\lambda|M^*| - |\hat{M}^*|) \max_{e \in E(H)} w(e) + |\hat{M}^*|x}{|\overline{M^*}| \min_{e \in E(H)} w(e) + |\hat{M}^*|x}. \end{aligned} \tag{2}$$

Since $\lambda(|M^*| - |\hat{M}^*|) \max_{e \in E(H)} w(e) \geq |\overline{M^*}| \min_{e \in E(H)} w(e)$ for $\lambda \geq 1$, Eq (2) is non-increasing as x increases. Taking $x = \min_{e \in E(H)} w(e)$, Eq (2) becomes

$$\frac{w(M)}{w(M^*)} \leq \frac{\lambda|M^*| \max_{e \in E(H)} w(e)}{|M^*| \min_{e \in E(H)} w(e)} - \frac{|\hat{M}^*|(\max_{e \in E(H)} w(e) - \min_{e \in E(H)} w(e))}{|M^*| \min_{e \in E(H)} w(e)}.$$

From this and $|\hat{M}^*|(\lambda+1)/\lambda \geq |M^*|$ (by Lemma 2),

$$\begin{aligned} \frac{w(M)}{w(M^*)} &\leq \lambda\mu - \frac{\lambda|\hat{M}^*|(\max_{e \in E(H)} w(e) - \min_{e \in E(H)} w(e))}{|\hat{M}^*|(\lambda+1) \min_{e \in E(H)} w(e)} \\ &= \lambda\mu - \frac{\lambda(\max_{e \in E(H)} w(e) - \min_{e \in E(H)} w(e))}{(\lambda+1) \min_{e \in E(H)} w(e)} \\ &= \lambda\mu - \frac{\lambda \cdot \max_{e \in E(H)} w(e)}{(\lambda+1) \min_{e \in E(H)} w(e)} + \frac{\lambda \cdot \min_{e \in E(H)} w(e)}{(\lambda+1) min_{e \in E(H)} w(e)} \\ &= \frac{\lambda^2\mu + \lambda\mu - \lambda\mu + \lambda}{\lambda+1} = \frac{\lambda^2\mu + \lambda}{\lambda+1}. \end{aligned}$$

If $M = M_2$, $\frac{w(M_2)}{w(M^*)} \leq \frac{\lambda^2\mu+\lambda}{\lambda+1}$ since $w(M_2) \leq w(M_1)$. Therefore, the theorem holds. □

Algorithm GreedyMinNum. Let $H(U \cup V, E, w)$ be the hypergraph constructed for the an instance of the MTRS-minNum problem.

- Initially, $M = \emptyset$. In each iteration, select e in $E(H)$ with maximum $|R(e)|$, $M = M \cup \{e\}$, and remove all vertices of e and all edges of $\cup_{u \in e} E(u)$. Continue to the next iteration until $R(M) = R$.

Algorithm GreedyMinNum has at most $|R|$ iterations to compute M. By a similar analysis of Lemma 1, Algorithm GreedyMinNum computes a feasible solution M to the MTRS-minNum problem. We show that GreedyMinNum is $\frac{\lambda+2}{2}$-approximate if Assumption 2 is satisfied. The complete analysis is given in [15].

Theorem 3. *Let M be a solution found by Algorithm GreedyMinNum and M^* be an optimal solution to the MTRS-minNum problem. Then $\frac{|M|}{|M^*|} \leq \frac{\lambda+2}{2}$ for $\lambda \geq 2$, under Assumption 2.*

Algorithm LS. Algorithms GreedyMinDist and GreedyMinNum achieve better approximation ratios than the trivial upper bounds – $\lambda\mu$ for MTRS-minDist and λ for MTRS-minNum with Assumption 1. We also give a local search algorithm (Algorithm LS) for MTRS-minNum with Assumption 1 (details in [15]).

4 Clustering

Computing all feasible matches and the hypergraph $H(U \cup V, E, w)$ can be time-consuming when the problem instance size is large. Furthermore, a designated driver may need to travel a long distance to serve a rider in order to satisfy Assumption 2. To address these issues, we propose a clustering algorithm (called **CL**) to partition the drivers and riders into groups, where each group contains a limited number of drivers and riders. Given the set $\mathit{\Gamma}$ of personal drivers and the set R of riders in an instance of MTRS, $\mathit{\Gamma} \cup R$ is partitioned into disjoint groups $\mathcal{C} = \{C_1, C_2, \dots, C_l\}$ such that $C_i \cap C_j = \emptyset$ for all $i \neq j$ and $\cup_{1 \leq i \leq l} C_i = \mathit{\Gamma} \cup R$. Each C_i is called a *cluster*, and denoted by $\mathit{\Gamma}(C_i)$ and $R(C_i)$ are the sets of drivers and riders contained in C_i, respectively. We then allocate designated drivers to clusters. For each cluster, the feasible matches and hypergraph are computed independently.

Algorithm CL uses a graph to partition $\mathit{\Gamma} \cup R$, where each element of $\mathit{\Gamma} \cup R$ is represented by a node in the graph. A driver (rider) who specifies FM match type is called an FM-driver (FM-rider), and similarly for LM-driver and LM-rider. Let $\mathit{\Gamma}(FM)$ and $\mathit{\Gamma}(LM)$ be the sets of personal FM-drivers and LM-drivers, resp. Let $R(FM)$ and $R(LM)$ be the sets of FM-riders and LM-riders, resp. Each driver η_u has a driver interval window $[a_u, b_u] \subseteq [\alpha_u, \beta_u]$, where α_u is the earliest departure time and β_u is the latest arrival time of η_u. Each rider r_v has a rider interval window $[a_v, b_v] \subseteq [\alpha_v, \beta_v]$. A driver η_u and a rider r_v are connected by an edge in the graph if η_u and r_v have the same match type and their interval windows satisfy some overlapping conditions (details omitted). Algorithm CL has two phases, partially described below (see [15] for full details).

4.1 Construction Phase (I)

Clusters are first created for FM match type and then for LM match type, independently. There are three major steps to construct the clusters.

(I.A) Typically, the spatial area (physical road-map) can be viewed as a large rectangle. We partition the large rectangle into $m_1 \times m_2$ smaller rectangles called *cells*, where $1 \leq m_1$ and $1 \leq m_2$ are input parameters (integers) to Algorithm CL. Each cell is denoted by $g(x, y)$ for $1 \leq x \leq m_1$ and $1 \leq y \leq m_2$, where $g(1, 1)$ is the top left corner. Two cells $g(x, y)$ and $g(x', y')$ are *adjacent* if

$|x - x'| + |y - y'| \leq 2$ (at most two cells away). For each cell $g(x, y)$, we create a bin $B(x, y)$.

(I.B1) Distribute riders of $R(FM)$ and drivers of $\Gamma(FM)$ to each bin $B(x, y)$ if their origins are in cell $g(x, y)$. Let $R(x, y)$ and $\Gamma(x, y)$ be the sets of riders and drivers in bin $B(x, y)$, respectively. For each cell $g(x, y)$, the $m_1 \times m_2$ cells are divided into 8 non-overlapping sectors, 4 diagonal sectors, 2 horizontal and 2 vertical sectors, all centered at $g(x, y)$.

(I.B2) Riders and drivers with similar travel direction (determined by their bins and sectors) and overlapping time windows are put into the same cluster. Let $R(sec_j) \subseteq R(x, y)$ be the set of riders and $\Gamma(sec_j) \subseteq \Gamma(x, y)$ be the set of drivers with destinations in sector sec_j, $1 \leq j \leq 8$, respectively. For each $1 \leq j \leq 8$, find a subset R'_j of $R(sec_j)$ such that for every $r_v \in R(sec_j) \setminus R'_j$, $[a_v, b_v]$ overlaps with $[a_u, b_u]$ for some rider $r_u \in R(sec_j)$. Let C_o be the origin cell and C_d be the destination sector for cluster C. Then, clusters are created as follows.

- For each rider $r_u \in R'_j$, create a FM-cluster $C(r_u)$ with $C_o = g(x, y)$ and $C_d = sec_j$, and add r_u to $C(r_u)$. To add riders and drivers to $C(r_u)$, we first construct an interval graph $I_j(U \cup V, E)$ as follows. $U(I_j) = R'_j$ and $V(I_j) = (R(sec_j) \setminus R'_j) \cup \Gamma(sec_j)$. For any $u \in U(I_j)$ and $v \cup V(I_j)$, add an edge $\{u, v\}$ to $E(I_j)$ if $[a_u, b_u]$ and $[a_v, b_v]$ overlap. For every node u in I_j, let $\deg_{I_j}(u)$ be the degree of u.
- For every rider node and driver node $v \in V(I_j)$ with $\deg(v) = 1$, add the rider/driver corresponding to v to the corresponding cluster $C(r_u)$, where $\{u, v\} \in N(u)$, and remove v from I_j. Then in each iteration, select the cluster $C(r_u)$ with the smallest $|R(C(r_u))|$ ($\deg_{I_j}(u)$ as a tiebreaker). Pick the node v adjacent to u with smallest $\deg_{I_j}(v)$, add v to $C(r_u)$, and remove v from I_j. Repeat this until all riders and drivers are clustered. For every isolated node v of $V(I_j)$, create a cluster $C(\eta_v)$, which will be used in the refinement phase.

(I.C) Repeat Step (I.B) symmetrically for $R(LM)$ and $D(LM)$ (i.e., treat destinations as origins and vice versa).

4.2 Refinement Phase (II)

The input to the second phase of Algorithm CL is the set $\mathcal{C}$ of clusters constructed in phase (I). For brevity, let $|C| = |R(C) \cup \Gamma(C)|$ be the *size* of C for any cluster C of $\mathcal{C}$. Large clusters are divided into multiple smaller clusters, while small clusters are merged with nearby clusters. Further, we try to keep $\frac{|R(C)|}{|\Gamma(C)|}$ to be within some reasonable range for each cluster $C \in \mathcal{C}$.

(II.A) Small clusters are merged with their adjacent clusters. Let s_{min} be an input parameter to Algorithm CL (denoting the threshold of small cluster size) and $C_{min} = \{C \in \mathcal{C} \mid |C| < s_{min}\}$. Construct a graph $G_C(U \cup V, E)$ for clusters of $\mathcal{C}$, where $U(G_C) = C_{min}$ and $V(G_C) = \mathcal{C} \setminus C_{min}$. Two clusters C and C' are *adjacent* if $C_d = C'_d$, C_o and C'_o are adjacent, and interval windows satisfy some overlapping conditions. Adjacent clusters in $U(G_C)$ and $V(G_C)$ are connected, and adjacent clusters within $U(G_C)$ are connected. Process each cluster C of C_{min}, starting with the smallest size. *Merge* C with an adjacent cluster C' by moving all riders and all drivers of C to C', and remove C from C_{min}. Contract the two vertices corresponding to C and C' in G_C. Repeat until C_{min} is empty, which is the end of this step.

(II.B) Given two input parameters τ_1 and τ_2 to Algorithm CL, where $0 < \tau_1 \leq |R|/|\Gamma|$ and $\tau_2 \geq 1$, a cluster $C \in \mathcal{C}$ is *balanced* if $\tau_1 \leq \frac{|R(C)|}{|\Gamma(C)|} \leq \frac{\tau_2|R|}{|\Gamma|}$; otherwise, C is *imbalanced.* Let $C_{imb} = C_l \cup C_h$, where $C_l = \{C \in \mathcal{C} \mid \frac{|R(C)|}{|\Gamma(C)|} < \tau_1\}$ and $C_h = \{\frac{|R(C)|}{|\Gamma(C)|} > \frac{\tau_2|R|}{|\Gamma|}\}$ for the resulting set $\mathcal{C}$ of clusters after Step (II.A). For each cluster $C \in C_{imb}$, define an *imbalanced value* f_C as:

$$f_C = \begin{cases} \frac{|\Gamma(C)|}{|R(C)|}, & \text{if } C \in C_l, \\ \frac{|R(C)|}{|\Gamma(C)|}, & \text{if } C \in C_h. \end{cases}$$

Construct a graph $G_C(U \cup V, E)$ for clusters of $\mathcal{C}$, where $U(G_C) = C_{imb}$ and $V(G_C) = \mathcal{C} \setminus C_{imb}$. Create edges for clusters between $U(G_C)$ and $C' \in V(G_C)$ and edges for clusters between $U(G_C)$ and $U(G_C)$ as described in Step (II.A). Process each cluster C of C_{imb}, starting with the most imbalanced. Select a cluster C' in G_C that is adjacent to C such that merging C with C' reduces f_C the most. Merge cluster C with C' and remove C from C_{imb}. Contract the two vertices corresponding to C and C' in G_C. Repeat until C_{imb} is empty, which is the end of this step.

(II.C) Finally, large clusters are divided. Let s_{max} be an input parameter to Algorithm CL (denoting the threshold of large cluster size), $\mathcal{C}$ be the set of clusters obtained after Step (II.B), and $C_{max} = \{C \in \mathcal{C} \mid |C| > s_{max}\}$. The algorithm processes each $C \in C_{max}$ one by one. C is heuristically partitioned into $C_1, C_2, \ldots, C_z$, where $z = \lfloor |C_i|/s_{max} \rfloor$, such that $\sum_{1 \leq i \leq z} |R(C_i)| = |R(C)|$, $\sum_{1 \leq i \leq z} |\Gamma(C_i)| = |\Gamma(C)|$, $|C_i| \approx |C|/z$ and $|R(C_i)|/|\Gamma(C_i)| \approx |R(C)|/|\Gamma(C)|$ for every $1 \leq j \leq z$. After the partitioning, $C_1, \ldots, C_z$ are included into $\mathcal{C}$, and they do not need to be checked again.

5 Experiment

We perform a computational study to evaluate our system and algorithms. Using datasets and surveys on transit and travel patterns in Chicago, we create a simulation environment that generates rider requests and driver offers. A summary of the experimental results for MTRS-minDist and MTRS-minNum is presented here. The simulation and all algorithms were implemented in Java and executed on Ryzen 5 7600 with 20 GB RAM. All ILP formulations were solved using CPLEX v12. The complete experiment setup and results is presented in [15].

MTRS-minDist Results. The overall results are shown in Table 1. Exact, labeled as '240' and '720', indicates that the Exact ILP is allowed to run for 240 and 720 seconds, respectively, to find a solution. The average incurred travel distance per assigned driver is calculated as the total incurred travel distance of assigned drivers divided by the total number of assigned drivers. The average travel time per rider is reduced by about 38.6%, while the average incurred travel distance per driver is less than 9.5 km (for Exact). With clustering, the average travel time per rider is reduced by about 38.7%, while the average incurred travel distance per driver is 10.6 km (for Exact) with 56% and 70% running time reduction for Exact-240 and Exact-720, respectively.

MTRS-MinNum Results. We first compute an assignment Π_Γ that assigns a maximal set $R' \subseteq R$ to drivers of Γ. Two algorithms described in [14], exact ILP (labeled as *ILP-P* here) and *ImpGreedy*, are used to compute assignment Π_Γ. Then, we use each of LS, GreedyMinNum (*Greedy* for short), and Exact to compute an assignment Π_Δ such that all of $R \setminus R'$ are assigned to drivers of Δ. The results are shown in Table 2. The average travel time per rider is reduced by about 38.5%, while 3179 drivers (99.99% of Γ and 7.9% of Δ) are required for Exact under ILP-P. With clustering for Exact under ILP-P, the average travel time per rider is reduced by about 39%, while 3730 drivers (95.9% of Γ and 16.4% of Δ) are required to serve all riders, achieving a 69% reduction in running time. Greedy outperforms LS slightly in both solution quality and running time on average. This is mainly because LS computes a solution with only Assumption 1, which requires it to find a feasible solution M in H first, and improvements made upon M are restricted.

Table 1. Results of all algorithms. Distance is in kilometers and time is in minute.

Avg public transit duration per rider	47.41 minutes		
Total number of drivers generated	2571 personal and 7700 designated		
Total number of riders generated	7700		
Without Clustering	GreedyMinDist	Exact-240	Exact-720
Avg incurred travel dist per assigned driver	11.69	9.34	8.72
Avg transit time saved per rider	18.28	18.29	18.32
Avg running time per interval	17.72	24	34.75
With Clustering	GreedyMinDist	Exact (no time limit)	
Avg incurred travel dist per assigned driver	13.98	10.64	
Avg transit time saved per rider	18.21	18.38	
Avg running time per interval	10.29	10.61	

Table 2. Results of all algorithms. Distance is in kilometers and time is in minute.

	ILP-P			ImpGreedy		
Without Clustering	LS	Greedy	Exact	LS	Greedy	Exact
Total # of assigned drivers	3324	3253	3179	3438	3359	3250
Total # of assigned personal drivers	2570	2570	2570	2452	2452	2452
Total # of assigned designated drivers	754	683	609	986	907	798
Avg transit time saved per rider	18.33	18.3	18.26	18.28	18.23	18.21
Avg running time per interval	14.63	14.58	15.9	14.8	14.1	15.67
With Clustering	LS	Greedy	Exact	LS	Greedy	Exact
Total # of assigned drivers	3885	3823	3732	3810	3756	3657
Total # of assigned personal drivers	2466	2466	2466	2270	2270	2270
Total # of assigned designated drivers	1419	1357	1266	1540	1486	1387
Avg transit time saved per rider	18.55	18.47	18.48	18.49	18.47	18.40
Avg running time per interval	4.94	4.93	4.98	5.41	5.4	5.27

6 Conclusion

The proposed system integrates public transit with ridesharing to enhance FM and LM commutes, prioritizing ridesharing routes to improve convenience, reduce commute times, and minimize traffic impacts (while potentially improving traffic conditions). We give a hypergraph-based ILP exact algorithm and approximation algorithms, along with a clustering algorithm that significantly reduces the time required to compute the hypergraph. Computational results show that our system and algorithms are effective in improving FM/LM commutes.

It will be interesting to see if the approximation ratios of GreedyMinDist and GreedyMinNum can be improved by adapting some local improvement approach, such as incorporating with Algorithm LS or local search algorithms for set packing problems. The clustering algorithm improves the running time significantly, but it also decreases the solution quality notably. It is worth to improve the clustering algorithm so that it can also lead to solution with quality comparable to non-clustering, or to show that it is NP-hard to do so.

Acknowledgments. The authors thank anonymous reviewers for their constructive comments. This work was partially supported by Canada NSERC Discovery Grants.

References

1. Ahuja, R., Magnanti, T., Orlin, J.: Network Flows: Theory, Algorithms, and Applications. Prentice Hall, Upper Saddle river (1993)

2. Alonso-González, M., Liu, T., Cats, O., Oort, N.V., Hoogendoorn, S.: The potential of demand-responsive transport as a complement to public transport: an assessment framework and an empirical evaluation. Transp. Res. Rec. **2672**(8), 879–889 (2018)
3. Bei, X., Zhang, S.: Algorithms for trip-vehicle assignment in ride-sharing. In: Proceedings of the AAAI Conference on Artificial Intelligence, vol. 32, no. 1 (2018)
4. Blad, K., de Almeida Correia, G.H., van Nes, R., Anne Annema, J.: A methodology to determine suitable locations for regional shared mobility hubs. Case Stud. Transp. Policy **10**(3), 1904–1916 (2022)
5. Bürstlein, J., López, D., Farooq, B.: Exploring first-mile on-demand transit solutions for North American suburbia: a case study of Markham, Canada. Transport. Res. Part A: Policy Pract. **153**, 261–283 (2021)
6. Carrese, F., et al.: The integration of shared autonomous vehicles in public transportation services: a systematic review. Sustainability **15**(17) (2023)
7. Chandra, B., Halldórsson, M.: Greedy local improvement and weighted set packing approximation. J. Algor. **39**(2), 223–240 (2001)
8. Chen, Y., Wang, H.: Pricing for a last-mile transportation system. Transport. Res. Part B: Methodol. **107**, 57–69 (2018)
9. Fielbaum, A.: Strategic public transport design using autonomous vehicles and other new technologies. Int. J. Intell. Transp. Syst. Res. **18**, 193–191 (2020)
10. Fielbaum, A., Jara-Díaz, S., Alonso-Mora, J.: Beyond the last mile: different spatial strategies to integrate on-demand services into public transport in a simplified city. Public Transp. **16**, 855–892 (2024)
11. Fielbaum, A., Tirachini, A., Alonso-Mora, J.: Improving public transportation via line-based integration of on-demand ridepooling. Transport. Res. Part A: Policy Pract. **190**, 104289 (2024)
12. Fürer, M., Yu, H.: Approximating the k-set packing problem by local improvements. In: Fouilhoux, P., Gouveia, L., Mahjoub, A.R., Paschos, V.T. (eds.) ISCO 2014. LNCS, vol. 8596, pp. 408–420. Springer, Cham (2014). https://doi.org/10.1007/978-3-319-09174-7_35
13. Gu, Q., Liang, J.: Multimodal transportation with ridesharing of personal vehicles. In: 32nd International Symposium on Algorithms and Computation (ISAAC 2021). LIPIcs, vol. 212, pp. 39:1–39:16. Schloss Dagstuhl – Leibniz-Zentrum für Informatik (2021)
14. Gu, Q.P., Liang, J.L.: Algorithms and computational study on a transportation system integrating public transit and ridesharing of personal vehicles. Comput. Oper. Res. **164**, 106529 (2024)
15. Gu, Q.P., Liang, J.L.: Exact set packing in multimodal transportation with ridesharing system for first/last mile (2025). https://doi.org/10.48550/arXiv.2505.01989. arXiv:2505.01989
16. Hall, J.D., Palsson, C., Price, J.: Is uber a substitute or complement for public transit? J. Urban Econ. **108**, 36–50 (2018)
17. Huang, Y., Kockelman, K.M., Garikapati, V.: Shared automated vehicle fleet operations for first-mile last-mile transit connections with dynamic pooling. Comput. Environ. Urban Syst. **92**, 101730 (2022)
18. Javed, T., Ahmed, A., Raman, V., Alquaity, A., Johansson, B.: Combustion-based transportation in a carbon-constrained world—a review. In: Agarwal, R.A., Agarwal, A.K., Gupta, T., Sharma, N. (eds.) Pollutants from Energy Sources. EES, pp. 7–34. Springer, Singapore (2019). https://doi.org/10.1007/978-981-13-3281-4_2

search efficiency is constrained by imperfect sensing and mobility limitations. One such application arises in autonomous robotic search-and-rescue missions, where drones or ground robots must locate missing persons in uncertain environments. High-speed traversal allows for rapid coverage of large areas, but the probability of detecting a small target may decrease unless slower, more careful searches are employed. Another application is in environmental monitoring and hazardous material detection, where mobile sensors must scan an area for pollutants, leaks, or radioactive sources. Fast movement enables broad sweeps, but sensors often have detection latencies or require sufficient exposure time for reliable readings. Similar challenges arise in underwater exploration, where autonomous underwater vehicles (AUVs) must detect submerged objects with sonar systems that may exhibit probabilistic detection characteristics depending on speed and environmental conditions. In cybersecurity and computer networks, our model also relates to intrusion detection systems (IDS) scanning for malicious activity. A security algorithm may switch between fast, coarse-grained sweeps of a network and slower, more detailed analyses of suspicious areas, reflecting the trade-offs between speed and detection reliability in anomaly detection. Beyond these practical applications, our study contributes to the broader theoretical understanding of search and exploration in adversarial and uncertain environments. By characterizing competitive strategies for different parameter settings, we provide insights into the fundamental limits of probabilistic detection in search problems.

The contributions of this paper are:

1. We analyze the problem in the extreme case where $p = 0$ (no detection at speed 1), providing an algorithm with tight competitive bounds.
2. We consider the other extreme case where $v = 0$ (no slow speed), providing an algorithm and its competitive ratio.
3. For the case where $p \in (0, 1)$ and $v \in (0, 1)$, we propose a hybrid algorithm.

The rest of the paper is organized as follows. In Sect. 2, we review related work on the linear search problem and its variants. In Sect. 3, we present an algorithm and its competitive ratio for the case where $v = 0$ (i.e., the robot must always move at speed 1). In Sect. 4, we present an algorithm and its competitive ratio for the case where $p = 0$ (i.e., the robot can only detect the target when moving at speed v). In Sect. 5, we discuss the general case where $p, v \in (0, 1)$ and propose a hybrid algorithm. Finally, we conclude in Sect. 6.

2 Related Work

The linear search problem, also known as the cow-path problem, was first studied by Beck [2] and Bellman [3] in the 1960s. The classical formulation involves an agent searching for a stationary target on an infinite line while minimizing the competitive ratio—the worst-case ratio of the agent's travel distance to the minimum distance required by an omniscient searcher. Over the decades, numerous variations have been investigated, including multi-agent group search and evacuation [10], moving targets [7], faulty agents [11], and search-and-rescue [5]. A recent survey [13] discusses the impact of prior knowledge on the competitive ratio in linear search.

A few studies have explored variations of the problem where the agent's speed is non-uniform. In [9], a search model is examined in which an agent can switch between a faster movement mode (without detection capability) and a slower mode that enables detection. This model is closely related to our work when $v = 1$ and $p = 0$, but differs in that it considers multi-agent search on a ray rather than an infinite line. Another study [12] investigates terrain-dependent speeds, where an agent's velocity depends on its location and direction of travel.

The challenge of uncertain detection has been analyzed in multiple settings. The competitive bound for the $v = 0$ case was previously established in [1] (Sect. 8.6.2), although a matching lower bound remains an open problem. The study in [4] examines the case of probabilistic detection on a ray, demonstrating that a non-monotonic strategy is optimal. Similarly, [6] explores a setting in which an agent can detect a target only when operating in the correct one of multiple searching modes.

To the best of our knowledge, our study is the first to consider a search model that simultaneously incorporates variable speeds and probabilistic detection on an infinite line.

3 The Fast-Only Case

In this section, we consider the case where $v = 0$ and so the agent must always move at the high speed 1. We start by presenting an algorithm with a competitive ratio of $\frac{8}{p} + \frac{p}{2-p}$, Algorithm 1.

Algorithm 1. Fast Approach

1: $a \leftarrow \frac{2}{2-p}$, $i \leftarrow 0$ ▷ Expansion Ratio, Round
2: **while** target not found **do**
3: move to $(-1)^{i+1}a^i$ and back to 0 at speed 1
4: $i \leftarrow i + 1$ ▷ Increment Round
5: **end while**

A version of the following theorem was previously proven in [1] (Sect. 8.6.2). For the sake of completeness, a proof can be found in the full version of the paper [8].

Theorem 1. *Algorithm 1 has an expected competitive ratio of* $\frac{8}{p} + \frac{p}{2-p}$.

Observe that, as expected, when $p = 1$, the competitive ratio is 9 (consistent with the original linear search problem).

4 The Slow-Only Case

In this section, we consider the case where $p = 0$ and so the agent cannot detect the target when moving at speed 1. For this case, we present an algorithm where the agent always moves at speed v, except over sections of the line that have already been explored. The algorithm is presented in Algorithm 2.

Algorithm 2. Slow Approach

$a \leftarrow 1 + \sqrt{\frac{2v^2}{v(1+v)}}$, $i \leftarrow 0$ ▷ Expansion Ratio, Round
while target not found **do**
  move to $(-1)^{i+1}a^{i-2}$ at speed 1 ▷ Traverse previously-explored regions quickly
  move to $(-1)^{i+1}a^{i}$ at speed v ▷ Traverse new regions slowly
  move back to 0 at speed 1
  $i \leftarrow i + 1$ ▷ Increment Round
end while

Theorem 2. *Algorithm 2 has a competitive ratio of* $3 + 2\sqrt{2 + \frac{2}{v}} + \frac{2}{v}$.

Proof. Observe that the agent executing Algorithm 2 essentially runs in rounds, where each round i consists of the agent moving from the origin to a^i and back to the origin. Let T_i denote the time it takes for the agent to complete round i. Observe that $T_0 = 2$ and $T_1 = 2a$, T_i for every $i \geq 2$ can be written:

$$T_i = a^{i-2} + \frac{a^i - a^{i-2}}{v} + a^i$$

Note that, because the agent moves at speed v over unexplored sections, it is guaranteed to detect the target the first time it passes through its location. Let us first consider the case where the target is detected in the first or second round. If the target is detected in the first round, then the competitive ratio is clearly 1 and the theorem holds. If the target is detected in the second round, then the target must be located at position $d \in (1, a^2]$. The expected competitive ratio is then:

$$\frac{2+d}{d} = 1 + \frac{1}{d}$$

which is maximized as $d \to 1$. Thus, the competitive ratio is at most 2 and the theorem holds.

Now let us consider the case where the target is detected in round $i + 2$. The target, then, must be located at position $d \in (a^i, a^{i+2}]$ and the expected competitive ratio is

$$\frac{2 + 2a + \sum_{j=2}^{i+1} T_j + a^i + \frac{d - a^i}{v}}{d}.$$

which, after substituting and evaluating the geometric series, simplifies to:

$$\frac{1}{d}\left(\frac{d-a^i}{v}+a^i+\frac{\left(a^{i+2}-1\right)\left(a^2(1+v)+v-1\right)}{(a-1)a^2v}\right).$$

Observe the above is decreasing with respect to d. Then, since $d < a^i$, the competitive ratio is maximized as $d \to a^i$. Thus, the competitive ratio is at most:

$$\begin{aligned}\lim_{d\to a^i}\frac{1}{d}&\left(\frac{d-a^i}{v}+a^i+\frac{\left(a^{i+2}-1\right)\left(a^2(1+v)+v-1\right)}{(a-1)a^2v}\right)\\ &=1+\frac{(a^{i+2}-1)(a^2(1+v)+v-1)}{a^{i+2}(a-1)v}\end{aligned}$$

Since the above is increasing with respect to i, the competitive ratio is maximized as $i \to \infty$. Thus, the competitive ratio is at most:

$$\lim_{i\to\infty} 1+\frac{(a^{i+2}-1)(a^2(1+v)+v-1)}{a^{i+2}(a-1)v}=1+\frac{a^2(1+v)+v-1}{a^2(a-1)v}$$

Substituting $a = 1+\sqrt{\frac{2v^2}{v(1+v)}}$ yields result stated in the theorem. □

Again, as expected, when $v = 1$, the competitive ratio is 9 (consistent with the original linear search problem).

Theorem 3. *If* $p = 0$*, there exists no algorithm with a competitive ratio better than* $3 + 2\sqrt{2+\frac{2}{v}}+\frac{2}{v}$.

Proof. Let $l(t) \geq 0$ and $r(t) \geq 0$ denote the leftmost and rightmost points such that the entire interval $[l(t), r(t)]$ has been explored (traversed by the agent at the slow speed v at least once) at time t. We will refer to the interval $[l(t), r(t)]$ as the *explored region*, $l(t)$ as the *left boundary* of the explored region, and $r(t)$ as the *right boundary* of the explored region.

Then, let $\beta_t = \inf_{t'>t}\frac{t'}{|l(t')|+|r(t')|}$. Clearly if $t_1 < t_2$ then $\beta_{t_1} \geq \beta_{t_2}$. Furthermore, $\beta_t \geq 1/v$ for all t since the maximum speed at which the agent can explore a region is v. Now let $\beta = \lim_{t\to\infty}\beta_t$. By the definition of the limit infimum, there must exist a finite time t such that for all $\beta \leq \frac{t_1}{|l(t_1)|+|r(t_1)|}$ for all $t' > t$ and there must exist infinitely many t' such that the (without loss of generality) the right boundary of the explored region $r(t') = \frac{t'}{\beta}$ (touches the the line $t = \beta x$).

Consider such a time t_i. Then, we can define x_j for $j < i$ as the furthest point from the origin the agent could have traveled on the opposite side of the line at time t_{j+1} and t_j as the minimum time the agent could have reached x_j ($t_j = \beta x_j$ if x_j is on the right side of the line and $t_j = -\beta x_j$ if x_j is on the left side of the line). See Fig. 1 for an illustration.

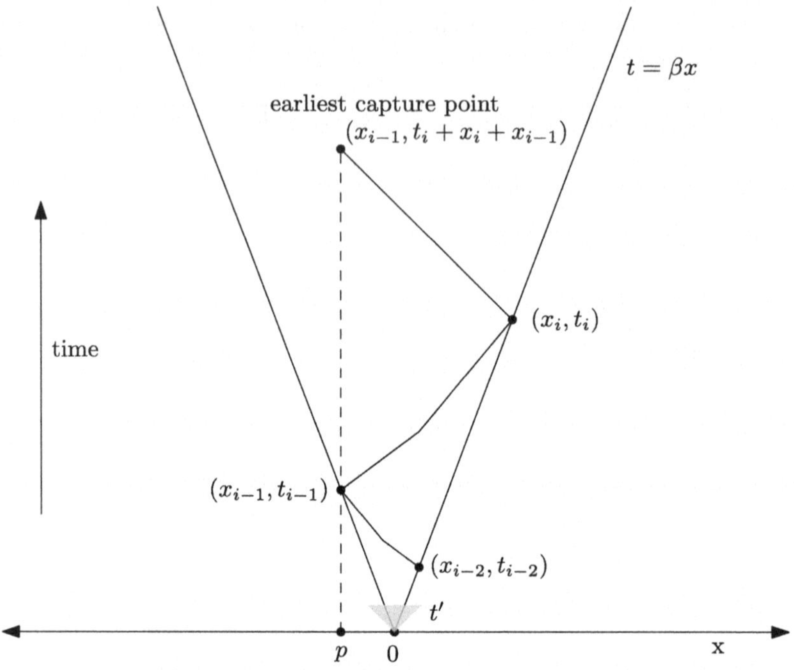

Fig. 1. Cone of exploration for the agent.

Observe, then, that t_i can be bounded recursively:

$$t_i \geq t_{i-1} + \frac{|x_i| - |x_{i-1}|}{v} + |x_i| + |x_{i-1}|$$

$$t_i \geq t_{i-1} + \frac{1}{\beta}\left(\frac{t_i - t_{i-1}}{v} + t_i + t_{i-1}\right)$$

Using trivial base-case bounds of $t_0 \geq 0$ and $t_1 \geq 1$, this recurrence can be solved[1] to show that:

$$t_i \geq \frac{\beta v - 1}{x}\left(\left(\frac{v(1+\beta)+x}{2(\beta v - 1)}\right)^i - \left(\frac{v(1+\beta)-x}{2(\beta v - 1)}\right)^i\right)$$

where $x = \sqrt{4 + v(6\beta v + (1+\beta^2)v - 4(1+\beta))}$.

Consider the case where the target is located just to the left of x_{i-1} so that it could not have been detected before time t_i. Then the earliest time the target could have been detected is at time $t_i + |x_i| + |x_{i-1}| + \epsilon$ for some arbitrarily small

[1] We solved this recurrence using Mathematica. The notebook code is open-source and available at https://github.com/kubishi/code_multi_speed_cow.

$\epsilon > 0$. Then, the competitive ratio can be written:

$$\begin{aligned}
\text{CR} &= \sup_{\epsilon,i} \frac{t_i + |x_i| + |x_{i-1}| + \epsilon}{|x_{i-1}| + \epsilon} = \sup_{\epsilon,i} \left[1 + \frac{t_i + |x_i|}{|x_{i-1}| + \epsilon}\right] \\
&= \sup_i \left[1 + \frac{t_i + |x_i|}{|x_{i-1}|}\right] = \sup_i \left[1 + (1+\beta)\frac{t_i}{t_{i-1}}\right] \\
&= \sup_i \left[1 + \frac{(1+\beta)(v(1+\beta)+x)}{2(\beta v - 1)\left(1 + \frac{2x}{v(1+\beta)-x-(v(1+\beta)+x)^i(v(1+\beta)-x)^{1-i}}\right)}\right] \qquad (1) \\
&= \sup_i \left[1 + \frac{(1+\beta)(v(1+\beta)+x)}{2(\beta v - 1)\left(1 + \frac{2x}{(v(1+\beta)-x)\left(1-\left(\frac{v(1+\beta)+x}{v(1+\beta)-x}\right)^i\right)}\right)}\right]. \qquad (2)
\end{aligned}$$

Equation 1 results from substituting t_i and t_{i-1} into the competitive ratio. Equation 2 follows since $v(1+\beta) - x$ (and therefore $v(1+\beta) + x$) is always positive. Observe that, since $\frac{v(1+\beta)+x}{v(1+\beta)-x} > 1$, (2) is decreasing with respect to i. Thus, the competitive ratio is maximized as $i \to \infty$. Then, the competitive ratio is at most

$$1 + \frac{(1+\beta)(v(1+\beta)+x)}{2(\beta v - 1)\left(1 + \frac{2x}{(v(1+\beta)-x)}\right)}.$$

Substituting $x = \sqrt{4 + v(6\beta v + (1+\beta^2)v - \ 4(1+\beta))}$ yields

$$1 + \frac{(1+\beta)\left(v(1+\beta) + \sqrt{4 + v(6\beta v + (1+\beta^2)v - 4(1+\beta))}\right)}{2(\beta v - 1)}.$$

Using standard minimization techniques, it can be verified that the minimum is achieved with $\beta = 1 + 2/v$. Substituting this value of β yields the result stated in the theorem. □

5 The General Case

Clearly, if both $v = 0$ and $p = 0$, the agent will never find the target and if both $v = 1$ and $p = 1$ the problem is equivalent to the original cow-path problem (and Algorithms 1 and 2 are equivalent and optimal). The only interesting case left, then, is when $v \in (0,1)$ and $p \in (0,1)$. In this case, the agent must decide when to move at speed 1 and when to move at speed v.

5.1 Moving Fast or Slow

The first thing to note about this case is that both Algorithm 1 and 2 (and their competitive ratios) remain valid for this model. A first question, then, is to determine for what values of p and v either algorithm is better than the other. Let CR_{Fast} and CR_{Slow} denote the expected competitive ratios of Algorithm 1 and Algorithm 2, respectively. Then, we have the following result:

$$CR_{Fast} < CR_{Slow}$$

$$3 + 2\sqrt{2 + \frac{2}{v}} + \frac{2}{v} < \frac{8}{p} + \frac{p}{2-p}$$

$$v < \frac{2p}{8 - p(1-p)^2 - 2\sqrt{p(8 + p^2(2-p))}} \tag{3}$$

Thus, a simple algorithm with a competitive ratio of $\max\{CR_{Fast}, CR_{Slow}\}$ is to run Algorithm 1 if Inequality (3) holds and Algorithm 2 otherwise. Figure 4 shows the values of p and v for which Algorithm 1 is better/worse than Algorithm 2.

5.2 A Hybrid Approach

In this section, we introduce a hybrid search strategy that balances the trade-offs between the slow and fast search approaches. The algorithm follows the same structure as the slow search algorithm but incorporates an additional "scouting" step. After reaching the turnaround point in each round, instead of immediately returning to the origin, the agent briefly continues forward at the fast speed before heading back.

The intuition behind this approach is that it disrupts the worst-case scenario of the slow search, where the target is placed just beyond the turnaround point. In the standard slow search algorithm, the agent would only reach such a target in a later round, leading to a high competitive ratio. However, by scouting ahead at high speed, the hybrid algorithm has a chance to detect the target much earlier, potentially improving performance in these cases. The challenge is to carefully balance the length of the scouting phase so that it provides meaningful improvement without excessively increasing the total search time for each round.

Algorithm 3. Hybrid Approach

inputs: expansion ratio a and scout-ahead ratio b
$i \leftarrow 0$ ▷ Round
while target not found **do**
  move to $(-1)^{i+1}a^{i-2}$ at speed 1 ▷ Traverse previously-explored regions quickly
  move to $(-1)^{i+1}a^{i}$ at speed v ▷ Traverse new regions slowly
  move to $(-1)^{i+1}\left(a^{i} + b(a^{i+2} - a^{i})\right)$ at speed 1 ▷ Scout ahead at the fast speed
  move back to 0 at speed 1
  $i \leftarrow i + 1$ ▷ Increment Round
end while

First, observe that Algorithm 3 is equivalent to Algorithm 2 if $b \leftarrow 0$. Thus, the expected competitive ratio of Algorithm 3 is at most the expected competitive ratio of Algorithm 2. Unfortunately, the expected competitive ratio of Algorithm 3 is difficult to compute.

Theorem 4. *The expected competitive ratio of Algorithm 3 with expansion ratio a and scout-ahead ratio b is* $\max\{CR_1, CR_2\}$ *where* CR_1 *is*

$$\frac{1}{(a^2-1)\,b+1}\left[\frac{(a^2-1)\,b}{v}+\frac{a^2\left(a^2\left(2\left(a^2-1\right)bv+v+1\right)+v-1\right)}{(a-1)a^2v}+1\right]$$

and CR_2 *is*

$$(a+1)\left[2a^2b(p-1)^2+a\left(\frac{1}{a-1}-2b(p-1)p\right)\right.$$
$$\left.+\frac{(p-2)p(v-1)}{av}+p(2b+p-2)+\frac{(p-1)^2}{v}\right].$$

Proof. Suppose the target is located at position $d \in (a^i, a^{i+2}]$. Then there are a few cases to consider:

1: The target is $d \geq a^i + b(a^{i+2} - a^i)$. In this case, the target is detected during round $i+2$.
2: The target is $d < a^i + b(a^{i+2} - a^i)$. In this case, the target may be detected during round i.
  2.a: The target is detected during round i while the agent is "scouting ahead".
  2.b: The target is detected during round $i+2$.

Let T_i denote the time it takes for the agent to complete round i:

$$T_i = a^{i-2} + \frac{a^i - a^{i-2}}{v} + 2b(a^{i+2} - a^i) + a^i$$

For case 1, the expected competitive ratio is:

$$\sup_{i,d\in(a^i+b(a^{i+2}-a^i),a^{i+2}]} \frac{\sum_{j=0}^{i+1} T_j + a^i + \frac{d-a^i}{v}}{d} \tag{4}$$

Observe that (4) is decreasing with respect to d. Then, since $d < a^i + b(a^{i+2} - a^i)$, the competitive ratio is maximized as $d \to a^i + b(a^{i+2} - a^i)$. The competitive ratio, then, can be written:

$$\sup_i \frac{\frac{(a^2-1)b}{v} + \frac{(a^2-a^{-i})(a^2(2(a^2-1)bv+v+1)+v-1)}{(a-1)a^2v} + 1}{(a^2-1)\,b+1}$$

Since the above is increasing with respect to i, the competitive ratio is maximized as $i \to \infty$. Thus, the competitive ratio is at most:

$$\frac{1}{(a^2-1)\,b+1}\left[\frac{(a^2-1)\,b}{v}+\frac{a^2\left(a^2\left(2\left(a^2-1\right)bv+v+1\right)+v-1\right)}{(a-1)a^2v}+1\right] \tag{5}$$

For case 2, the agent detects the target during round i while scouting ahead on the way to $a^i + b(a^{i+2} - a^i)$ with probability p and on the way back to the origin with probability $p(1-p)$. With probability $(1-p)^2$, the agent does not detect the target while scouting ahead and detects it only during round $i+2$. The expected competitive ratio is:

$$\sup_{i, d \in (a^i, a^i + b(a^{i+2} - a^i))} \left[p \left(\sum_{j=0}^{i-1} T_j + a^i + \frac{d - a^i}{v} \right) /d + \right.$$
$$p(1-p) \left(\sum_{j=0}^{i+1} T_j + a^{i+2} + \frac{d - a^{i+2}}{v} \right) /d$$
$$\left. (1-p)^2 \left(\sum_{j=0}^{i+1} T_j + a^{i+2} + \frac{d - a^{i+2}}{v} \right) /d \right] \tag{6}$$

Observe again that (6) is decreasing with respect to d. Then, since $d < a^i + b(a^{i+2} - a^i)$, the competitive ratio is maximized as $d \to a^i$. Then, (6) can be simplified to:

$$\frac{1}{a^2(a-1)v} \left[\frac{1 - v - a^2(1 + v + 2(a-1)^2 bv)}{a^i} + \right.$$
$$a(1+a)\Big((a-1)\big[a(1-p)^2 + (2-p)p\big] +$$
$$(a^2 + 2(a-1)a^3 b + 2p - 2a(a + b + a^2(2a-3)b)p$$
$$\left. + (a-1)(1 + a + 2(a-1)a^2 b)p^2)v \Big) \right] \tag{7}$$

Observe now that (7) is decreasing with respect to i since

$$1 - v - a^2(1 + v + 2(a-1)^2 bv) < 1 - v - a^2 < 0$$

when $a > 1, b \in [0,1], v \in (0,1)$. Thus, the competitive ratio is maximized as $i \to \infty$.

$$(a+1) \left[2a^2 b(p-1)^2 + a \left(\frac{1}{a-1} - 2b(p-1)p \right) \right.$$
$$\left. + \frac{(p-2)p(v-1)}{av} + p(2b + p - 2) + \frac{(p-1)^2}{v} \right] \tag{8}$$

The expected competitive ratio is then the maximum of (5) and (8). □

Unfortunately, minimizing the expected competitive ratio of the Hybrid Approach analytically with respect to both a and b has proven difficult. Figure 2 shows the numerically optimized values of the expansion ratio a and the scout-ahead ratio b for different combinations of p and v. As expected, the value of b

increases with p, reflecting the increased benefit of scouting ahead when detection at high speed is more reliable. The expansion ratio a also varies with both parameters, reflecting the trade-off between coverage and redundancy in uncertain detection regimes.

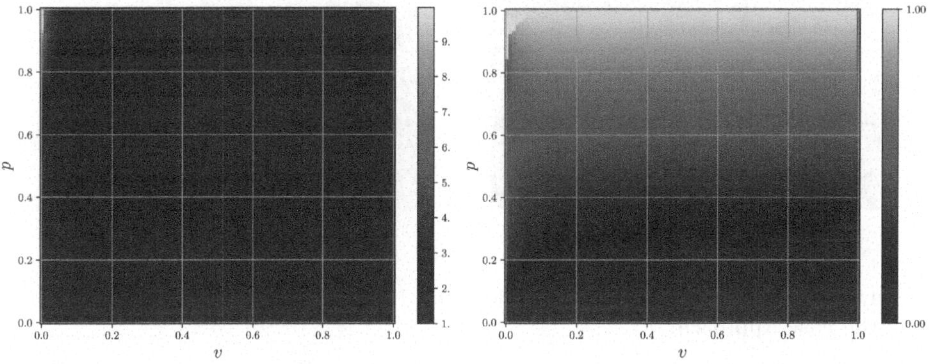

Fig. 2. The numerically computed best expansion ratio a (left) and scout-ahead ratio b (right) values for different values of p and v.

Figure 3 illustrates the expected competitive ratio of the best-performing algorithm between the Fast Approach and the Hybrid Approach[2]. Figure 3 shows the improvement of the Hybrid Approach over the Slow Approach, demonstrating its advantage. Finally, Fig. 4 shows the regions in the p-v plane where the Slow Approach and the Hybrid Approach outperform the Fast Approach. The

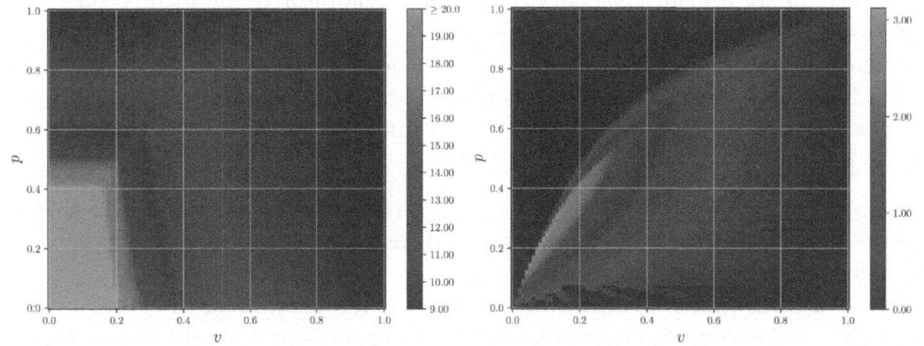

Fig. 3. Competitive ratio (left) and difference in expected competitive ratios (right) of Algorithm 2 and Algorithm 3, showing how much the hybrid approach improves on the slow approach.

[2] The code used to generate all figures is open-source and available at https://github.com/kubishi/code_multi_speed_cow.

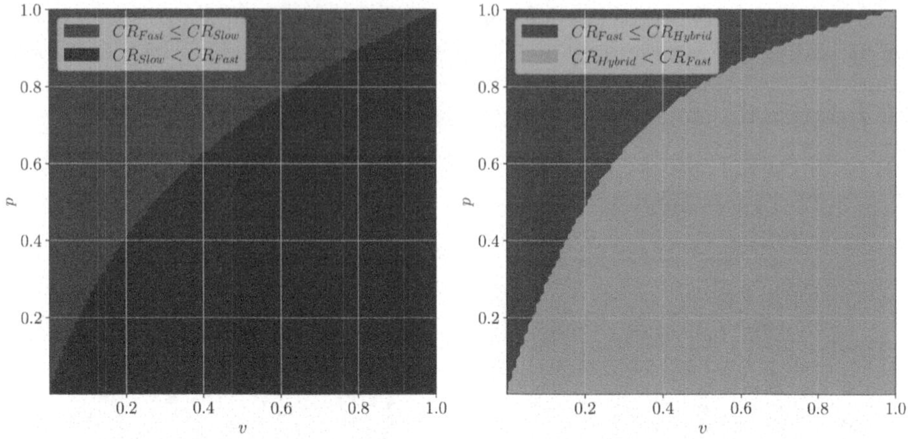

Fig. 4. Values for p and v where the Slow and Hybrid Approaches are better/worse than the Fast Approach.

Hybrid Approach never performs worse than the Slow Approach and expands the parameter space in which it surpasses the Fast Approach. This highlights the effectiveness of a hybrid strategy that balances exploration speed with targeted scouting.

6 Conclusion

In this paper, we introduced and analyzed a novel variant of the linear search (cow-path) problem, where a searching agent balances probabilistic detection with variable movement speeds. We presented competitive algorithms for three key cases: (1) when the agent cannot detect the target at high speed ($p = 0$), (2) when the agent cannot move at a slow speed ($v = 0$), and (3) the general case where both p and v are within $(0, 1)$.

For the $p = 0$ case, we derived an optimal algorithm with a competitive ratio of $3 + 2\sqrt{2 + \frac{2}{v}} + \frac{2}{v}$. For the $v = 0$ case, we established an algorithm with a competitive ratio of $\frac{8}{p} + \frac{p}{2-p}$. In the general case, we demonstrated that a hybrid algorithm leveraging both slow and fast movement strategies improves upon existing methods.

Several directions remain for future work. First, tightening the lower bound for the $v = 0$ case remains an open question. Additionally, it would be interesting to explore models where the detection probability depends continuously on speed (i.e., $p(v)$), or where speed and sensing capabilities degrade over time. Extensions to higher-dimensional settings or bounded domains (e.g., finite intervals or graphs) may reveal different trade-offs. Finally, considering online variants where p or v is initially unknown and must be learned adaptively could connect this problem to learning-augmented algorithms and online optimization.

References

1. Alpern, S., Gal, S.: The theory of search games and rendezvous. In: International Series in Operations Research and Management Science, vol. 55. Kluwer (2003)
2. Beck, A.: On the linear search problem. Israel J. Math. **2**(4), 221–228 (1964). https://doi.org/10.1007/BF02759737
3. Bellman, R.: Problem 63-9, an optimal search. SIAM Rev. **5**(3), 274–274 (1963). http://www.jstor.org/stable/2027629
4. Bonato, A., Georgiou, K., MacRury, C., Prałat, P.: Probabilistically faulty searching on a half-line. In: Kohayakawa, Y., Miyazawa, F.K. (eds.) LATIN 2021. LNCS, vol. 12118, pp. 168–180. Springer, Cham (2020). https://doi.org/10.1007/978-3-030-61792-9_14
5. Coleman, J., Cheng, L., Krishnamachari, B.: Search and rescue on the line. In: Rajsbaum, S., Balliu, A., Daymude, J.J., Olivetti, D. (eds.) Structural Information and Communication Complexity, pp. 297–316. Springer, Cham (2023). https://doi.org/10.1007/978-3-031-32733-9_13
6. Coleman, J., Ivanov, D., Kranakis, E., Krizanc, D., Morales-Ponce, O.: Multimodal search on the line. In: Structural Information and Communication Complexity (2025)
7. Coleman, J., Kranakis, E., Krizanc, D., Morales-Ponce, O.: Line Search for an Oblivious Moving Target. In: Hillel, E., Palmieri, R., Rivière, E. (eds.) 26th International Conference on Principles of Distributed Systems (OPODIS 2022). Leibniz International Proceedings in Informatics (LIPIcs), vol. 253, pp. 12:1–12:19. Schloss Dagstuhl – Leibniz-Zentrum für Informatik, Dagstuhl (2023). https://doi.org/10.4230/LIPIcs.OPODIS.2022.12. https://drops.dagstuhl.de/entities/document/10.4230/LIPIcs.OPODIS.2022.12
8. Coleman, J., Ponce, O.M.: Linear Search with Probabilistic Detection and Variable Speeds (2025)
9. Czyzowicz, J., Gasieniec, L., Georgiou, K., Kranakis, E., MacQuarrie, F.: The beachcombers' problem: walking and searching with mobile robots. Theor. Comput. Sci. **608**, 201–218 (2015). https://doi.org/10.1016/J.TCS.2015.09.011
10. Czyzowicz, J., Georgiou, K., Kranakis, E.: Group Search and Evacuation, pp. 335–370. Springer, Cham (2019). https://doi.org/10.1007/978-3-030-11072-7_14
11. Czyzowicz, J., Kranakis, E., Krizanc, D., Narayanan, L., Opatrny, J.: Search on a line with faulty robots. In: Proceedings of the 2016 ACM Symposium on Principles of Distributed Computing, PODC '16, pp. 405–414. Association for Computing Machinery, New York (2016). https://doi.org/10.1145/2933057.2933102
12. Czyzowicz, J., Kranakis, E., Krizanc, D., Narayanan, L., Opatrny, J., Shende, S.: Linear search with terrain-dependent speeds. In: Fotakis, D., Pagourtzis, A., Paschos, V.T. (eds.) CIAC 2017. LNCS, vol. 10236, pp. 430–441. Springer, Cham (2017). https://doi.org/10.1007/978-3-319-57586-5_36
13. Kranakis, E.: Invited paper: a survey of the impact of knowledge on the competitive ratio in linear search. In: Masuzawa, T., Katayama, Y., Kakugawa, H., Nakamura, J., Kim, Y. (eds.) Stabilization, Safety, and Security of Distributed Systems, pp. 23–38. Springer, Cham (2025). https://doi.org/10.1007/978-3-031-74498-3_2

Streaming Algorithms for Scheduling Jobs with Priorities

Bin Fu[1], Yumei Huo[2], and Hairong Zhao[3(✉)]

[1] University of Texas Rio Grande Valley, Edinburg, TX 78539, USA
bin.fu@utrgv.edu
[2] College of Staten Island, CUNY, Staten Island, NY 10314, USA
yumei.huo@csi.cuny.edu
[3] Purdue University Northwest, Hammond, IN 46323, USA
hairong@purdue.edu

Abstract. We study the classical makespan minimization problem on parallel machines where jobs have bounded priority h. We develop the first one-pass streaming approximation scheme when all jobs' processing times differ no more than a constant factor c and the number of jobs n is at least $\frac{3mhc}{2\epsilon}$ where m is the number of machines. The algorithm is then extended to the more general problem where the largest αn jobs have no more than c factor difference for some constant α, $0 < \alpha \leq 1$. For both problems, our algorithms can generate a "sketch schedule" for the applications where not only an approximate value, but also a schedule associated with the approximate value is needed. The algorithms can be applied and extended to the makespan minimization scheduling subject to precedence constraints where the precedence graph has bounded depth.

Keywords: Streaming algorithms · makespan minimization · job priorities

1 Introduction

This paper aims at designing streaming algorithms for the classical parallel machine scheduling problem, where jobs have varying priorities. Given a set of n jobs, each job j, $1 \leq j \leq n$, has a processing time p_j and a priority pr_j. All jobs are available for processing at time 0 and can be scheduled on any of the m identical machines. At any given time, only one job can be scheduled on each machine, and once a job starts, it cannot be interrupted. A smaller value of pr_j indicates a higher priority. Jobs with higher priority must be scheduled before jobs with lower priority are completed. Given a schedule S, let C_j represent the completion time of job j in S. The makespan of the schedule S is defined as $C_{max} = \max_{1 \leq j \leq n} C_j$ and the objective is to minimize the makespan. This scheduling problem, which aims to achieve load balancing while adhering to priority constraints, has significant applications in fields such as computer

H. Fernau and B. Zhu (Eds.): IWOCA 2025, LNCS 15885, pp. 446–458, 2025.
https://doi.org/10.1007/978-3-031-98740-3_32

science (e.g., operating systems, client-server networks, and cloud computing) as well as in manufacturing and service industries.

In this paper, we design streaming algorithms for the makespan minimization problem. Unlike the traditional data model, where all input data can be stored in memory and accessed at any time, the data streaming model processes data as a sequence of items that can only be examined in a limited number of passes (typically just one). The goal of streaming algorithms is to provide good approximate solutions without storing the entire input. These algorithms differ from online algorithms in that online algorithms must act immediately upon receiving each data item, without any knowledge of future data. In contrast, streaming algorithms can store partial or compressed input data in limited memory and delay action until all the data is received. Online algorithms can be seen as a special type of streaming algorithms, where a job needs to be scheduled immediately once it arrives. For more results on online scheduling algorithms related to the problems we study, readers can refer to the survey paper [5]. In this work, we focus on general streaming algorithms that requires sublinear space with respect to the input size n.

In this paper, we focus on the problem with a constant number of distinct priorities, denoted as h. We use smaller values to represent high priorities. That is, if $pr_{j_1} < pr_{j_2}$, then job j_2 must be scheduled after job j_1 is completed. Furthermore, we assume that the jobs have constrained processing times. We first consider the case where the processing times of jobs vary within a constant factor c, i.e., $p_{max} \leq c \cdot p_{min}$ where p_{max} and p_{min} are the maximum and minimum processing time, respectively. Using the three-field notation, we denote this problem as $P \mid pr_j \leq h, p_{max} \leq c \cdot p_{min} \mid C_{max}$. Next, we consider a more general case where the largest αn jobs have no more than a c-factor difference for some constant $0 < \alpha \leq 1$. For a given set of n jobs, let $[j]$ be the j-th largest job. Thus, $p_{[1]}$ and $p_{[n]}$ represent the largest and smallest job, respectively. We denote this more generalized problem as $P \mid pr_j \leq h, p_{[1]} \leq c \cdot p_{[\alpha n]} \mid C_{max}$. Using partition problem, one can easily show that the problem $P \mid p_{max} \leq c \cdot p_{min} \mid C_{max}$ is NP-hard, which implies that both of our problems are also NP-hard. We will present streaming algorithms for both problems, where the input stream consists of a sequence of jobs, each described by pairs (p_j, pr_j), representing the processing time and priority of job j for $1 \leq j \leq n$.

It should be noted that jobs with priorities studied in this paper can be viewed as a special case of jobs with precedence constraints. In this case, jobs with lower priority depend on all jobs with higher priorities, meaning that all higher-priority jobs can be considered as the predecessors of the lower-priority ones. In contrast, for general precedence constraints, each job depends only on a specific set of predecessors and can be scheduled immediately once all its predecessors have completed. For a precedence graph $G = (V, E)$, where V is a set of vertices representing the jobs and E is a set of directed arcs representing the precedence constraints among the jobs, let dp_j be the job j's depth which is the number of jobs on the longest directed path ending at j in G. The scheduling problem where the precedence graph has a bounded depth h, can be denoted as

$P \mid prec, dp_j \leq h, p_{[1]} \leq c \cdot p_{[\alpha n]} \mid C_{max}$. We will demonstrate that the streaming algorithm for $P \mid pr_j \leq h, p_{[1]} \leq c \cdot p_{[\alpha n]} \mid C_{max}$ can be applied or extended to solve $P \mid prec, dp_j \leq h, p_{[1]} \leq c \cdot p_{[\alpha n]} \mid C_{max}$ depending on the format of the stream input.

Literature Review. For the problem of makespan minimization problem subject to job priorities, jobs must be scheduled in groups based on their priorities. The optimal schedule consists of the optimal schedule for each group of jobs with a specific priority, which is the same as the classical makespan minimization scheduling problem. Much research has been conducted on the classical makespan minimization scheduling problem where n jobs are scheduled on m machines. The makespan minimization problem has long been known to be NP-hard. The List Scheduling rule and Longest Processing Time rule provide approximation ratios of $2 - \frac{1}{m}$ and $\frac{4}{3} - \frac{1}{3m}$, respectively ([8] and [9]). Additionally, approximation schemes have been developed when the number of machines m is fixed [11] and when m is arbitrary [10].

For arbitrary precedence, when $m = 2$ and jobs have unit processing time, i.e., Coffman and Graham [2] gave an optimal polynomial time algorithm for $P_2 \mid prec, p_j = 1 \mid C_{max}$ in 1972. In 1978, Lenstra and Kan [12] showed that $P \mid prec, p_j = 1 \mid C_{max}$, is strongly NP-hard. However, the complexity of the problem $P_3 \mid prec, p_j = 1 \mid C_{max}$ remains open. Graham [8] showed that list scheduling is a $(2 - \frac{1}{m})$-approximation for $P \mid prec \mid C_{max}$. Most recently, Levey and Rothvoss [13] proved that for any fixed m and arbitrary precedence constraint, when jobs have equal processing time, a $(1 + \epsilon)$-approximation can be obtained in polynomial time. For the precedence graph with bounded height h, in 1978 Lenstra and Kan [12] showed that for arbitrary number of machines m, the problem is NP-hard even if the jobs have unit processing time; in 1984, Doleva and Warmuth [4] developed an optimal algorithm that runs in $O(n^{h(m-1)+1})$ time.

Regarding streaming algorithms, effective streaming algorithms have been developed for a variety of problems in statistics, optimization, and graph algorithms (see surveys by Muthukrishnan [15] and McGregor [14]). However, relatively little research has been conducted in the area of scheduling and operations research in general.

Beigel and Fu [1] developed a randomized streaming approximation scheme for the bin packing problem. Cormode and Veselý [3] developed a streaming algorithm for vector bin packing in d dimensions with approximation ratio $(d+\epsilon)$. For the related vector scheduling problem, they also gave a streaming algorithm with a $(2-1/m+\epsilon)$-approximation, where m is the number of identical machines. More recently, Fu, Huo and Zhao studied streaming algorithms for multitasking scheduling with shared processing, targeting objectives such as makespan [6] and total completion time [7].

New Contributions. Assuming that the number of jobs n is large enough, we first design a streaming approximation scheme for the problem $P \mid pr_j \leq$

$h, p_{max} \leq c \cdot p_{min} \mid C_{max}$. Next, we extend the algorithm to handle the more general case, where the largest αn jobs have no more than a c-factor difference, for some constant α (with $0 < \alpha \leq 1$), i.e., $P \mid pr_j \leq h, p_{[1]} \leq c \cdot p_{[\alpha n]} \mid C_{max}$. Both algorithms can generate a "sketch schedule" for applications where not only an approximate value but also the associated schedule is required. A real schedule can be constructed based on the "sketch schedule" when the stream input is scanned in a second pass.

Our streaming algorithm for $P \mid pr_j \leq h, p_{[1]} \leq c \cdot p_{[\alpha n]} \mid C_{max}$ can be applied and extended to the scheduling problem with precedence constraints when the precedence graph has bounded depth (denoted as $P \mid prec, dp_j \leq h, p_{[1]} \leq c \cdot p_{[\alpha n]} \mid C_{max}$), depending on how the precedence constraints are given in the input stream. Note that these approximation results complement the inapproximability results of the studied problems. In the strong NP-hardness proof of $P \mid prec, p_j = 1 \mid C_{max}$ in [12], the scheduling instance created from the instance of clique problem has a precedence graph of height 3, and there is a schedule of the $n = 3m$ jobs with makespan of 3 if and only if there is a solution to the clique instance. This implies when $m = \frac{n}{3}$, even if $h = 3$ and $c = 1$, the problem $P \mid prec, dp_j \leq h, p_{[1]} \leq c \cdot p_{[\alpha n]} \mid C_{max}$ cannot be approximated within a factor of $\frac{4}{3}$ in polynomial time unless P=NP. Our results show that when m is a little bit smaller, i.e., upper bounded by n times a factor that depends on ϵ, h and α, then the problem admits polynomial time approximation schemes. For example, if $m \leq \frac{n}{15}$, $h = 3$ and $c = 1$, then there is a polynomial time 1.3-approximation for $P \mid prec, dp_j \leq h, p_{max} \leq c \cdot p_{min} \mid C_{max}$.

2 Streaming Approximation Schemes

In this section, we develop streaming algorithms for our problems. Both of our algorithms follow a common framework. At the conceptual level, the framework has two stages:

Stage 1: Generate and store a sketch of the input while reading the input stream.

Stage 2: Compute an approximation of the optimal value based on the sketch of the input.

Roughly speaking, the sketch of the input is a summary of the input jobs which requires only sublinear space. Instead of storing the actual processing times of the jobs, we use the rounded processing time. That is, if the processing time of a job is in the range of $[(1+\delta)^u, (1+\delta)^{u+1})$ for some parameter δ, we round it up to $(1+\delta)^{u+1}$. We store the number of jobs whose processing time falls into each range for each depth. We then use the rounded processing times to obtain an approximation of the optimal makespan. A formal definition of the sketch for our problems is given below.

Definition 1. *For a given parameter δ, and an instance of the problem $P \mid pr_j \leq h, p_{max} \leq c \cdot p_{min} \mid C_{max}$ or $P \mid pr_j \leq h, p_{[1]} \leq c \cdot p_{[\alpha n]} \mid C_{max}$, the* ***sketch***

of input *with respect to* δ, *denoted as* $SKJ_\delta = \{(d, u, n_{d,u})\}$, *consists of a set of tuples,* $(d, u, n_{d,u})$, *where* $n_{d,u}$ *is the number of jobs with the priority* d *and the processing time in the range of* $[(1+\delta)^u, (1+\delta)^{u+1})$.

The size of the sketch, which is the number of tuples $(d, u, u_{d,u})$, depends on the number of priorities, the range of processing times, and the approximation error ϵ. In the following subsections, we will first present our streaming algorithm for $P \mid pr_j \leq h, p_{max} \leq c \cdot p_{min} \mid C_{max}$. Then, we generalize the algorithm for problem $P \mid pr_j \leq h, p_{[1]} \leq c \cdot p_{[\alpha n]} \mid C_{max}$. We will see that the difference between the algorithms mainly lies in Stage 1, which involves generating the sketch from the input stream. Finally, for applications where not only an approximate value, but also a schedule associated with the approximate value is needed, our framework can generate a "sketch of schedule" which is a succinct description of a schedule that achieves the approximate value. As illustrated in the Sect. 2.3, we can use the "sketch of schedule" to easily generate a real schedule when the jobs are read in second pass.

2.1 $P \mid pr_j \leq h, p_{max} \leq c \cdot p_{min} \mid C_{max}$

We study the problem assuming c and h are known. The jobs are given via stream and each job j is described by a pair (p_j, pr_j), where p_j and pr_j are job j's processing time and priority, respectively. Without loss of generality, we can assume $p_j \in [1, c]$ in this case. The algorithm is straightforward: first scan the jobs from the stream and generate the sketch of the input, $SKJ_\delta = \{(d, u, n_{d,u})\}$; then for each priority d, $1 \leq d \leq h$ (h represents the lowest priority), compute the length of the time interval where all the jobs with the priority d can be feasibly scheduled, and finally return the total length of these intervals. The complete algorithm is given in Streaming-Algorithm1.

Theorem 1. *For any* ϵ, *when* $n \geq \frac{3hc}{2\epsilon} m$, *Streaming-Algorithm1 is a one-pass streaming approximation scheme for* $P \mid pr_j \leq h, p_{max} \leq c \cdot p_{min} \mid C_{max}$ *that uses* $O(\frac{h \log c}{\epsilon})$ *space,* $O(1)$ *update time for each job in the stream, and* $O(\frac{h \log c}{\epsilon})$ *time to return the approximate makespan.*

Proof. First we consider the complexity. The space complexity is dominated by the sketch for which we can use a two dimensional array of size $h \cdot k = O(\frac{h \log c}{\epsilon})$. It is easy to see that the update time for each job is $O(1)$. Finally, it takes $O(h \cdot k) = O(\frac{h \log c}{\epsilon})$ time to compute and return the approximate value A.

Now we consider the approximation ratio of the algorithm. Let I be the input instance, and let I' be the instance corresponding to the sketch SKJ_δ which consists of $n_{d,u}$ jobs that have processing time rp_u for each d, u. Alternatively, we can also view I' being obtained from I by rounding up the processing time of each job. Let C^*_{max} and C'_{max} be the optimal makespan for the instance I and I', respectively. It is easy to see that $C^*_{max} \leq C'_{max} \leq (1+\delta)C^*_{max}$. In the following, we prove that the returned value from Streaming-Algorithm1, A, satisfies the inequality, $C^*_{max} \leq A \leq (1+\epsilon)C^*_{max}$.

Algorithm . Streaming-Algorithm1

Input: Parameters ϵ, m, c and h
  Stream input: (p_j, pr_j), $1 \leq j \leq n$.
Output: An approximate value of the optimal makespan

let $\delta = \frac{\epsilon}{3}$, $k = \lfloor \log_{1+\delta} c \rfloor$
// read the input stream and generate the sketch of the input SKJ_δ:
  initialize $SKJ_\delta = \{(d, u, n_{d,u}) : 1 \leq d \leq h, 0 \leq u \leq k, n_{d,u} = 0\}$
  for each job j with (p_j, pr_j) in the stream input **do**
    update $n_{d,u} = n_{d,u} + 1$ where $d = pr_j$ and $u = \lfloor \log_{1+\delta} p_j \rfloor$
  end for
// compute the approximate makespan:
  for each u, $0 \leq u < k$
    let $rp_u = (1+\delta)^{u+1}$
  let $rp_k = c$
  for each d, $1 \leq d \leq h$ **do**
    let $A_d = \frac{1}{m} \sum_{u=0}^{k} (n_{d,u} \cdot rp_u)$
  end for
  $A = \sum_{d=1}^{h} (\lfloor A_d \rfloor + c)$
return A

We first show that $A \geq C'_{max}$ which implies $A > C^*_{max}$. Consider a schedule S for the jobs of instance I' which schedules the jobs as follows: the jobs with each priority d are scheduled using list scheduling rule (i.e. schedule the jobs one by one in the given order to the machine that is available at the earliest time), and jobs with priority $d+1$ can start only after all jobs with depth d complete. It is easy to see that in S the jobs with the priority d are scheduled into an interval of length at most $\lfloor \frac{1}{m} \sum_{u=0}^{k} (n_{d,u} \cdot rp_u) \rfloor + c = \lfloor A_d \rfloor + c$. Therefore, the makespan of the feasible schedule S is at most $\sum_{d=1}^{h} (\lfloor A_d \rfloor + c) = A$, which implies that $A \geq C'_{max}$, where C'_{max} is the optimal makespan for the instance I'.

On the other hand, it is obvious that $\sum_{d=1}^{h} A_d = \sum_{d=1}^{h} (\frac{1}{m} \sum_{u=0}^{k} (n_{d,u} \cdot rp_u))$ is a lower bound of C'_{max}, and $C^*_{max} \leq C'_{max} \leq (1+\delta) C^*_{max}$. Thus,

$$A = \sum_{d=1}^{h} (\lfloor A_d \rfloor + c) \leq \left(\sum_{d=1}^{h} A_d \right) + h \cdot c \leq C'_{max} + h \cdot c \leq (1+\delta) C^*_{max} + h \cdot c.$$

Since $C^*_{max} \geq \frac{n}{m}$, when $n \geq \frac{3hc}{2\epsilon} m$, we have $h \cdot c \leq \frac{2\epsilon}{3} \frac{n}{m} \leq \frac{2\epsilon}{3} C^*_{max}$. Therefore,

$$A \leq (1+\delta) C^*_{max} + \frac{2\epsilon}{3} C^*_{max} = (1 + \frac{\epsilon}{3}) C^*_{max} + \frac{2\epsilon}{3} C^*_{max} = (1+\epsilon) C^*_{max}.$$

In summary, we have $C^*_{max} \leq C'_{max} \leq A \leq (1+\epsilon) C^*_{max}$, and this completes the proof. ■

Theorem 1 shows that Stream-Algorithm1 only takes constant time to read and process each job in the stream input, and then constant time and constant space to return an approximation of the makespan if all parameters are known. In some cases, however, we may not know the exact value of c, but we are given an estimate $\hat{c}$ of the parameter c. In these cases, we can still apply Stream-Algorithm1 by using $\hat{c}$. As long as $\frac{\hat{c}}{c}$ is a constant, we still have a $(1 + \epsilon)$-approximation with the same space and time complexity.

2.2 $P \mid pr_j \leq h, p_{[1]} \leq c \cdot p_{[\alpha n]} \mid C_{max}$

In this subsection, we study the problem that the processing times among the jobs can vary widely, but the processing times of the top αn jobs are within a factor of c. Without loss of generality, we assume p_{max} is not known until all input is read.

The first stage of the algorithm is to efficiently generate the sketch of the input while we scan the jobs from the input stream. The details of this step are given by the procedure **Compute-Input-Sketch**. The algorithm uses a balanced binary search tree to store the sketch of the jobs that have been read so far. Each node of the tree stores a tuple $(d, u, n_{d,u})$ in the current sketch where (d, u) is the key. Given two tuples (d_1, u_1, n_{d_1,u_1}) and (d_2, u_2, n_{d_2,u_2}), we say $(d_1, u_1, n_{d_1,u_1}) < (d_2, u_2, n_{d_2,u_2})$ if 1) $u_1 < u_2$, or 2) $u_1 = u_2$ and $d_1 < d_2$. To ensure that the size of the sketch is bounded, we exclude jobs with processing time less than $\frac{p_{max}}{n^2}$ from the sketch. Specifically, $SKJ_\delta = \{(d, u, n_{d,u}) : u_- \leq u \leq u_+,\ 1 \leq d \leq h\}$, where $u_- = \lfloor \log_{1+\delta} \frac{p_{max}}{n^2} \rfloor$, and $u_+ = \lfloor \log_{1+\delta} p_{max} \rfloor$. The sketch is updated dynamically as more jobs are read in the stream.

Once we get the sketch, for each priority value d, $1 \leq d \leq h$, we compute the length of the time interval where all the jobs with the priority value d can be feasibly scheduled, and finally return the total length of these intervals. We present our algorithm formally in Streaming-Algorithm2.

Theorem 2. *When* $n \geq \frac{3m(h+1)c}{2\alpha\epsilon}$, *there is a streaming approximation scheme for the problem* $P \mid pr_j \leq h, p_{[1]} \leq c \cdot p_{[\alpha n]} \mid C_{max}$ *that uses* $O(\log n)$ *space, takes* $O(\log \frac{h}{\epsilon} + \log\log n)$ *update time for each job in the stream, and* $O(\frac{h}{\epsilon} \log n)$ *time to return the approximate value.*

Proof. We first analyze the approximation ratio. We use similar notations as in the proof of Theorem 1. Let I be the given instance. Let I' be the instance obtained from I by rounding up all the processing times that are greater than or equal to $\frac{p_{max}}{n^2}$, i.e. for each job j in I, if $p_j \geq \frac{p_{max}}{n^2}$, we round it up to $rp_u = (1 + \delta)^{u+1}$ where $u = \lfloor \log_{1+\delta} p_j \rfloor$; otherwise, we keep it same as before. Let C^*_{max} and C'_{max} be the optimal makespan for I and I' respectively. Let I'' be the instance corresponding to the sketch SKJ_δ. Apparently I'' can be obtained from I' by removing the small jobs whose processing time is less than $\frac{p_{max}}{n^2}$. Let C''_{max} be the optimal makespan for I''. Then we have $C''_{max} \geq \sum_{d=1}^{h} A_d$. It is easy to see that $C''_{max} \leq C'_{max} \leq (1+\delta)C^*_{max}$, and $C^*_{max} \leq C'_{max} \leq C''_{max} + n \cdot \frac{p_{max}}{n^2} = C''_{max} + \frac{p_{max}}{n}$.

We can construct a feasible schedule S for I'' using list scheduling rule to schedule the jobs group by group based on their priority values starting with

Algorithm . Compute-Input-Sketch

Input: Parameters δ, n
    Stream input: (p_j, pr_j), $1 \leq j \leq n$.
Output: p_{max}, h, SKJ_δ, u_-, u_+

create an empty balanced binary search tree T
initialize $p_{max} = 1$, $h = 1$
for each job j with (p_j, pr_j) in the stream input **do**
    let $u = \lfloor \log_{1+\delta} p_j \rfloor$
    if $p_j \geq \frac{p_{max}}{n^2}$ **then**
        if $p_{max} < p_j$, $p_{max} = p_j$
        if $h < pr_j$, set $h = pr_j$
        if node $(dp_j, u, n_{pr_j,u})$ exists in the tree **then**
            update the node with $n_{pr_j,u} = n_{pr_j,u} + 1$
        else
            insert a new node $(pr_j, u, 1)$
            let $(d', u', n_{d',u'})$ be the smallest node in tree T
            if $u' < \log_{1+\delta} \frac{p_{max}}{n^2}$, delete node $(d', u', n_{d',u'})$ from the tree
        end if
    end if
end for
let $u_- = \lfloor \log_{1+\delta} \frac{p_{max}}{n^2} \rfloor$, $u_+ = \lfloor \log_{1+\delta} p_{max} \rfloor$
$SKJ_\delta = \{(d, u, n_{d,u}) : 1 \leq d \leq h, u_- \leq u \leq u_+\}$ where $(d, u, n_{d,u})$ is a node in T

the group that has the priority value $d = 1$. To get a schedule for all jobs in I' based on S, for each priority value d, we can simply insert into S all the small jobs with the priority value d onto the first machine after all jobs of group d finish and before the first big job of group $d+1$ starts. Let the new schedule be S'. Apparently the makespan of S' is at least C^*_{max} and at most $A = \sum_{d=1}^{h} (\lfloor A_d \rfloor + p_{max}) + \lceil \frac{p_{max}}{n} \rceil$. Thus, we have $A \geq C^*_{max}$ and

$$A = \left(\sum_{d=1}^{h} (\lfloor A_d \rfloor + p_{max}) \right) + \lceil \tfrac{p_{max}}{n} \rceil \leq C''_{max} + h \cdot p_{max} + \lceil \tfrac{p_{max}}{n} \rceil. \quad (1)$$

Since the largest αn jobs have no more than c factor difference, each of the largest αn jobs has processing time at least $\frac{p_{max}}{c}$. Thus, we have

$$C^*_{max} \geq \alpha \cdot n \cdot \tfrac{p_{max}}{c} \cdot \tfrac{1}{m} = \tfrac{\alpha n}{c \cdot m} p_{max},$$

Algorithm . Streaming-Algorithm2

Input: Parameters ϵ, m, n
Stream input: (p_j, pr_j), $1 \le j \le n$.
Output: An approximate value of the optimal makespan

1: let $\delta = \frac{\epsilon}{3}$
2: call Compute-Input-Sketch to get SKJ_δ, p_{max}, u_-, u_+ and h
3: for each $u_- \le u < u_+$
4: let $rp_u = (1+\delta)^{u+1}$
5: let $rp_{u_+} = p_{max}$
6: for each d, $1 \le d \le h$
7: let $A_d = \frac{1}{m}\sum_{u=u_-}^{u_+}(n_{d,u} \cdot rp_u)$
8: let $A = \sum_{d=1}^{h}(\lfloor A_d \rfloor + p_{max})$
9: return A

which implies $p_{max} \le \frac{c \cdot m}{\alpha \cdot n} C^*_{max}$. If we plug this into inequality (1), we get

$$\begin{aligned} A &\le C''_{max} + h \cdot p_{max} + \lceil \tfrac{p_{max}}{n} \rceil \\ &\le C''_{max} + (h+1) \cdot p_{max} \\ &\le C''_{max} + (h+1) \cdot \tfrac{c \cdot m}{\alpha \cdot n} C^*_{max} \\ &\le (1+\delta) C^*_{max} + \tfrac{(h+1) \cdot c \cdot m}{\alpha n} C^*_{max} \\ &\le (1 + \delta + \tfrac{(h+1) \cdot c \cdot m}{\alpha n}) C^*_{max} \\ &\le (1 + \tfrac{\epsilon}{3} + \tfrac{(h+1) \cdot c \cdot m}{\alpha n}) C^*_{max}. \end{aligned}$$

When $n \ge \frac{3m(h+1)c}{2\alpha\epsilon}$ we have $A \le (1 + \frac{\epsilon}{3} + \frac{2\epsilon}{3}) C^*_{max} = (1+\epsilon) C^*_{max}$.

Now we consider the complexity. The space complexity is dominated by the tree. As the way that the tree is implemented, each time a node is inserted into the tree, if there is a node $(d, u, n_{d,u})$ such that $u < \log_{1+\delta} \frac{p_{max}}{n^2}$, the smallest such node will be deleted from the tree. In this way, the number of nodes in the tree is at most $h \cdot \log_{1+\delta} n^2 = O(\frac{h}{\epsilon} \log n)$. Therefore, Streaming-Algorithm2, uses only $O(\log n)$ space to store the tree T, which is sublinear in the size of stream input.

For each job in the stream input, at most two tree operations are needed, and thus the update time for processing each job is $O(\log(\frac{h}{\epsilon} \log n) = O(\log(\frac{h}{\epsilon}) + \log\log n)$ time. After reading all the jobs, the computation of the approximation is bounded by the size of the sketch ■

2.3 The Sketch of Schedule

The streaming algorithms we have presented so far return an approximate value of the optimal makespan. This may be sufficient for some scheduling and planning applications. However, in many other applications, it would be desirable to have a schedule whose makespan is the approximate value. In the traditional data model, a schedule explicitly or implicitly specifies for each job on which machine and in what time interval it is scheduled. This means we need at least $\Omega(n)$ time complexity and space complexity to describe a schedule. For the big data model, we introduce the concept of **sketch of schedule** which is a condensed description of a schedule using only sublinear space.

Definition 2. *For the problems* $P \mid pr_j \leq h, p_{max} \leq c \cdot p_{max} \mid C_{max}$ *and* $P \mid pr_j \leq h, p_{[1]} \leq c \cdot p_{[\alpha n]} \mid C_{max}$, *the* **sketch of a schedule**, *denoted as* $SKS = \{t_d : 1 \leq d \leq h\}$, *describes a feasible schedule such that all the jobs of priority* d *can be scheduled during the interval* $[t_{d-1}, t_d)$ *for* $1 \leq d \leq h$ *where* $t_0 = 0$.

For the problem $P \mid pr_j \leq h, p_{max} \leq c \cdot p_{min} \mid C_{max}$, by the proof of Theorem 1 we know all the jobs of priority d can be feasibly scheduled during an interval of length $\lfloor A_d \rfloor + c$, which implies that $SKS = \{t_d : t_d = \sum_{k=1}^{d} (\lfloor A_d \rfloor + c), 1 \leq d \leq h\}$ is the sketch of a schedule whose makespan is at most $(1+\epsilon)$ times the optimal, and the sketch that can be computed by Streaming-Algorithm1.

Based on the sketch of schedule, if we scan all the jobs in the second time, we can easily generate a feasible schedule using the following procedure:

Procedure SketchtoSchedule: For each priority d, we maintain the index of the current machine $curM_d$ (Initially, $curM_d = 1$), and the completion time of the last job on this machine, $curT_d$ (Initially, $curT_d = t_{d-1}$). For each job (p_j, pr_j) in the job stream input, let $d = pr_j$, we try to schedule the job at $curT_d$ on machine $curM_d$. If a job cannot complete before t_d, we start scheduling the job at $curT_d = t_{d-1}$ on the next machine $curM_d = curM_d + 1$.

The total space needed in this procedure is $O(d)$, and the time to schedule each job is $O(1)$.

Theorem 3. *For* $P \mid pr_j \leq h, p_{min} \leq c \cdot p_{max} \mid C_{max}$, *given any* $0 < \epsilon < 1$, *when* $n \geq \frac{3hc}{2\epsilon} m$ *Streaming-Algorithm1 can compute a sketch* $SKS = \{t_d : 1 \leq d \leq h\}$ *where* $t_d = \sum_{k=1}^{d} (\lfloor A_d \rfloor + c)$ *for all* $1 \leq d \leq h$; *based on* SKS, *Algorithm SketchToSchedule can generate, with* $O(1)$ *time for each job, a feasible schedule with the makespan at most* $(1+\epsilon)$ *times the optimal makespan.*

Similarly, Streaming-Algorithm2 can compute a sketch of schedule for $P \mid pr_j \leq h, p_{[1]} \leq c \cdot p_{[\alpha n]} \mid C_{max}$. For this problem, the small jobs with the priority d are not considered when we calculate A_d, so the sketch of schedule is slightly different from previous problem. Specifically, in this case, the sketch of a schedule is given by $SKS = \{t_d : t_d = t_{d-1} + (\lfloor A_d \rfloor + p_{max} + \lceil \frac{p_{max}}{n} \rceil), 1 \leq d \leq h\}$, $t_0 = 0$. We have the following theorem.

Theorem 4. *For the problem* $P \mid pr_j \leq h, p_{[1]} \leq c \cdot p_{[\alpha n]} \mid C_{max}$, *given any* $0 < \epsilon < 1$, *when* $n \geq \frac{3m(h+1)c}{2\alpha\epsilon}$, *Streaming-Algorithm2 can compute a sketch of the schedule* $SKS = \{t_d : t_d = t_{d-1} + (\lfloor A_d \rfloor + p_{max} + \lceil \frac{p_{max}}{n} \rceil), 1 \leq d \leq h\}$, $t_0 = 0$; *and based on* SKS, *Algorithm SketchToSchedule can generate, with* $O(1)$ *time for each job, a feasible schedule with the makespan at most* $(1+\epsilon)$ *times the optimal makespan.*

Proof. From the proof of Theorem 2, we know that the interval with the length $\lfloor A_d \rfloor + p_{max}$ can feasibly fit in all the jobs with priority d whose process time is at least $\frac{p_{max}}{n^2}$. To ensure that all other jobs with priority d, whose processing time is less than $\frac{p_{\max}}{n^2}$, can also be accommodated, we add to the interval an additional length of $\lceil n \cdot \frac{p_{max}}{n^2} \rceil = \lceil \frac{p_{max}}{n} \rceil$. Hence, $SKS = \{t_d : t_d = t_{d-1} + (\lfloor A_d \rfloor + p_{max} + \lceil \frac{p_{max}}{n} \rceil), 1 \leq d \leq h\}$, $t_0 = 0$, describes a feasible schedule such that all the jobs of priority d can be scheduled during the interval $[t_{d-1}, t_d]$. Based on the sketch SKS, we can use Algorithm SketchToSchedule to generate a feasible schedule with the makespan at most

$$t_h = \sum_{d=1}^{h} (\lfloor A_d \rfloor + p_{max} + \lceil \tfrac{p_{max}}{n} \rceil) \leq \left(\sum_{d=1}^{h} \lfloor A_d \rfloor \right) + (h+1) p_{max}.$$

From the proof of Theorem 2, we know $t_h \leq C''_{max} + (h+1)p_{max} \leq (1+\epsilon)C^*_{max}$. ∎

Theorem 5. *For the problems* $P \mid pr_j \leq h, p_{[min]} \leq c \cdot p_{max} \mid C_{max}$, *and* $P \mid pr_j \leq h, p_{[1]} \leq c \cdot p_{[\alpha n]} \mid C_{max}$, *when* $n \geq \frac{3mhc}{2\epsilon}$ *and* $n \geq \frac{3m(h+1)c}{2\alpha\epsilon}$, *respectively, there exist streaming approximation schemes that can return a* $(1+\epsilon)$*-approximation and a sketch of a schedule in the first pass, and in the second pass output a feasible schedule with the approximate value.*

2.4 $P \mid prec, dp_j \leq h, p_{[1]} \leq c \cdot p_{[\alpha n]} \mid C_{max}$

In this subsection, we further extend our results to the more general precedence constrained scheduling problem where the precedence graph has bounded depth h, that is, $P \mid prec, dp_j \leq h, p_{[1]} \leq c \cdot p_{[\alpha n]} \mid C_{max}$. It is easy to see that if the jobs' depth are known and the stream input consists of a sequence of jobs, each described by pairs, (p_j, dp_j) for job j, $1 \leq j \leq n$, where p_j and dp_j are job j's processing time and depth in the precedence graph, our streaming algorithm for $P \mid dp_j \leq h, p_{[1]} \leq c \cdot p_{[\alpha n]} \mid C_{max}$ can be directly applied to $P \mid prec, dp_j \leq h, p_{[1]} \leq c \cdot p_{[\alpha n]} \mid C_{max}$.

For the scenario where the depths of the jobs are not provided directly, we consider the case that the stream input consists of all the jobs (j, p_j) in arbitrary order, followed by all the arcs $\langle i, j \rangle$ of the precedence graph in topological order. Our algorithm can be extended and turned into a semi-streaming algorithm in that the needed space is linear in number of jobs, but sublinear in the number of edges which is usually quadratic in n. In addition to the binary search tree, we need to use an array B to store the jobs information: for each job j with processing time p_j, we maintain a pair (dp_j, u_j), where dp_j represents its current depth, and $u_j = \lfloor \log_{1+\delta} p_j \rfloor$. Similarly, we exclude jobs with processing times less than $\frac{p_{max}}{n^2}$ from the sketch. Specifically, $SKJ_\delta = \{(d, u, n_{d,u}) : u_- \leq u \leq u_+, 1 \leq d \leq h\}$, where $u_- = \lfloor \log_{1+\delta} \frac{p_{max}}{n^2} \rfloor$, and $u_+ = \lfloor \log_{1+\delta} p_{max} \rfloor$. The sketch and the depths of the jobs are updated dynamically as more jobs and arcs of the precedence graph are read in the stream.

When a job (j, p_j) arrives in the stream input, we update job j's entry in the array B such that $dp_j = 1$ and $u_j = \lfloor \log_{1+\delta} p_j \rfloor$. If $p_j \geq \frac{p_{max}}{n^2}$, we update the tree as follows. If the node with the key (dp_j, u_j) does not exist in the tree, create and insert a node $(dp_j, u_j, 1)$ into the tree, otherwise update the node $(dp_j, u_j, n_{dp_j,u_j})$ by increasing n_{dp_j,u_j} by one. If needed, we update p_{max} and then remove the smallest node $(d', u', n_{d',u'})$ with $u' < \log_{1+\delta} \frac{p_{max}}{n^2}$ from the tree.

When an arc $\langle i, j \rangle$ arrives in the stream input, we access job i's and j's entries in the array to obtain (d_i, u_i) and (d_j, u_j), and compute $dp_j = \max(d_j, d_i + 1)$. if $dp_j > d_j$, update j's entry in the array B; if in addition $u_j \geq u_-$, that is, the processing time of j is big enough, we update the node (d_j, u_j, n_{d_j,u_j}) by setting $n_{d_j,u_j} = n_{d_j,u_j} - 1$ and delete this node if n_{d_j,u_j} becomes 0; and then update the node $(dp_j, u_j, n_{dp_j,u_j})$ by setting $n_{dp_j,u_j} = n_{dp_j u_j} + 1$ or insert a new node if the node with the key (dp_j, u_j) does not exist in the tree. Once the sketch of the input is generated, we can run the same algorithm to compute the approximate value of the makespan for the problem. And we have the following theorem:

Regarding the practicality of the stream input formats, we note that a preprocessing of the precedence graph at the data center or customer end is possible and thus the depth of jobs or a topological sorting order the graph can be provided in the stream input. If neither the jobs' depth nor the topological sorting order of the graph is known, that is, the stream input consists of a sequence of job and arcs in arbitrary order, it is impossible to store sufficient information for a graph within a space sublinear in the size of stream input: missing the precedence constraint between any pair of jobs would result in an infeasible schedule.

3 Conclusion

In this work, we developed streaming algorithms for the parallel machine makespan minimization scheduling problems with job priority constraint, $P \mid pr_j \leq h, p_{max} \leq c \cdot p_{min} \mid C_{max}$ and $P \mid pr_j \leq h, p_{[1]} \leq c \cdot p_{[\alpha n]} \mid C_{max}$. The streaming approximation algorithms can be applied and extended to the scheduling problems $P \mid prec, dp_j \leq h, p_{[1]} \leq c \cdot p_{[\alpha n]} \mid C_{max}$ for two types of stream input, which complement the in-approximability result of the classical scheduling problem $P \mid prec, p_j = 1 \mid C_{max}$.

Besides the approximate value of the optimal makespan, our algorithms can also compute the "sketch of schedule", which can be used to generate a schedule when the jobs are read in the second pass. For our studied problems, it is also interesting to design streaming approximation algorithms for other various precedence constraints and other performance criteria including total completion time, maximum tardiness, etc.

References

1. Beigel, R., Fu, B.: A dense hierarchy of sublinear time approximation schemes for bin packing. In: Proceedings of the 6th International Frontiers in Algorithmics, and Proceedings of the 8th International Conference on Algorithmic Aspects in Information and Management. FAW-AAIM'12, pp. 172–181. Springer-Verlag, Beijing (2012). isbn: 9783642296994
2. Coffman, E.G., Graham, R.L.: Optimal scheduling for two-processor systems. Acta Informatica **1**(3), 200–213 (1972)
3. Cormode, G., Veselý, P.: Streaming algorithms for bin packing and vector scheduling. Theory Comput. Syst. **65**, 916–942 (2021)
4. Dolev, D., Warmuth, M.K.: Scheduling precedence graphs of bounded height. J. Algor. **5**(1), 48–59 (1984)
5. Dwibedy, D., Mohanty, R.: Online list scheduling formakespan minimization: a review of the state-of-the-art results, research challenges and open problems. SIGACT News **53**(2), 84–105 (2022). issn: 0163-5700. https://doi.org/10.1145/3544979.3544993
6. Bin, F., Huo, Y., Zhao, H.: Streaming algorithms for multitasking scheduling with shared processing. Disc. Appl. Math. **320**, 346–355 (2022)
7. Fu, B., Huo, Y., Zhao, H.: Streaming approximation scheme for minimizing total completion time on parallel machines subject to varying processing capacity. Theor. Comput. Sci. **961**, 113937 (2023). issn: 0304-3975
8. Graham, R.L.: Bounds for certain multiprocessing anomalies. Bell Syst. Tech. J. **45**, 1563–1581 (1966)
9. Graham, R.L.: Bounds on multiprocessing timing anomalies. SIAM J. Appl. Math. **17**, 416–429 (1969)
10. Hochbaum, D.S., Shmoys, D.B.: Using dual approximation algorithms for scheduling problems theoretical and practical results. J. ACM **34**(1), 144–162 (1987)
11. Horowitz, E., Sahni, S.: Exact and approximate algorithms for scheduling nonidentical processors. J. ACM **23**, 317–327 (1976)
12. Lenstra, J.K., Kan, A.: Complexity of scheduling under precedence constraints. Oper. Res. **26**(1), 22–35 (1978)
13. Levey, E., Rothvoss, T.: A (1+epsilon)-approximation for makespan scheduling with precedence constraints using LP hierarchies. SIAM J. Comput. **50**(3), STOC16-201–STOC16-217 (2021)
14. McGregor, A.: Graph stream algorithms: a survey. SIGMOD Rec. **43**(1), 9–20 (2014)
15. Muthukrishnan, S.: Data streams: algorithms and applications. Found. Trends Theor. Comput. Sci. **1**(2), 117–236 (2005)

Author Index

H. Fernau and B. Zhu (Eds.): IWOCA 2025, LNCS 15885, pp. 459–460, 2025.
https://doi.org/10.1007/978-3-031-98740-3